T0213236

Mechanical Engineering Series

Series Editor

Francis A. Kulacki, Department of Mechanical Engineering, University of Minnesota, Minneapolis, MN, USA

The Mechanical Engineering Series presents advanced level treatment of topics on the cutting edge of mechanical engineering. Designed for use by students, researchers and practicing engineers, the series presents modern developments in mechanical engineering and its innovative applications in applied mechanics, bioengineering, dynamic systems and control, energy, energy conversion and energy systems, fluid mechanics and fluid machinery, heat and mass transfer, manufacturing science and technology, mechanical design, mechanics of materials, micro- and nano-science technology, thermal physics, tribology, and vibration and acoustics. The series features graduate-level texts, professional books, and research monographs in key engineering science concentrations.

More information about this series at http://www.springer.com/series/1161

Giancarlo Genta · Lorenzo Morello

The Automotive Chassis

Volume 1: Components Design

Second Edition

Springer

Giancarlo Genta
Politecnico di Torino
Turin, Italy

Lorenzo Morello
Politecnico di Torino
Turin, Italy

ISSN 0941-5122 ISSN 2192-063X (electronic)
Mechanical Engineering Series
ISBN 978-3-030-35637-8 ISBN 978-3-030-35635-4 (eBook)
https://doi.org/10.1007/978-3-030-35635-4

This Springer imprint is published by the registered company Springer Nature Switzerland AG
The registered company address is: Gewerbestrasse 11, 6330 Cham, Switzerland

Foreword

Each book—even one that at first glance might seem like a "cold" university engineering text—tells a fascinating story of experience and knowledge.

When I was invited to write an introduction to this volume, based on the experiences of a great professor and a very experienced industrial manager, I felt the pleasant feeling of a puzzle just completed: the text has, in fact, achieved the important goal of helping readers understand what it really means to move from the concept to the creation of a new car, from design to assembly.

To describe the impact of the text, it is sufficient to highlight how the authors' passion for creating this manual—so much so that it has become an international point of reference in the design of chassis—coincides substantially with the enthusiasm of the thousands of people who work every day at Fiat Chrysler Automobiles with the aim of conceiving and creating cars that are increasingly innovative.

Before they even start studying on these pages, I would like students to be aware that creating a new car—or even just contributing to its birth—is a fascinating job, made up of successive joints of creative and technical skills, which requires an extraordinary commitment. The same commitment needed to lay the foundations for new engineers to grow and make their contribution to creating ever more cutting-edge cars.

Flipping through the pages of the book and investigating the various design steps, it is clear that the authors have achieved an important objective: to give due importance to the fact that a university text must not only tell the theory on how to build new cars, but also describe in a coherent and comprehensive way the content of a car starting from the manufacturer's point of view, sometimes very different from the theory, too often the only subject in university texts.

And in telling this small story of effective integration between the world of study and the world of work, I rediscover a bit of the experiences I lived first by studying Engineering at the University and then working in Fiat Chrysler Automobiles: I find out how much effort and dedication both paths have taken. But, above all, how much satisfaction can they bring.

The approach used by the authors shows the indissoluble link between the academic system and the professional paths that a car company like FCA is able to offer, making it well understood that "knowing how to do" is very different from "doing" but that the two voices together are a winning combination, an essential method to always keep up with the times and make a difference in a context in which knowledge remains the fundamental competitive advantage.

Torino, Italy
June 2019

Daniele Chiari
Head of Product Planning &
Institutional Relations, FCA, Emea

Foreword to the Second Edition

It is a great pleasure and honor for me to write the foreword to the second edition of *The Automotive Chassis*.

First of all, I want to express the gratitude that I have for the authors, who have been great masters of my education: Professor Genta as my Tutor at Politecnico di Torino and Professor Morello as Head of Engineering at Fiat Auto. Their innovative, methodical, rational approach, and their effort to promote and develop the technical competence have helped to form my core values and beliefs.

The two books of *The Automotive Chassis* represent an exceptional masterpiece that has been useful in these years to the engineering students and also to the automotive engineers of my generation. Thanks to this work, we have further developed our knowledge of the most complex and fascinating area of the vehicle where the real technical competence of the engineer is tested.

In this second edition, the first book maintains its robust and structured approach to "Components Design" and the second book on "Systems Design" has been further enriched and updated according to the rapid growth of our car industry toward NEVs and Autonomy. With these actions, *The Automotive Chassis* will continue its role of spreading the chassis engineering culture in our fascinating automotive world.

Shangai, China
April 2019

Giorgio Cornacchia
Head of APAC Product
Development at FCA

Preface

This book is the result of a double decades-long experience: from one side a teaching experience of courses such as Vehicle Mechanics, Vehicle System Design, Chassis design, and more to students of Engineering, from the other side from the design praxis of vehicle and chassis components in a large automotive company. This book is primarily addressed to students of Automotive engineering and secondarily to all technicians and designers working in this field. It also addressed to all people enthusiast of cars that are looking for a technical guide.

The tradition and the diversity of disciplines involved in road vehicle design lead us to divide the vehicle into three main subsystems: the engine, the body, and the chassis.

The chassis isn't today a visible subsystem anymore, tangible as a result of a certain part of the fabrication process, while engine and body are; chassis components are assembled, as a matter of fact, directly on the body. For this reason, the function of the chassis cannot be assessed separately from the rest of the car.

As we will see better, reading the chapters dedicated in the first and in the second part of this book, to the historical evolution, the situation was completely different in the past; in the first cars the *chassis* was defined as a real self-moving subassembly that included the following:

- a structure, usually a ladder *framework*, able to carry on all the remaining components of the vehicle;
- the *suspensions* for the mechanical linkage of wheels with the framework;
- the *wheels* completed with tires;
- the *steering system* to change wheel angles accordingly to the vehicle path;
- the *brake system* to reduce the speed or to stop the vehicle;
- the *transmission* to apply the engine torque to the driving wheels.

This group of components, after the engine assembly, was able to move autonomously; this happened at least in many experimental tests, where the body was simulated with a ballast and during the fabrication process, to move the chassis from the shop of the carmaker to that of the body maker.

Customers often bought from the carmaker a chassis to be completed later on by a body maker, according to their desire and specification.

On contemporary vehicles, this particular architecture and function is only provided for industrial vehicles, with the exception of buses where the structure, even if built by some body maker, participates with the chassis framework to the total stiffness, such as a kind of unitized body.

On almost every car, the chassis structure cannot be separated from the body as being part of its floor (*platform*); sometime some auxiliary framework is also added to interface suspensions or power train to the body and to enable their pre-assembly on the side of the main assembly line.

Nevertheless, tradition and some particular technical aspect of these components have justified the development of a particular discipline within vehicle engineering; as a consequence, almost all car manufacturers have a technical organization addressed to the chassis, separated from those addressed to the body or to the engine.

A new reason has been added in recent times to justify a different discipline and a specific organization and is the setting up of the so-called *technological platforms*: the modern trend of the market calls for an unprecedented product diversification, never reached in the past; sometimes marketing expert calls this phenomenon fragmentation.

This high diversification couldn't be sustained with acceptable production cost without a strong cross standardization of non-visible or of non-specific part of a certain model.

This situation has been very well known since years to all industrial vehicle manufacturers. The term platform implying the underbody and the front side members, with the addition of the adjective technological, describes a set of components substantially equal to the former chassis; the particular technical and scientific issues, the different development cycle, and the longer economic life have reinforced the specificity of engineers that are dedicated to this car subsystem.

The contents of this book are divided into five parts, organized into two volumes.

The first volume describes main chassis subsystems in two parts.

The first part describes the main components of the chassis from the tire to the chassis structure, including wheels, suspension, steering, and braking systems, not forgetting the control systems that show an increasing importance, due to the diffusion of active and automatic systems.

The second part is addressed to the transmission and to the related components; the complexity of this topic justifies a separated presentation.

It should be noticed that, by many car manufacturers, the engineering and production organization dedicated to this subsystem are integrated into the power train organization, instead of the chassis organization. This has obviously no influence on the technical contents of this book and can be justified by the standardization issues and by the life cycle of this component, in certain aspects more similar to the engine than to the chassis.

The explanation approach of the chassis components assumes the existence of a general knowledge of the mechanical components that can be gathered through a conventional machine design course. Topics that can be found on a non-specific course are not treated. In particular, gears design in the second part will not be approached exhaustively, as well as shafts, bearing, and seals design.

Nevertheless, in many parts of this book, design and testing knowledge that are usually not approached in general purpose design courses are introduced and discussed.

We also decided to spend two chapters on the historical evolution of the automotive product; they should enable the reader to appreciate the technical progress of the car in its first 120 years of life. In the opinion of the authors, this subject is a useful technical training and proves to be sometimes useful for inspiration too.

Only architectures that are typical to the most diffused road vehicles will be considered: cars with some mention to industrial vehicles, without considering other applications as motor bicycles, tractors, or earth moving machines and quadricycles.

The second volume is divided into three parts and is entirely addressed to the chassis as a system, putting in evidence the contribution of the chassis to the vehicle performance, as perceived by the customer and as imposed by the legislation rules.

The third part is dedicated to an outline of the functions that the vehicle is expected to perform, of the customers' expectations and to the legislation.

In the fourth part, the influence of the chassis design for the vehicle performance is explained. Particularly, the longitudinal, transversal, and vertical dynamics are explained, with its influence on speed, acceleration, consumption, breaking capacity, and maneuverability (or handling) and comfort.

The fifth part is addressed to mathematical models of the chassis and, more in general, of the vehicle. As known, car engineers take more and more advantage from mathematical models of virtual prototypes and perform numerical testing of prototypes before they are available for physical tests.

Even if mathematical models are based upon calculation codes that are prepared by specialist, and are available on the market, we think to be necessary to supply the students with a clear idea of the methods at the base of these codes and on the approximations that these codes imply. The purpose of this part isn't to enable specialists to built up their models, but to suggest a correct and responsible usage of their results.

The two books are completed by five appendices.

The first appendix recalls some notion on system dynamics, useful to understand the setting up of mathematical models that are introduced in the fourth and fifth parts.

The second appendix is dedicated to two-wheeled vehicles. The study of two-wheeled vehicles, for some aspects more complicated than for four-wheeled vehicles, is very particular and has nothing to do with cars; in addition to that, industries that produce motorcycles are very well separated from the car industry.

Nevertheless, there are disciplines common to the two worlds, due to the fact that both vehicles use pneumatic tires as interface with the ground; some knowledge exchange between the two vehicle engineers could be of mutual benefit.

The third appendix is dedicated to the particular issues that should be faced when vehicles on wheels will be developed for planets or environments different as the earth. Starting from the only vehicle of this kind that was developed for the Apollo Project, similarities and differences between conventional vehicles and those that in the future could be utilized for interplanetary exploration.

The fourth appendix analyzes some mathematical approach, sometimes very simplified to interpret the motion of cars after the impact due to an accident.

The last appendix reports the main data of vehicles of different kinds that are used in some explanatory example in the book; these data could also enable the student to practice their skills on exercises with a minimum of realism.

Turin, Italy

Giancarlo Genta
Lorenzo Morello

Acknowledgements

The authors wish to thank Fiat Research Center for having made possible the preparation of these two volumes, not only by supporting the cost of this work, but also by supplying a lot of technical material that contributed to update the content of these books and to orient them to practical applications.

Particularly, the authors appreciated the many suggestions and information they received from Isabella Anna Albe Camuffo, Kamel Bel Knani, Roberto Cappo, Paolo Mario Coeli, Silvio Data, Roberto Puppini, and Giuseppe Rovera.

The first volume of this work has, in addition, benefited of the lecture notes prepared by Fiat Research Center, to sustain the teaching activity of the courses of Vehicle System Design, Chassis Design, and Automotive Transmission Design, within the course of Automotive Engineering of the Politecnico of Turin and of the Master in Automotive Engineering of the Federico II University of Naples.

The authors' gratitude must also be shown to the Companies that have supplied part of the material used for the illustrations, mainly in the first volume; in alphabetical order, we remember: Audi, Fiat Chrysler Automobiles (FCA, formerly Fiat Auto), Getrag, Honda, Iveco, Marelli, Mercedes, Shaeffers, and Valeo. Without their contribution, this book haven't been complete and, for the time being, topical.

Particular thanks are conveyed to Donatella Biffignandi of the Automobile Museum of Turin for the help and material supplied for the preparation of the historical sections.

About the Authors

Giancarlo Genta got a degree in aeronautical engineering in 1970 and in aerospace engineering in 1971 at the Politecnico of Turin. He started immediately after his career at the Politecnico as Assistant of Machine Design and Technologies.

He has been Visiting Professor of Astronautical Propulsion Systems since 1976 and of Vehicle Mechanics since 1977 and, more recently, of Vehicle System Design at the course of Mechanical Engineering and Automotive Engineering.

He was appointed Associate Professor of Aeronautical Engines Design in 1983, at the Aerospace Engineering School of the Politecnico of Turin; he was appointed full professor of the same course in 1990.

He was elected Director of the Mechanical Engineering Department of the Politecnico from 1989 to 1995. He has been holding the course of Applied Stress Analysis II for the Master of Science of the University of Illinois at the Politecnico of Turin.

He also held many courses in Italy and abroad, in the frame of development cooperation projects, in Kenya (2 years), Somalia (6 months), India (1 month), and at the Bureau International du Travail.

He has been Honorary Member of the Academy of Sciences of Turin, since 1996, and of the International Academy of Astronautics, since 1999; he was elected full member of the same Academy in 2006.

He coordinates the Research Doctorate in Mechatronics, since 1997.

He performed research activities, mainly in the field of Machine Design, particularly on static and dynamic structural analysis.

He studied the magnetic suspension of rotating parts, the vehicle dynamics, and the related control systems; he was one of the promoters of the Interdepartmental Laboratory on Mechatronics, where he performs research activities on magnetic bearings, moving robots, and vehicle mechanics.

He is author of more than 270 scientific publications, covering many aspects of mechanical design, published by Italian, English, and American magazines or presented in Congresses.

He wrote textbooks of Vehicle Mechanics (published in Italian and English), adopted as reference in some Italian and American University. He also wrote monographs on composite materials design, on the storage of energy on flywheels (published in English and translated in Russian), on Rotating Systems Dynamics, and of popular books on space exploration.

Lorenzo Morello got his degree on Mechanical Automotive Engineering in 1968, at the Politecnico of Turin.

He started immediately after his career at the Politecnico as Assistant of Machine Design and Technologies.

He left the Politecnico in 1971 and started a new activity at a branch of Fiat dedicated to vehicles studies that will be joined to the new Research Center in 1976. He participates in the development of some car and of experimental prototypes for the ESV US Program. He also developed some mathematical model for vehicle suspension and road holding simulation.

Starting from 1973 he was involved on an ample project for the development of mathematical models of the vehicle, to address the product policies of the company to face the first energy crisis; as part of this activity, he started the development of a new automatic transmission for reduced fuel consumption and of a small direct injection diesel engine to be used on automobiles.

He was appointed manager of the chassis department of the Vehicle Research Unit and has been coordinating the development of many research prototypes, such as electric cars, off-road vehicle, trucks, and buses.

He was appointed manager of the same Research Unit in 1977 and has been leading a group of about 100 design engineers, dedicated to the development of prototypes; a new urban bus with unitized thin steel sheet body, with spot welded joints, a commercial vehicle that will start production later on, a small light weight urban car, under contract of the National Research Council, and a hybrid car, under contract of the US Department of Energy, were developed in this period of time.

He took the responsibility of the Engines Research Unit in 1980; this group, of about 200 people, was mainly addressed to the development of new car engines. He has been managing the development of many petrol engines, according to the principle of high turbulence fast combustion, a car direct injection diesel engine, many turbocharged prechamber diesel engines, a modular two-cylinder car engine, and many other modified prototypes.

He was appointed Director of Products development in 1983; this position includes all applied research activities on Vehicle Products of Fiat Group. The Division included about 400 people, addressed to power train, chassis, and bodies studies and also to prototype's construction.

He joined Fiat Auto in 1983 to take the responsibility of development of some new car petrol engines and of the direct injection diesel (the first in the world for automobile application). He was appointed Director for Power Train Engineering in 1987; the objective of this group was to develop all engines produced by the Fiat Auto brands; the most important activity of this time is the development of the new engine family, to be produced in Pratola Serra and including more than 20 different engines.

He returned, at the end of his career, to vehicle development in 1994, as director for Vehicle Engineering; the group was addressed to the design and test of bodies, chassis components, electric and electronic systems, wind tunnels, safety center, and other facilities.

He retired in 1999 and started a new activity as consultant to the strategic planning of Elasis, a new company of the Fiat Group, entirely dedicated to vehicle applied research.

Together with Fiat Research Center he participated in the planning of some courses of the new Faculty on Automotive Engineering of the Politecnico of Turin and of the preparation of the related lecture notes.

He was Contract Professor of Vehicle System Design and has been Contract Professor of Automotive Transmissions Design since many years, at the Politecnico of Turin and the University of Naples; he also published a textbook on this last subject and many articles and books about car technology evolution. He is coauthor of *The Automotive Body*, also published by Springer in the Mechanical Engineering Series.

Acronyms and Symbols

Acronyms

Some of these acronyms are of general use, while others are used as a sort of trademark of a particular manufacturer. Moreover, many started as trademarks and then became of common usage.

4WD	4 Wheel drive
4WS	4 Wheel Steering
AAA	American Automobile Association
AAC	Adaptive Cruise Control
ABS	Antilock Braking System
AC	Alternating Current
AD	Additive Manufacturing
ADAS	Advanced Driving Assistance System
ADAV	Advanced Driving Assistance Vehicle
AEB	Autonomous Emergency Braking
AEBS	Advanced Emergency Braking System
AFC	Alkaline Fuel Cells
AMoD	Advanced Mobility on Demand
ARC	Active Roll Control
ASR	Anti Spin Regulator
BEV	Battery Electric Vehicle
BMS	Battery Management Systems
CACC	Cooperative Adaptive Cruise Control
CAD	Computer-Aided Design
CAE	Computer-Aided Engineering
CAM	Computer-Aided Manufacturing
CPA	Centrifugal (or rotating) Pendulum Absorber
CVT	Continuously Variable Transmission
DARPA	Defence Advanced Research Projects Agency

DC	Direct Current
DMFC	Direct Methanol Fuel Cells
DOE	(American) Department Of Energy
EBD	Electronic Brake Distributor
ECU	Electronic Control Units
EM	Electric Motor
EMF	Electromotive Force
EPS	Electric Power Steering
EREV	Extended-Range Electric Vehicles
ESV	Experimentally Safety Vehicle
FAA	Federal Axiation Administration
FCW	Forward Collision Warning
FESS	Flywheel Energy Storage System
HEV	Hybrid Electric Vehicle
HV	Hybrid Vehicle
ICE	Internal Combustion Engine
ICT	Information and Communication Technologies
IPTS	Inductive Power Transfer System
KERS	Kinetic Energy Recovery Storage
LiDAR	Light Detection And Ranging
Li-ion	Lithium-ion (Li-ion)
LiPh	Lithium-iron-phosphate
LiPo	Lithium ion-polymer
LiS	Lithium-Sulfur
LPG	Liquified Petroleum Gas
LRR	Long-Range Radar
MCFC	Molten Carbonate Fuel Cells
MEMS	Micro Electromechanical System
MHEV	Mild Hybrid Electric Vehicles
MoD	Mobility on Demand
MTA	Mechanical Transmission Automatized
NaS	Sodium-Sulfur
NASA	National Aeronautics and Space Administration
NEDC	New European Driving Cycle
NHTSA	National Highway Safety Administration
NiCd	Nickel-Cadmium
NiFe	Nickel-iron
NiMH	Nickel-Metal Hydride
NiZn	Nickel-Zinc
NVH	Noise, Vibrations and Harshness
OLEV	On-Line Electric Vehicle
PAFC	Phosphoric Acid Fuel Cells
PAV	Personal Air Vehicles
PEMFC	Proton Exchange Membrane Fuel Cells
PMBM	Permanent Magnets Brushless Motors

PUMA	Personal Urban Mobility and Accessibility
RDE	Real Driving Emissions Test
REEV	Range-Extended Electric Vehicle
RPEV	Roadway-Powered Electric Vehicle
SAE	Society of Automotive Engineers
SOFC	Solid Oxide Fuel Cells
SRR	Short-Range Radar
SUV	Sport Utility Vehicle
TCS	Traction Control Systems
TMR	Triple Modular Redundancy
ToF	Time of Flight
UAV	Unmanned Aerial Vehicles
UE	European Union
UNECE	United Nations Economic Commission for Europe
UPM	Urban Personal Mobility
V2I	Vehicle to Infrastructure (communications)
V2V	Vehicle to Vehicle (communications)
V2x	Vehicle to any other thing (communications)
VDC	Vehicle Dynamics Control
WHO	World Health Organization
WLTP	Worldwide Harmonized Light Vehicles Test Procedure
WPTS	Wireless Power Transfer System
ZEBRA	Zero Emissions Batteries Research Activity
ZEV	Zero Emission Vehicle

Symbols

a	Acceleration; generic distance; distance between center of gravity and front axle
b	Generic distance; distance between center of gravity and rear axle
c	Viscous damping coefficient; specific heat
d	Generic distance, diameter
e	Base of natural logarithms
f	Rolling coefficient; friction coefficient
f_0	Rolling coefficient at zero speed
f	Force vector
g	Gravity acceleration
h	Wheel deflection
h_G	Center of gravity height on the ground
k	Stiffness
l	Wheelbase; length
m	Mass
p	Pressure

r	Radius
s	Stopping distance, thickness
t	Temperature; time; track
u	Displacement vector
v	Slipping speed
z	Teeth number
A	Area
C	Cornering stiffness; damping coefficient
C_γ	Camber stiffness
C_0	Cohesiveness
E	Energy; Young modulus
F	Force
G	Shear modulus
H	Thermal convection coefficient
I	Area moment of inertia
J	Quadratic mass moment
K	Rolling resistance coefficient; stiffness; thermal conductivity
K	Stiffness matrix
M	Moment
M_f	Braking moment
M_m	Engine moment
M_z	Self-aligning moment
P	Power; tire vertical stiffness; force
P_d	Power at the wheel
P_m	Power at the engine
P_n	Required power
Q	Thermal flux
R	Undeformed wheel radius; path radius
R_e	Rolling radius
R_l	Loaded radius
S	Surface
T	Temperature, force
V	Speed; volume
W	Weight
α	Sideslip angle; road side inclination; angle
α_t	Road transverse inclination angle
γ	Camber angle
δ	Steering angle
ε	Toe-in, -out; brake efficiency; deformation
η	Efficiency
θ	Angle; pitch angle
μ	Torque transmission ratio; adherence coefficient
μ_p	Max friction coefficient
μ_x	Longitudinal friction coefficient

μ_{x_p}	Max longitudinal friction coefficient
μ_{x_s}	Slip longitudinal friction coefficient
μ_y	Transversal friction coefficient
μ_{y_p}	Max transversal friction coefficient
μ_{y_s}	Slip transversal friction coefficient
ν	Speed transmission ratio; kinematic viscosity
ρ	Density
σ	Normal pressure; slip
τ	Transversal pressure; transmission ratio
φ	Angle; roll angle, friction angle
ω	Pulsation; frequency
Φ	Diameter
Π	Tire torsion stiffness
χ	Torsion stiffness
Ω	Angular speed

Contents

Part I
Wheels, Structures and Mechanisms

Introduction

The first part of this book is dedicated to the study of parts (they should better be called subsystems, because they are often complex systems, as suspensions) that constitute the chassis. Their main function is to allow a suitable exchange of force with the ground, in order to obtain the desired vehicle speed and path.

With reference to the coordinates system that will be defined as vehicle system in the fourth part, the forces exchanged with the ground can be classified as follows:

- forces perpendicular to the ground (vertical for the motion on a plane road): in steady-state conditions, they can be considered as constant, but because of the obstacles on the road, they are variable and they are relevant to the comfort issues of vehicle passengers;
- longitudinal forces: they are mainly due to the propulsion (engine and transmission) and braking systems and they are relevant to the vehicle speed control; and
- transversal forces: they are due to wheel steering angles and they are relevant to road holding and stability.

All these forces are acting on tires, that because of their deformable structure make the dynamic behavior of the vehicle on rubber wheels more similar to a floating or flying vehicle than a rail vehicle.

Chassis technologies can be defined as mature, but saying so we wouldn't like to underestimate the various ongoing evolutions, mainly regarding the application of controlled or active systems, based upon electronic and informatic technologies.

As a matter of fact, the fast evolution of automotive electronics in terms of performance and cost had and will have a big influence on improving active safety and comfort of vehicles.

A last aspect that shouldn't be forgot is that in many markets the development and production of chassis systems are leaving car manufacturers in favor of part manufacturers that are becoming specialists in their business.

This is true, since many years, for brake systems, steering systems, and tires, and is becoming to be true for suspensions and transmissions. In this situation, it is very important for both those that will address their career to car manufacturing or to part manufacturing to develop a good system understanding; the development of these components is virtually impossible if separated from that of the vehicle.

As we will see, chassis components have received a quick evolution during these last years: almost all cars feature today radial tires with low aspect ratio (radial dimension is much smaller as transverse dimension) and need suspension with very precise elasto-kinematic behavior. Mc Pherson and double wishbone suspensions share the market as far as the front axle is concerned, while a significant percentage of the rear axles is featuring multilink suspensions.

It is quite unlikely that the kinematic configuration will receive new innovations; the same can be said for the steering system where the wide diffusion of assistance systems has almost standardized the rack and pinion configuration.

A similar situation can be seen for car brake systems where disc brakes are widely diffused with exceptions for the rear axle of economy cars that conserve the drum solution.

New developments are instead expected for electronic control systems and the related fields of sensors and actuators, where electromechanical actuators will give more opportunities for performance improvements.

Electronic control systems have initially entered the marked as add-on devices.

The case of the brake antilock system (ABS) is typical: it made possible a significant performance improvement to the brake system at the cost of new and sophisticated components (the electronic control system, the wheel speed sensors, the valve group able to regulate the pressure on the brake actuators of the wheel independently of the pedal pressure).

The introduction of this system was initially gradual, but afterward it reached high volumes, with consistent cost reduction and, now, as a consequence, the diffusion is nearly total. In parallel, the system performance was improved offering new possibilities, either in the field of cost reduction (i.e., giving the possibility to incorporate the brake distribution valve function and the power function at no cost), or in the field of functions, where, with the addition of some sensor, also the vehicle dynamic control has been obtained.

A similar story can be told for power steering systems, initially totally hydraulic; the addition of electronic controls allowed a better regulation of the power assistance pressure, reducing the sensitivity of the steering wheel torque to the vehicle speed.

The present trend consists in the substitution of the hydraulic electronic system with an electric electronic system: the power assistance is coming from a controlled electric motor; from this opportunity comes the possibility of having an active steering system that can improve vehicle performance while avoiding a sudden obstacle.

In a probable future, all actuators could become electric, with cost reduction and increased performance; the further step could be to avoid any mechanical linkage between pilot controls (pedal, steering wheel, etc.) and actuators.

This goal has been already reached for the engine, where throttle position or fuel injection quantity is no more controlled by the accelerator pedal mechanically, but through a *drive-by-wire* system. We can easily foresee for the future a *brake-by-wire* system or a *steer-by-wire* system.

The next step, now discussion topic in many technical congresses, is the *corner-by-wire* that is a wheel-suspension group (*corner*) with total electric actuation (driving, braking and steering functions); a system like this could have a significant result on vehicle performance and architecture.

Similar evolution processes are also present in the suspension field; a first step is the application of the electronic control to the damping properties of shock absorbers and to the position of the body relative to the ground while the vehicle is standing still (*trim*); a possible evolution could lead to a suspension where the body position is controlled also dynamically. This possible achievement could simplify the elasto-kinematic requirements of the suspension.

We think that these and other examples could offer a view of the possible fields of the chassis evolution.

After the chapter dedicated to the historical evolution, the most diffused configurations for chassis components will be described. The following components will be considered.

Wheels and Tires

Tires will not be studied from the standpoint of their product and process design techniques, useful to determine their performance. They will be studied almost as a *black box* examining their static and dynamic response that is the base of the vehicle static and dynamic response.

A good knowledge of the tire performance is fundamental for an effective communication between vehicle and tire specialists.

Suspensions

While studying suspensions, the main kinematic schemes will be considered and their influence on the working angles of the tires, on vehicle roll and pitch. The most important suspension components will be described, as the main elastic elements, the secondary elastic elements, and the damping elements.

Steering System

A main mechanism of the steering system will be studied and their mechanical properties; the main components will be described as the steering box and the most important power assistance systems.

Brake System

The most important brake types will be introduced and their actuation and power assistance systems. The industrial vehicle brake systems will be described separately, because of the different actuation systems (pneumatic instead of hydraulic power).

Control Systems

As far as chassis control systems are concerned, this volume will describe sensors and actuators in use and the technical target that these systems should reach mainly on vehicle dynamics; the most diffused control strategies will be also described that the different systems adopt, while the interaction between control system and dynamic behavior of the vehicle will be afforded in the second volume.

Chassis Structures

Although this topic could better be tackled in a book dedicated to body design, this chapter will outline the integration of the chassis functions into the body structure and will offer a short description of the main types of auxiliary frameworks in use on unitized bodies. A short description of industrial vehicle frameworks is also offered.

Chapter 1
Historical Evolution

1.1 Introduction

It is always arduous to decide if historical notes should come before or after the description and the interpretation of the state of the art.

In favor of the second alternative there the fact that readers should already have understood the motivations that lead a design decision. In favor of the first alternative, we have chosen, there is the fact that readers can situate the state of the art in its context and can better appreciate the effort that has been made by those people that have contributed to the state of the art.

The unavoidable drawback of this approach is that some considerations are only superficially introduced on the historic section and are again introduced in a better detail later on; readers will forgive us for repeating some concept. We suggest that readers that are very interested in the chassis history will read again the historic notes at the end of the volume.

We will start in this chapter from suspensions and steering system, to describe later wheels, tires, brake system and structures. This fact is only due to a larger impact that suspensions and steering system have on the vehicle architecture and the consequent evolution; steering system will be described together with suspensions because these two systems are indissoluble from a designer point of view.

The primary function of a vehicle suspension is to isolate in the best way the body, the sprung mass, from the disturbances coming from the road uneven surface. To obtain this result, the wheels and the masses integral with them, the so called unsprung masses, are connected to the body with mechanical linkages that allow their relative motion, mainly in the vertical direction; according to this direction forces are transmitted to the body trough elastic and damping elements.

The elastic characteristics of tires are also contributing to suspension quality.

Because the path of the wheel motion is not strictly vertical with reference to the car body, but can show also components in the other two directions, a second function for the suspension can be identified: it should guide the wheel, during its

G. Genta and L. Morello, *The Automotive Chassis*, Mechanical Engineering Series,
https://doi.org/10.1007/978-3-030-35635-4_1

displacement (suspension deflection) in order to avoid undesired motions, in terms of:

- steering angle of the wheel (*toe-in*, *toe-out*), that can modify vehicle path;
- wheel camber angle, that can affect the cornering stiffness of the tire;
- track variation, because can influence negatively the tire duration;
- wheelbase variation, because can cause resonances with traction and braking force application.

These displacements, we will define as secondary, cannot be completely eliminated, but must be accurately designed while defining the kinematic linkages and the stiffness of the reaction points (elasto-kinematics); in fact a design variation of these parameters can positively affect the vehicle dynamic behavior.

This topic will be discussed completely in the suspension chapter.

To tell in advance what will be discussed later on, we can say that:

- the toe angle of the wheel can correct the vehicle understeering (and over steering) behavior and improve the vehicle stability during braking;
- a camber angle equal but opposite to the roll angle, or, in other words, a constant perpendicularity of the wheel to the ground can allow the maximum exploitation of the tire cornering stiffness;
- an appropriate wheelbase variation can improve the contribution of the wheel to absorb the effect of obstacles, with positive results for comfort; in addition to that the suspension can be designed to have *anti dive* and *anti squat* characteristics, minimizing the pitch angle variation as a consequence of driving and braking forces application.

It shouldn't be forgot that four wheels vehicles are, as far as the exchange of forces with the ground is concerned, simply hyperstatic systems; therefore a last function of the suspension is enabling to design forces exchanged with the ground and to make the wheel contact possible also if the road isn't flat.

Many of these considerations, that we will better explain later on, were not known at the beginning of the motor era and have been developed quite recently, as compared with the hundred years of life of the automotive product, because of the increase of the vehicle speed and the improvements of the road conditions.

The first two sections of this chapter trace the evolution of the mechanical linkages of the suspension and of the steering system in an attempt to understand, according to documents and drawing of that time, the motivations and the ideas of the engineers that developed them.

1.2 The Rigid Axle Mechanical Linkages

We include under this title the articulated and elastic systems of both suspension and steering mechanism.

Fig. 1.1 This coach built around 1650 shows the existence of a suspension; the sprungmass includes the passengers compartment only and is connected to the unsprung mass with four leaf springs with leader belts (Automobile Museum of Turin)

The suspension function was already known in the XVI century. Coach bodies were suspended through a leaf spring set fit to the chassis framework, a rigid structure bearing wheel hubs. The free end of the springs was connected to the body through leader belts; Fig. 1.1 offers an interesting example of this kind of configuration in a coach dated about 1650. The steel leaf springs, present in this vehicle, were introduced during this time; formerly, they were made of wood. There is no component explicitly dedicated to damping suspension oscillations; the internal friction of leaf springs and of belts should have been enough to reach the expected comfort.

Secondary motions weren't present, because wheels were each other rigidly connected; the system was isostatic because of the play of front steering axle.

Elliot, an English wheelwright, was credited with the invention of the single rigid axle suspension, using semi elliptical steel leaf springs; also in this case the front axle can steer on a pivot in the middle of the axle.

This suspension system, in use on coaches and carriages, was also adopted in the first steam road vehicles, in the nineteenth century, before of the internal combustion engine. Figure 1.2 shows an example of this kind.

The carriage steering system showed the inconvenience of reducing the roll-over stability of the vehicle, while turning; when lateral centrifugal forces were applied, the roll-over line (it can be obtained joining the two contact points with the ground of the wheels of the same side) was shifted closer to the centre of the vehicle.

Fig. 1.2 The Bordino's steam coach was built in 1854. The suspension system is strictly derived from that of a horse carriage. The rear axle is shaped as a crankshaft, where the connecting rods of the two cylinders are directly working (Automobile museum of Turin)

In 1810 Längensberger envisaged a steering system where the front axle was always parallel to the rear axle but the front wheel hub only were articulated to the axle using a king-pin; the two stub axles were connected through track-rod arms and a track rod, shaping up an articulated parallelogram. Using this device the vehicle roll over stability was unaffected in turns.

In 1818 a patent was filed in London in name of Ackermann; this patent described the law the steering angles of the two wheels should follow in order to have wheels rolling correctly with their symmetry plane containing the local speed vector. This invention didn't find immediate practical application because there was no real need on carriages to have a steering system different as the turntable steering and there was no idea about how to satisfy this law in a simple way.

Jeantaud, again a wheelwright, proposed in 1878 a mechanism perfected from the idea of Längensberger, where the two track-rod arms were slightly inclined to the middle of the vehicle in such a way as to have their axis crossing near the middle of the rear axle.

This mechanism obtains with acceptable approximation the Ackermann's low and is today still in use on steering rigid axles.

Independently from the Jeantaud's idea, the solution presented by Bollée in his Mancelle, again in 1878, was not very far away from the Ackermann's low. We like to remember this steering system, because it is probably one of the first bound to an automobile: Fig. 1.3 shows the scheme of the front axle; it should be noticed that it is an independent suspension with transversal leaf springs, equivalent to the double wishbone mechanism.

We shouldn't, in fact, think that independent wheel suspensions were invented in the '940s, at the end of an era that saw only rigid axle with leaf springs; independent

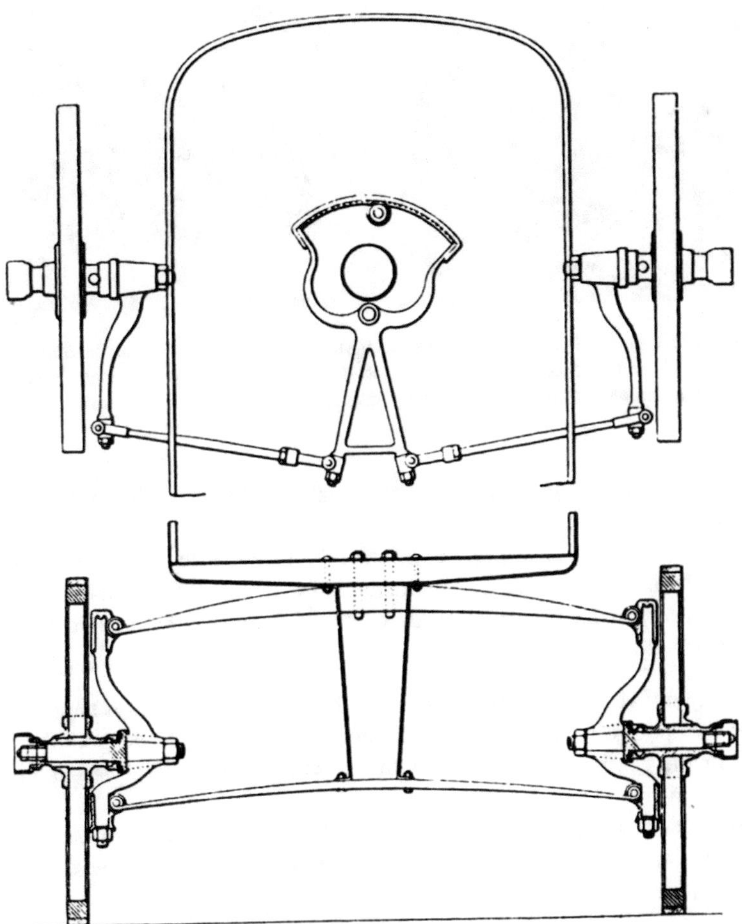

Fig. 1.3 Drawing of the front steering axle of the Mancelle of Bollée, in 1878; it should be noticed the independent double wishbone suspension with transverse double leaf springs

wheel suspensions are sporadically present also in the first cars. Nevertheless we should think that the advantages obtained on the existing roads and at those speeds were negligible as compared with the enormous design complications.

The first cars with internal combustion engine, the tricycle of Benz, in 1886 and the *Stahlradwagen* (German word for steel wheel car) of Daimler, in 1889 (see Fig. 1.4) give no evidence of a great attention to the suspension system.

The first is in fact a tricycle with a single non suspended front steering wheel; the second is a four-wheeled vehicle with no suspension, where the front axle is balancing on a central horizontal pivot, that makes the system isostatic. We can assume that the undoubted genius of these two precursors was concentrated on developing a reduced

Fig. 1.4 The first cars with internal combustion engine, the tricycle of Benz and the *Stahlradwagen* of Daimler give no evidence of a great attention to the suspension system. Both have no front suspension and the second features no suspensions at all with a suspended seat only

weight engine with a significant, but light, quantity of energy on board, instead on the passengers comfort.

Let us see now the most important features of rigid axle suspensions with leaf springs: axle linkages are in this case integrated with the elastic element.

As far as suspensions are concerned, we should we aware of two typical problems of suspended axles.

- If the engine is part of the sprung mass (and this has almost always happened with some exception), it is necessary to develop a mechanism to connect the engine shaft or the gearbox output shaft with the wheels, that have a variable relative position.
- Because the steering control (steering wheel or steering bar) must be at reach of the driver, again mechanisms must be developed that connect stub axles, without affecting the steering angles with the suspension bouncing motion.

The solution of this two problems should have committed first cars designers, particularly as far as the steering system is concerned; the multiplicity of solutions of the first cars give evidence of the difficulty met in finding an adequate technical solution to the problem.

On the Benz car the problem is bypassed, considering that there is no front suspension. On Daimler car the steering control is fixed to the balancing axle near the pivot point.

As far as the transmission is concerned, on the Benz car a leather belt gearbox is applied, that compensates for the center line variation of the pulleys with a spring-loaded moving tensioner.

This device is common to other car of this period of time and has the advantage of integrating transmission functions and gearbox functions; the gearbox function is obtained by using a number of couples of pulleys with a shiftable belt.

Almost in the same years the idea of using a tooth wheel gearbox was born, in connection with a chain transmission; the Fiat 3 $\frac{1}{2}$ HP of 1899 could be taken as an example of this concept, shown in the phantom view of Fig. 1.5. The sprocket

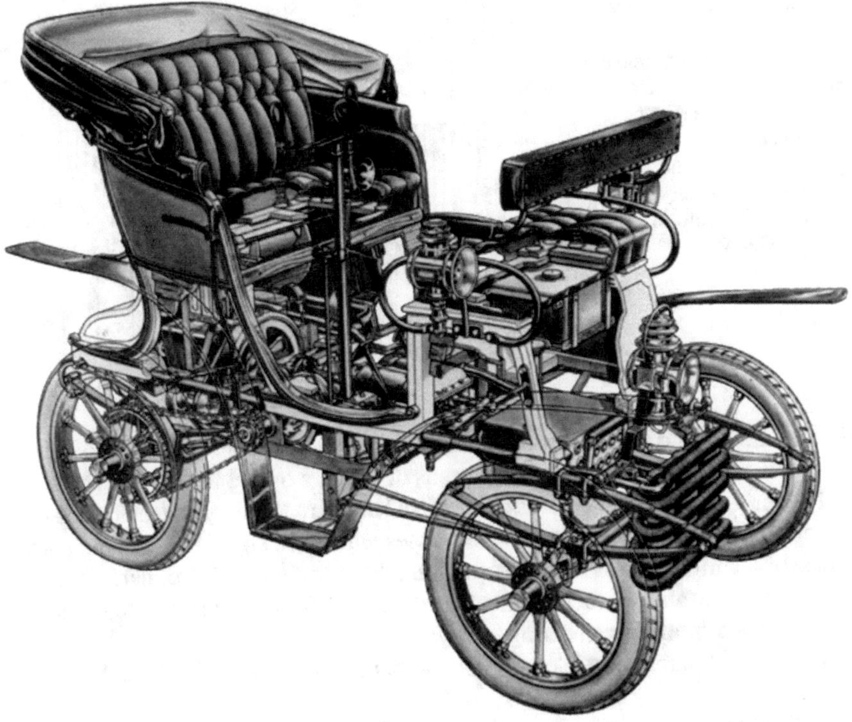

Fig. 1.5 The Fiat 3 $\frac{1}{2}$ HP of 1899 features a chain and sprocket transmission, where the sprocket is set in the center of curvature of the path of motion of the rear axle. It can be noticed on the same car the Jeantaud steering system

axis was fixed to the sprung mass, almost in center of curvature of the suspension path, described in the bouncing motion. This design detail allowed the axle to move without affecting the chain length.

This transmission and suspension architecture will be widely applied in the first twenty years of the car history.

On the same car can be noticed the Jeanteaud steering mechanism connected to the steering bar through a longitudinal linkage; also in this case the linkage knuckle mounted on the sprung mass was positioned in the center of curvature of the motion of the axle in its bouncing movement, to avoid undesirable change of path while riding on uneven roads.

Almost in the same years, in 1898, De Dion and Bouton introduced the propeller shaft transmission with universal joints. It is still in use, almost unchanged in its basic elements, in the today's cars with front engine and rear drive; it solves in the best way any problem connected to the suspension motion.

Leaf springs, as we have seen, were already known in the XVII century; they reached at the beginning of the XX century a satisfactory level of fabrication technology and application know-how. Leaf springs integrate in a single element the

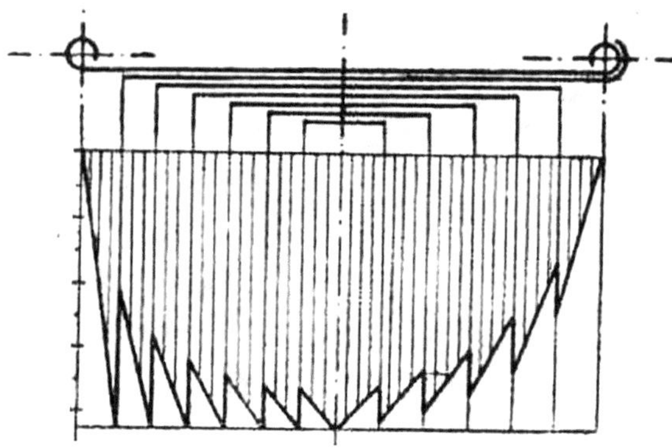

Fig. 1.6 On the automotive engineering manual of Baudry de Saunier of 1900 is exposed the theory of a leaf spring with constant leaf thickness and length reduced by constant steps. The structure is practically equivalent to a uniform resistance body

elastic function and the structure function; the elasto-kinematic performance of this component was fully adequate to the needs of the cars of that time.

The spring is made with leaves of different length, reduced by constant steps from the top to the bottom; this assembly simulates with good approximation the performance of a uniform stress structure, granting the maximum value of deformation at a given stress level. This fact is rationalized in Fig. 1.6, coming from one of the first manual of automotive engineering.

The same structure, flexible on a vertical plane can be quite stiff in an horizontal plane.

The spring was mounted to the chassis with a fixed eye on the mother leaf and with a moving eye articulated to a swinging shackle or with a sliding element; this device allowed the spring to change his length because of the deflection. The fixed eye was on the same side as the fixed point of transmission and steering system, typically behind the front axle and in front of the rear axle.

The suspension architecture could assume many different variants, according to objectives of comfort, cost and weight of the unsprung mass. Leaf springs have been named after their shape, with reference to ideal elliptical shape, made of two mirror like leaf springs; with reference to Fig. 1.5 front springs are called elliptical, rear ones are called semi elliptical.

The most diffused solution featured two semi elliptical springs for each axle; on luxury cars was some time adopted the $\frac{3}{4}$ of ellipse solution (Fig. 1.7, top left) where the side beam of the chassis structure was fit directly to a $\frac{1}{4}$ of ellipse spring. This solution allows an increased flexibility at a higher cost and weight of the unsprung mass.

Again on luxury cars a 3 semi elliptical spring solution was also adopted for the rear axle; the function is comparable to the of the $\frac{3}{4}$ of ellipse arrangement, where the two $\frac{1}{4}$ of ellipse springs are integrated in a single element, fit in the middle to the

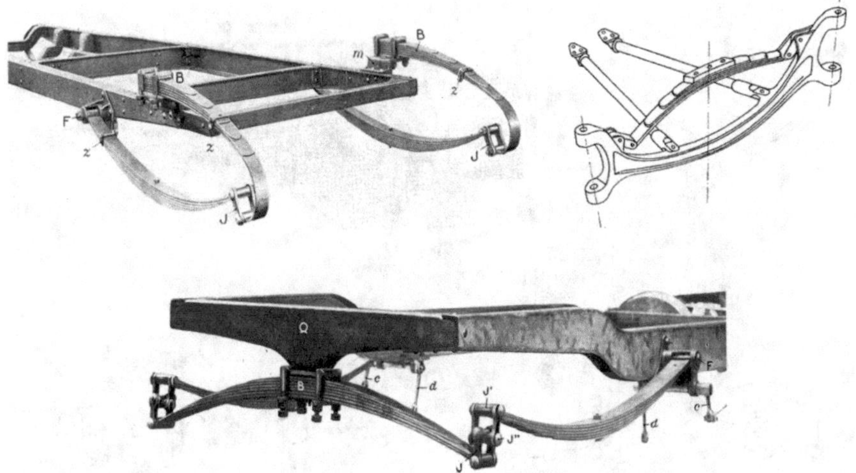

Fig. 1.7 Some application schemes for leaf springs. On the top left is shown a $\frac{3}{4}$ of ellipse solution; on the right the Ford scheme with a single spring for each axle; at the bottom a 3 leaf spring solution equivalent to the $\frac{3}{4}$ of ellipse one

structure of the chassis. In this case, swinging shackles are made with a universal joint (Fig. 1.7, bottom).

A particular cantilever spring, shown in Fig. 1.7 at the top right, was developed and diffused by Ford. A single semi elliptical spring is fit to a cross beam of the chassis structure, at the mid section; it performs like two $\frac{1}{4}$ of ellipse cantilever springs, mounted on the axle; in this case, two swinging shackles must be used. Undesired motion of the axle in the longitudinal direction is avoided with a couple of dedicated thrust beams.

A convenient suspension should not only have a deformable elastic member, but also a damping element; otherwise a permanent oscillation of the sprung mass could take place as a result of an external force input. This cannot really take place in the reality, because of the internal friction; nevertheless too many oscillations could be produced around the static position.

With leaf spring this problem wasn't considered for long time, because of the high value of the internal friction due to the relative motion of lives; on the contrary, every effort was made to improve leaves lubrication in order to avoid annoying squeaks and permanent sticking, because of oxidation. Nevertheless additional dampers were not considered for long time. The first shock absorbers appeared in the '910s, being considered initially as accessories to be installed after the sale, for demanding costumers with attitude to sport driving.

Many solutions were considered. Figure 1.8 shows the most interesting. The first four are friction shock absorbers; the first and the third work on the extension stroke only, while the second and the fourth work in the compression stroke too.

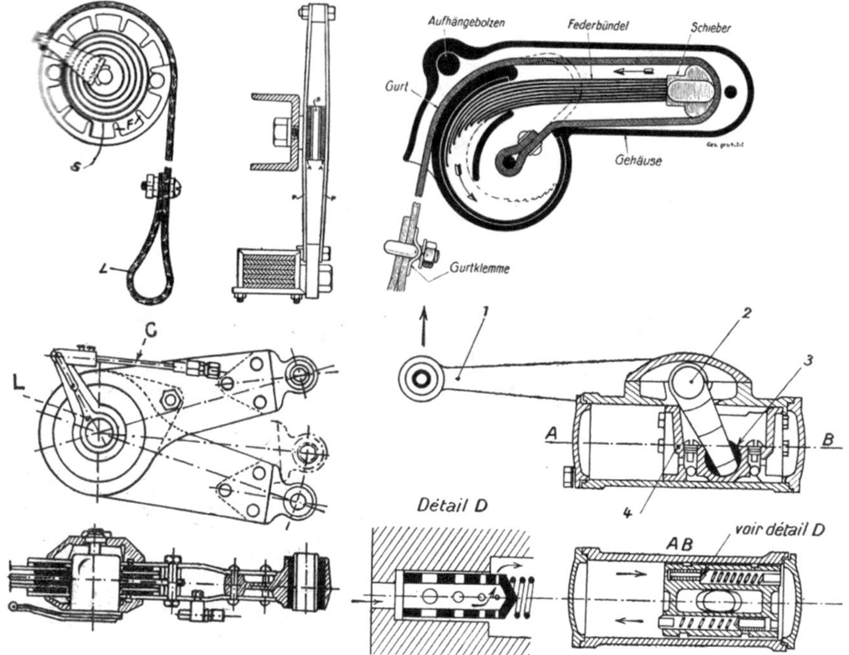

Fig. 1.8 Some type of shock absorber, suitable for after market application. The first three in the first raw and the first in the second row work on mechanical friction; the last of these can be also adjusted. The fifth example refers to a hydraulic shock absorber

Many engineers thought that damping forces were noxious in the compression stroke because increased the value of forces applied to the sprung mass; other engineers thought the reduction of danger of oscillation was the most important priority. Only later, the importance of having a bilateral force proportional to suspension compression or extension speed was understood and theorized.

A friction shock absorber shows some property negative to comfort. In fact, a dry friction force allows suspension motion only over a certain threshold value of force. It can happen that suspensions are always blocked, by roads with small unevenness or by macadam pavement, and the only working elastic members are tires. For this reason the fourth type of shock absorber features an adjustment screw suitable to change the pressure force on the friction discs; this screw can be adjusted by the driver from the inside of the car, having the possibility to reduce damping on good road, or low speed and increasing it in the opposite conditions: a first example of controlled suspension!

The last set of figures represents a hydraulic shock absorber, to be applied on the side beam of the chassis structure. This shock absorber works bilaterally with progressive forces depending on the suspension extension or compression speed. Modern telescopic shock absorbers implement this principle in a simpler way; they were introduced from the end of the ‘920s.

Coming back to leaf springs, we shouldn't forget that their success was also bound to the reliability of the fabrication process of steel leaves. Material microstructure could be improved easily by plane forging; a rupture of a leave was never catastrophic, because leaves broke down one at a time and the consequent vehicle trim variation was easily detected. The consequent repair was simple and could be done by a competent smith. The arrival of coil springs was delayed by these facts; as a matter of fact leaf springs were initially applied also to independent suspension.

With the increase of vehicle speed some critical aspects of the rigid axle started to appear: the most important problems were weight, secondary deformations and *shimmy*.

The axle weight was remarkable; the amount of the unsprung mass as compared with the relatively high vertical flexibility of tires could cause on certain road axle hopping, very concerning in turns and never suitable for a good ride comfort.

Secondary deformations were caused by braking (S deformation) or by vehicle roll (different elongations of the two suspension of the same axle). The first type of motion could cause very annoying resonances of the entire driveline while starting up or braking on a gravel road or on a dirty road.

The roll motion of the body, due to the centrifugal force in a turn, causes a different longitudinal motion of the two points where leaf springs are joined to the axle; said in a different way, the rigid axle is steering, while the vehicle is turning. The architecture dictated by the transmission installation (fixed eye in front and swing shackle in the rear side) cause the vehicle to oversteer or to increase the path curvature, because of the deformation; this fact could cause instability or spinning around also in curves that today appear not demanding.

The scheme on the upper side of Fig. 1.9 illustrates this behavior for a Rolls-Royce rear suspension; the advantage is that while driving on an asymmetric obstacle a self aligning steering action can be provided.

To avoid this inconvenience some rigid axle suspensions showed two longitudinal linkages that modified the cinematic behavior of the axle; an example is given on Fig. 1.10, where two rods have the double function of avoiding the S deformation and of giving the axle an understeering behavior.

The previous Fig. 1.9 explains us, looking at its lower part, the *shimmy* phenomenon, typical of rigid steering axles; riding over an asymmetric obstacle imposes a certain roll speed to the axle. Because wheels are rotating and they had in old cars a significant inertia moment, a gyroscopic torque is applied to the steering mechanism; the elasticity of the steering wheel driveline (some elasticity was required to limit steering wheel vibration cause by the engine) could induce very annoying oscillations. Since obstacles are mainly asymmetric this phenomenon was very well known; it was solved on steering rigid axle by using additional shock absorbers, to be applied to the track rod.

A single big advantage of the rigid axle was highlighted after the diffusion of independent suspensions; on rigid axle the body roll doesn't affect the camber angle of the wheel, because the will is always perpendicular to the surface of the road.

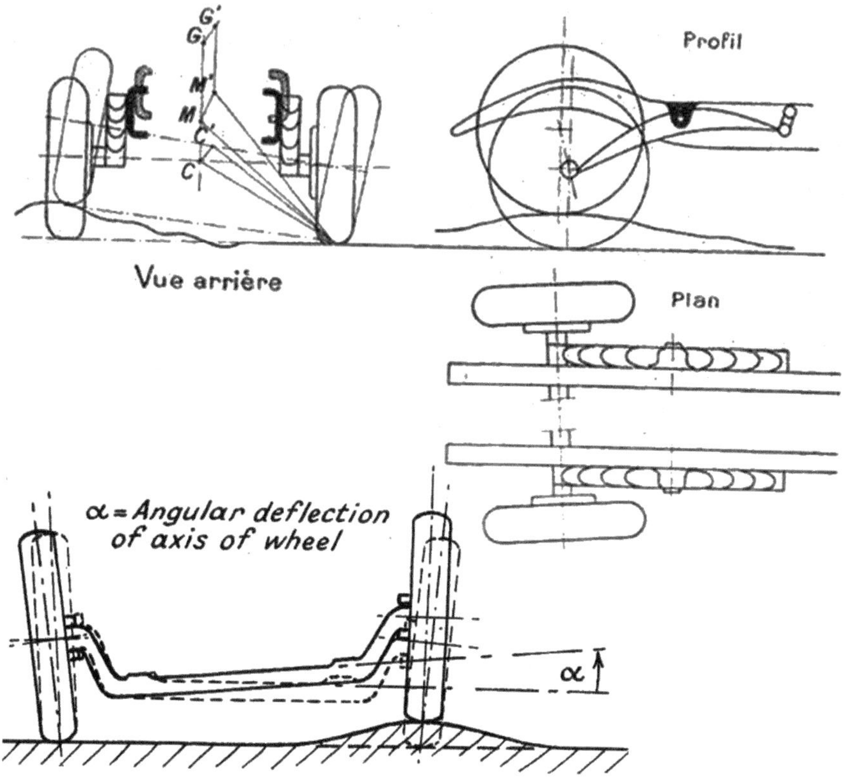

Fig. 1.9 The upper scheme illustrate the axle steering of a Rolls-Royce suspension while negotiating asimmetric obstacles or in a turn. The lower scheme justifies the phenomenon of the *shimmy*

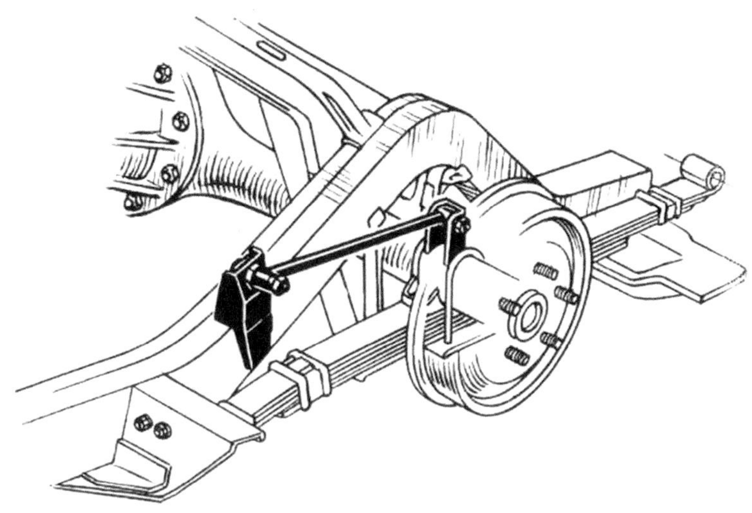

Fig. 1.10 Example of a rigid axle with leaf springs, showing additional linkages, to reduce axle steering and S deformation

1.3 The Independent Suspension Mechanical Linkages

Rigid axle suspensions with leaf springs dominated the market for long time and they started to be substituted by independent suspension on the '930s only; nevertheless many independent suspensions examples are older than this date, starting from the very first, the already cited Mancelle.

Let us consider steering suspension at first.

Many example are present of suspensions where the wheel hubs are guided by vertical tubes; this solution was perfected and produced by Lancia.

Probably the first application of this kind was presented in the Stephens of 1898, shown on Fig. 1.11, where the front wheels are guided by telescopic steering fork, inspired by a bicycle; they are sustained by a transversal leaf spring fit in the middle to the body.

This architecture was considered later by Sizaire and Naudin and Decauville that many authors quote as precursor to Lancia. In any case we should doubt that these precursors had a clear view of the advantages they could claim.

It is interesting to note that many of the first advertisement on independent suspensions were showing the wheel while riding on an asymmetric obstacle and never on a curve road. That fact suggests thinking that the new architecture was developed to avoid the shimmy phenomenon and no attention was paid to the camber variation caused by roll angle; this fact also explains why independent suspensions were applied to steering axles at first.

Fig. 1.11 The vertical tubes suspension was probably introduced for the first time by Stephens in 1898

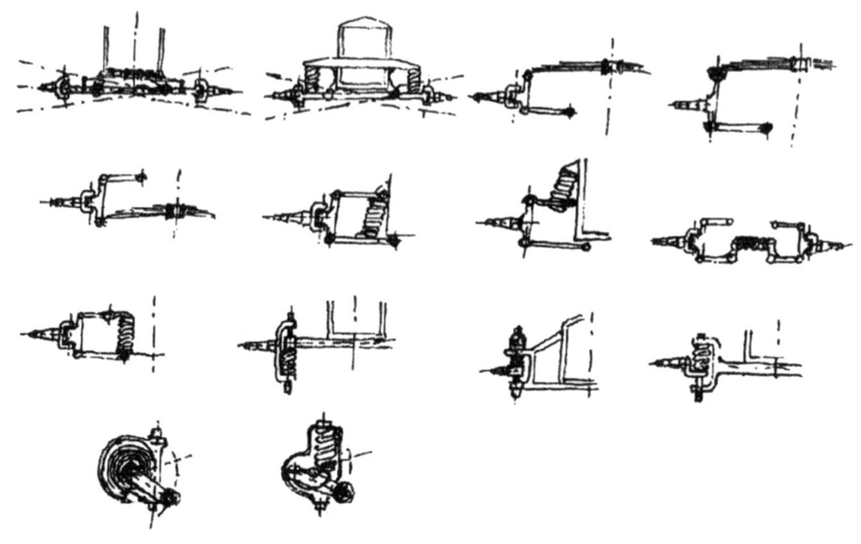

Fig. 1.12 This table is an evidence of the alternatives that were examined by Falchetto while designing the front suspensions of the Lancia Lambda. The architecture considered are almost all those existing on present cars

The problem of camber angle variations was studied and solved only later in the '940s, when practical speed increased significantly.

An other thrust to change, we shouldn't forget, were the advantages obtainable for the entire vehicle architecture. The front rigid axle, with its relevant lateral extension, couldn't be set to close to the ground; over the axle a certain clearance should be provided to allow for suspension compression stroke. As a consequence the engine had to be positioned at a relevant height on the ground; the front part of the body, the hood started just behind the front axle and had a relevant vertical dimension.

The independent wheel suspension made this layout no more necessary; the engine could be set down and forwards, between the suspension arms, with advantages of vehicle length and weight, with reduction on the height of the center of gravity and consequent improvements in speed because of streamlined aerodynamic shapes.

Lets return to front independent suspensions with vertical tubes. Figure 1.12 shows a set of sketches drafted by Falchetto, the vehicle engineer which, under the guidance of Vincenzo Lancia, faced the problem of introducing an independent wheel suspension on the future Lambda.

The solutions considered in that occasion were almost all those that were developed in the following years; the choice was in favour of the third on the third row from the top, probably because of the problem of shimmy, but there is no trace of this decision.

Fig. 1.13 The front view of the Lancia Lambda shows the front suspension and the gate structure of connection to the body (Automobile Museum of Turin)

Figure 1.13 shows a picture of the final vehicle launched in 1922; the most important feature of this design, as we saw not completely new, was the solution of the problem of an efficient lubrication of the sliding tubes by integrating, in a sealed element, guide, spring and hydraulic shock absorber.

The elastic element is now a coil spring; this is probably the first application were the car weight is entirely sustained by a spring of this kind.

An interesting variation of this suspension family was presented by Cottin-Degouttes on a car shown on Fig. 1.14 and introduced in 1927; the vertical tubes suspension was modified, but we don't know the reasons for this decision.

The tubes are slightly inclined with the top closer to the center of the car, while the elastic element is again a cross leaf spring, like in the Stephens. The leaf spring is articulated to the sliding strut; because the path of the tip of the spring is about a circle and the strut it coupled to the tube with spherical element, we can assume that the wheel can partially recover the camber angle variations caused by the body roll.

The kinematic properties of this suspension can be similar to a double wishbone suspension; but this kind of architecture didn't have an industrial follow-up.

A different scheme of front independent wheel suspension is attributed to Dubonnet, an important independent car designer, and was exploited by many car manu-

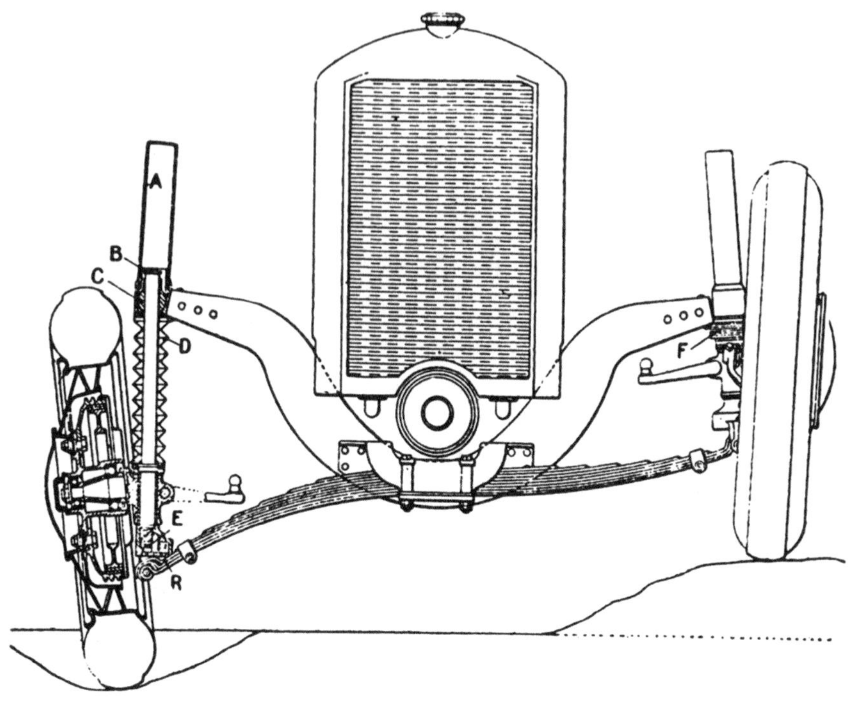

Fig. 1.14 This front suspension was presented by Cottin-Deguttes in 1927. No other examples of this kind were developed

facturers, among them by Fiat that applied this solution to production starting from 1935.

The suspension includes a sealed bearing element (cartridge) that integrates the helical spring in the same oil of the shock absorber, as shown on the detail on the far right of Fig. 1.15; spring and damper work on a finger crossing through the cartridge. This finger is connected to one of the arms of a double wishbone mechanism.

The cartridge can be integrated on the suspension in two different ways. As we can see on the left of the same figure, the cartridge can be mounted on a non suspended rigid axle through a king-pin; in this case the entire suspension is steering with the wheel and the steering mechanism is the same as for a rigid steering axle, being the motion of wheels completely independent. The kinematic behavior is almost the same as a vertical tube suspension, but the spring and shock absorber unit is easier to be manufactured. The architecture is similar to that of a double trailing arms suspension.

The second alternative is shown on the right of the same figure, in this case the cartridge is flanged to a front cross member of the chassis structure and the suspension arms can swing but not steer; the wheel strut is mounted with knuckles

Fig. 1.15 Two version of independent front suspensions designed by Dubonnet and produced by Fiat, starting from 1935. The suspension includes a bearing sealed cartridge, integrating the elastic element and the shock absorber

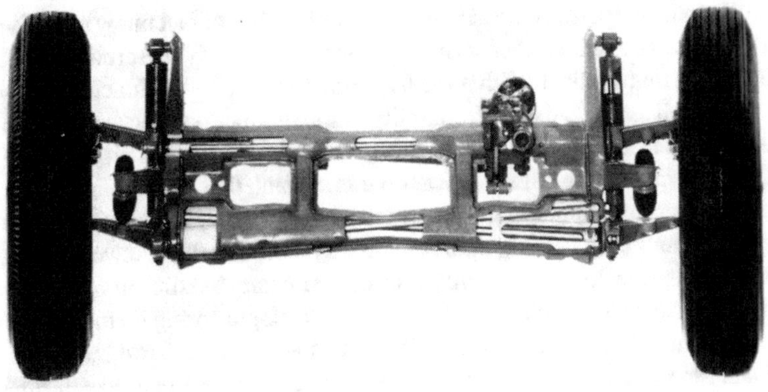

Fig. 1.16 The Porsche suspension is made with two double trailing arms elements; it was developed in 1931 and produced by Volkwagen till the '970s

to the oscillating arms. As a double wishbone suspension the steering mechanism is different and the track rod is substituted by an articulated system.

A suspension similar to the first scheme of double trailing arms is attributed to Porsche and is shown in Fig. 1.16; it was developed in 1931 and applied till the '970s on the Volkswagen Beetle and other car of this company.

The two articulated parallelograms are mounted at the end of a double tube structure building up also the front cross member of the platform.

Upper arms are flanged to the same torsion bar, contained in the upper tube; this bar limits the roll angle. Lower arms are flanged to two different torsion bars, each of them being flanged at the other end to the tubular structure; they act as elastic elements. The dimensions of this suspension are therefore very contained in the three directions.

When these suspensions were very effective in reducing shimmy, they showed later on an undesired feature: they caused the wheel to have, during turns, a camber angle equal to the roll angle of the body.

Bearing in mind the roll angle, due to the centrifugal force, is leaning the body to the outside of the curve, the wheels are assuming a camber angle such as to reduce the cornering stiffness of the tire. It would be better if the camber variation would be opposite, but the ideal behavior is to have no variation in every condition.

To improve this undesired feature double wishbone suspensions were developed with arms of unequal length; the shorter upper arm is increasing the camber angle when the suspension is compressing and decreasing when the suspension is extending. The value of camber angle is such as to compensate partially for the effect of the roll angle.

One of the first examples of this concept has been introduced by Studebaker in 1939 and followed by many car manufacturers till present days. A drawing of this suspension is reported in Fig. 1.17.

We would like to point out that the elastic element is integrated with the lower arm and is still made with a transversal leaf spring; as a matter of fact many manufacturers weren't so keen to abandon leaf springs. Reasons were many: first reliability, but we can also see in this application that the leaf spring integrates the function of two elements: arms and spring. Also the reliability of the production process was important and last but not least the existing investments for high volume of production.

There is an interesting patent of Fiat on this subject, that was applied to cars with rear engine, starting from 1955; it provided that the leaf spring was fit to the body through two symmetric bearing points. With a suitable distance between the bearing points it is possible to obtain a different elastic characteristic on symmetric and asymmetric suspension strokes; in this way with a single spring element is possible to obtain the function of the main elastic element an of the anti roll bar.

We can also observe in Fig. 1.17 that the Jeantaud mechanism has disappeared, which was present on previous steering systems; the track rod cannot be applied anymore, because the distance between the articulation points of track rod arms is changing with the suspension stroke.

Before the coming of the rack steering system, applied starting from the '960s, a small articulated parallelogram was employed whose rods were articulated to the body and were rotated by the steering box; to a suitable point of these rods were connected the two steering linkages.

Also this mechanism approximates with some error the ideal Ackermann law; the presence of sideslip angles always increasing, because of the increasing speed, makes this error less important.

The theorization of the behavior of tires under the application of side forces and the deriving concept of sideslip angle was also developed during these years.

The double wishbone suspensions with different length arms are diffusing quite rapidly in the following years and became almost totally applied to the front axles during the '960s.

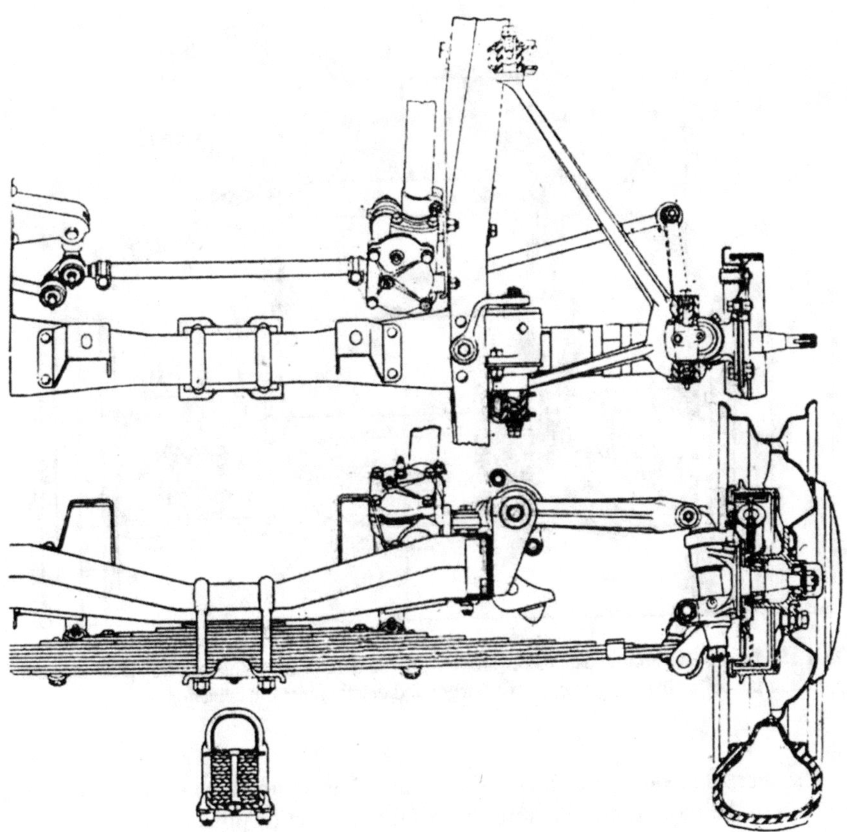

Fig. 1.17 Front suspension made by Studebaker in 1939. The mechanism is double wishbone type and features an upper arm of reduced length. The lower arm integrates the elastic function being made with a leaf spring

Figure 1.18 represents a suspension with stamped steel low thickness steel arms and coil springs launched by Fiat in 1950 on the 1400 and applied with minor modification on smaller models in the following years.

McPherson, a design engineer of American Ford , introduced this front suspension in 1947 that was called after his name (Fig. 1.19); it can be considered as a double wishbone suspension with different length arms, where the length of the upper arm is infinite.

It shouldn't be confused with a simple simplification of the double wishbone suspension for cost reduction; this suspension later contributed to the diffusion of modern front wheel driven cars, because the avoidance of the upper arm left the necessary space for the engine transversal installation. This suspension had a rapid diffusion starting from the end of the '960s not only on front wheel driven cars born in those years, but also on many cars with longitudinal engine.

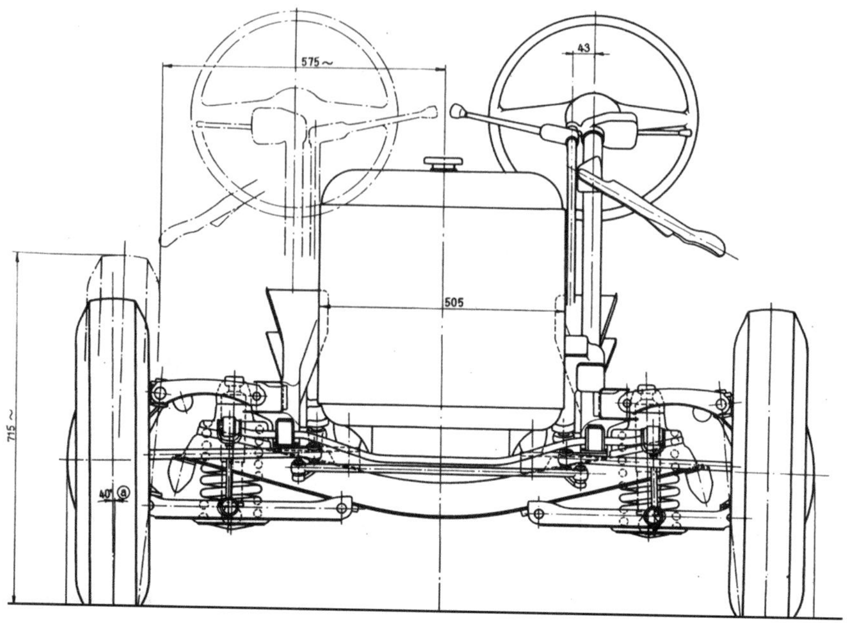

Fig. 1.18 Double wishbone suspension adopted on Fiat cars in 1950; this solution is almost totally applied in the following years because of fairly good elastokinematic behavior

The second advantage of this suspension is that it simplifies the body structure because of the more rational distribution of the connection points.

The reduction of kinematic performance, as compared with the double wishbone solution is not very relevant and, therefore this solution is also applied to contemporary sport cars.

Honda in the '980s introduced a last evolution to the front suspension architecture. It conceived a swan-necked wheel strut (see Fig. 1.20), which allowed the double wishbone suspension to be also installed on transversal front wheel driven cars. The descendants of the Mc Pherson and Honda suspension are sharing now the present market, with the introduction of many improved details the we will drop for sake of simplicity.

The rear suspension history should be much more complicate; the attempt to identify an evolution trend risk to oversimplify the explanation.

It should, first of all, be pointed out that the rigid axle, receiving elastic member more sophisticated than the leaf springs and additional linkages, had a very long life either on the front driven cars and on the rear wheel driven cars. On front wheel driven cars the weight of the axle was not relevant; the simplified function, because of the absence of the differential and final drive, in fact, allowed the use of tubular structure quite light.

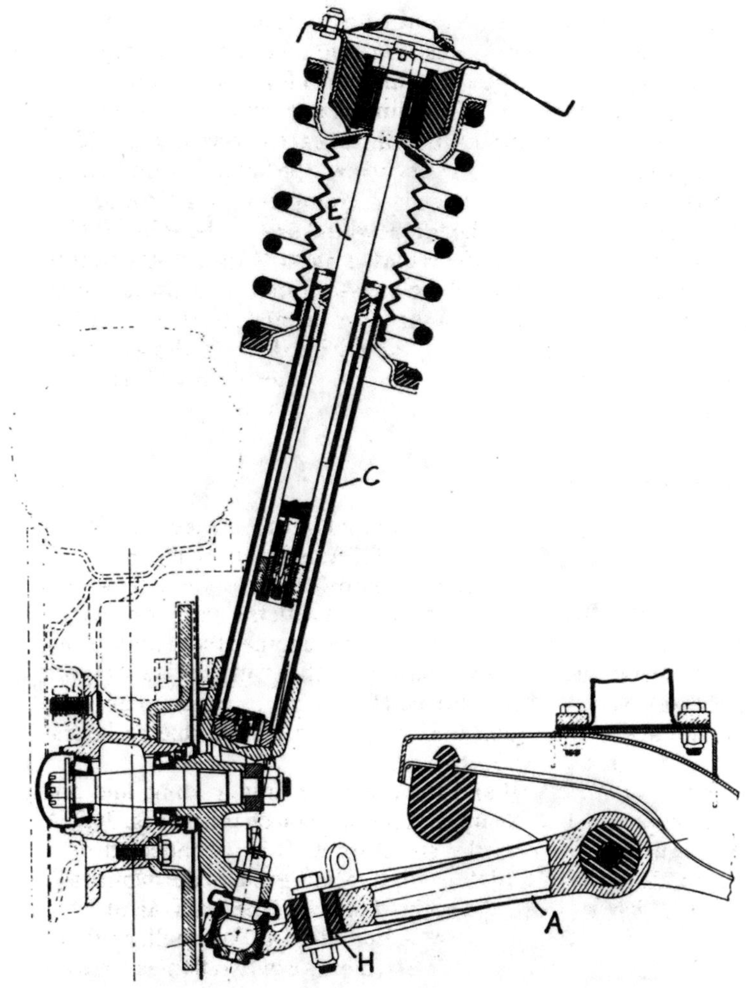

Fig. 1.19 McPherson introduced on American Ford cars this kind of suspension starting from 1947. The kinematic equivalent of this suspension is a double wishbone suspension where the upper arm has infinite length

On rear wheel driven cars, particularly on luxury and sport cars adopting the rigid axle, the weight was reduced with suspended differentials and a good kinematic behavior was obtained with coil spring and more complicated linkages.

An example of rigid rear axle particularly elegant is given by Alfa Romeo. This scheme was adopted on different cars starting from the '970s. It can be considered as an improvement of the De Dion-Bouton suspension. Figure 1.21 shows this example.

A triangular structure building up the rigid axle is linked to a spherical joint in the front, which determines a precise position for the roll axis; suspension stroke

Fig. 1.20 Honda front suspension with high double wishbone; high is referred to the position of the upper arm as compared with the conventional case

Fig. 1.21 An example of rigid rear axle particularly elegant is given by Alfa Romeo. This scheme was adopted on different cars starting from the '970s. It can be considered as an improvement of the De Dion-Bouton suspension

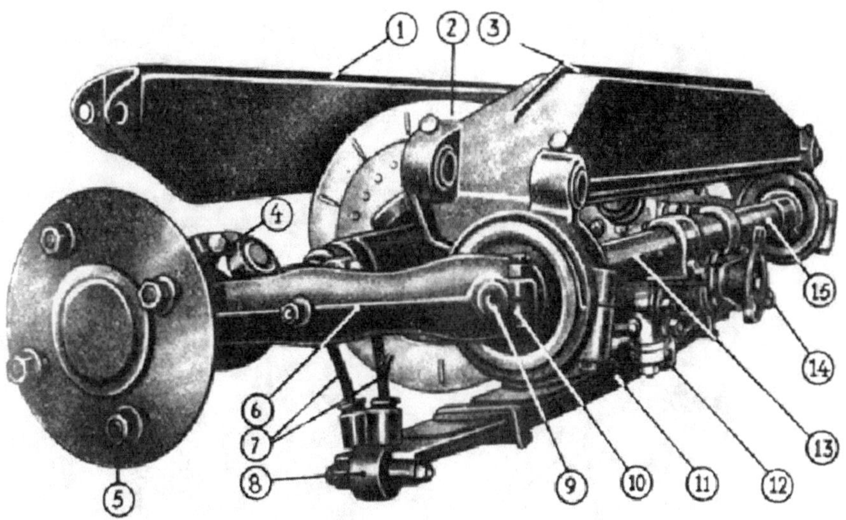

Fig. 1.22 One of the first application of an independent rear suspension with trailed arms appeared on the Lancia Aprilia in 1937

and body roll do not affect axle steering thanks to the guidance given by a Watt mechanism in the back.

The need to obtain a regular and large space for the trunk, imposed the development of rear suspension different from the rigid axle.

One of the first application of an independent rear suspension with trailing arms appeared on the Lancia Aprilia in 1937. This suspension, presented on Fig. 1.22, is adapted for a rear wheel drive and consists of two longitudinal arms with the articulation point in front of the wheel; in this case the elastic element is still a transversal leaf spring with the addition of a transversal torsion bar.

Being an independent suspension, the differential and final drive is suspended to the body. This kind of suspension, in connection with the application of coil springs, had a considerable diffusion, lasting on present cars, essentially with front wheel drive.

It enjoys the advantage of a reduced space for linkages and elastic system; the disadvantage is the modest performance and the weight of the trailing arms.

The rear wheel driven cars with rear engine had a significant diffusion between the '940s and the '960s; today they are no more applied, making an exception for some niche sport cars.

The reason for the diffusion of this architecture was the very good interior roominess with contained exterior dimensions, thanks to the absence of the propeller shaft, when economic and reliable component for the front wheel drive were not available yet.

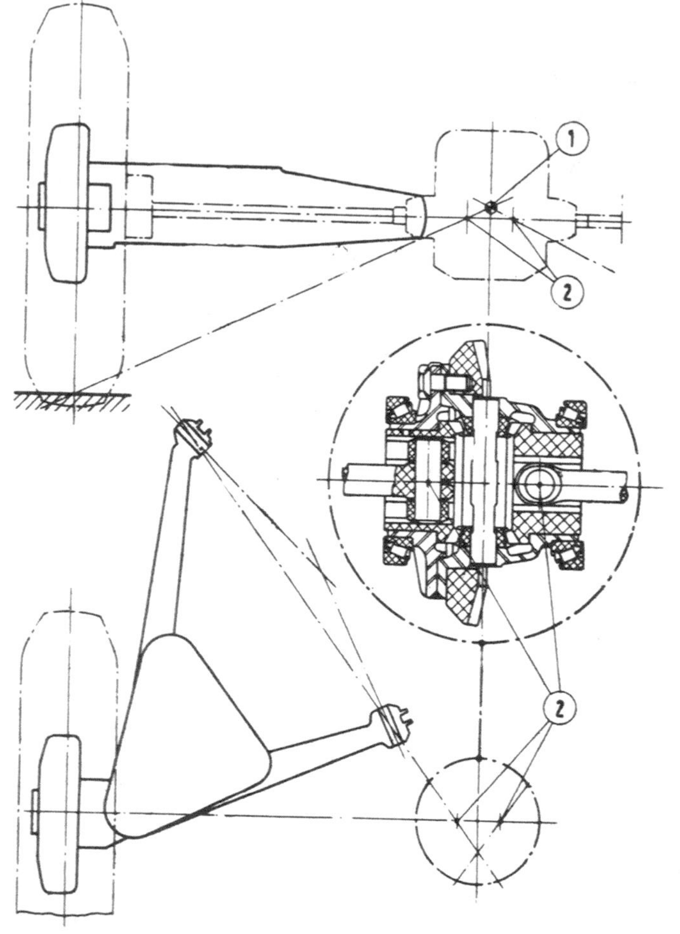

Fig. 1.23 The rigid axle couldn't be applied for compatibility with the rear powertrain. On the Fiat 600 of 1955 the semi trailing arm suspension was chosen because it allowed to use very simple constant velocity joints

The rigid axle couldn't be applied because of the short distance between power train and axle; the so called semi trailing arms were usually selected, because of the compatibility with cheap constant velocity joints.

On Fig. 1.23 is shown a scheme of the Fiat 600 of 1955. The traced dotted lines show how it was possible to have the suspension roll axis crossing the differential; this condition is mandatory to avoid the application of sliding constant velocity joints.

The camber angle relative to the ground doesn't change because of the roll of the body, but features remarkable variations because of payload variations; the look of the unloaded car is characterized by a sensible positive camber (wheel mean planes cross under the ground). In this conditions rod holding is quite approximate. In

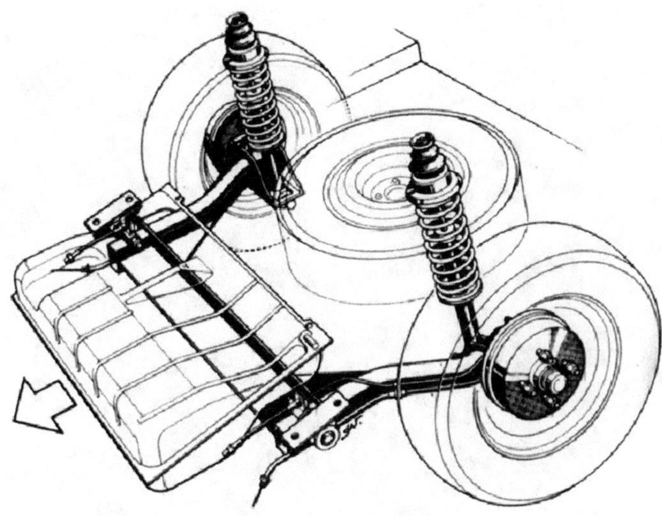

Fig. 1.24 In 1969 Volkswagen conceives a new rear suspension quite suitable to front wheel driven cars: the so called semirigid axle or twist axle; in short time it will become one of the most diffused architecture

addition to that on bumpy roads the track is changing continuously, because of the suspension stroke, leading the tires to premature wear.

This kind of suspension was also adopted with many rear wheel driven cars with front engine, because of its limited height; today is abandoned.

In 1969 Volkswagen conceives a new rear suspension quite suitable to front wheel driven cars: the so called semirigid axle or twist axle; in short time it will become one of the most diffused architecture.

Figure 1.24 shows the original application and demonstrates quite clearly the advantages given on the car; tank an spare wheel find their place between the arms and the cross member and leave the trunk bottom low, flat and wide.

From this point of view only, trailing arms are slightly better. They don't allow any camber angle variation relative to the body, during suspension stroke; the wheel wells steel sheets can be closer to the tire profile and leave more space to the trunk at a given car track. Their elasto-kinematic behavior is lower, as far as the cornering stiffness of tires is concerned, as we already explained.

On twist axles the gate structure bearing wheels is characterized by a cross beam made with a steel bar with open cross section, like a horizontal U, for example, open to the back. With this feature the gate is flexible to torsion (differential suspension strokes), but is stiff to flexion (side forces). Wheels are, therefore, very effectively guided and don't change their camber angle respective to the ground very much during body roll.

We can conclude this historic overview of car suspensions mentioning the multilink architecture, which represents the top evolution of suspension mechanism. It

Fig. 1.25 Multilink rear suspension were applied for the first time to mass production cars by Mercedes in 1982

was introduced during the '960s on competition and sport cars, but it was applied for the first time on mass production by Mercedes in 1982; it is represented on the drawing on Fig. 1.25.

If we consider the suspension from a geometric point of view only, we can say that the suspension mechanism must leave to the wheel one degree of freedom only, the stroke, not considering the wheel rotation around his hub. If we consider cinematic linkages with spheric joints only, we can apply a maximum of five linkages to reduce the six degree of freedom of a free body in the space to one. The multilink suspension features five linkages to obtain the maximum number of adjustable parameters to approximate the ideal behavior in the best way.

With this arrangement we can minimize camber variation with reference to the ground, we can also induce too angle variations able to optimize steering behavior and longitudinal flexibility, for optimum comfort, can be obtained with no drawback on steering.

The multilink suspension family is very large; it is now applied to almost all rear wheel driven cars with front engine; many application are already present on large and medium front wheel driven cars.

The independent wheel suspension chapter covers the last sixty years of the car history; but evolution hasn't stopped. If it would be difficult to think about new configuration of mechanisms many efforts are made on the field of controls: but this topic will be covered on chapters dedicated to the state of the art.

1.4 Wheels and Tires

In consideration of the importance of wheels on cars, we will extend our historic outline to much more ancient times.

For hundreds of thousands of years man did live without using any particular means of transportation. When he had to move an object, he simply lifted and carried it, if he was strong enough. If the object was too heavy, he arranged to drag it along. It is well possible that occasionally branches or other round objects were slipped under the load to reduce friction, but no evidence of this practice remains.

With the Neolithic revolution the need for transportation greatly increased and, on the other hand, the practice of taming animals opened new perspectives. The development of agriculture caused the need to transport seeds to the field and crops back to the homestead and the number of objects which were considered important and which men needed to carry with them increased as a result of the new exigencies of village life.

It is given for sure that sleighs were used in northern Europe before 5000 B.C., and their use at that time in other places can be inferred. Sleighs and drags can actually be used for transportation not only on snow and ice, but also on grassland (American Indians used the *travois* well in the XIX century) and even in deserts and sometimes on rocks.

It is impossible to state when a drag was for the first time mounted on a pair of wheels or who operated this technical revolution. Ancient wheels were made mainly of wood, and little direct archeological evidence could remain.

About 3500 B.C. the potter's wheel was introduced in order to produce pots with axial symmetry. The use of the potter's wheel can be inferred from the marks left on pots made with it. The supporting wheel for vehicles is thought to have been originated about the same time.

The most ancient evidence of a wheeled vehicle is from a pictogram on a tablet from Inanna temple in Erech, Mesopotamia. Such document dates back to slightly later than 3500 B.C., and includes a small sketch of a cart with four wheels, together with that of a drag (Fig. 1.26a).

The vehicle shown in Fig. 1.26b has two features typical of all vehicles of its times for more than a thousand years: The wheels are discs made from three planks of wood and the animals are harnessed to a central shaft. This uniformity of types of wheels and of driving systems, particularly if compared with the great variety of vehicle structures, has led to the opinion that the wheel was *invented*, or better, developed, in a certain place and then started a slow diffusion in all the ancient world.

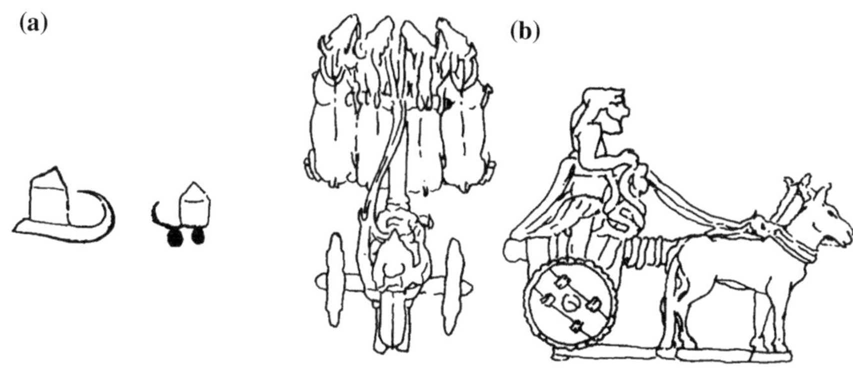

Fig. 1.26 **a** Pictogram on a tablet from Inanna temple in Erech, Mesopotamia. Such document dates back to slightly later than 3500 B.C., and includes a small sketch of a cart with four wheels, together with that of a drag. **b** Copper model of a war cart, driven by four onagers, found in the tomb of Tell Agrab, from the third millennium B.C.

In various places where it was introduced, the local type of sleigh was adapted to the new vehicle, by using the standard wheels and harness.

The place where the wheel was first developed is not known, but it is possible to infer that it was in Southern Mesopotamia, where the wheel was for sure used about 3500 B.C. The diffusion of the wheel was quite slow. Evidence of its use dates from 3000 B.C. in Elam and Assyria, 2500 B.C. in Central Asia and Indus Valley, 2250 B.C. in northern Mesopotamia, 2000 B.C. in southern Russia and Crete, 1800 B.C. in Anatolia, 1600 B.C. in Egypt and Palestine, 1500 B.C. in Greece and Georgia, 1300 B.C. in China and about 1000 B.C. in northern Italy. Some centuries later it reached northern Europe.

It is not possible to understand from ancient pictures whether the axle did turn together with the wheels or was stationary. The fact that the central hole of the wheel disc was round has little meaning as a circular hole can also be explained by the ease of construction. It is likely that both solutions were used, as there are still people who use those primitive technologies nowadays.

It is however likely that the wheel did not derive from the roller: The types of wheels used would rule that out, and it is likely that, in the mind of the ancient wheel maker, the wheel and the roller had little in common.

The need to build lighter wheels for war chariots probably led to the development of the spoked wheel, which is much more efficient for such use. The wheel with spokes was first used probably at about 2000 B.C. and by 1600 B.C. it reached its fully developed form, particularly in Egypt. The central part of a wheel of that type (with 8 spokes) is shown in Fig. 1.27a. It is a part of a chariot dating back to 1350 B.C., found in a tomb near Thebes. The spokes are fitted in the hub; the felloe is usually built in various parts but some examples of felloes in one piece, bent in a circular shape, were found as well.

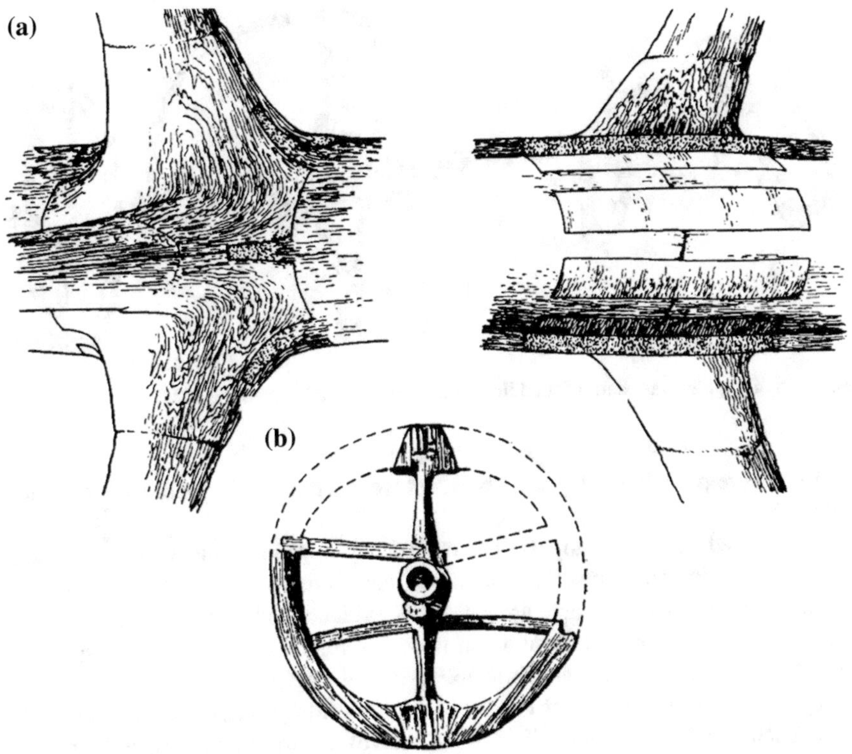

Fig. 1.27 **a** Cross section and view of the center part of a wheel of an Egyptian cart from 1350 B.C. found in a tomb near Tebe. **b** Wheel found in Mercurago, on the River Po flat, probably dating to about 1000 B.C.

A wheel which seems to be a stage in the evolution between the disc and the spoked wheel is shown in Fig. 1.27b. The time this wheel was built has however led one to think that it was simply an attempt to copy a spoked wheel by a wheelwright used to disc wheels; but wheels of the same type are represented on more ancient Greek paintings.

In many cases, even in very ancient times, wheels had a hoop or tire or at least some device to strengthen the rim. Some disc wheels have a wooden rim, in one or more pieces. Sometimes the rim of the wheel is inlaid with copper nails, to reduce wear or perhaps to keep in position a leather tire. Certainly many Egyptian war chariots had wheels covered with leather. In some pictures, even very ancient ones, something which looks like a metal tire can be seen. The evidence of such practice is however much more recent, dating back to about 1000 B.C. These metal tires were built in various parts, welded together and then shrink-fit to the wheel.

As it was already stated, only after animals were tamed could wheeled vehicles be propelled in a proper way. In Mesopotamia both transportation vehicles and war

Fig. 1.28 Egyptian war chariot found in a tomb near Tebe (XV century B.C.)

chariots were pulled by onagers. Also oxen were without any doubt used for transportation.

The spoked wheel, that appeared about 2000 B.C., was accompanied by the use of horses to drive war chariots. It is not known where horses were first tamed and used for that purpose, but the scarce archeological evidence indicates that it should have happened in north-east Persia, and that from that region the use of horses spread to the whole ancient world, from China to Egypt and Europe.

The structure of an Egyptian chariot of the XV century B.C. is shown in Fig. 1.28. It represents without doubt the best of the state of the art of its times, and remained unchanged for centuries.

The progress from the Sumerian vehicle shown in Fig. 1.26b is great, and if the greater power of the two horses compared with that of the onagers is considered, it is easy to understand why some historians ascribed to the use of this weapon the expansion of the Hittites in Anatolia, the Achaei in Greece and of the Hyksos who in the XVIII Century B.C. invaded Egypt, teaching the new technology to the Egyptians.

Chariots became obsolete when the knowledge of riding became wide-spread. Donkeys were used as pack animals and for transportation of persons in the third millennium B.C. and surely also horses were occasionally used in the same way.

In Europe only the Celtic tribes continued to use war chariots, which at the beginning were carried north of the Alps by Etruscans. Celtic wheelwrights learned the art of building wheeled vehicles and made significant progress.

The remains of the wagon found at Dejbjerg are shown in Fig. 1.29a. It is the first example of a wagon with steering on the front axle, but it can be considered an articulated vehicle made by two chariots. It is however unlikely that the solution was actually used for transportation; it looks more an insulated solution, for ceremonial (burial) purposes. At any rate it incorporated other interesting features, like the wooden roller bearings in the hubs.

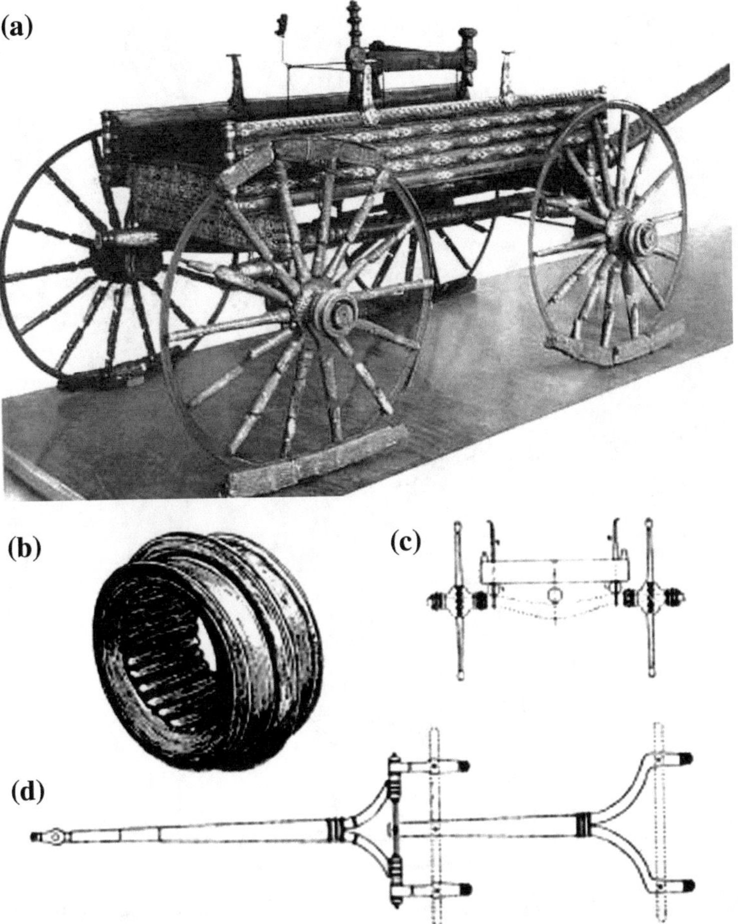

Fig. 1.29 Celtic carriage from the I century B.C. found near Dejbjerg: **a** picture, **b** wooden roller bearings in the wheel hub, **c** cross section, **d** view

The wheel shows construction details we will find on more modern vehicles: particularly interesting are the roller bearings, completely made of wood, in wheel hubs, shown on detail b.

Wheels commonly applied to carriages and coaches at the end of nineteenth century and on many of the first cars were the so called artillery wheels; the hub of this wheel is shown on the upper part of Fig. 1.30.

The wheel is made by wooden spokes, whose base is shaped like a sector of the hub. They are clamped by two metal flanges which build up also the outer ring of the bearing; the rim is made of wooden arches pressure fitted in a steel ring which forces the spokes into the hub. Spokes are perpendicular to the hub with a certain inclination according to a bevel structure.

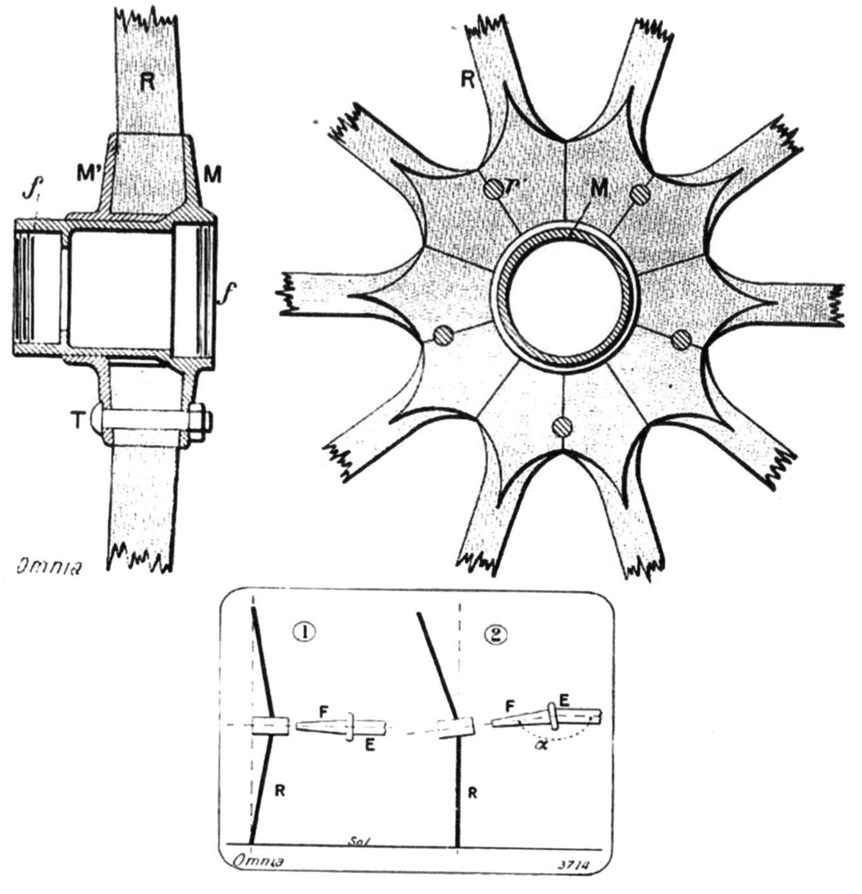

Fig. 1.30 So called artillery wheels were common on carriages and on the first cars. Wheel spokes are perpendicular to the wheel axis but are slightly inclined, according to a bell shape. This shape required the wheel to be mounted with a certain inclination, respect to the vehicle, to have spokes working correctly. This angle was called camber angle (*carrossage* in french)

This kind of design increase the lateral robustness of the wheel and gives to the wheel the necessary radial flexibility to allow for rim hot pressure fitting.

This shape required a suitable inclination of the wheel rotation axis to have spokes working correctly under the vehicle weight; this angle (lower sketch) was called camber angle (*carrossage* in french).

On metallic wheel this kind of inclination didn't have anymore reason to exist, but the name remained.

On this kind of wheel a solid rubber rim could be mounted. Other application showed a cycle type wheel with solid rubber tire, like the cars of Daimler and Benz, shown on Fig. 1.4.

Fig. 1.31 Evolution of the car tire wheel. The first two in the upper row are made of wood with metallic reinforcements, the second one has dismountable rim. The third is the Sankey wheel, entirely made of steel and separable from the hub, solution chosen also on present wheels. On second raw two solution are shown with cycle type spokes, where the second has a rapid dismounting fit. Finally the Michelin disc wheel scheme, used also on present cars

In Fig. 1.31 many of the fundamental evolution stages of the pneumatic wheel are reported; we will speak of the tire itself later on.

The first wheel on the upper raw, in use till the '910s, were similar to the artillery wheels, we have referred about; instead of the metallic rolling rim, they presented a channel ring able to receive the pneumatic tire. These wheels cannot be easily dismounted form the hub; initially, a punctured tire was changed only directly on the lifted car, with enormous difficulty.

The separable channel rim, shown on the second figure makes the tire change easier, having one or more spare rims with inflated tire on board.

First solid metal wheels appeared in the '920s; the mechanic resistance was impressively increased. On this occasion the wheel was modified in order to be separable from the hub through a bolted flange, the so called Sankey wheel. Punctured wheel substitution was made quite easier.

Cycle type spoke wheels were applied till the '950s for their low weight, especially on sport and luxury cars; the first type shown present a conventional flange hub, while the second features a rapid dismounting Rouge Whitworth joint. This is characterized by a spline coupling, between wheel and hub, held in place by a big butterfly nut; this feature became a style sign of fine cars.

Fig. 1.32 Spoke wheel with Rouge Whitworth nut; spokes are arranged in two rows. The first inside raw has spokes with only radial inclination; these spokes transmit braking and driving torque and their inclination is motivated by. The second outside row provides for spokes on a bevel surface to carry on lateral forces

In comparison with the first car of Daimler and Benz, we can observe on more modern cars, like on Fig. 1.32, a different spokes arrangement, now on two rows, with different geometric configuration.

The first raw, inside with reference to the car, has spokes with only radial inclination; these spokes transmit braking and driving torque and their inclination is motivated by the fact that the spoke can only transmit forces along its axis.

The second outside row provides for spokes on a bevel surface and their orientation is suitable, together with those of the first raw to carry on lateral forces applied to the wheel.

Coming back to Fig. 1.31 we see, as a last example, the Michelin wheel; this wheel includes a stamped steel disc, welded to the channel rim; this solution, after a number of improvement, is still in use.

Is it is now unconceivable our everyday life without a car, it would be unconceivable a car without the existence of a tire.

The tire invention isn't as old as the wheel invention, but is, anyway, older as the automobile invention.

Thomson is remembered as the inventor of tires, described in an English patent of 1845; the invention was finalized to improving the ride comfort of coaches and to reduce their rolling resistance; the initial proposal provided for a leather sewed lining and a cloth tube treated with rubber.

This idea wasn't put to practice, but was proposed again by Dunlop, who filed a similar patent in 1888, without knowledge of Thomson's priority; this time bicycles could take profit of this invention that started to be applied extensively. As happened for the legal dispute of Otto, regarding the internal combustion engine, Dunlop patents were invalidated in 1890, because of Thomson priority; going on with the comparison of the two cases, this fact didn't affect Dunlop's business negatively, but was beneficial to those who wanted enter the same market and contributed to a rapid development of this technology.

Only in 1980 semi tubular tires were introduced; they could be adapted to a channel rim by a steel cord circle and allowed an easy dismounting.

Michelin Brothers started to diffuse dismountable tires on cars, starting the first time on the Paris-Bordeaux race of 1895.

This could be thought as the starting point of the tire life on automobiles, receiving many technology contributions by several manufactures.

A base invention for the diffusion of tires was rubber vulcanization, after a patent of Goodyear of 1844. Before the invention of this chemical treatment, probably discovered by chance, rubber didn't have stable shape and wore-out very quickly because of the temperature.

Many further improvements allowed to reach present performance; we recall the application of synthetic fibers to tire clothes, initially of line and cotton, made of rayon in the '930s, later of polyesters and of kevlar, from the '970s. A second major evolution refers to the radial texture of clothes, formerly diagonal, starting from the '960s.

Other improvements concern the chemistry of rubber and the introduction of synthetic elastomers; on this subject we will not supply further details.

A last evolution regards the geometric dimension of tires, characterized by the channel diameter D, by the tire section width W and by the so called *aspect ratio*, we will explain later, as ratio between the radial dimension of the tire and its section width; aspect ratio wasn't standardized, but was set around 100%; later, to improved comfort tires became wider and the need to reduce car height and weight reduced the tire diameter. Aspect ratios started to reduce.

Figure 1.33 shows the evolution of these dimensions for a series of Fiat cars, with curb mass in the range of 1000 kg; the rim diameter decreases from 25″ in 1910 to

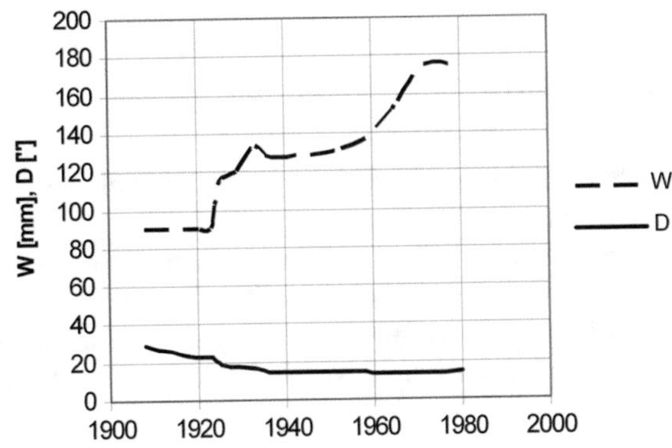

Fig. 1.33 Evolutionary trend of tires dimensions on a car with about 1000 kg of mass; the evolution is shown by the rim diameter D (″) and by section dimension W (mm)

Fig. 1.34 *Tweel*: elastic non pneumatic wheel recently presented by Michelin

13″ in 1970, increasing slightly in the following years; the section width doubles from an initial value of 90 mm to about 180 mm on modern cars.

The aspect ratio, not shown in this diagram had a stepped evolution; from an initial value of about 100% has gone to 80% in the ’940s, to descend again to 50% in the ’970s; in these years tires have conserved their outside diameter, increasing gradually the rim diameter.

On more recent cars outside diameter tend to increase.

We should finally point out that from the beginning of motor history many patents on *elastic wheels* have been filed, where elasticity was obtained my mechanical elastic means, rather than by the elasticity of the tire; they are often called *non pneumatic tires*. Diffusion of tires has made this kind of wheel to disappear for about one century, except as for the wheel of the LRV of the Apollo Project (see Appendix C). Quite recently Michelin has presented an elastic non pneumatic wheel, called *tweel* (contraction of tire and wheel) that could introduce a real innovation in the future (Fig. 1.34).

1.5 Brakes

Brake evolution is bound to top speed growth and to traffic density increase, that impose the need of stopping the car in short space; it is reasonable to think that braking first cars in their contemporary traffic wasn’t a serious concern.

First cars show brakes that work on the external surface of the wheel with moving shoes, like on horse drawn carriages and on railway cars; this very simple device was no more suitable for tired wheels, because of the potential wear.

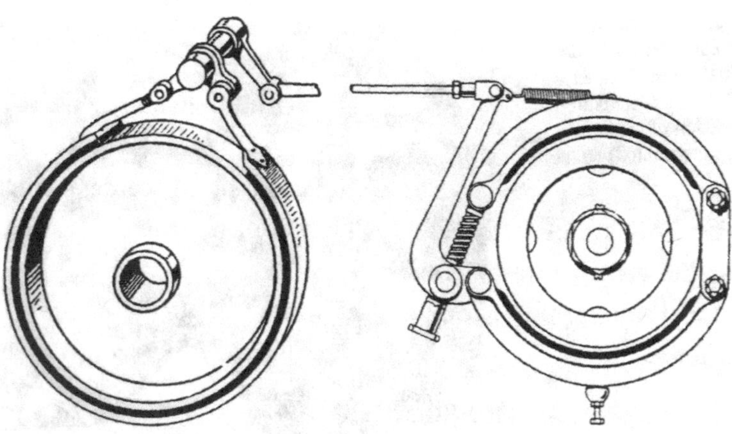

Fig. 1.35 Band brakes and external shoes brakes, to be applied on transmission line before the differential and final drive

For this reason external shoe brakes and band brakes, like those shown on Fig. 1.35 were introduced; they worked on the transmission line only. They choice was motivated by the need to exploit the differential for braking force uniform distribution too, very difficult to be achieved with mechanical linkages control. They were also almost insensitive to machining tolerances.

Brake liners were missing, but some example of *Ferodo* liners, synthetic material developed by Frood were already present on coaches shoe brakes; this practice diffused on large scale in the '920s.

Drum brakes were developed later on and they were applied on rear wheels only; they were more difficult to be machined but showed a superior capability of dissipating heat generated during long braking distances. Band and shoe brakes were initially not abandoned, but they continued to be applied on the transmission as supplementary or parking brake.

Disc brakes were introduced by Jaguar in series production in 1953.

Drum brakes were controlled by pedal and were usually designed to stop the vehicle; the transmission brake, vice versa, was controlled by a hand lever which could be stopped in the desired position and was usually designed for continuous braking on downhill slopes and a parking brake.

It should be noticed that position and function of driver controls was standardized informally only in the '930s and the solution initially adopted could be very different; a system that received a certain diffusion which appears unjustified today, provided for a combined control of brake and clutch, to avoid stalling the engine at the end of braking.

The application of brakes to front axle started only in 1910, thanks to a luxury car manufacturer: Isotta Fraschini.

Fig. 1.36 Completely mechanical control of front wheel brakes; the contact point of the cam, actuating the expansion of shoes is near the king pin axis of the wheel

This innovation didn't regard the brakes per se, similar to other rear brakes, nor the discovery of their usefulness, but the control system for steering wheels, shown on Fig. 1.36.

Brake controls were in fact mechanical, as already shown on Fig. 1.35 and were made by linkages and cables put in tension by pedal; in drum brakes the internal shoes were expanded by a cam. What should be avoided, at any rate, was that wheel steering could affect braking; the patent of Isotta Fraschini consisted in rounding the cam in such a way as contact points were always on the king-pin axis.

In this way brake control was insensitive to steering; this patent was overridden by other mechanical solutions and four wheel braking was state of the art at the end of the '920s.

What we said about steering motion, can be also applied to the suspension motion of both axles.

The distance between the articulation points of tie rods had not to change, over a certain threshold during the suspension stroke; if this had happened brake could be actuated by road bumps or tie rods could have been broken by excessive pull.

For this reason, as we can see on Fig. 1.37 articulations point between levers and tie beams had to stay on center of instantaneous rotation of axles, during suspension stroke that were located nearby the fixed eyes of leaf springs.

Tie rods must have adjustments for their length, to recover for wear of shoes; it should be noticed that also distribution of braking force is influenced by the length of rods.

These problems were completely solved by the application of hydraulic controls; this kind of control spread during the '930s and was proposed by Lockeed and introduced on a car for the first time by Duesemberg in 1921.

Another important evolution for braking systems was the application of auxiliary power, to increase the driver's thrust on the hydraulic lines; this device was introduced

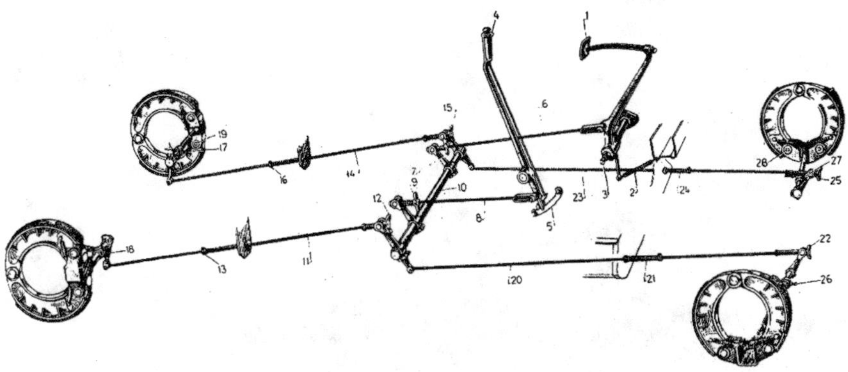

Fig. 1.37 Mechanical control for four wheels braking system; articulations point between levers and tie rods had to stay on center of instantaneous rotation of axles, during suspension stroke that were located nearby the fixed eyes of leaf springs

using the engine vacuum as energy source and was applied for the first time on a Ford car in 1954; it allowed also weakest drivers to brake efficiently.

Already in 1910 Rolls-Royce applied on its cars a completely mechanical power brake, showed on the drawing of Fig. 1.38.

The mechanical energy necessary to operate brakes is derived by the engine through a power take-off; this energy is transferred to the braking system through a clutch kept constantly sliding in controlled way by the brake pedal. The energy coming from the propeller shaft is contributing to slow down the car too.

1.6 Chassis Frame

To speak of chassis structure we must say something about car body manufacturers.

At the dawn of the motor industry era, people who dedicated their financial and intellectual efforts to this product preferred, understandably, to concentrate on the most peculiar problems of cars development that are, particularly, engine, transmission, suspension and steering system. In other words they became *chassis* manufacturer, according to a french name.

The body, according to technical rules known at the end of the nineteenth century was not considered a critical technology and was totally imported by coaches and carriages, as it was; many wheelwrights became also car body manufacturer.

This fact caused an initial separation of these two kinds of industry.

Chassis manufacturers used metallic materials mainly and their plants were able to cast, to stamp and to machine; due to the precision of couplings they worked according to drawings and produced small series.

Body manufacturers used mainly wooden structures and they worked as carpenters, using fixtures and tools to produce single pieces, often with no drawings. This

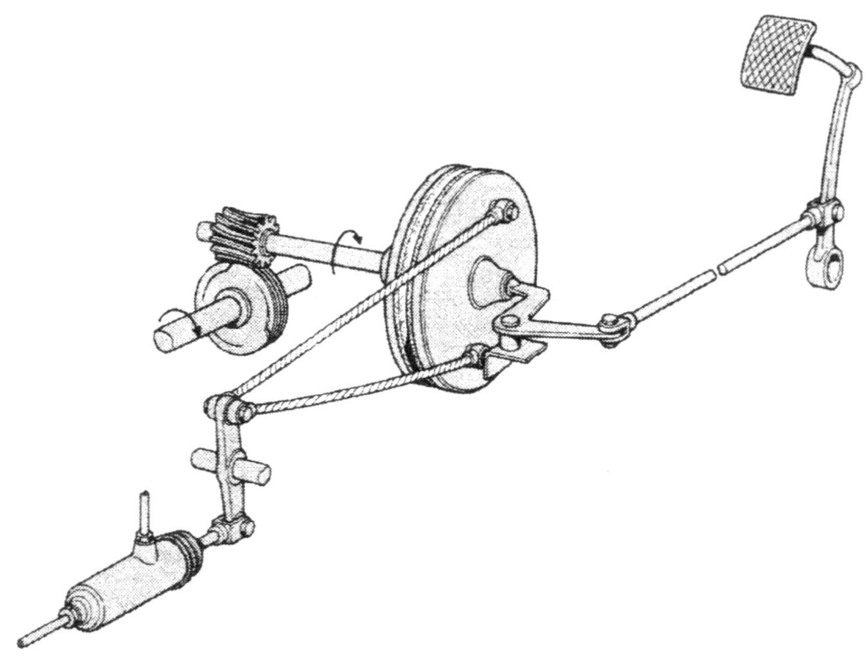

Fig. 1.38 Mechanical power brake adopted by Rolls-Royce; the necessary energy derives from the propeller shaft and is transferred to the braking system by a clutch, operated by the brakes pedal

tradition came from the fact that wood was much more suitable than metal, to obtain curved shapes that already followed elaborated aesthetic rules.

Also painting, vital to surface protection and nice appearance, was again in favour of wood, considering the existing oil paints and varnishes and justified a complete separation of the chassis and body fabrication to avoid damages during the complicate assembly process of mechanical parts.

It should be remembered that a complete treatment of a body required more than 400 h of work; to this time the drying time of any layer had to be added.

The chassis had, therefore an important technological value, required by the industry organization. It allowed the car manufacturer to produce a finished product that could be delivered to the body work, relieving the car manufacturer of any responsibility; it could also be driven, in fact, to be tested and to show the customer its performance.

The final customer bought, therefore, a chassis that was put in charge of a coach builder to have the car completed; also example were available where coach builders bought chassis on their own and sold finished cars and car manufacturers that had permanent agreement with coach builders to produce and sell complete cars.

This was the situation in Europe at the beginning of the first world war.

After the world mass production started and many manufacturers introduced their internal body shop. This fact determined also a gradual transition from wood to steel, with intermediate solutions, we could define as hybrid, with wooden skeleton and external panels made of wood or partially of steel sheets.

Synthetic enamels shortened the painting time of an order of magnitude and made possible a better integration of body and chassis fabrication.

The organization of the production cycle didn't change very much either because of tradition and because customers still requested special bodies that had to be assembled on a finished chassis.

The integration of chassis and body manufacturing was mainly developed in the United States and imitated in Europe by major manufacturers, starting from the '920s.

The frame which is the bearing structure of the chassis had to carry on all mechanical components, the complete body and the payloads. In addition the frame was the assembly support of all chassis components, including the engine, to obtain a rational organization of the fabrication process of the car.

Frames of first cars were made either of wood or of steel; this last solution was far more applied. Figure 1.39 shows an example of a complete chassis of a car of the beginning of the twentieths century, where the frame can be easily identified.

The frame is made by a *ladder structure*, showing two longitudinal *side beams* with particular shape.

In the front side the distance between the two beams is reduced to allow for the necessary space to steer the front axle wheels, while the rear side is larger to better install the car body; in this area the distance between side beams is determined by the track and by the dimension of the transmission.

Front and rear ends are often curved on the side view, to better match the leaf springs shape and allow for suspension compression stroke. The curvature of latest cars was increased to reduce floor car height between the wheels.

Side beams are connected by a number of *cross members*, shaping the ladder structure.

Side beams and cross members are bolted in our example, but in other cases the were heat riveted and, later, lap welded.

This kind of structure has been maintained for years and is still in use on many industrial vehicles; the choice of bolts or rivets instead of welding was imposed by the dimensions of the structure, such as to induce unacceptable residual stresses after welding, because of the quantity of applied heat.

Torsional stiffness of this kind of frame is not particularly high, because it is determined by the cross section of side beams, almost always of open type to make the assembly with cross members and chassis components easier. Also bending stiffness is low because of the limitations on side beams height.

Chassis frame deformations became an important issue because of their effect on body; the problem became more evident by increasing of maximum speed.

The body was made by a wooden skeleton structure which was quite stiff after the completion of fabrication, considering the carpentry joint that could be applied; but these joints couldn't withstand the task of making the entire car stiff.

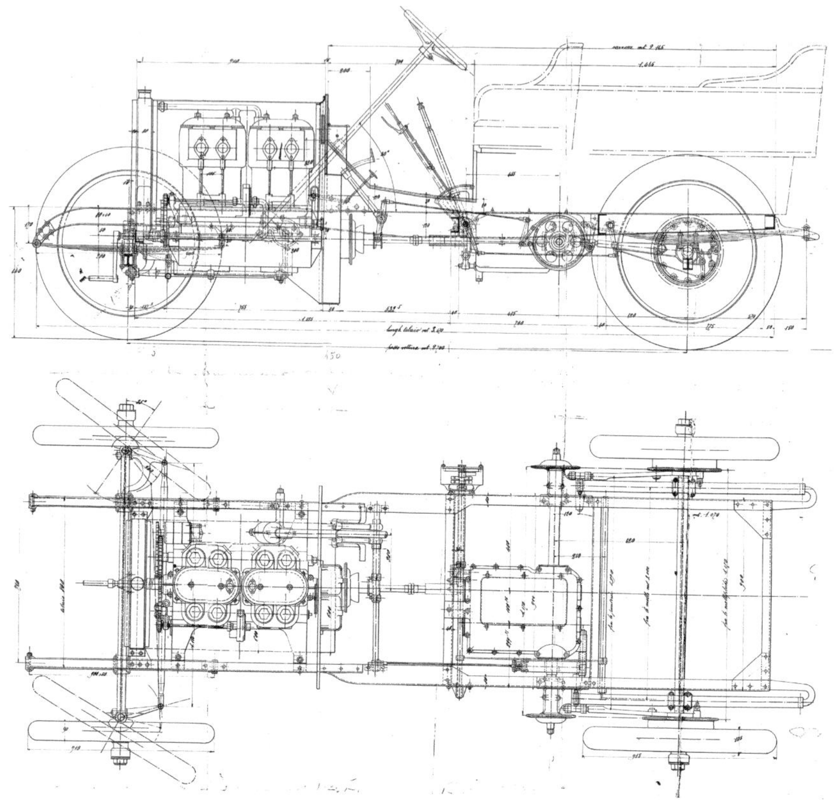

Fig. 1.39 Complete chassis of a Fiat car of 1905. The chassis structure made of a ladder frame is similar to that of many cars of those years

Chassis deformations applied to a stiffer body resulted in loosening or ruptures of the body skeleton joints; these facts built up plays between body parts and, therefore, annoying squeaks and rattles.

Body joints were perfected abandoning carpentry solutions in favour of more flexible metal joints that avoided contact between parts; to this purpose also fabric structure were developed, according to Weymann's patents, coming from the aeronautical technology of those years, that allowed for a very flexible, but silent body.

The final solution came by using steel sheets, at least for outside body panels; on one side the increased flexibility better matched chassis deformations; on the other side it could contribute to the entire structure stiffness, being able to carry on an increased stress.

Chassis frames of some car from the '930s changed also aspect and showed more elaborated schemes, as can be seen on Fig. 1.40.

We can see at first on this figure that frame elements are made with stamped parts. Side and cross members, for their complicate shape are deep-drafted using dedicated

Fig. 1.40 Central beam frame structure of Fiat 1500 of 1935; the structure is made of steel sheet pieces stamped and riveted each to the other

stamps; on previous example, on Fig. 1.39, same components were cut from profiled steel bars and were bent in open stamps.

The advantage of this innovation is a better control of structure weight, because every part of the frame is designed according to local stress; the disadvantage is obviously the higher amount of dedicated tool investments.

Instead of the already shown ladder frame, we have now a cross frame, where side beams are shaped to build an X; cross members instead of connecting side members shape up cantilever beams to carry on the floor and the body. The two side members are joined together in the middle building up a tubular structure much torsionally stiffer; the scheme of the frame allows the floor to be in a lower position with benefit for the car height.

Front cross member is a cast element bolted to the frame, able to receive the independent suspension shown on Fig. 1.15.

Most joints are riveted.

This kind of chassis is connected to the body through robust bolts. The body was made by stamped steel panels, invested on a wooden skeleton; the assembly is not particularly stiff but it should contribute to the total torsional stiffness.

A further step is shown on the Volkswagen Beetle, almost contemporary of the cars previously shown, but produced for longer time; the chassis structure is shown on Fig. 1.41.

In this case the structure is made by a central beam to which floor and front and rear cross members are welded, to carry on front suspension (shown on Fig. 1.16) and rear suspension and powertrain.

The floor is bolted to an integral full steel sheet body completely welded; the solution can't be called a unitized body, but its structural behavior can be comparable when assembly is completed.

This kind of chassis structure, called chassis *platform*, was still in use in many cars till the '950s.

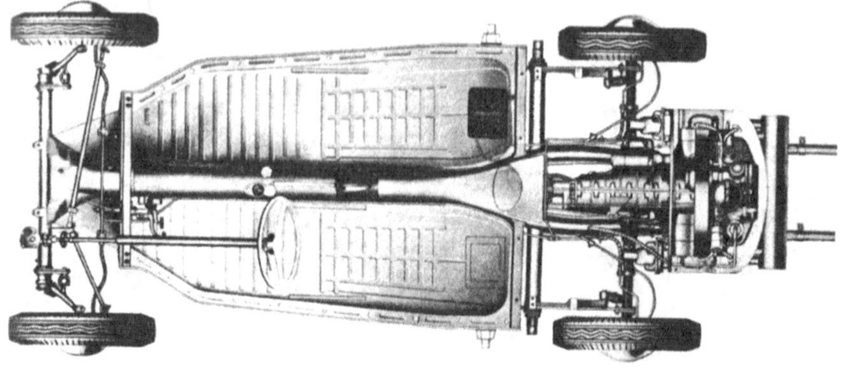

Fig. 1.41 Platform chassis structure of Volkswagen Beetle of 1936. This structure is separated from the body during the assembly process of the chassis, but after body assembly shapes up a closed structure, whose torsional stiffness can be compared to that of a unitized body

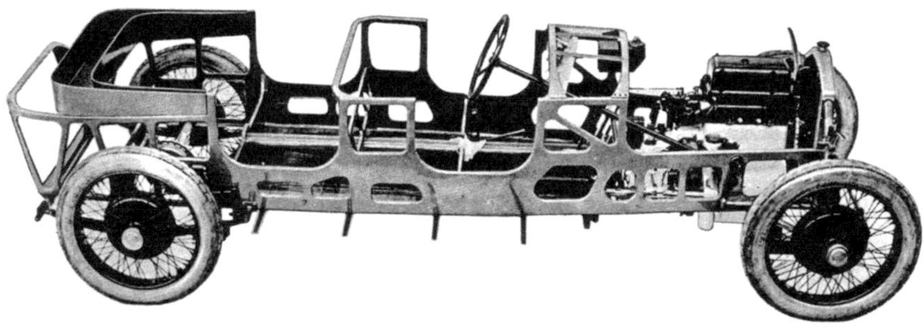

Fig. 1.42 Complete chassis with parts of the unitized body of the Lancia Lambda of 1922. This picture has only communication purpose; in fact mechanical components were assembled to the structure only after the completion and painting of the body

It shows the advantage of being still compatible with the assembly process organization already described; this solution was obviously suitable to produce different car variants on the same chassis.

The unitized body was considered to obtain optimum structural performance with a reduced weight; the considerable distance, existing in a sedan car, between chassis side members and longitudinal roof reinforcements allows to have a very stiff structure, if these elements are connected with vertical members (*pillars*) sufficiently stiff and well joined to longitudinal and cross members.

In addition to this body panels, if shaped for this purpose, can contribute to the structure stiffness by limiting angular deformations of pillars side beams and cross members.

The first car to try this way was the Lancia Lambda of 1922, whose chassis is shown on Fig. 1.42.

Fig. 1.43 Phantom view of the unitized body of the Fiat 1400 of 1950; darkened areas put in evidence body structure. Chassis is no more existing (as physical subassembly) and all components (powertrain, suspension, steering system, etc.) are bolted to the body

The body is made by rived panels put in such a way as to build up boxed areas; many parts of the structure are over side beams, integrating seats, fire wall and trunk and contributing impressively to the total stiffness.

Front suspension is connected with a gate structure surrounding the cooling radiator, as already shown on Fig. 1.13.

We are not able to comment quantitatively on structural performance of this solution, but the result on volumes layout of this car is impressive and allows the modern and slender shape characteristic of this car as compared with contemporary ones.

This structure could fit a convertible kind of body; the sedan type didn't take advantage of this structure, featuring a Weymann's semi hard top.

The crucial step was made about fifteen years later when spot welding was developed and deep stamped steel sheets were available; it should be noticed the every cut and bent of the Lambda structure is made by open and general purpose stamping tools.

One of the first examples in Europe of this technology developed by Budd in the United States was the Fiat 1400 of 1950, represented on Fig. 1.43.

All body panels are shaped in such a way as to build boxed beams in the junction areas; these boxed beams shape up a space frame carrying on all structural tasks.

For example the space frame carrying on the power train is made by two members shaped like an U (in the upper side) that together with the wheels side walls build up closed section beams.

The chassis, as physical subassembly of certain mechanical components, is no more existing and its separation from the car is no more possible; nor is possible to identify its structural function. Nevertheless the structural function of the body is very well evident by looking to the darkened elements of the phantom view.

The production process is completely different; the body is stamped, welded and painted and chassis components, grouped by small sub assemblies are assembled only afterwards.

The role of body manufacturers has been completely revolutionized.

Chapter 2
Wheels and Tires

2.1 Description

Vehicle wheels have essentially two functions:

- to carry on the vehicle weight by exchanging vertical forces with the road surface;
- to exchange with the road surface longitudinal and side forces, able to move the vehicle and to control its path.

As we have seen in the previous chapter the first function was already present on antique wheels that had to carry weights but only occasionally were demanded to brake the vehicle or to withstand significant side forces.

With speed increasing the bearing capacity for side and longitudinal forces becomes more and more important.

Road vehicle wheels include two elements: the rim and the tire.

2.1.1 Rim Characteristics

A peculiar characteristic of a rim is the possibility to allow a fast and simple substitution of the tire. A mean average tire life should be set in the range between 30,000 and 60,000 km, according to the heaviness of the mission. In addition, tires can be punctured and require immediate substitution.

These facts justify the particular shape of the rim and the fact that its dimensions are standardized in order to be exchanged with other of different manufacturers.

The wheel is made by a *disc* and a *flange*; these are shown on the right side of Fig. 2.1; on the same figure can be seen the undeformed section of the tire (on the upper side) and the deformed section (on the lower side) of the contact.

Disc and flange are usually integral; the wheel is fixed to its hub with bolts.

Flanges show a particular shape suitable to keep the tire in place. The flange is characterized by the *diameter d* and shows a dropped center useful to assemble and

G. Genta and L. Morello, *The Automotive Chassis*, Mechanical Engineering Series,
https://doi.org/10.1007/978-3-030-35635-4_2

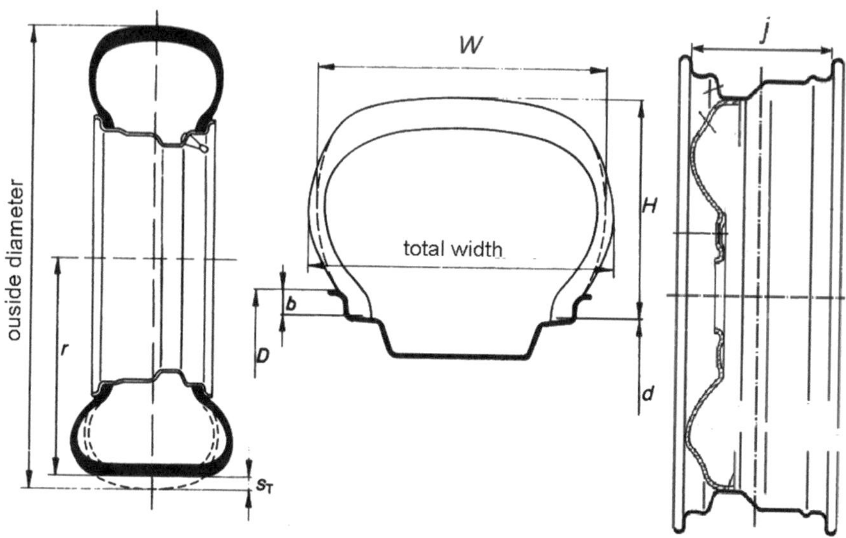

Fig. 2.1 Schematic section of a complete wheel including rim and tire and their main dimensions

disassemble the tire; in fact the bead cable of the tire is quite inextensible and must be set initially in the dropped center from one side, to cross the flange and to be finally installed.

The j dimension between the two flanks of the flange determines the mounting width of the rim; dimensions d and j are usually always measured in inches and must be identical to those of the tire.

The wheel size measurement is usually written like in the following example:

$$5\frac{1}{2}\ J \times 15,$$

to show a rim with a j dimension of $5\frac{1}{2}''$ (139.7 mm) and a d diameter of $15''$ (381 mm); J letter is showing the type of adopted rim profile, almost always J type.

Groups of letters can be added to this indication to show the *hump* type of the flange; the hump is the shape of the area matching with the internal diameter of the tire; a particular tapered shape is used to improve the air sealing for tubeless tire or to avoid sudden tire removal after puncturing. No letter stands for conical rim for tube tires.

Rims can be made by stamped and welded steel or by cast aluminum or magnesium; this last alternative, initially addressed to weight reduction, is also justified by aesthetic reasons.

The rim disc is characterized by windows suitable to improve brake cooling down and by a central hole able to sustain the wheel before bolts tightening; wheels fit the disc in a tapered area to increase disc elasticity; this has the function of reducing the statistical pull dispersion after tightening.

Rim discs show usually a bell shape addressed to give a correct position to the symmetry plane of the tire (*equator* plane) with respect to the car and to allow the necessary space for hubs and brakes installation.

Today many cars can be fitted with tires of different size, according to the engine power of the version and of customer desire; usually more performing versions have a wider tire with the same rolling radius. As a consequence the aspect ratio is reduced to increase cornering stiffness with compromise on comfort.

The rim disc shape is usually such as to maintain the outside profile of the tire in the same position.

2.1.2 Tire Characteristics

The rigid structure of the wheel is surrounded by a flexible element made by the tire and its *tube*, used to maintain the inflation pressure. This last can be missing, as in *tubeless* tires, where the tire is hermetically fitted to the wheel; this more costly solution is today preferred because of the improvement on safety, since punctured tubeless tires are slower in loosing their pressure.

The tire is a quite complex composite structure, where many layers of rubberized fabric (*plies*) with reinforcements cords; the orientation of warp and weave and of reinforcement cords is peculiar to give a tire its mechanical design characteristics.

The angle between the direction of the cords with respect to the circumferential direction of the tires is usually referred to as the *crown angle*. As a general rule, plies with a low value of the crown angle enhance the handling characteristics of the vehicle, while those with a high value, up to 90°, of the crown angle enhance ride comfort.

Each tire is designated by a group of letters and numbers, like the following example:

$$175/65\ R\ 14\ 82\ T.$$

We report the meaning of this designation with reference to Fig. 2.1.

- The first figure (175) shows the dimension W, usually measured in mm; it is separated by a bar from the following figure. Because the tire is a deformable structure, this measure must be referred to an undeformed situation with correct inflation pressure and no load applied.
- The second figure (65) refers to the *aspect ratio* of the tire, given by the ratio H/W between the radial height and the width; usually aspect ratios are expressed as a percentage: in the given example H is 65% of W. If this figure is omitted it should be assumed to be 80%, for long time a standardized value.
- The following letter shows the type of tire plies; R stands for radial, otherwise it is omitted.
- The third figure is showing the rim diameter in inches.

Table 2.1 Letters showing the maximum allowed speed of a tire

Speed (km/h)	80	130	150	160	170	180	190	210	240	270
Letter	F	M	P	Q	R	S	T	H	V	W

- The fourth figure in the *load factor*, which determines the allowed vertical load at a certain assigned inflation pressure; this figure is the row order of a standardization table where maximum allowed vertical loads are shown as a function of the inflation pressure; this figure has therefore no physical meaning.
- The final letter indicates the maximum allowed speed for the tire, as shown on Table 2.1.

Other no more up-to-date standardized designation survive on non recent publications; for instance:

$$P\ 225/50\ SR\ 15.$$

In this case, the first letter stands for vehicle category (P is for cars), the first two figures indicate tire width in millimeters and aspect ratio, the following letters shows the maximum allowed speed and the next letter the tire type (R for radial, B for belted, D for cross-ply); the last figure is the rim diameter in inches.

Independently of their application, the structure of a tire can be made according to different fundamental schemes; these schemes regarding the tire *carcass*, that is the organization of plies, can be classified according to two categories: *cross* or *conventional ply* and *radial ply*. A radial ply tire is shown on Fig. 2.2; an intermediate category could be that of *belted tires*.

On cross ply tires reinforcement cords are crossing the equator plane with a crown angle of 35°–40°; on radial tires some plies running perpendicular to the circumferential direction (carcass plies, with a crown angle of 90°) are surrounded by other plies (belt plies) with a much smaller angle, in the range of 15°. On cross ply tires, plies are extending from bead to bead, while on radial tires belt plies are limited to the region that comes into contact with the ground, while the carcass plies keep the sides of the tire into shape.

Conventional tires have a less vulnerable structure in the flanks but are more flexible in the belt region; on the contrary radial tire are very flexible and vulnerable in the flanks, being stiffer in the belt region: this arrangement improves cornering and comfort behavior with penalty on the flank vulnerability (shocks against sidewalk steps).

For their better characteristics radial tires are the most common type; conventional tire are in use for more demanding applications such as off-road driving.

Tires main function is to distribute the vertical load on a surface which should be large enough to absorb road irregularities. It is vital that the flexibility, according to different directions, is distributed in a certain way: on radial tires, for example, vertical deformation is high, because of high deformation of the sides, while circumferential deformation is low, because of the almost straight direction of belt cords. Superior

Fig. 2.2 Radial tire structure: there are plies with circumferential cords (belt) and plies with radial cords (flanks) and two reinforcement cables in the bead for the part of the tire contacting the rim

comfort behavior and high capability of side force generation are a direct consequence of this flexibility distribution.

Particularly relevant to the tire behavior is the *tread* of the tire, essentially made by vulcanized filled rubber; it is the contact surface with the ground and determines friction of the tire-ground coupling. Treads are featuring sculptures of different designs; circumferential and transversal grooves have the essential purpose of making water to drain easier from the contact area, when the road is wet and improve friction on incoherent soils; thinner transversal grooves improve friction on slippery ground. If the vehicle drive was limited to well paved and dry roads, sculptures would be unnecessary, as for *slick* tires.

2.1.3 Wheel Reference System

While studying forces between tire and ground, it is useful to use the wheel reference system $X'Y'Z'$ shown on Fig. 2.3. The system origin is the center of the contact patch, defined as the theoretical contact point between the undeformed equatorial plane and the ground; X' axis is the intersection of such plane with the ground and its positive

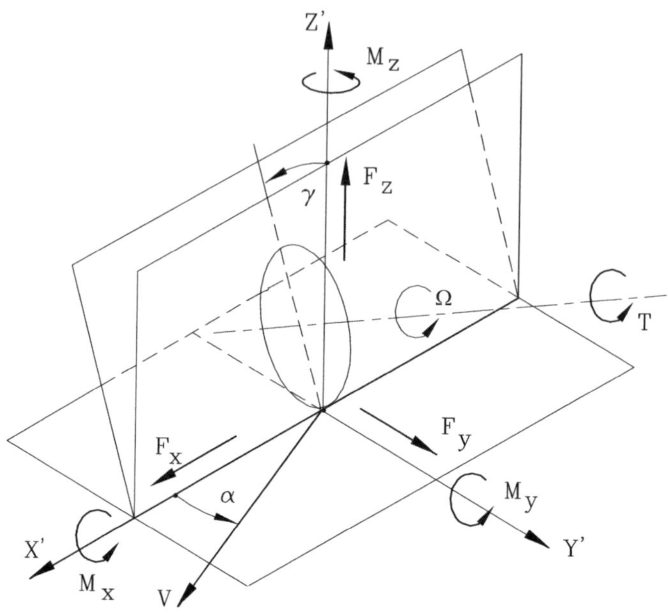

Fig. 2.3 Reference system used to study forces exchanged between tire and ground. Definition of positive direction of forces, moments and side slip angle

direction is the same as for vehicle speed. Z' axis is perpendicular to the ground[1] and, as a consequence Y' axis lies on the ground and is pointing to the left.

The total force is supposed to be applied in the center of contact and can be seen as three components acting along X', Y' and Z' axis, as *vertical force* F_x, *lateral force* F_y and *normal force* F_z. In the same way the total moment can be broken down in three components that are the *overturning moment* M_x, the *rolling resistance moment* M_y and the *self aligning moment* M_z. A *wheel torque* T can be applied by the vehicle along the tire rotation axis.

The angle between the $X'Z'$ plane and the direction of the wheel hub speed is called *wheel side slip angle* α, while the angle between the $X'Z'$ plane and the wheel equatorial plane is called *inclination* or *camber angle* γ. Positive directions for angles are shown on the same figure; in particular the inclination angle is positive when the highest part of the wheel is oriented to the right of the driver (in other words, left wheel has a positive inclination angle if the top is pointing to the car interior). While

[1]The terminology we are going to use is that suggested by SAE (Society of Automotive Engineers) in his document J670e Vehicle Dynamics Terminology, published in 1952 and revised on July 1976. Nevertheless in some cases the Authors have decided not to follow some suggestion, as for the positive direction of Z' axis, which, in this document, is pointing downwards.

the inclination angle is referred to the road, the camber angle is usually referred to the car and has positive directions opposite for the two vehicle sides.

In this book we will assume that inclination and camber angle are coincident.[2]

2.2 Tire Operation

While explaining tire operation, we should identify two different situations:

- prepared ground, when the tire is in contact with paved or concrete grounds;
- unprepared ground, when the tire is in contact with natural grounds or dirty roads.

The first situation is typical of on-road driving, while the second one is typical of off-road driving; the physical phenomenon discriminating the two situations is that ground deformations should be considered or neglected: ground deformation are neglected on dry paved roads.

2.2.1 *On-road Driving*

In this case two different aspects are to be considered:

- the *adhesion* between rubber and ground; thanks to this phenomenon, tires can exchange with the ground also forces which are contained in the contact patch plane;
- the *elasticity* of the tire structure, which gives to the tire certain absorption capabilities as far as road asperities are concerned and are the main reason for having tires slipping in the two direction of the contact patch, when longitudinal and lateral forces are applied.

Rubber-Ground Adhesion

We define adhesion the result of physical phenomena that allow a specimen of rubber set on the ground and pressed with a certain vertical force to withstand forces contained in the ground plane, without any relative motion.

Adhesion is caused by two different phenomena, which have, anyway, similar results:

- *physical adhesion* and
- *local deformation*.

We can put in evidence the physical adhesion phenomenon considering a piece of rubber pressed on a completely smooth and rigid ground. Rubber molecules and

[2]This is a further deviation from SAE recommendations.

ground molecules apply certain attraction forces; these attraction forces refer to distances in the range between 10 and 10,000 Å (equivalent to 0.001–0.01 μm). Impurities present between the two contact surfaces limit the adhesion phenomenon to a certain number of adhesion sites, on both surfaces.

Adhesion forces will develop between adhesive sites. If we now imagine to apply a lateral force to our rubber specimen, it will be balanced by adhesion force.

This force will be maintained till a certain distance is reached between the adhesive sites and will become zero, when the distance will be greater than a certain value; when the adhesion force is destroyed, molecules layers are instantaneously left free from stress and they start oscillating in the material. Because of internal material damping this fact will dissipate a certain amount of energy; the dissipated energy divided by the displacement between rubber specimen and ground will result in reaction force, opposing this displacement.

Adhesion force is therefore controlled by the following parameters:

- surface energy of contacting materials;
- damping properties of those materials, where rubber plays the most important role; they are controlled by temperatures, bound to relative speed;
- deformation of contacting surfaces, because of lateral forces which can cause also instability as in case of *stick and slip*.

Adhesion force measurement can be made by pressing a rubber specimen on a perfectly polished mirror-like surface.

It should be noticed as adhesion is bound to material damping properties.

Local deformations are, instead, caused by road asperities; if we don't take into account adhesion, as it could happen putting our rubber specimen on a perfectly lubricated but rough ground, we will have areas where rubber is deforming and afterwards is loosing its deformation.

Again, mechanical work dissipated by damping will build up forces along the contact surface.

Also now friction and damping are bound each other.

Local deformation forces can be measured by having a rubber carpet specimen sliding on a low resistance plane, such as a plane equipped with rollers or spheres free to rotate.

Orientatively adhesion force is about 70% of the total friction force in case of rubber on dry paved road.

These phenomena change their mechanism radically in case of watered ground; we again distinguish three fundamental cases.

- Water layer thickness is so high to establish a permanent lubricated meatus between tire and ground (*aquaplaning*); in this case adhesion and local deformations cannot take place and tangential forces can be calculated, according to the lubrication theory, starting from liquid viscosity.
- Water layer thickness isn't such as to establish permanent lubrication, but enough to avoid adhesion forces; in this case we can still have local deformation forces if the ground is rough.

- Water is completely removed from the contact patch area and in this case the behavior of rubber is as it was explained.

This last case is what we would like to obtain by mains of tire grooves and draining pavement.

In a real case it can happen that the first part of the patch will receive a lubrication hydrodynamic lift, when the areas behind are in intermediate or dry conditions.

We define *friction coefficient* μ the ratio between the lateral force and the vertical force; as forces, it can be broken down in longitudinal friction coefficient, the ratio:

$$\mu_x = \frac{F_x}{F_z} \tag{2.1}$$

between force along the X' axis and vertical force along the Z' axis and longitudinal friction coefficient, the ratio:

$$\mu_y = \frac{F_y}{F_z} \tag{2.2}$$

between force along the Y' axis and vertical force along the Z' axis.

Friction coefficient can be defined globally, as we did, or locally as ratio between tangential and vertical pressures.

Elastic Behavior

The tire can be compared to an elastic deforming structure with internal damping.

Without inflation pressure tire rigidity will be modest and not adequate to practical use. On the contrary, by neglecting mechanical contribution of cloths and cords, the tire could be thought as a perfect membrane with internal constant pressure and where a certain vertical load is applied on a part of its structure.

This membrane could withstand vertical load only by its internal pressure; tire surface should flatten for an area as big as to generate a pressure-area product able to equal the vertical force. This surface is the *contact patch* of the tire.

Contact patch area is decreasing if the inflation pressure is increasing.

If the tire is rolling previously deformed patch areas must return round, while new material ha to be flattened as to withstand vertical force.

It can be imagined that this process is implying energy dissipation, because of rubber damping properties. These damping properties are not only due to the internal material damping, but also to the relative motion that must take place between cloths, cords and rubber, to allow for local bending deformation.

If the ground isn't rigid also the anelastic ground deformation is playing a role in determining roll resistance.

The membrane model can easily justify why lateral and longitudinal forces can induce local deformation of the contact patch; what is more difficult is to understand what is going on when lateral and longitudinal forces are applied to a rolling wheel.

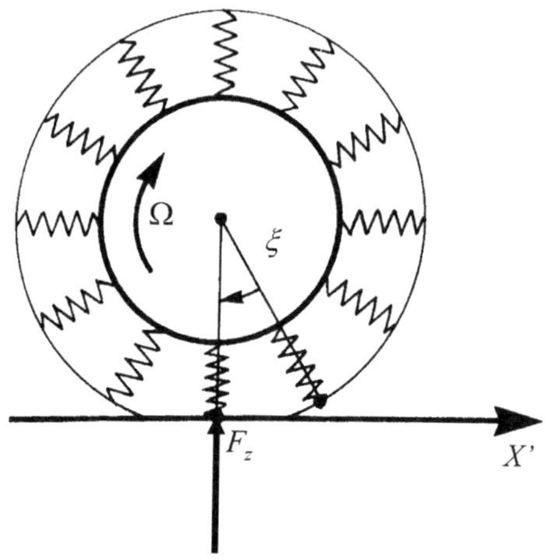

Fig. 2.4 "Brush" simplified model of a tire; the tire is reduced to a discrete number of springs, featuring vertical, lateral and longitudinal flexibility and bearing a specimen of tire rubber on their tip

To predict, only in a qualitative way, what will occur we will use a simpler model, represented on Fig. 2.4.

The tire is reduced to a discrete number of springs, featuring vertical, lateral and longitudinal flexibility and bearing a specimen of tire rubber on their tip.

Let us assume, for sake of simplicity, that the number of springs is such as there is a position where one only of them is contacting the ground; they are fit on a rigid rim with a uniform angular displacement ξ.

Spring tips are free to move, according to applied forces, with no connection with near springs. Simplifications can be removed, without affecting the nature of the results we are going to study. We will call this model, extremely simplified, *brush model*.

Under the application of the vertical force, contacting spring will assume a certain radial deformation, as in a real tire; if we associate to the spring a suitable damping coefficient, the model is also able to simulate rolling resistance, proportional to the mechanical work dissipated by any contacting spring, when leaving the contact patch, because of rolling.

In addition to that, rolling radius, because of the applied vertical force will assume a value smaller as the unloaded value.

Let us assume that the tire is withstanding a longitudinal force F_x in driving direction (in the middle of Fig. 2.5) or in braking direction (at the right of Fig. 2.5); according to the proposed model an angular deformation ζ, proportional to the applied force will be applied to the contacting spring and by the tire.

It is easy to understand that if the tire has a rolling speed Ω, each contacting spring will returned straight, when leaving the contact patch, while a new spring will assume the angular displacement ζ. To the initial tire speed Ω a new deformation speed will be added, equal to the ratio between the angular deformation ζ and the time ξ/Ω, needed to put the following spring in contact with the ground.

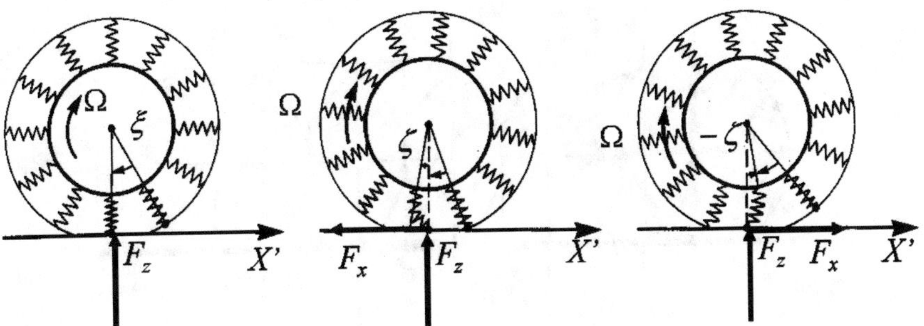

Fig. 2.5 Brush model: static representation of radial and circumferential deformations of a tire, with contemporary application of a vertical and of braking or driving force

Under the action of a driving force the wheel should roll faster as when rolling free of force; on the contrary, if the wheel is braking, it should roll slower as when rolling free.

If our model was accurate speed variation should be proportional to applied longitudinal force; this speed variation is called *longitudinal slip speed.*

When braking or driving force should go over the friction limit, a new added speed would be generated depending only on mass characteristics; over this value longitudinal forces cannot increase anymore.

Let us assume, finally, that our tire is withstanding vertical and lateral force, due, for example, to a centrifugal force applied to the vehicle in a bend. In this case, the spring in contact with the ground will assume a lateral displacement in the Y' direction. The phenomenon is shown on Fig. 2.6.

If the wheel is rolling, each spring will contact the ground as shown on the upper part of this figure, relative to the side view, on a certain point of the X' axis.

The initial contact point will receive a lateral displacement Δy, shown on the lower part of the same figure; this displacement again is proportional to the lateral force; when next spring will contact the ground it will assume the same displacement, that will be added to the wheel hub. As a result the wheel hub will receive a lateral motion, whose speed will be given by the ratio between this displacement and the time needed to have the next spring in contact with the ground, that is ξ/Ω.

If the model would be accurate, because deformation is proportional to the force, we would expect the wheel hub to deviate its path from the initial one of an angle α, proportional to the lateral force; this deviation is called side slip angle. It should be better called wheel side slip angle, to be not confused with the vehicle side slip angle, we will explain later on.

When lateral force should go over the friction value of rubber, again the added speed will be only controlled by masses and the side force could not have any further increase.

The brush model has given us the possibility to understand and predict in a very rough way the concepts of longitudinal slip and side slip, we will examine in a better detail in the following paragraphs.

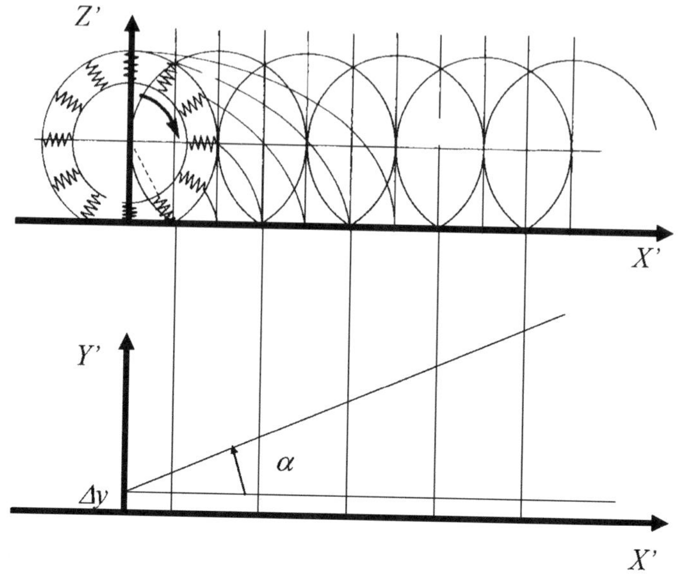

Fig. 2.6 Brush model; graphic representation of a rolling tire with contemporary application of vertical and longitudinal force

2.2.2 *Off-road Driving*

On off-road driving, besides of the phenomena we have described, which remain unchanged also in this case, ground deformations should be taken into account and they could predominate in certain situations.

Some kind of vehicle is designed not only to be driven on paved roads, but also on natural grounds, on country and forest roads, with their surface only roughly levelled. Prevailing factors limiting vehicle mobility are the deformation of the surface and the shape of the ground which can interfere with the mechanical parts of the chassis.

We will discuss in the following paragraphs about ground deformation only. In fact, also on a level road, the mechanical behavior of an untreated ground can have a severe impact on vehicle motion. The shape has a purely geometric impact on geometric chassis design, concerning:

- wheelbase,
- track,
- wheel diameter,
- available suspension stroke,
- clearance of the chassis from the ground and
- attack angles.

We will refer to natural soils and to fresh snow, which for certain reasons shows a similar behavior.

Natural soils are the last result of the effect of wind and water erosion on the rocks from which they have been originated and are made by tamped down powders.

Mechanical properties of soil are determined by the solid incoherent particles which are its main component; main particles characteristics are their granulometry, their apparent density and the content of water or *humidity*. We distinguish between *apparent* and *real density* to take into account for the space separating contiguous particles. The first is defined by the ratio between weight and volume of an average lump cut from the ground and measured as it is; the second represent the limit that can be obtained by compressing this lump in order to eliminate any space between particles; in other words it is the average density of single particles.

The forces that can be exchanged by near particles and that oppose deformations caused by external forces that may be applied, are cohesive forces which are usually increasing as particle size, or *granulometry*, is decreasing; in fact, if granulometry is decreasing the contact surface between near particles is increasing.

The most relevant contribution to soil deformation is given by making particles to slide along their contact surface; the higher the cohesion force the lower will be deformation. Cohesion forces are also influenced by the presence of water trapped between near particles.

On the other side a significant contribution to limit particles displacement and consequent soil deformation is the friction force at the contact between particles; friction coefficient too is influenced by the presence of water.

For low contents of humidity, soils feature a good value of mechanical resistance which is decreasing as the quantity of trapped water is increasing.

The *liquid limit* takes place when water is so abundant as to completely eliminate the cohesive effect between particles and therefore the shear resistance of the soil; this limit is expressed by the quantity of water divided by the dry residual.

The *plastic limit* is instead reached, when the soil is loosing is capability to be shaped, being too brittle and, also in this case, is expressed by the quantity of trapped water, divided by the dry residual.

The difference between the to said limits is called *plasticity index*, while the ratio between the actual water content of a given soil and the quantity referred to the plastic limit, is called *relative content of water* of the soil.

For our purposes we propose the following classification of soil types.

Cohesive Soils

Cohesive soils contain variable quantities of clay, mixed with very fine granules of sedimentary rock, with a mean diameter between 0.05 and 0.10 mm. They can be classified in *lightly clayey*, *clayey soils* and *clays*, according to the quantity of clay. In these soils cohesion is particularly relevant and is dominated by relative humidity; for this purpose solid, plastic and fluid states are identified.

Relative humidity is therefore an important parameter to describe the mechanical properties of cohesive soils; these soils are stable only if relative humidity is in the field between 0.4 and 0.65; over those values mechanical properties deteriorate severely. A peculiar characteristic of these soils is that of draining water with high difficulty; vehicle mobility will result particularly difficult during rain season and ice melt.

Cohesive soils occupy a large part of the earth surface.

Sandy Soils

In sandy soils particles have higher dimensions and clay is practically missing; a possible classification of these soils can take into account particles diameter and apparent density. The mechanical resistance to deformation of sandy soils increases initially as the water content increases, while it decreases rapidly at higher contents; as a difference with clayey soils, they have a very good draining capacity: they are diffused on see and river shores and in some deserts.

Swampy Soils

Swampy soils are characterized by the high content of peat, where most diffused constituent are residuals of animal and vegetal organic matter, at the end of an anaerobic decomposition process; they can be classified in *solid swampy soils*, when pit lies directly on a rock layer, *saprofelic soils*, when the peat lies over a layer of materials in decomposition and *floating soils* when peat floats over the water.

These grounds are diffused in Central Europe and Siberia.

Snowy Soils

Snow is particularly abundant during winter season of the Northern Hemisphere. Vehicle mobility on snow depends mainly on snow layer thickness, on his density, on temperature and on the snow structure. Fresh snow density lies between 0.075 and 0.20 g/cm^3, while seasoned and compacted snow density is between 0.30 and 0.40 g/cm^3; the process of transformation of fresh to seasoned snow is irreversible.

Soil Deformation Because of Vertical Load

Vertical load application to soils originate deformation; we will take into consideration the application of uniform loads, as we could ideally have by an infinite flexible tire (membrane model). Let us consider Fig. 2.7, which represents in a qualitative way what could happen by pressing a pad, applying a pressure p, defined as the ratio between the applied load and the pad surface.

This pad simulates the effect of the contact patch of a tire, pressed against the soil by the vehicle weight.

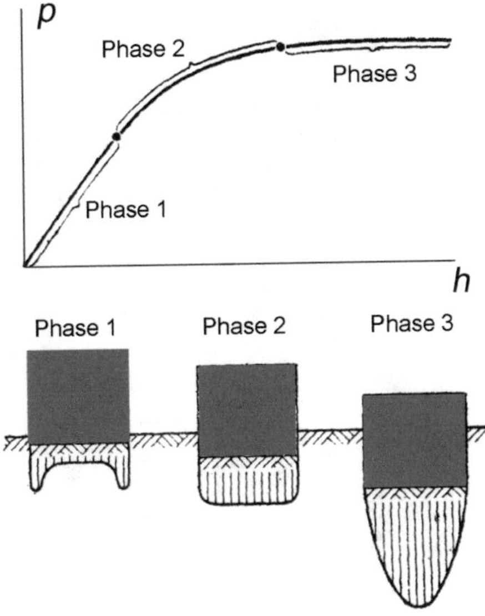

Fig. 2.7 Main phases of deformation of an ideal homogeneous soil, expressed by sinking h as a function of applied pressure p. Lower diagrams shows qualitatively the distribution of stress in the soil

There is no difference in this description considering the different types of soils we have classified.

By relatively small pressure, the soil is cut along the perimeter of the pad and is tamped down; pressure reaction of soil concentrates on the edge of the cut and is increasing as soil cohesivity is increasing. A nucleus of tamped down soil is building up below the pad and moves in the direction of the applied load, loading lower layers of soil. In a first phase deformation is almost proportional to applied pressure; soil reaction pressure is shown qualitatively below the diagram, showing the corresponding part of the $h(p)$ diagram.

As pressure is increased more significant volumes of soil reach plasticity, as is shown in the central picture, relative to phase 2. When the entire volume of soil displaced by the pad has reached the plastic condition it reacts with the same pressure at any value of sinking. The position and extension of phase 2 depends on initial soil constipation.

Figure 2.8 shows some diagram of sinking h as function of applied pressure p, for some kind of soil, characterized by the qualitative behavior we have described.

Diagram A refers to a cohesive soil (a: humid soil, b: plastic soil); diagram B refers to a sandy soil (a: loose sand, about 200 mm thick; b: compacted layer); diagram C refers to a layer of peat (a: pressing pad of 4 m^2 of surface; b: pressing pad of about 0.4 m^2); diagram D refers to a snowy ground (a: fresh snow, density of 0.15 g/cm^2, b: compacted snow, density of 0.20 g/cm^2).

Diagram C shows haw the surface of the pad can influence the value of sinking at the same pressure. In particular the bigger the surface, the higher the sinking; this happens because bigger surfaces increase the amount of soil interested by costipation.

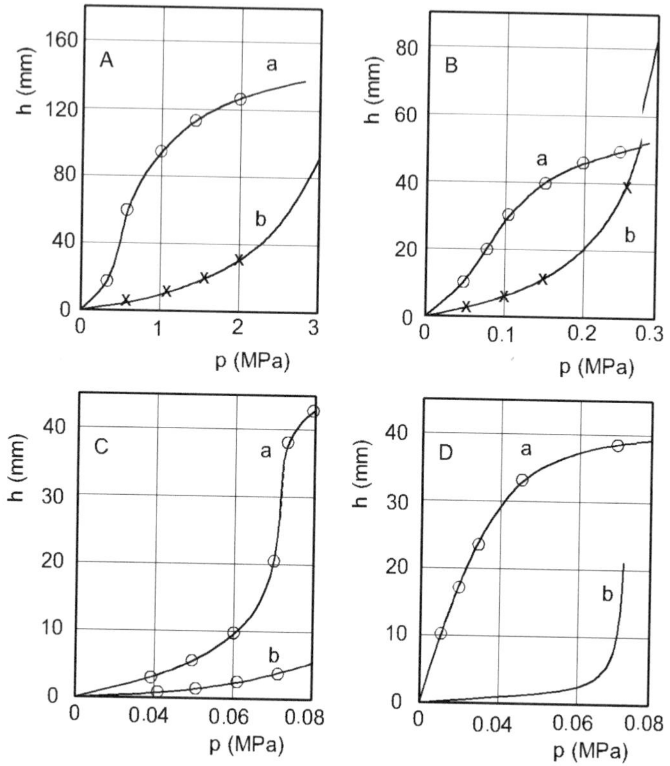

Fig. 2.8 Diagram of sinking h as function of applied pressure p, for some kind of soil. Diagram **A** refers to a cohesive soil (a: humid soil, b: plastic soil); diagram **B** refers to a sandy soil (a: loose sand, about 200 mm thick; b: compacted layer); diagram **C** refers to a layer of peat (a: pressing piston of 4 m^2 of surface; b: pressing piston of about 0.4 m^2); diagram **D** refers to a snowy ground (a: fresh snow, density of 0.15 g/cm^2, b: compacted snow, density of 0.20 g/cm^2)

Please take note of the different order of magnitude for pressure of cohesive soils, which demonstrates their superior mechanical properties; considering a tire with a patch pressure of about 0.2 MPa, we can draw immediately first approximation conclusions about possibility of driving a vehicle on different kind of soil.

Only for the case of fresh snow, we show on Fig. 2.9 some sinking curves for different average dimension d of the pad ($d = \sqrt[2]{A}$, where A is the pad area). Curve a corresponds to a pressure $p = 0.1$ MPa; curve b to $p = 0.08$ MPa, curve c to $p = 0.06$ MPa and curve d to $p = 0.04$ MPa.

Similar diagram shapes can be found for different kind of soil, even if, sometimes, there is a minimum point for low values of d.

In the first part of the curve in Fig. 2.7a (phase 1), where the soil behaves as an elastic material, it is possible to define a coefficient of proportionality, sometimes called the stiffness or the *bedding number* of the soil of the soil

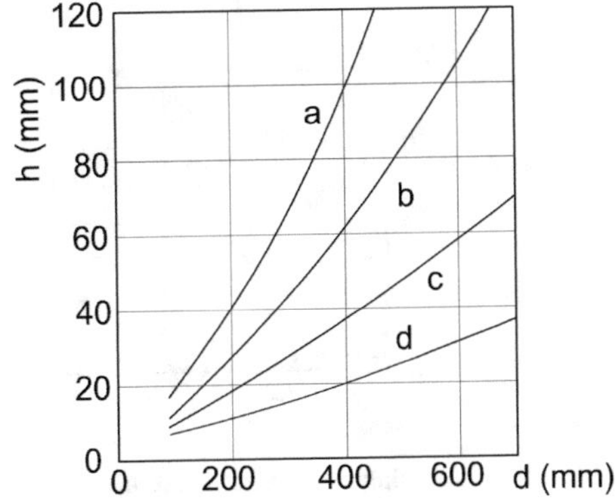

Fig. 2.9 Effect of the average piston dimension d on sinking h, for different value of pressure p, on snow (curve a: $p = 0.1$ MPa; curve b: $p = 0.08$ MPa; curve c: $p = 0.06$ MPa; curve d: $p = 0.04$ MPa)

$$k = \frac{p}{h} \,. \tag{2.3}$$

The stiffness k does not depend only on the elastic characteristics of the ground (its Young's modulus E and its Poisson's ratio ν), but also on the area and the shape of the loading pad.

Consider a pad exerting a constant pressure p on the soil. In Phase 1 the sinking h reduces to the elastic sinking h_e defined as

$$h_0 = \frac{p}{k} \,. \tag{2.4}$$

The sinking h in the subsequent phases 2 and 3 may be expressed as

$$h = \frac{h_e p_0}{p_0 - p} = h_e \frac{1}{1 - p/p_0} \,, \tag{2.5}$$

where h and h_0 have the meaning of effective deformation and of contribution, to total deformation, of elasticity, interpreted by previous formula; p_0, *plasticity pressure*, is the value of pressure over which plasticity takes place, as in phase 3. Obviously this formula is valid for $p \geqslant p_0$ only.

Soil Deformation Because of Shear Forces

Simultaneous presence of side forces applied to the pad causes an increase of sinking at the same pressure; some experimental behaviors are reported on Fig. 2.10.

Diagrams report pad sinking h due to contemporary presence of normal pressure p and shear pressure τ, for a sandy soil (Figure A) and a clayey soil (Figure B). Curves a, b and c correspond to pressure p of 0.02, 0.03 and 0.04 MPa.

The tangential force the pad can exert on the ground is limited by cohesive and friction phenomena occurring at the soil-pad interface. If the pad dopes not sink into

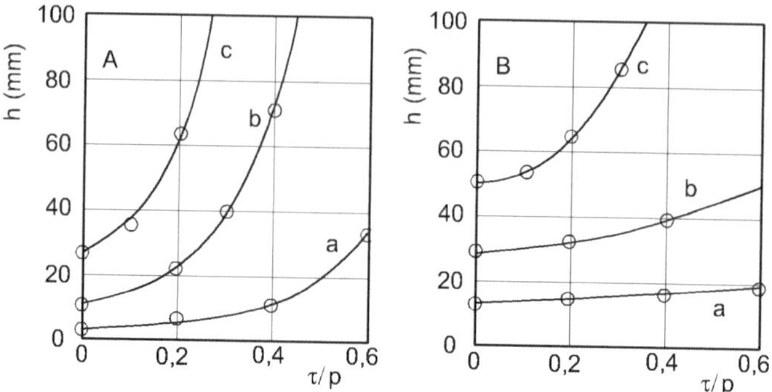

Fig. 2.10 Piston sinking h due to contemporary presence of normal pressure p and shear pressure τ, for a sandy soil (figure **A**) and a clayey soil (figure **B**). Curves a, b and c correspond to pressure p of 0.02, 0.03 and 0.04 MPa

the ground, the maximum tangential force is

$$F_{t_{max}} = AC_0 + F_n \tan \Phi_0 \, , \tag{2.6}$$

where C_0 is the contribution of cohesion, or *cohesivity* and Φ_0 is the *friction angle* between particles; it can be practically measured by evaluating the semi-aperture of the cone which approximates the shape of a pile of this soil at the loose state.

The shear resistance of the considered soil τ_o is then

$$\tau_0 = \frac{F_{t_{max}}}{A} = C_0 + p \tan \Phi_0 \, . \tag{2.7}$$

If the pad sinks into the ground, a further component of the force, the *bulldozing* force, must be added to those due to cohesivity and friction.

Mechanical Properties of Different Soils

Figure 2.11 shows the main parameters of identification of the mechanical behavior of a natural ground:

- elastic modulus E,
- plasticity pressure p_0,
- friction angle Φ_0,
- cohesivity C_0,

to calculate sinking h as a function of normal pressure p and shear pressure τ. In addition, τ_0 allows to predict the maximum longitudinal force (driving or braking) that the ground can withstand.

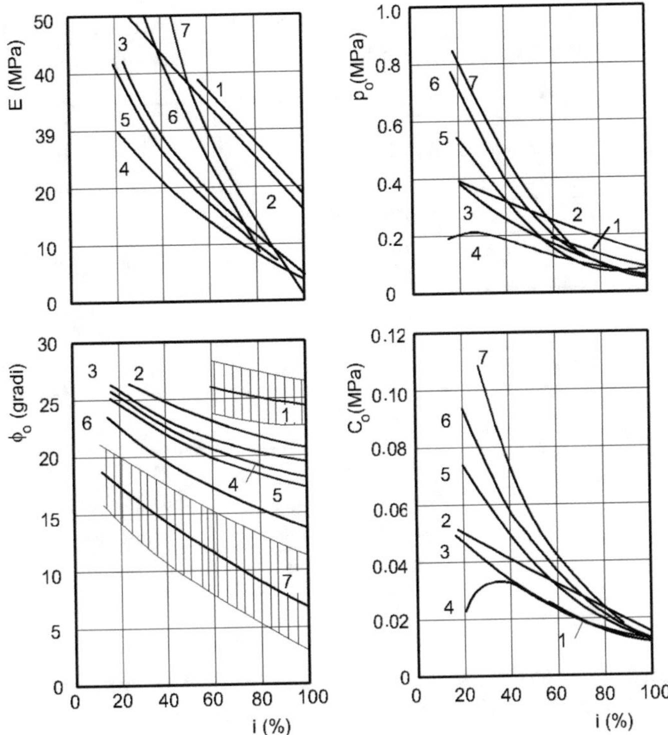

Fig. 2.11 Main identification parameters of the behavior of a natural ground, as a function of the relative content of water i; curves with number 1 refers to a sandy loose soil, with 7 to a clayey soil, while numbers from 2 to 6 refer to sandy soils with increasing sand content

These parameters are function of the relative content of water i; curves with number 1 refer to a sandy loose soil, with 7 to a clayey soil, while numbers from 2 to 6 refer to sandy soils with increasing sand content.

The deterioration of these parameters for increasing water content can be noticed; as a reference a ploughed and hoed soil in Northern Europe can contain till 100% of water at the end of ice melt and 30% of water in summer season.

Also the higher cohesivity of clayey soils and the higher friction angle of sandy soils can be noticed.

Table 2.2 illustrates the main mechanical properties of snow, as a function of its state and external temperature.

What we have reported allows to solve the following problems:

- to understand if a certain ground can withstand vehicle wheel pressure, without interference with the chassis;
- to predict the sinking of the contact patch;
- to define the maximum longitudinal or side force, by calculation of the shear resistance τ_0.

Table 2.2 Main mechanical properties of snow, as a function of its state and external temperature

T (°C)	Physical state	Fresh	Compacted by wind	Compressed
–	Density (g/cm^3)	0.2	0.4	0.6
−5	Cohesivity (MPa)	0.05	0.07	0.19
−10	Cohesivity (MPa)	0.06	0.09	0.22
−5	Friction angle (°)	18	22	31
−10	Friction angle (°)	19	23	26
−5	Elastic modulus (MPa)	0.4–0.6	1.5–2.0	~7.5
−10	Elastic modulus (MPa)	0.6–0.8	2.5–3.0	~10

Table 2.3 Values for f_0 for different kinds of natural grounds

Type of road	f_0
Flat snowy road	0.025–0.040
Natural flat ground	0.08–0.16
Flat sandy ground	0.15–0.30

The contributions to tire patch displacements (vertical deformation, longitudinal and side slips) will be described in the following paragraphs; such contributions are relevant on paved roads, but can be neglected on natural grounds.

In addition, it is necessary to add some consideration on rolling resistance. A tire, rolling on unpaved roads or natural grounds, is submitted to the following forces.

- Resistance due to elastic hysteresis of the tire; it is the same we will describe on next paragraphs and is caused by the energy dissipated by the deformation of the contact patch.
- Resistance due to elastic hysteresis of soil; it is caused by the mechanical work of plastic deformation, we have described.
- *Bulldozing resistance* due to the fact that a certain amount of soil is moved in front of the contact patch of the tire.

The resistance due to tire hysteresis is slightly different as that measured on paved grounds, because the patch flattening is lower; other contributions could be calculated starting from the physical properties of the soil in question: they can increase rolling resistance of an order of magnitude.

Table 2.3 reports some values of rolling resistance f_0[3] for most common types of natural ground. We invite to compare those values with those reported on Table 2.4 for paved roads.

The following chapters will refer to the tire on paved roads only.

[3]Rolling resistance coefficient f_0 will be defined later on.

Table 2.4 Values of f_0 for different kinds of roads

Type and conditions of road	f_0
Very good concrete	0.008–0.010
Very good tarmac	0.010–0.0125
Average concrete	0.010–0.015
Very good paved road	0.015
Very good Mc Adam	0.013–0.016
Average tarmac	0.018
Bad concrete	0.020
Good paved road	0.020
Average Mc Adam	0.018–0.023
Bad tarmac	0.023
Dusty Mc Adam	0.023–0.028
Good stone paved road	0.033–0.055
Very good dirty road	0.045
Bad stone paved road	0.085

2.3 Rolling Radius

Consider a wheel rolling on a level road with no braking or tractive moment applied to it and its mean plane perpendicular to the road. While the relationship between the angular velocity Ω and the forward speed V of a rolling rigid wheel of radius R is simply

$$V = \Omega R,$$

for a pneumatic tire an effective rolling radius R_e can be defined as the ratio between V and Ω:

$$R_e = V/\Omega\,. \tag{2.8}$$

The effective *rolling radius* is defined as the radius of a rigid wheel which travels and rotates at the same speed of the pneumatic wheel.

The wheel-road contact is far from being a point-contact and the tread band is compliant also in circumferential direction; as a consequence radius R_e coincides neither with the loaded radius R_l nor with its unloaded radius R and the centre of instantaneous rotation is therefore not coincident with the centre of contact A (Fig. 2.12).

Owing to longitudinal deformations of the tread band, the peripheral velocity of any point of the tread varies periodically. When it gets near to the point in which it enters the contact zone it slows down and consequently a circumferential compression results. In the contact zone there is very limited sliding between tire and road.

The peripheral velocity of the tread (relative to the wheel centre) in that zone coincides with the velocity of the centre of the wheel V. After leaving the contact zone, the tread regains its initial length and its peripheral velocity ΩR is restored.

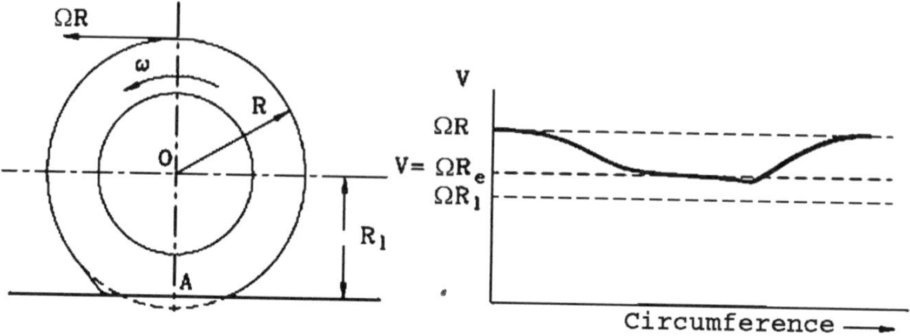

Fig. 2.12 Tire rolling on a flat road; geometric configuration and peripheral speed on contact area

As a consequence of this mechanism, the spin speed of a wheel with pneumatic tire is smaller than that of a rigid wheel with the same loaded radius R_l and travelling at the same speed:

$$R_l < R_e < R \ .$$

The centre of rotation of the wheel lies then under the surface of the road, at a short distance from it.

Owing to their lower vertical stiffness, radial tires have a lower loaded radius R_l than bias-ply tires with equal radius R but their effective rolling radius R_e is closer to the unloaded radius, as the tread is circumferentially stiffer.

For instance in a bias-ply tire R_e can be about 96% of R while R_l is 94% of it; in a radial tire R_e and R_l can be respectively 98% and 92% of R.

The effective rolling radius depends on many factors, some of which are determined by the tire as the type of structure, the wear of the tread, and by the working condition as inflation pressure, load, speed and others.

An increase of the vertical load F_z and a decrease of the inflation pressure p lead to similar results: A decrease of both R_l and R_e. With increasing speed, the tire expands under centrifugal forces, and consequently R, R_l and R_e increase. This effect is larger in bias-ply tires while, owing to the greater stiffness of the tread band, radial tires expand to a very limited, and usually negligible, extent. As it will be shown in the following sections, any tractive or braking torque applied to the wheel will cause strong variations of the effective rolling radius.

The diagram of rolling radius as function of speed for tires 155 D 15 with cross ply and 155 R 15 with radial ply is shown on Fig. 2.13. Diagrams are obtained by having the tire to roll inside a 3.8 m diameter steel drum. As it will we shows later any variation in tractive or braking force is affecting rolling radius.

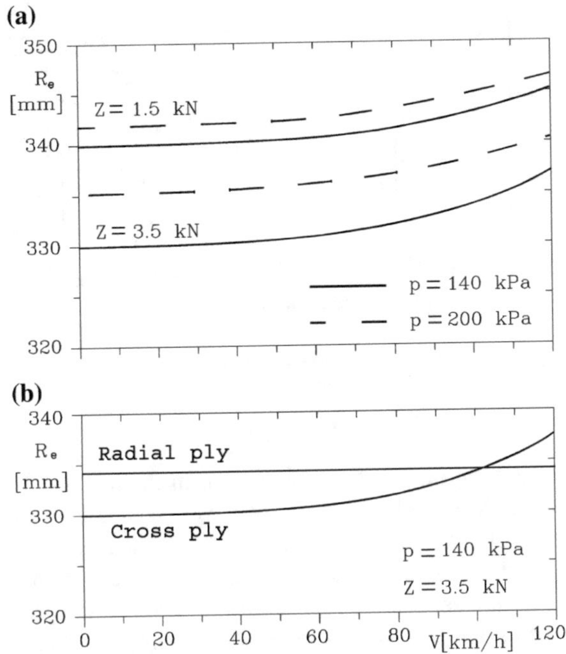

Fig. 2.13 Rolling ratio R_e as function of V for a cross ply tire 155 D 15 (**a**) and for a radial ply tire 155 R 15 (**b**)

2.4 Rolling Resistance

Consider a wheel rolling freely on a flat surface. If both the wheel and the road were perfectly undeformable, there would be no resistance and consequently no need to exert a tractive force. In the real world however perfectly rigid bodies do not exist and both the road and the wheel are subject to deformation in the contact zone.

During the motion new material enters continuously into this zone and is deformed, to spring back to its initial shape when leaving it. To produce this deformation it is necessary to spend some energy which is not completely recovered at the end of the contact zone due to the internal damping of the material.

This energy dissipation is what causes rolling resistance. It is then clear that it increases with increasing deformations and, mainly, with decreasing elastic return. A steel wheel on a steel rail has a lower rolling resistance than a pneumatic wheel and the motion on compliant soil causes greater resistance than that on a rigid surface. From this viewpoint, a wheel on compliant soil is always in the situation of a wheel that attempts to climb out of the pit it is digging by itself (Fig. 2.14a).

In the case of pneumatic tires rolling on tarmac or concrete, the deformations are almost only localized in the wheel and then the energy dissipated in the tires governs the phenomenon. Other mechanisms, like small sliding between road and

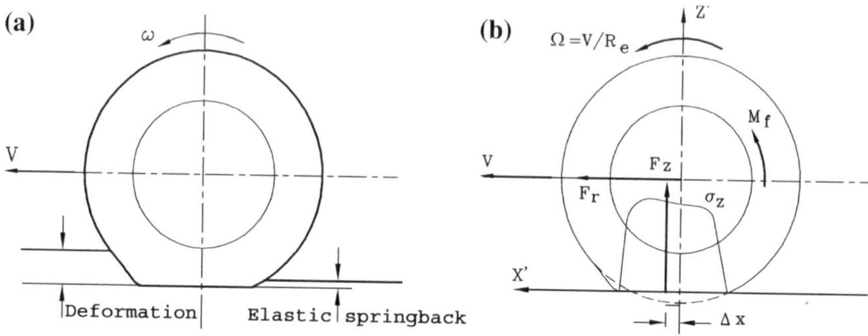

Fig. 2.14 **a** Rolling tire on a deformable surface: ground deformation and spring back. **b** Forces F_z and F_r and contact pressure σ_z in a rolling tire

wheel, aerodynamic drag on the disc and friction in the hub are responsible for a small contribution to the overall resistance, of the order of a few percent.

The distribution of the contact pressure, which at standstill was symmetrical with respect to the centre of the contact zone, becomes unsymmetrical when the wheel is rolling and the resultant F_z moves forward (Fig. 2.14b) producing a torque $M_y = -F_z \Delta x$ with respect to the rotation axis. Rolling resistance is due to this torque, together with the small contributions of the resistance in the hub and aerodynamic drag.

The two mentioned ways of seeing rolling drag are equivalent since resultant F_z is displaced forward to the centre of the contact owing to the energy dissipations occurring in the deformed parts of the wheel and possibly of the ground.

Rolling resistance is defined by the mentioned SAE document J670 as the force which must be applied to the wheel at the wheel centre with a line of action parallel to the X' axis so that its moment with respect to a line through the centre of tire contact and parallel to the spin axis of the wheel will balance the moment of the tire contact forces about this line.

Here this definition is used, with two small modifications.

First thing the force is changed in sign, so that a true resisting force, i.e. a force acting in a direction opposite to the speed, is obtained.

Second the aerodynamic drag moment on the wheel and the resisting torque applied to the hub are included, in order to include also these two effects in the overall rolling resistance.

To maintain a free wheel spinning a force at the wheel-ground contact is required and then some of the available traction is used: On the idle wheel, to supply a torque which counteracts the total moment M_y, and on the driving wheels which must supply a tractive force against the rolling resistance of the former.

On driving wheels the driving torque is directly applied through the driving shafts to overcome rolling resistance moment. Rolling resistance of driving wheels thus does not involve forces acting at the road-wheel contact and does not use any of the available traction.

This is particularly important in the motion on compliant ground, which is usually characterized by high rolling drag and low available traction: If all wheels are driving wheels, rolling drag can be overcome directly by the driving torque; if some of the wheels are idle, the traction the driving wheels can supply may not be sufficient to overcome the drag of idle wheels and motion may be impossible, even on level road.

Consider a free rolling wheel on level road with its mean plane coinciding with $X'Z'$-plane, i.e. with $\alpha = 0$, $\gamma = 0$ (Fig. 2.14). Assuming that no traction or braking moment other than moment M_f due to aerodynamic drag and bearing drag is applied to the wheel, the equilibrium equation in steady state rolling, solved in the rolling resistance F_r, is:

$$F_r = \frac{-F_z \Delta x + M_f}{R_l} . \tag{2.9}$$

It must be noted that both rolling resistance F_r and drag moment M_f are negative. In the case of driving wheels, moment M_f must be substituted by the difference:

$$M_m - |M_f| ,$$

between the driving and resistant moments.

If such difference is positive and greater than moment $F_z \Delta x$, force F_r is positive, i.e. the wheel is exerting a driving action.

Equation (2.9) is of limited practical use, as Δx and M_f are not easily determined. For practical purposes, rolling resistance is usually expressed as:

$$F_r = -f F_z , \tag{2.10}$$

where the rolling resistance coefficient f must be determined experimentally.

The minus sign in Eq. (2.10) comes from the fact that traditionally the rolling resistance coefficient is expressed by a positive number.

Coefficient f depends on many parameters, as the travelling speed V, the inflation pressure p, the normal force F_z, the size of the tire and of the contact zone, the structure and the material of the tire, the working temperature, the road conditions and, last but not least, the forces F_x and F_y exerted by the wheel.

Effect of Speed

The rolling resistance coefficient f generally increases with the speed V of the vehicle, at the beginning very slowly and then at an increased rate (Fig. 2.15).

The law $f(V)$ can be approximated by a polynomial expression of the type:

$$f = \sum_{i=0}^{n} f_i V^i . \tag{2.11}$$

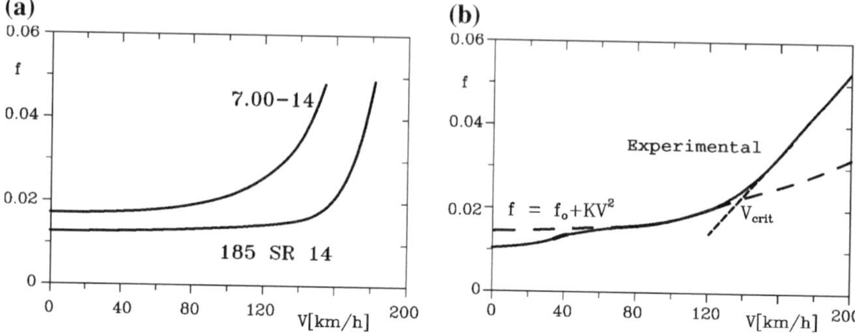

Fig. 2.15 Diagram of rolling resistance f as a function of speed V. **a** Measured on a cross ply tire 175 14 and on a radial tire 185 R 14; **b** experimental curve (radial tire 135 R14, inflated at 190 kPa, with vertical load of 340 kN) compared with Eq. (2.13)

Generally speaking, two terms of Eq. (2.11) are considered sufficient for approximating experimental data in a satisfactory way, at least up to the speed at which f starts to grow at a very high rate (Fig. 2.15). The following expressions can be used:

$$f = f_0 + kV \tag{2.12}$$

or:

$$f = f_0 + KV^2 \,. \tag{2.13}$$

The second is generally preferred; it will be used throughout this book. The values of f_0 and K must be measured on any particular tire; as an example the tire of Fig. 2.15b is characterized, in the test conditions reported, by the values: $f_0 = 0.013$, $K = 6.5 \times 10^{-6}$ s^2/m^2.

On more recent tires rubber mix and fillers have been developed able to reduce rolling resistance; values of $f_0 = 0.008$ can be reached.

The speed at which the curve $f(V)$ shows a sharp bend upward is generally said to be the *critical speed* of the tire. Its presence can be easily explained by vibratory phenomena which take place in the tire at high speed, as the ones clearly visible in the pictures of a tire rolling at high speed against a drum reported in Fig. 2.16. The standing waves which propagate along the circumference of the tire from the contact zone are clearly visible. The tread band vibrates both in its plane and in the direction of the axis of the wheel.

The increase of rolling resistance which is linked with the occurrence of standing waves is easily explained by the fact that their wavelength is not much different from the length of the contact zone (Fig. 2.17). In the trailing part of the contact zone the tread then has a tendency to lift from the ground, or at least to decrease its pressure on it. The pressures concentrate consequently in the leading zone of the contact and their resultant moves forward, with an increase of the moment $F_z \Delta x$.

The critical speed of the tire, i.e. the speed at which such vibrations become important, must be considered as the speed at which the tire stops working in a

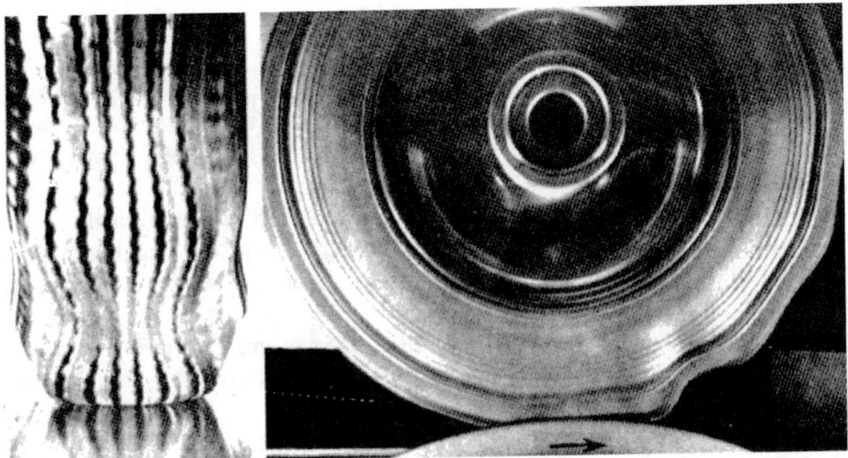

Fig. 2.16 Stationary waves on a rolling tire over its critical speed

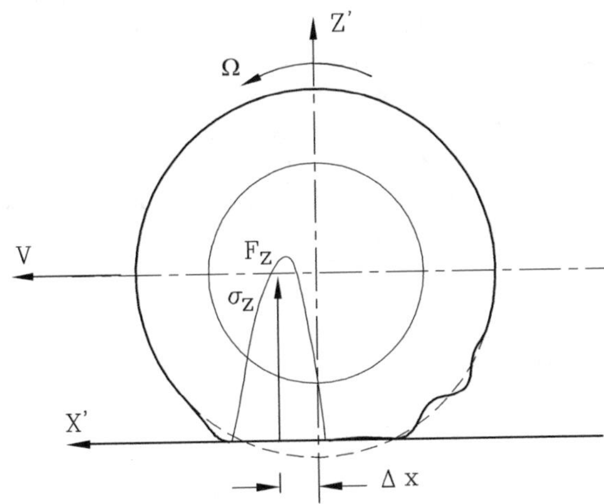

Fig. 2.17 High speed stationary waves (A. Morelli, *Costruzioni automobilistiche*, in *Enciclopedia dell'ingegneria*, ISEDI, Milano, 1972.)

regular way and consequently it should never be exceeded or even approached in the normal use of the vehicle. Above that speed strong overheating takes place; as most of the increase of the rolling power is converted into heat, the increase of temperature can quickly cause the destruction of the tire itself.

The critical speed is influenced by many parameters and it is one of the factors which must be taken into account in the choice of the tires for a particular vehicle.

Effect of Material Nature and Structure

The type of structure and the material used for the construction of the tire play an important role in determining the rolling resistance and the critical speed. Generally speaking, radial tires show a value of f about 20% lower than bias-ply types (Fig. 2.15) and a higher value of the critical speed.

Even between tires of the same type, it is possible to observe significant differences in rolling resistance, as it is possible to optimize the structure (number of plies, their orientation etc.) in order to obtain suitable characteristics for any particular application.

On Fig. 2.18 some typical curve is reported, as an example, referring to conventional cars, sport cars and race cars. It can be seen that critical speeds are quite different in the three cases.

Notable differences can be seen between car and industrial vehicle tires. These last present value of f_0 much lower, down to 0.005–0.008 and a limited increase of f with speed ($K \approx 0$), sometimes a slight reduction (Fig. 2.19).

The nature of the material used has a great importance, as different rubber compositions are characterized by different values of the internal damping and by different dependence of the latter with the loading frequency. Natural rubbers have generally low damping, if compared with synthetic rubbers. This results in lower rolling resistance but also in lower critical speed.

Also the type and quantity of the other chemical components added to rubber have an important influence on damping and consequently on rolling resistance. On Fig. 2.20 two curves are shown for same size tires one made by natural rubber, the other by synthetic one.

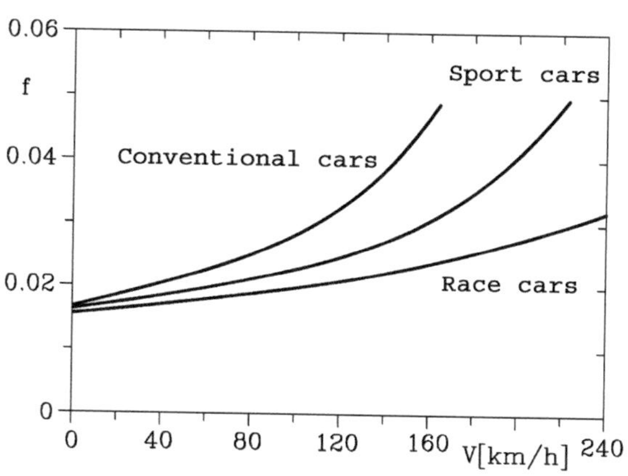

Fig. 2.18 Rolling resistance coefficient f and speed V for different purpose tires

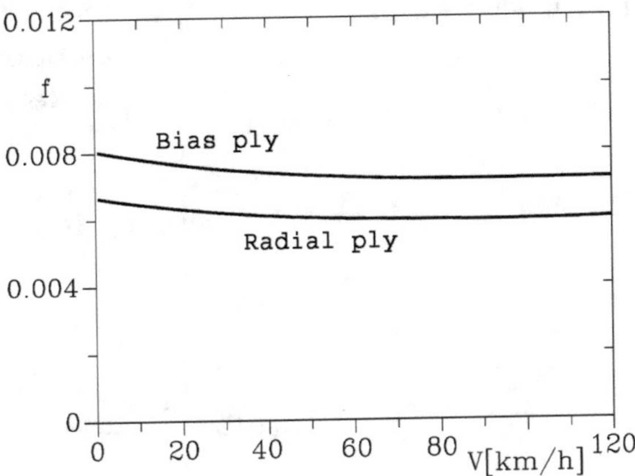

Fig. 2.19 Diagram of f with V on industrial vehicle tires 300 20, pressure $p =$ 750 kPa

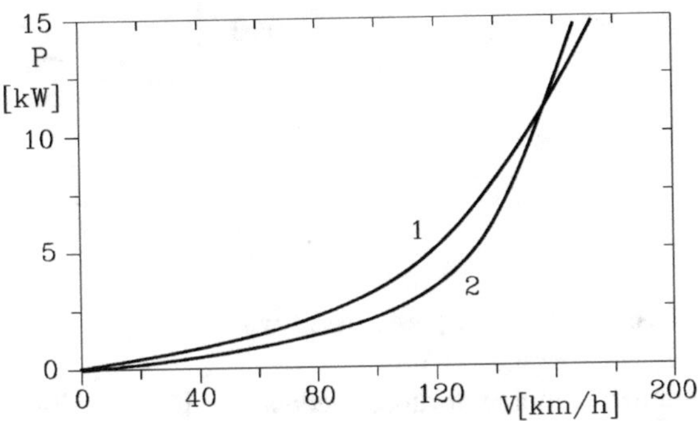

Fig. 2.20 Rolling power as a function of speed. Tires 135 14, $p = 175$ kPa, $F_z = 3{,}7$ kN. Curve 1: synthetic rubber; curve 2: natural rubber

Effect of Tread Wear

On Fig. 2.21 curves $f(V)$ are reported for two tires, cross ply and conventional, in new conditions. To simulate tread wear part of the tread has been artificially removed and tests repeated.

In the case of bias-ply tires the rolling resistance decreases with wear, and its behavior at high speed improves. This behavior can be ascribed to the fact that deformations are localized in a small zone surrounding the contact zone and consequently hysteresis losses take place mainly in the tread band. Also vibratory phenomena interest mainly the zone immediately near the tread band. A decrease of the vibrating mass has the consequence of increasing the natural frequency and hence the critical speed.

In radial tires rolling resistance decreases with wear, but the behavior at high speed can get worse. In radial tires deformations are more evenly distributed in the whole structure, as the stiffness of the sidewalls is low; with a decrease of the mass of the tread band, the centrifugal stiffening of the whole structure decreases, and vibratory phenomena become more important.

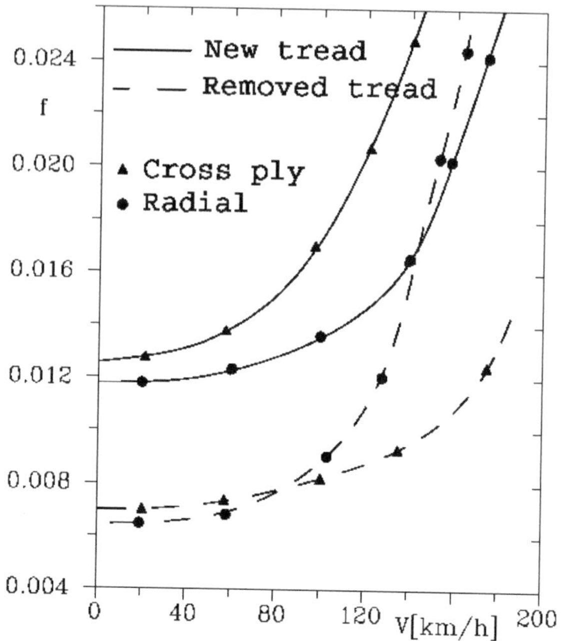

Fig. 2.21 Effect of wear on f

Effect of Operating Temperature

Internal damping of rubber decreases with increasing temperature and consequently rolling resistance, which is mainly due to hysteresis losses, decreases. Also that small part of rolling resistance which is due to localized sliding in the contact zone, decreases for the decrease of the friction coefficient.

The decrease of rolling resistance tends to stabilize the temperature of the tire as an increase of the temperature causes a decrease of power dissipation and consequently of the rate at which heat is generated within the tire.

Some curves $f(V)$ obtained at constant temperature are reported in Fig. 2.22a, while in Fig. 2.22b the rolling coefficient of the same tire maintained at each speed at the equilibrium temperature is shown. The curve is compared with that obtained by maintaining the tire at a temperature corresponding to low-speed running, as it occurs at start-up of the vehicle, before equilibrium temperature is reached.

The decrease in time of the rolling resistance and the increase of the temperature in a tire rolling at 185 km/h is shown in Fig. 2.22c while in Fig. 2.22d the equilibrium temperature is plotted as a function of the speed. The last two plots have been obtained on a Nylon 185 13 radial tire, with $F_z = 4$ kN and $p = 150$ kPa, rolling against a roller of 2.5 m diameter.

The temperature was measured using a thermocouple inset in the carcass and reflects the temperature of the material which is higher than that of the air. As the tests were performed on a roller, the increase of temperature and the resistance are somewhat higher than those which would be obtained on the road.

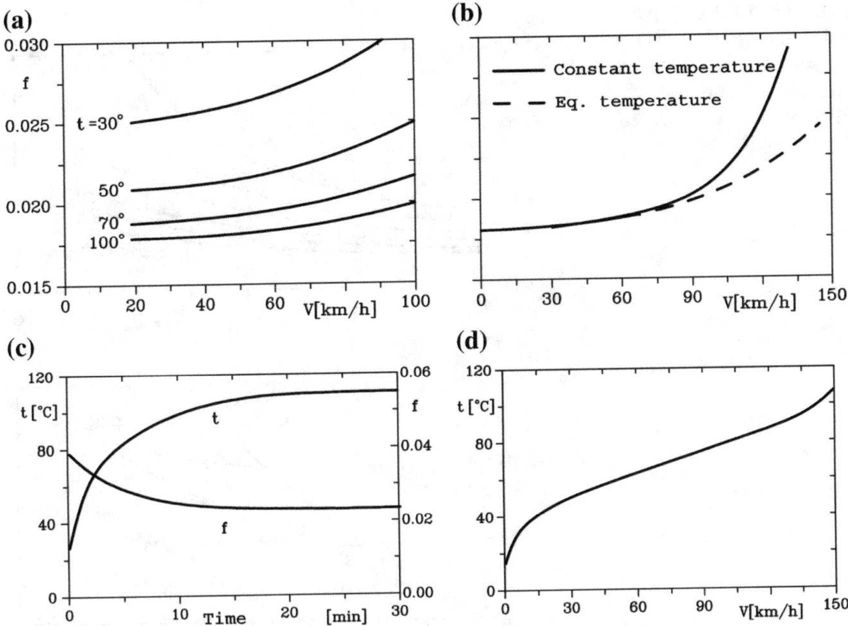

Fig. 2.22 **a** Rolling resistance coefficient at constant temperature. **b** Comparison between $f(V)$ at constant temperature and at equilibrium temperature for each speed. **c** Reduction of rolling resistance coefficient and temperature increase for increasing time at a speed of 185 km/h. **d** Equilibrium temperature in function of speed

Effect of Inflation Pressure and Vertical Load

Generally speaking an increase of inflation pressure or a reduction of the normal force F_z acting on the wheel cause the rolling resistance to decrease and the critical speed to increase. Also this effect tends to stabilize the temperature, as an increase of the latter causes an increase of the pressure which in turn decreases power dissipation and heat generation.

On Fig. 2.23a the $f(V)$ curves are reported, for different values of inflation pressure p. The behavior of the tire at different vertical loads F_z and at different pressures p is shown on Fig. 2.23b.

In order to take into account the influence of both load and pressure on the rolling resistance coefficient the following empirical formula suggested by the SAE can be used:

$$f = \frac{K'}{1000}\left(5.1 + \frac{5.5 \times 10^5 + 90F_z}{p} + \frac{1100 + 0.0388F_z}{p}V^2\right), \qquad (2.14)$$

where coefficient K' takes the value 1 for conventional tires and 0.8 for radial tires. The normal force F_z, the pressure p and the speed V must be expressed respectively

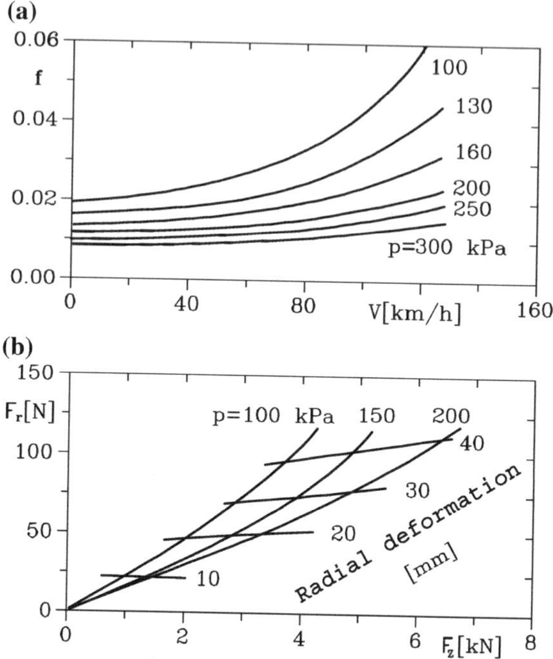

Fig. 2.23 **a** Effect of the inflation pressure on $f(V)$; **b** behavior of a 165 13 tire at different values of F_z and inflation pressure p. Speed $V = 30$ km/h

in N, N/m^2 (Pa) and m/s. The dependence of coefficient f on the speed V is the same as the one expressed by Eq. (2.13).

It must however be noted that for each tire the inflation pressure p is determined as a function of the force F_z by design considerations, and that it is impossible to increase the pressure in order to reduce the rolling resistance.

Effect of Tire Size

The two geometrical parameters which have more influence on the rolling resistance are the radius of the tire and the aspect ratio H/W. An increase of the former and a decrease of the latter cause a decrease of rolling resistance and an increase of the critical speed.

The decrease of the aspect ratio is favorable as it causes an increase of the stiffness of the sidewalls and decreases the deformation under load, which in turn lower the hysteresis losses. Such ratio can be as low as 0.4 on radial tires used on fast modern cars, while values of 0.7–0.8 were the most common in the past.

Figure 2.24 shows the rolling power dissipated by two tires of different size.

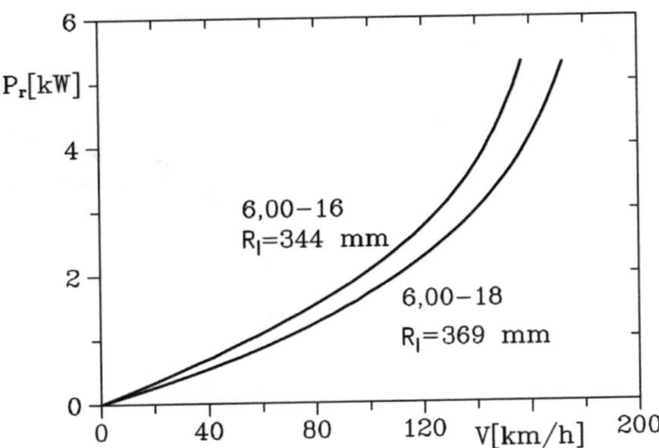

Fig. 2.24 Influence of rolling radius on rolling resistance power P_r. $F_z = 4.75\,\text{kN}$, $p = 250\,\text{kPa}$

Effect of Road

As a first approximation, the nature and conditions of the road are taken into account by choosing an appropriate value of f_0, i.e. by moving the curve $f(V)$ in the direction of the f axis. Some values of f_0 for different road types are reported in Table 2.4.

Effect of Road Wetness

Wetness of the road surface increases rolling resistance considerably. Everybody has experienced the sudden slow-down of a car when crossing a puddle at high speed and also many of the most energy-conscious drivers are aware of the fuel consumption increase of their car in wet weather. For certain tires and road surfaces in heavy rain, the rolling resistance increase may be as high as 50%.

Not only the force necessary to remove the layer of water under the contact patch but also a higher hysteresis loss due to lower tire temperature causes this increase. See Fig. 2.22 for reference; in the average a rolling resistance increases of about 1% for each degree of temperature decrease of the tire can be noticed.

Figures 2.25 and 2.26 show[4] the changes in rolling resistance f at different values for the car speed V because of the water thickness s, for two different tires, rolling on a tarmac ground; the figures refer to a standard P225/60R16 and a 195R14 M+S (for winter and light off road operation).

We can observe that:

- The increase in rolling resistance is increasing with water thickness;
- The thread design has influence on this increase with its different behavior on removing the water under the contact patch, as can be observed by comparing the standard tire with the M+S one. While the thread of the last is deeper, the number

[4]Ejsmont J., Sjögren L., Świeczko-Żurek B. and Ronowski G., *Influence of Road Wetness on Tire-Pavement Rolling Resistance*, Journal of Civil Engineering and Architecture, 9 (2015).

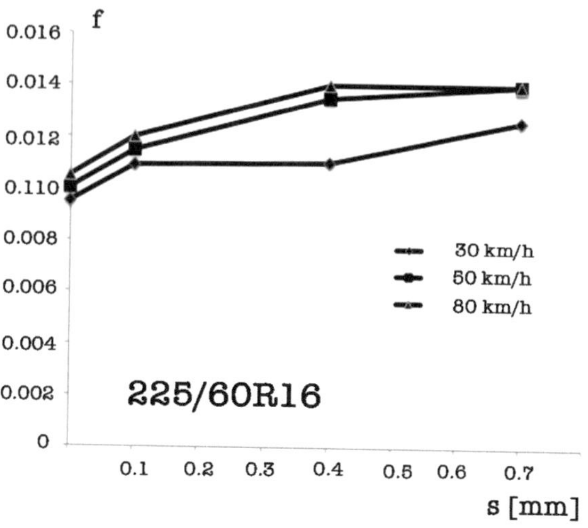

Fig. 2.25 Effect on rolling resistance of the water thickness *s* for a 225/60R16 tire on a wet tarmac ground

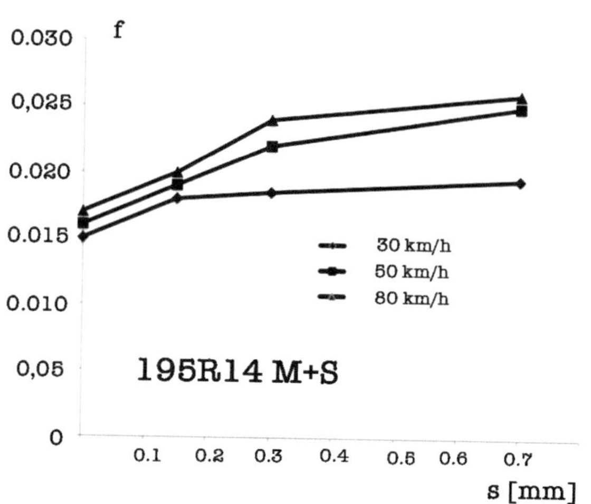

Fig. 2.26 Effect on rolling resistance of the water thickness *s* for a 195R14 M+S tire on a wet tarmac ground

and density of grooves in the thread pattern is higher on the first, which allows better water removal.

Parallel measurement showed that the water present on the road surface cooled the tires with a contribution on rolling resistance increase. Since in the atmospheric conditions existing during the tests tire temperature drop of 6–7 °C as compared with dry conditions, we can estimate an increases in rolling resistance by about 10% for this effect only.

However, the increase of rolling resistance cannot be explained by temperature drop alone. It appears that thicker water film prompt hydrodynamic effects that dissipate energy by water turbulences and create a kind of standing wave at the leading edge of the footprint which moves the vertical reaction force forward the theoretical contact point.

It was also demonstrated that the size of the inert material inside the tarmac has also a considerable influence on the water film thickness resulting in different rolling resistance. It is observed that surfaces with larger size can drain water from surface more easily, resulting in lower rolling resistance increase during very wet conditions.

Effect of Wheel Side Slip Angle

If the tire travels with a sideslip angle α, as it is the case any time it exerts a side force F_y or as a consequence of toe angle, a strong increase of rolling resistance can be expected. The force in the mean plane of the wheel increases but above all the transversal force F_y has a component which adds to the rolling resistance (Fig. 2.27). The rolling resistance is by definition the component of the force due to the road-tire contact directed as the velocity V; it can thus be expressed as:

$$F_r = F_x \cos(\alpha) + F_y \sin(\alpha) \; . \tag{2.15}$$

If the component in the plane of symmetry of the wheel F_x were independent of the sideslip angle and the cornering force F_y were linear with it ($F_y = -C\alpha$), for small values of α the rolling resistance would follow a quadratic law:

$$|F_r| = |F_x| + C\alpha^2 \; . \tag{2.16}$$

Effect of Camber Angle

If the mean plane of the wheel is not perpendicular to the ground a component of the aligning torque M_z (see Sect. 2.7) contributes to rolling resistance. Equation (2.9) becomes:

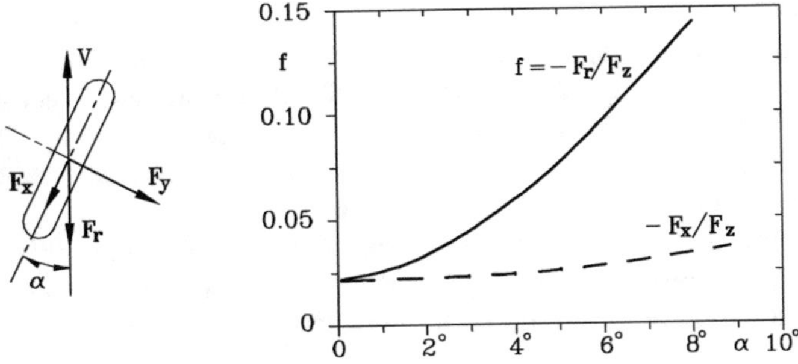

Fig. 2.27 Rolling resistance coefficient as function of the wheel side slip angle α. Tire 185 14, F_z = 4 kN, p = 170 kPa

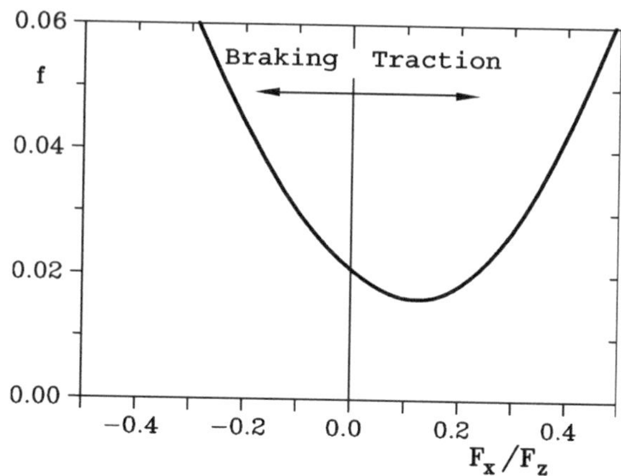

Fig. 2.28 Rolling resistance coefficient as function of braking and traction force

$$F_r = \frac{-F_z \Delta x \cos(\gamma) - M_z \sin(\gamma) + M_f}{R_l} \, . \tag{2.17}$$

This effect is usually very small, due to the fact that γ is usually small. It is however dependent on the sideslip angle α through the aligning torque M_z.

Effect of Longitudinal Force

Rolling resistance can also be defined when tractive or braking moments are applied to the wheel. In this case the power dissipated by rolling resistance $F_r V$ can be expressed as:

$$|F_r|V = \begin{cases} |F_f|V - |M_f|\Omega \text{ (braking)} \\ M_t|\Omega - |F_t|V \quad \text{(traction)} \end{cases} \tag{2.18}$$

where F_b, F_t, M_b and M_t are respectively the braking and tractive forces and moments. Equation (2.18) should be applied only in constant speed motion, since they do not include tractive (braking) moments needed to accelerate (decelerate) rotating parts.

The trend of the rolling resistance coefficient as a function of the longitudinal force F_x is shown in Fig. 2.28. The increase of rolling resistance is not negligible in case of strong longitudinal forces, particularly in the case of braking. This is due mainly to the fact that the generation of longitudinal forces is always accompanied by the presence of sliding in at least a part of the contact zone.

The minimum rolling resistance occurs when wheels exert a low driving force[5] and can take a value as low as 75–85% of that occurring in free rolling conditions. The fact that the rolling resistance decreases initially with the application of a driving

[5]D.J. Schuring, *Energy Loss of Pneumatic Tires Under Freely Rolling, Braking and Driving Conditions*, Tire Science and Technology, TSTCA, Vol. 4, No. 1, Feb. 1976, pp. 3–15.

force to increase steeply when the force increases would favor four wheel drive layouts in which all wheels work under moderate driving forces instead of having some wheels idle and some working with higher driving loads.

2.5 Static Forces

Consider a small portion of the tire-road contact area. The force per unit area exerted by the tire on the road can be decomposed into a component perpendicular to the road and a tangential component. The first is the contact pressure σ_z while the other can be further decomposed in the directions of X' and Y' axes giving way to the components τ_x and τ_y. The resultant of the distributions of σ_z, τ_x and τ_y are the already defined normal, longitudinal and lateral forces F_z, F_x and F_y respectively.

These distributions are not constant and are strongly influenced by many factors as tire structure, load, inflation pressure etc. Some typical results obtained on a stationary wheel which is exerting no force in the $X'Y'$ plane are reported in Fig. 2.29. At the centre of the contact, the contact pressure σ_z is close to that of the air in the tire at the centre of the contact area, while at the sides it is higher. If the wheel is not rolling the distribution is symmetrical with respect to $Y'Z'$ plane and the resultant passes through the centre of the contact. Tangential forces do not vanish locally even when no force is exerted in $X'Y'$ plane, i.e. when their resultant is zero. In such case components τ_x are directed towards the centre of the contact area and the tire acts to *compact* the ground towards the centre of the contact. Component τ_y has the effect of *stretching* the ground outward.

The force-deflection characteristics of tires depend on many factors, like travelling speed, pressure, wear and many other. A strong difference can be found for this

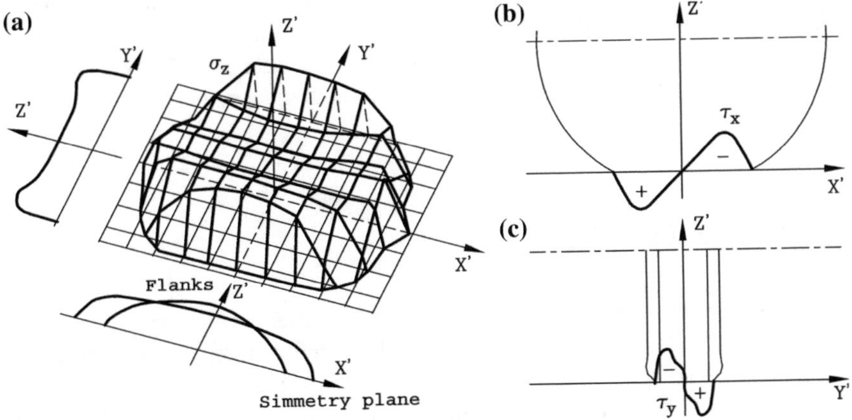

Fig. 2.29 Force distribution in contact patch. **a** Normal pressure σ_z. **b** Tangential pressure τ_x in the symmetry plane **c** τ_y and in the $Y'Z'$ plane

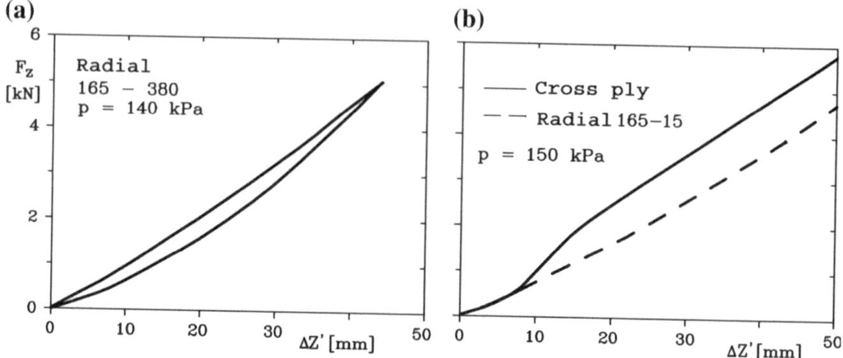

Fig. 2.30 Contact force F_z as function of deflection $\Delta Z'$ for some tires in static conditions (load applied and relieved slowly)

issue between bias-ply and radial tires, radial tires being less stiff at standstill in all directions.

The characteristics in direction perpendicular to the ground (force F_z versus deflection Z') for some tires are reported in Fig. 2.30. In Fig. 2.30a the curve obtained when removing the load has been reported together with that related to the application of the load. A hysteresis cycle can clearly be identified, denouncing the presence of damping in the motion along Z'-axis. This damping is usually at its maximum at standstill and decreases with rolling speed: The practice of neglecting it in the simulations of the motion of the vehicle is justified by this observation. In Fig. 2.30b the curves obtained for a radial and a bias-ply tire are compared.

A static tire rate can be defined as the tangent stiffness in any given equilibrium condition, i.e. at any given value of the load, inflation, pressure etc.

Similar plots can be obtained for forces in X' and Y' directions and moments about Z' axis versus the corresponding displacements of the centre of the wheel (Fig. 2.31). In all cases the plots show a nonlinear behavior and a hysteresis cycle; radial tires are generally less stiff than bias-ply tires of similar size. All characteristics are influenced by both the speed of rolling and the frequency of application of the force.

2.6 Longitudinal Force

Consider a pneumatic wheel rolling on level road. If a braking moment M_b is applied to it, the distributions of normal pressure and longitudinal forces which result from that application are qualitatively sketched in Fig. 2.32a. The tread band is circumferentially stretched in the zone which precedes the contact with the ground, while in free rolling the same part of the tire was compressed.

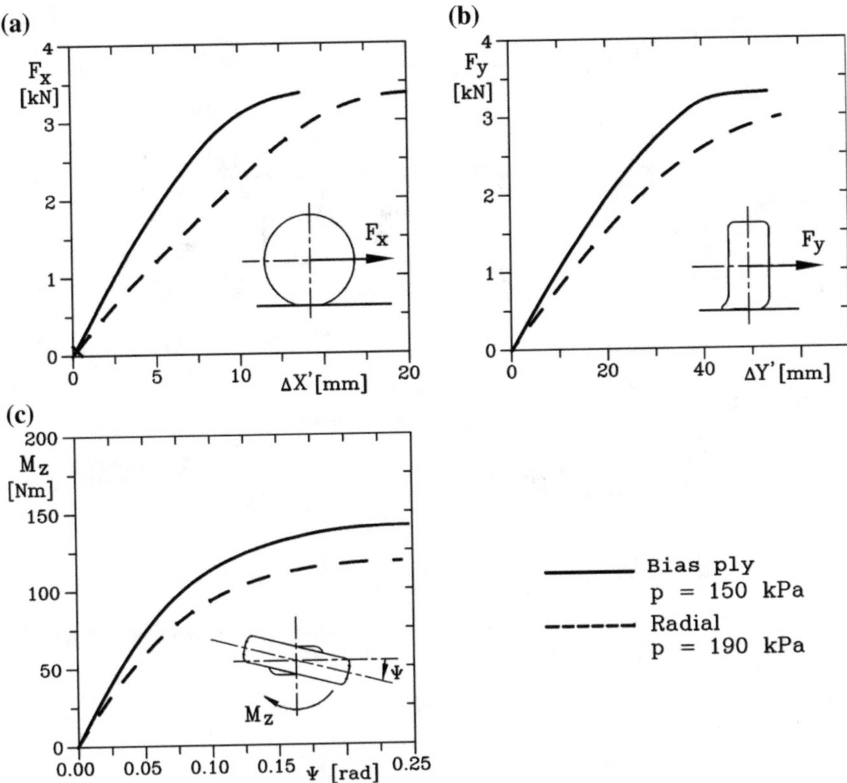

Fig. 2.31 versus deflections $\Delta X'$ and $\Delta Y'$ and rotation about Z' axis for a radial and a bias-ply tire at standstill (wheel non-rolling) and in static conditions (load applied and removed slowly)

The peripheral velocity of the tread band in the leading zone of the contact $\Omega R'_e$ is consequently higher than that (ΩR) of the undeformed wheel. The effective rolling radius R'_e, whose value R_e in free rolling was between R_l and R, grows towards R and, if M_b is large enough, becomes greater than R.

The instantaneous centre of rotation is consequently located under the road surface (Fig. 2.33). The angular velocity Ω of the wheel is lower than that characterizing free rolling in the same conditions ($\Omega_0 = V/R_e$). In such conditions it is possible to define a *longitudinal slip* as:

$$\sigma = \frac{\Omega}{\Omega_0} - 1 = \frac{v}{V} , \tag{2.19}$$

where v is the linear speed at which the contact zone moves on the ground. The longitudinal slip is often expressed as a percentage; in the present book however the definition of Eq. (2.19) will be strictly adhered to.

If instead of braking, the wheel is driving, the leading part of the contact zone is compressed instead of being stretched (Fig. 2.32b). The value of the effective rolling

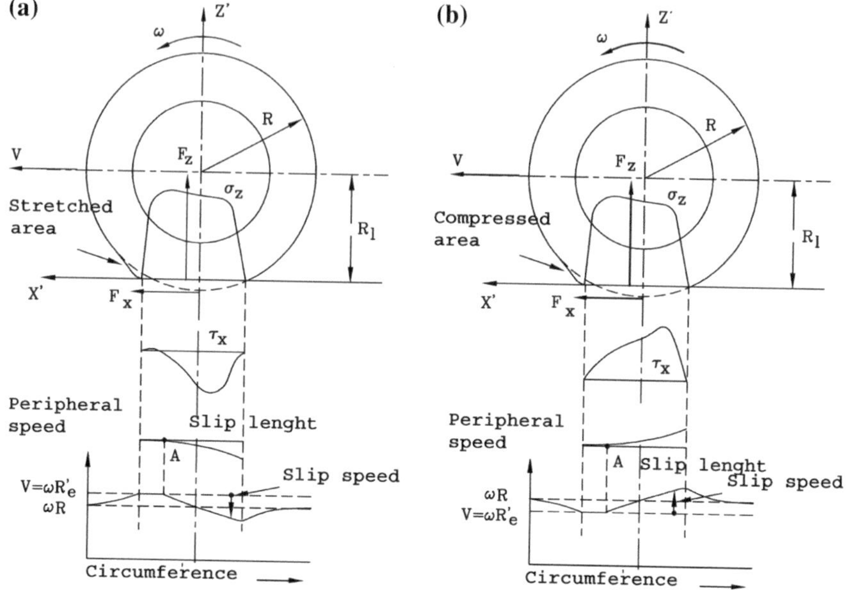

Fig. 2.32 Contact pressure distribution for a braking (**a**) and driving (**b**) wheel. The rolling radius R_e' is different as that by pure rolling R_e (force F_x lies on ground)

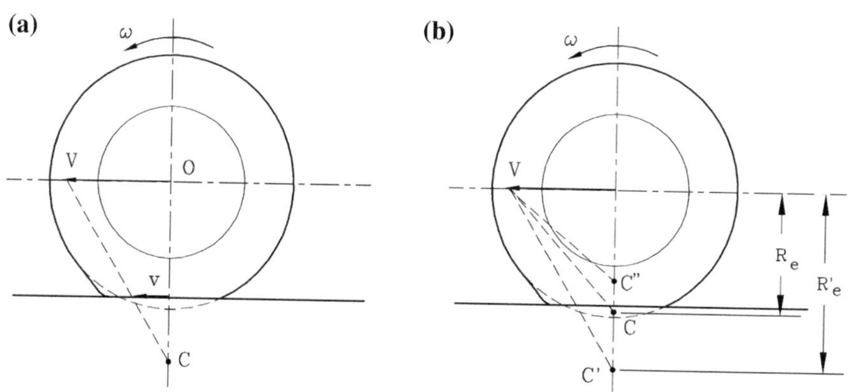

Fig. 2.33 **a** Braking wheel, center of instantaneous rotation and sleep speed. **b** Position of the instantaneous rotation center by pure rotation C, by braking C′ and by traction C″

radius R_e' is smaller than that characterizing free rolling and is usually smaller than R_l; the angular velocity of the wheel is greater than Ω_0.

The slip defined by Eq. (2.19) is positive for driving conditions and negative for braking.

The presence of the slip velocity[6] v does not mean however that there is an actual sliding of the contact zone as a whole. The peripheral velocity of the leading part of that zone is actually $V = \Omega R'_e$, and consequently in that zone no sliding can occur. The speed of the tread band starts to decrease (in braking, increase in driving) and sliding begins only at the point indicated in Fig. 2.32 as point A. The slip zone, which interests only a very limited part of the contact zone for small values of σ, gets larger with increasing slip and, at a certain value of that parameter, reaches the leading part of the contact zone and global sliding of the tire occurs (Fig. 2.34a).

The longitudinal force F_x the wheel exchanges with the road is a function of σ. It vanishes when $\sigma = 0$ (free rolling conditions)[7] to increase with almost linear law, for values of σ from about -0.25 to about 0.25:

$$F_x = C_\sigma \sigma \ , \tag{2.20}$$

where constant

$$C_\sigma = \left(\frac{\partial F_x}{\partial \sigma}\right)_{\sigma=0}$$

can be defined *slip stiffness* or *longitudinal stiffness* of the tire.

Outside this range, which depends on many factors, its absolute value decreases in braking up to the value $\sigma = -1$, which characterizes free sliding (locking of the wheel). Also in driving the force decreases above the stated range, but σ can have any positive value, up to infinity when the wheel spins while the vehicle is not moving.

As a first approximation, force F_x can be considered as roughly proportional to the load F_z, at equal value of σ. It is consequently useful to define a longitudinal force coefficient:

$$\mu_x = \frac{F_x}{F_z} \ . \tag{2.21}$$

The qualitative trend of such coefficient is reported against σ in Fig. 2.34b.

In the first part of the curve $F_x(\sigma)$, the slip stiffness is roughly proportional to the load and it is possible to define a slip stiffness coefficient $\partial C_\sigma / \partial F_z$, at least for limited variation of the load about a given value F_{z0}

$$C_\sigma = C_{\sigma 0} + \frac{\partial C_\sigma}{\partial F_z}(F_z - F_{z0}) \tag{2.22}$$

Two important values of μ can be identified on the curve both in braking and in driving: The peak value μ_p and the value μ_s which characterizes pure sliding. The first is referred to as driving traction coefficient when the wheel is exerting

---

[6]The slip velocity is defined by SAE Document J670 as $\Omega - \Omega_0$, i.e. the difference between the actual angular velocity and the angular velocity of a free rolling tire. Here a definition based on a linear velocity rather than an angular velocity is preferred: $v = R_e(\Omega - \Omega_0)$.

[7]Actually free rolling is characterized by a very small negative slip, corresponding to the rolling resistance. This is however usually neglected when plotting curves $F_x(\sigma)$.

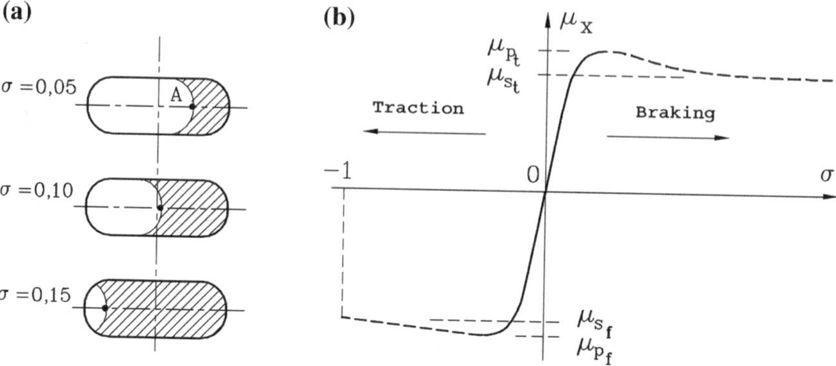

Fig. 2.34 **a** Slipping area at different values of slip σ. **b** Qualitative diagram of μ_x as function of longitudinal slip σ

a positive longitudinal force and as braking traction coefficient, which is usually reported in absolute value, in the opposite case. The second are the sliding driving traction coefficient and the sliding braking traction coefficient.

The part of the curve $\mu(\sigma)$ which lies beyond the range included by the two peak values, represented by a dashed line in Fig. 2.34b, is a zone of instability in the practical use of the vehicle. The equation of motion of a braking wheel is:

$$J\frac{d\Omega}{dt} = |F_x|R_l - |M_f| \ . \qquad (2.23)$$

If a decrease of σ at constant speed V leads to a decrease of the absolute value of μ_x, it causes a decrease of the absolute value of F_x, i.e. of the force which maintains the rotation of the wheel. If this decrease of $|F_x|$ is not quickly followed by a decrease of the braking moment $|M_b|$ (and it is unrealistic to assume that the driver can react fast enough to release the brakes) a further slowing down of the wheel takes place, which in turn causes a further reduction of $|F_x|$.

Actually, when the optimum value of σ which is characterized by μ_{p_b}, is exceeded, the wheel locks in a very short time. In order to prevent the locking of wheels, devices generally defined as antilock or antiskid systems are now widely used.

Such devices detect the deceleration of the wheel and, when it reaches a predetermined value, react decreasing the braking moment and avoiding the locking of the wheel. *Antilock* devices can operate on each wheel separately or, more often, on both wheels of an axle. Similarly, to avoid that a wheel slips under the effect of a driving torque applied to it, *antispin* devices limit the engine moment when the acceleration of the driving wheels exceeds a stated value.

The curves usually show a certain symmetry between the braking and driving conditions and often the maximum braking and driving forces are assumed to be equal. The values of function $\mu_x(\sigma)$ depend on a number of parameters, such as the type of tire, road conditions, speed, magnitude of the side force F_y exerted by the

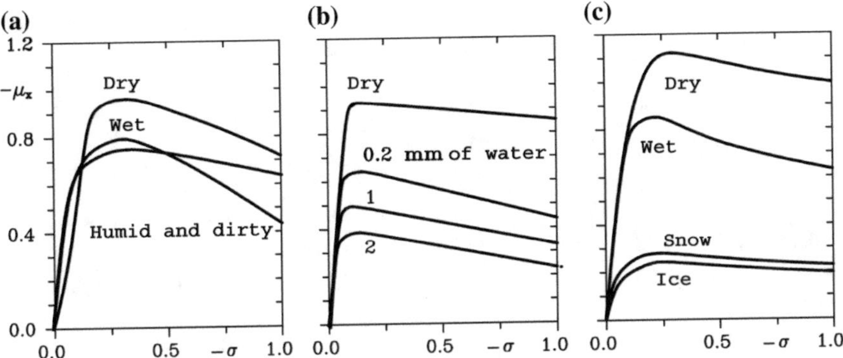

Fig. 2.35 Curves of $\mu_x(\sigma)$ obtained in different conditions

tire and many others. Moreover there is a significant difference between the curves obtained by different experimenters in conditions not exactly comparable. Some curves $\mu_x(\sigma)$ obtained in different conditions are reported in Fig. 2.35.

The maximum value of the longitudinal force decreases with increasing speed but this reduction is much influenced by operating conditions. Generally speaking, it is not very marked on dry road, while it is greater on wet surfaces. Also the difference between the maximum value (see Table 2.5) and the value related to sliding (Table 2.6) is more notable on wet roads (Fig. 2.36). Particularly dangerous conditions are encountered when the road is only partially wet and dirty: The behavior can vary from spot to spot and the value at slip can be very much different from the maximum one.

The values shown in the tables must be regarded only as average indications as they are much sensitive. Note that in good conditions the longitudinal force can be almost equal to the load acting on the tire or even larger; the values reported however refer to standard tires used on passenger vehicles. High performance tires, particularly those used on racing cars, show peak values of μ_x which can be as high as 1.5–1.8 but even these tires do not reach very high values of longitudinal force coefficient in sliding condition; the difference between μ_p and μ_s is even larger. To reach very high performances particular formulations of the rubber must be used, which are characterized by a strong wear and consequently are restricted to competition tires. Note that radial tires almost always show better performances than bias-ply types.

Tread wear has a great influence on the longitudinal forces, particularly at high speed. From Fig. 2.37 it can be assumed that the increase of μ_{p_b} due to wear can be, particularly at high speed, quite noticeable. It must however be noted that the figure refers to dry roads, as the presence of even a thin water layer on the road can change drastically these results.

When the road is wet, particularly if the water layer is thick, the tire can lift from the road surface as a result of hydrodynamic lift (*aquaplaning*). A thin layer of water can slip between the tire and the road thus reducing the contact area (Fig. 2.38). With increasing speed the area of the contact zone further reduces, until a complete lifting

Table 2.5 Values of μ_p for different tires at 30 km/h. Tires 145 15; 150 15 e 165 R 15; $F_z = 3$ kN, $p = 160$ kPa for cross ply tires (conv.) and 220 kPa for radial tires

Tire	Road					
	Concrete		Tarmac		Snow	Ice
	Dry	Wet	Dry	Wet		
Rad.	1.19	0.99	1.22	1.10	0.45	0.25
Conv.	1.13	0.84	1.02	1.07	0.27	0.24
Rad. (snow)	1.04–1.12	0.62–0.83	1.00–1.09	1.00–1.10	0.36–0.47	0.24–0.44
Conv. (snow)	0.86–1.02	0.59–0.70	0.81–0.89	0.78–1.02	0.41–0.48	0.29–0.37
Chains					0.60	0.40

Table 2.6 Values of μ_s for same tires of previous table

Tire	Road					
	Concrete		Tarmac		Snow	Ice
	Dry	Wet	Dry	Wet		
Rad.	0.95	0.73	1.03	0.90	0.43	0.16
Conv.	0.99	0.62	0.88	0.80	0.22	0.18
Rad. (snow)	0.88–1.00	0.50–0.61	0.87–0.99	0.77–0.93	0.35–0.45	0.22–0.41
Conv. (snow)	0.72–0.90	0.47–0.57	0.70–0.78	0.67–0.84	0.39–0.47	0.29–0.36

of the tire takes place. True hydrodynamic lubrication conditions can be said to exist in this case and consequently the force coefficient or, better, the friction coefficient as in this condition sliding usually occurs, reduces to very low values, of the order of 0.05.

In order to avoid hydroplaning, or at least to postpone its occurrence, it is mandatory to evacuate water from the contact zone as quickly and effectively as possible. This can be done in two distinct ways: By making the road surface permeable or by using deep groves in the tread in both circumferential and transversal direction in order to allow a high flow rate.

The assumption, in a way implicit in the definition of the longitudinal force coefficient, that longitudinal forces are proportional to the normal force acting on the wheel is only a crude approximation. Actually the longitudinal force coefficient decreases with increasing load as shown in Fig. 2.39.

The curves $\mu_x(\sigma)$ can be approximated by analytical expressions. One of the formulas which can be used in the range $-1 < \sigma < 1$ is:

$$\mu_x = A\left(1 - e^{-B\sigma}\right) + C\sigma^2 - D\sigma \,, \tag{2.24}$$

where:

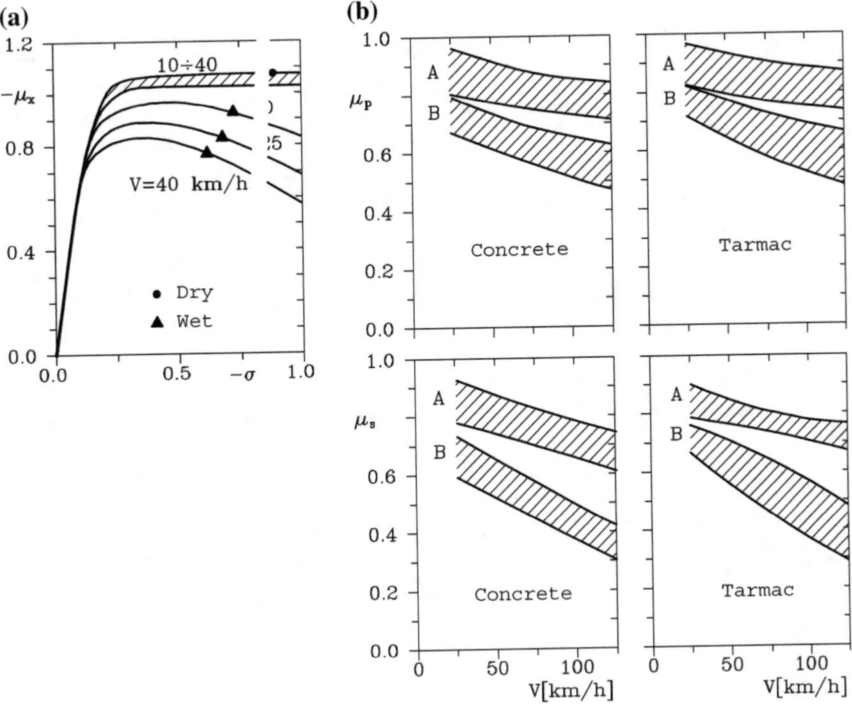

Fig. 2.36 Influence of car speed on μ_p and μ_s on wet and dry road. Cross and bias ply tires 165 15 e 150 R 15 with $F_z = 3.75\,\mathrm{kN}$

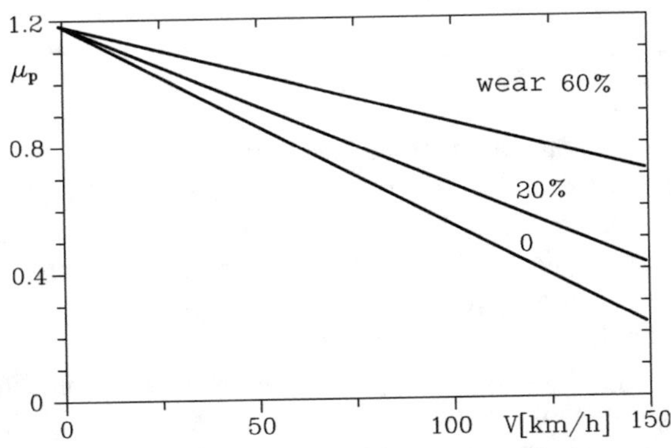

Fig. 2.37 Influence of tread wear on the peak value of longitudinal friction as function of speed

$$B = \left(\frac{K}{\alpha + d} \right)^{1/n}$$

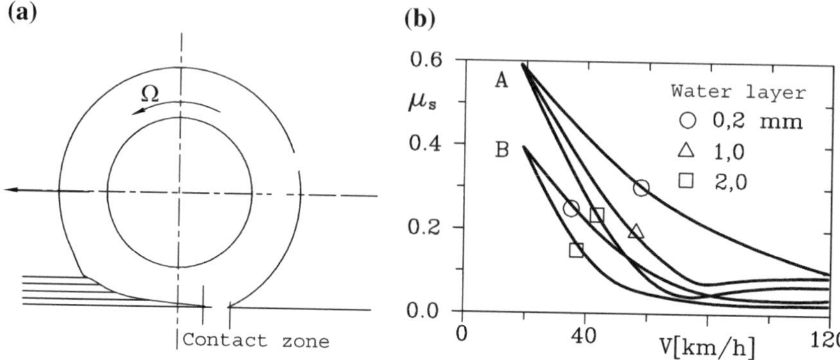

Fig. 2.38 Hydrodynamic lift of a wheel (*aquaplaning*). **a** Scheme; **b** diagram of μ_s in function of speed on a wet road. Tire 145-15 with tread (curve A) and without (curve B); $F_z = 3$ kN; $p = 150$ kPa

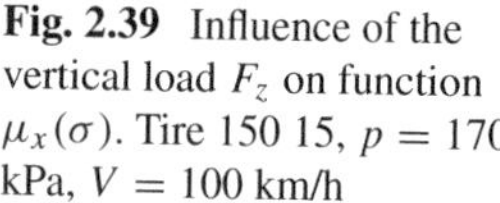

Fig. 2.39 Influence of the vertical load F_z on function $\mu_x(\sigma)$. Tire 150 15, $p = 170$ kPa, $V = 100$ km/h

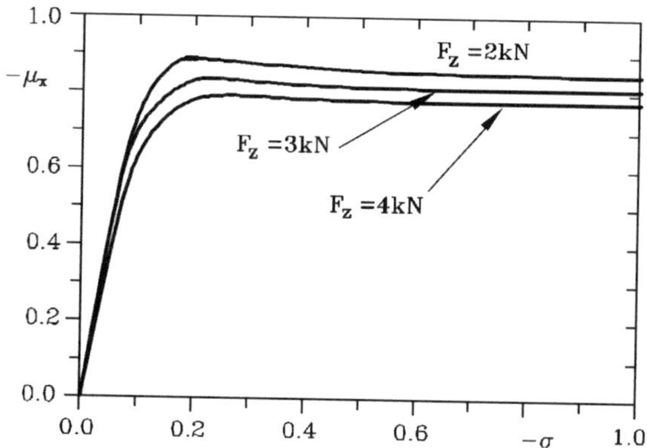

is a factor which takes into account the interaction between the longitudinal slip σ and the sideslip α (see Sect. 2.8). The derivative in the origin $(\partial\mu_x/\partial\sigma)_{\sigma=0}$ is simply $AB - D$. Coefficients A, C, D, K, d and n must be obtained from the experimental curves and have no physical meaning. They depend not only on the road conditions but also on the load. A curve $\mu_x(\sigma)$ for a radial tire 145/80 R 13 4.5J obtained through Eq. (2.24) is reported in Fig. 2.40a, curve A.

A very good approximation of longitudinal force F_x as a function of the slip σ can be obtained through the empirical equation introduced by Pacejka[8] and known as *magic formula*. Such mathematical expression allows one to express forces F_x and F_y and the aligning torque M_z as functions of the normal force F_z, of σ and of the sideslip (α) and camber (γ) angles.

The equation yielding the longitudinal force F_x, as a function of the slip σ, is:

[8]E. Bakker, L. Lidner, H.B. Pacejka, *Tire Modelling for Use in Vehicle Dynamics Studies*, SAE Paper 870421; E. Bakker, H.B. Pacejka, L. Lidner, *A New Tire Model with an Application in Vehicle Dynamics Studies*, SAE Paper 890087.

$$F_x = D \sin\left(C \arctan\left\{B(1-E)(\sigma + S_h) + E \arctan\left[B(\sigma + S_h)\right]\right\}\right) + S_v \,, \tag{2.25}$$

where B, C, D, E, S_v and S_h are six coefficients which depend on the load F_z and on angle γ. They must be obtained from experimental testing and do not have any direct physical meaning. In particular, S_v and S_h have been introduced to allow nonvanishing values of F_x when $\sigma = 0$.

Coefficient D yields directly the maximum value of F_x, apart from the effect of S_v. The product BCD gives the slope of the curve for $\sigma + S_h = 0$. The values of the coefficients are expressed as functions of a number of coefficients b_i which can be considered as characteristic of any specific tire, but depend also on road conditions and speed:

$$C = b_0 \qquad D = \mu_p F_z \,,$$

where for b_0 a value of 1.65 is suggested and:

$$\mu_p = b_1 F_z + b_2 \,, \qquad BCD = \left(b_3 F_z^2 + b_4 F_z\right) e^{-b_5 F_z} \,,$$
$$E = b_6 F_z^2 + b_7 F_z + b_8 \,, \quad S_h = b_9 F_z + b_{10} \,, \qquad S_v = 0 \,.$$

Note that product BCD is nothing else than the slip stiffness of the tire.

If a symmetrical behavior for positive and negative values of force F_x is accepted, this model can be used for both braking and driving. The curve is usually extended to braking beyond the point where $\sigma = -1$, to simulate a wheel rotating in backward direction while moving forward.

The coefficients introduced in Eq. (2.25) and the results obtained from it are usually expressed in non consistent units: Force F_z is in kN, longitudinal slip is expressed as a percentage and force F_x is in N.

A set of curves $F_x(\sigma)$ obtained for vertical loads $F_z = 2$, 4, 6 and 8 kN for a radial tire 205/60 VR 15 6J is shown in Fig. 2.40b.

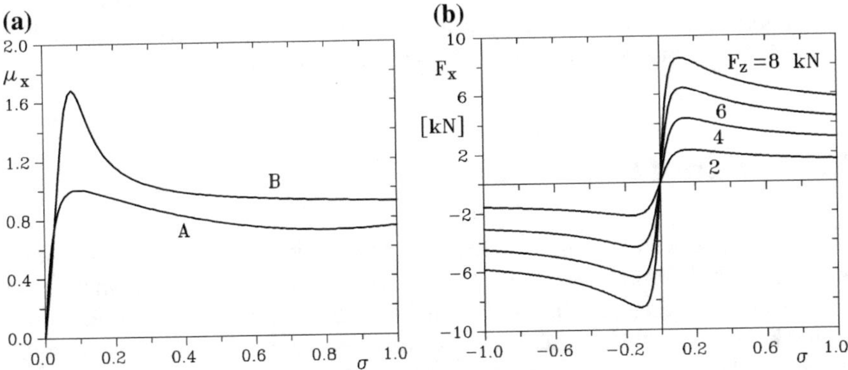

Fig. 2.40 **a** Curves $\mu_x(\sigma)$ for a 145 R 13 tire obtained with Eq. (2.24) (curve A) and for a 245/65 R 22.5 tire obtained from Fig. 2.25 (curve B). **b** Curves $F_x(\sigma)$ for some value of vertical load Fig. 2.25 for a 205/60 R 15 V tire

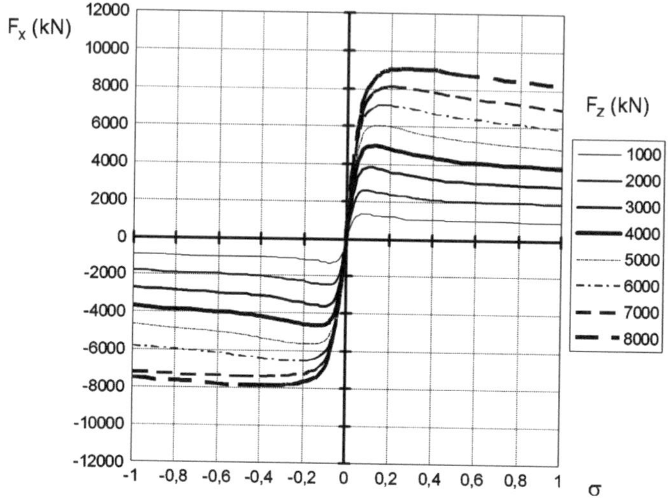

Fig. 2.41 Curves $F_x(\sigma)$ for a 195/65 R 15 tire obtained for different values of vertical load F_z

A curve $\mu_x(\sigma)$ for a high performance radial tire 245/65 R 22.5 obtained from Eq. (2.25), is also reported in Fig. 2.40 a, curve B. Note the very high peak value of the force coefficient.

Using the same equations, on Fig. 2.41 the curve $F_x(\sigma)$ is shown for a tire 195/65 R 15: it should be noted, in this case, the flat shape of the curve, after the peak, due to the high performance kind of rubber mix.

The importance of the model expressed by Eq. (2.25) is mainly linked to the fact that tire manufacturers are increasingly disclosing the performances of their tires in terms of magic formulae coefficients. If this trend will consolidate, the magic formula will prove to be a simple and accurate model for tire behavior and, which is even more important, one for which the data will be readily available.

Magic formulae have been modified by the Author in 1996, by the introduction of additional parameters; the new formulae include the modelling of the combined effect of longitudinal and side slip, as we will see on a dedicated paragraph.

2.7 Cornering Forces

In the previous section it was clear that a pneumatic tire can exert longitudinal forces only if deformations are present in the tread band and if the wheel has a nonvanishing longitudinal slip. In the same way the generation of cornering forces cannot be understood if no reference is made to the lateral deformations of the tire and to its sideslip angle: The generation of tangential forces in the road-wheel contact is directly linked with the compliance of the tire.

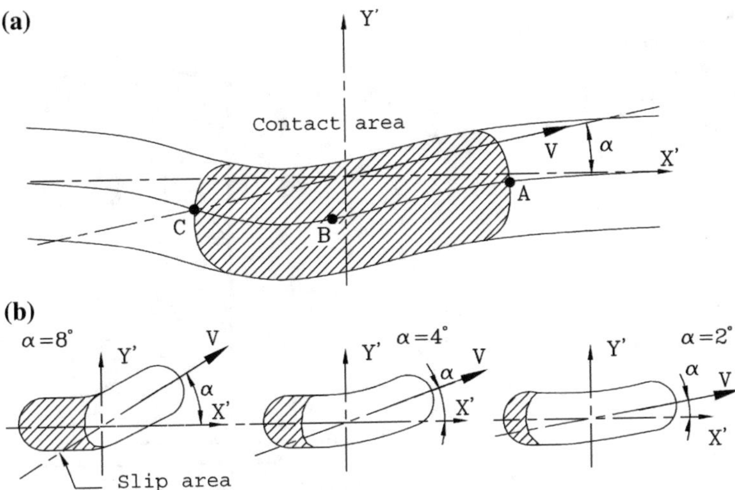

Fig. 2.42 Wheel-road contact when side slip angles are present. **a** Contact zone and path of a point of the tread on the equator plane; **b** contact zone and slip zone at different values of α (α is not in scale)

The fact that the wheel has a sideslip angle, i.e. is not in pure rolling, does not mean that in the contact zone the tire slips on the road: Also in this case, as seen for longitudinal forces, the compliance of the tire allows the tread to move, relatively to the centre of the wheel, with the same velocity as the ground. However some localized sliding between the wheel and the road can be present and, with increasing sideslip angle, they become more and more important, until the whole wheel is in actual, macroscopic sliding.

If the velocity of the centre of the wheel does not lie in its mean plane, i.e. if the wheel travels with a sideslip angle, the shape of the contact zone is quite distorted (Fig. 2.42). Consider a point belonging to the mean plane on the tread band. Upon approaching the contact zone it tends to move in a direction parallel to the velocity V, relatively to the centre of the wheel, and consequently goes out of the mean plane.

After touching the ground at point A, it continues following the direction of the velocity V (for an observer fixed to the ground, it remains still) until it reaches point B. At that point, the elastic forces pulling it towards the mean plane are strong enough to overcome those due to the friction on the road, forcing it to slide on the road and to deviate from its path. This sliding continues for the remaining part of the contact zone, until point C is reached. The contact zone can thus be divided into two parts: A leading zone in which no sliding occurs and a trailing one in which the tread slips towards the mean plane. This second zone grows with the sideslip angle (Fig. 2.42b), until it invades the whole contact zone and the wheel actually slips on the ground.

The lateral deformations of the tire are plotted in a qualitative way in Fig. 2.43, together with the distribution of σ_z, τ_y, and of the lateral velocity. The resultant F_y of the distribution of side forces is not applied at the centre of the contact zone but at a

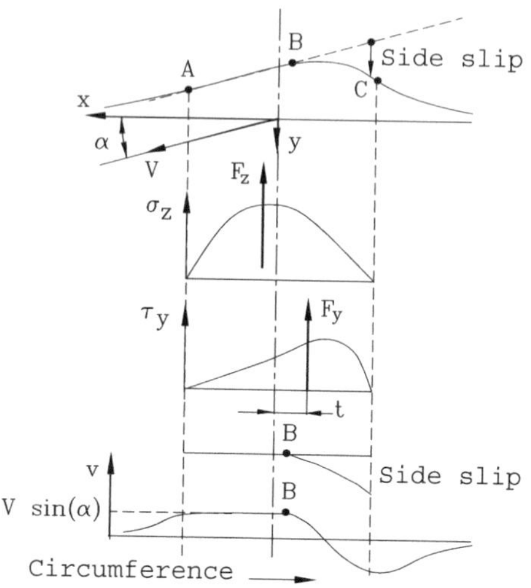

Fig. 2.43 Lateral deformation, distribution of pressures σ_z and τ_y, slip and lateral speed in a cornering tire

point which is located behind it at a distance t. Such distance is defined as *pneumatic trail*.

The moment $M_z = F_y t$ is the *aligning moment* as it tends to force the mean plane of the wheel towards the direction of the velocity V. The absolute value of the side force F_y grows almost linearly at first as α increases, then, when the limit conditions of sliding are approached, in a slower way. Eventually it remains constant, or decreases slightly, when sliding conditions are reached.

The side force F_y is plotted as a function of α for the cases of a radial and a bias-ply tire in Fig. 2.44a. Radial tires show a stiffer behavior than bias-ply ones for what side forces are concerned, as they require smaller sideslip angles to produce the same side force.

With increasing sideslip angle, τ_y is more evenly distributed and the pneumatic trail decreases. The aligning moment is consequently the product of a force which increases with α and a distance which decreases; its trend is consequently of the type shown in Fig. 2.44b. At high values of α, M_z can change direction, as is shown in the figure.

The side force coefficient

$$\mu_y = \frac{F_y}{F_z}$$

is often used. Its maximum value, usually defined as lateral traction coefficient , is written as μ_{y_p} and the value taken in sliding conditions as μ_{y_s}.

Both force F_y and moment M_z depend on many factors, besides the angle α, as normal force F_z, speed, pressure p, road conditions etc. Diagrams of cornering forces F_y, of self aligning moment M_z and of trail t are shown on Fig. 2.45 as function of the side slip angle α and of vertical load F_z. The figure refers to a 145 D 15 tire, inflated at 180 kPa at a speed of 50 km/h.

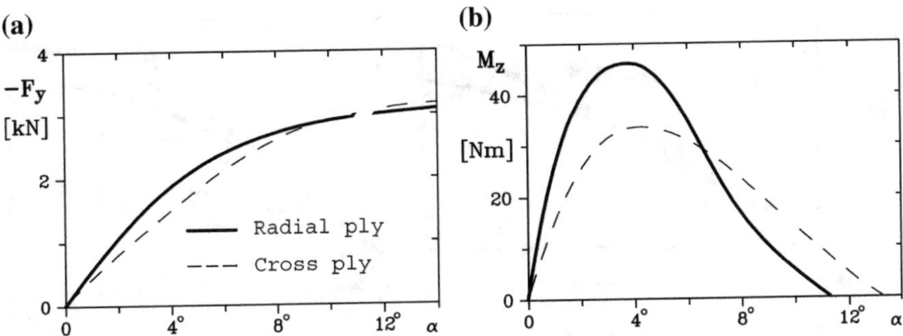

Fig. 2.44 Lateral force F_y and self aligning moment M_z for different tires of the same size 145 D 13 and 145 R 13; $F_z = 3\,\text{kN}$, $p = 170\,\text{kPa}$; $V = 40\,\text{km/h}$

It must be observed that all curves are to be taken as correct only as far as their shape is concerned, but they are less precise as far as numeric values are concerned; they depend, in fact, on the kind of test machine, used for measurement, which can never simulate the on-road behavior correctly. See the following paragraph dedicated to test machines.

The diagrams we are going to show cannot be never taken as a substitution of experimental date gathered on real cars, but they are a trend indicator for the interpretation of experimental results.

The lateral behavior of the tire can be summarized in a single diagram, the so-called *Gough diagram*, in which the side force F_y is plotted against the self aligning torque M_z with F_z, α and t as parameters. The Gough diagrams for two different tires are shown in Fig. 2.46.

Alternatively, the *carpet plot* can be drawn; the force is plotted against the slip angle and the various curves at different F_z are just superimposed, translated by a quantity proportional to the normal force with respect to each other. Also the aligning torque can be plotted in this way. Carpet plots are shown in Fig. 2.62.

At increasing speed, the curve $F_y(\alpha)$ lowers, mainly in the part corresponding to the higher values of the sideslip angle. The linear part remains almost unchanged (Fig. 2.47). Also the pneumatic trail t decreases with increasing speed and consequently the aligning torque shows a decrease which is more marked than that of the side force.

The decrease of F_y and of M_z as speed increases is higher on slippery roads, as we can see on Fig. 2.48 where the same curves as in Fig. 2.47 at two different speeds on dry and wet road are shown.

As far as hydrodynamic lifting (aquaplaning) is concerned, the same considerations seen for the longitudinal force F_x can be repeated for the side force F_y. The decrease of the aligning torque which goes together with the decrease of the side force should warn the driver of the approaching of the loss of traction. A poorly designed tire however could maintain strong aligning moments even in conditions of reduced transversal traction, leading the driver into error.

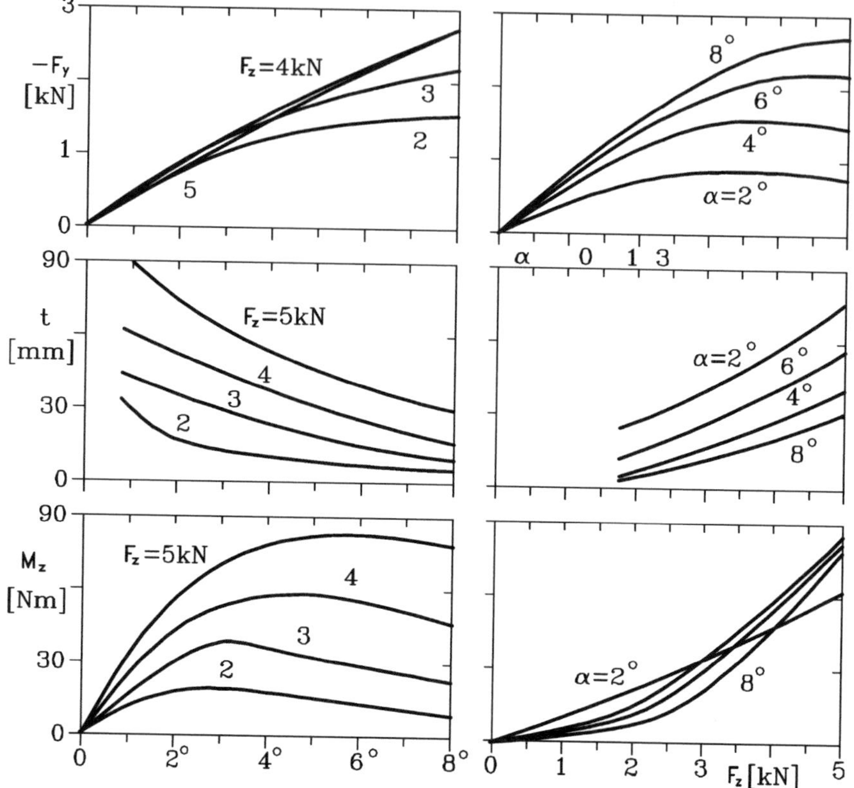

Fig. 2.45 Diagrams of F_y, M_z and t as function of F_z and α. Tire 145 D 15, p = 180 kPa, V = 50 km/h

Finally, on Fig. 2.49 we show a carpet plot for cornering forces for three tires 195/65 R 15, 205/55 R16 and 225/45 R 17 proposed for different version of the same car; on Fig. 2.50 the carpet plot for self aligning torque for the first and the last tire.

It should be remembered that even if tire width are different the free rolling radius of both tires is the same and equal to 317 mm, being used with the same wheel house.

The effect of increased width, combined with the different aspect ratio, can be seen on cornering stiffness; stiffer tires are justified for higher cornering force at the same side slip angle, as on sport version. It can be noticed that the effect on self aligning moment is not positive; this result can be corrected with a different king-pin geometry.

The presence of a camber angle produces, even if no slip angle is present, a lateral force. It is usually said camber thrust or camber force, as distinct from cornering force, due to sideslip angle alone. The camber force added to the cornering force give the total side or lateral force. The camber force is usually far smaller than the cornering force, at least at equal values of angles α and γ. It depends on the load F_z,

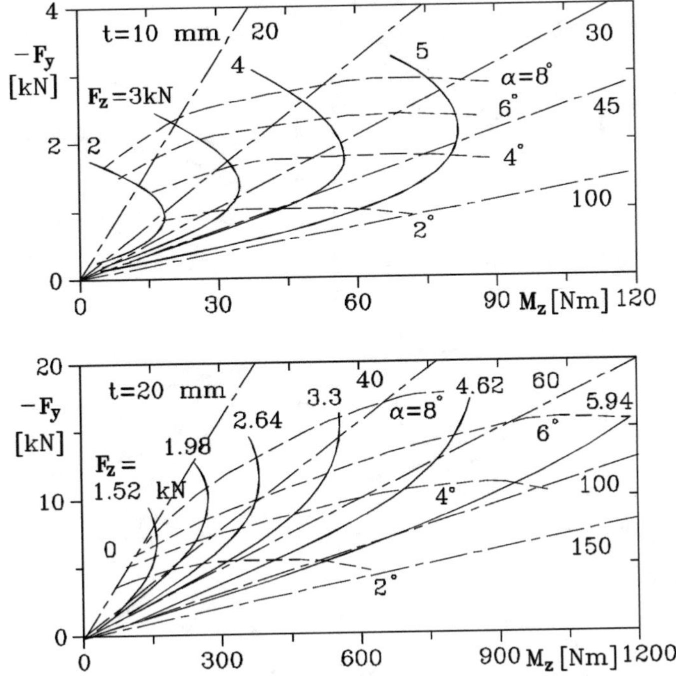

Fig. 2.46 Gough diagrams. **a** Radial ply tire for a car 145 R 16; $p = 180\,\text{kPa}$, $V = 50\,\text{km/h}$; **b** cross ply tire for an industrial vehicle 300 20, $p = 675\,\text{kPa}$

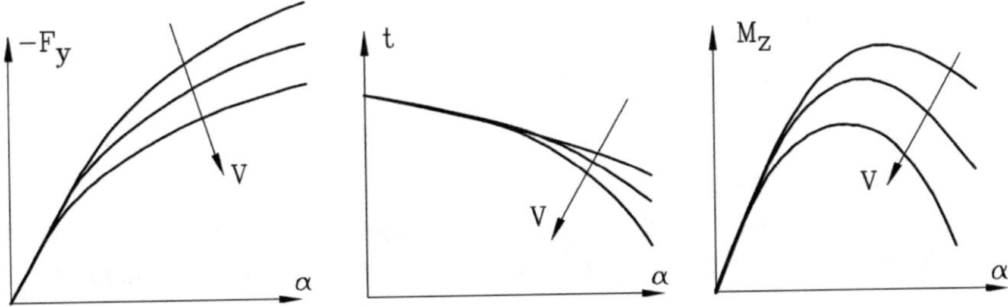

Fig. 2.47 Qualitative curves of $F_y(\alpha)$, $M_z(\alpha)$ and $t(\alpha)$ at different speeds

is practically linear with it (Fig. 2.51), and is strongly dependent on the type of tire considered.

The camber thrust is usually applied in a point which leads the centre of the contact zone, producing a small moment M_{z_γ}. It is usually neglected, owing to its very small value. Bias-ply tires usually produce greater camber thrusts and moments than radial ones.

Usually both sideslip and camber are simultaneously present. As an example, the effect of camber on the force due to sideslip for two different values of the load is

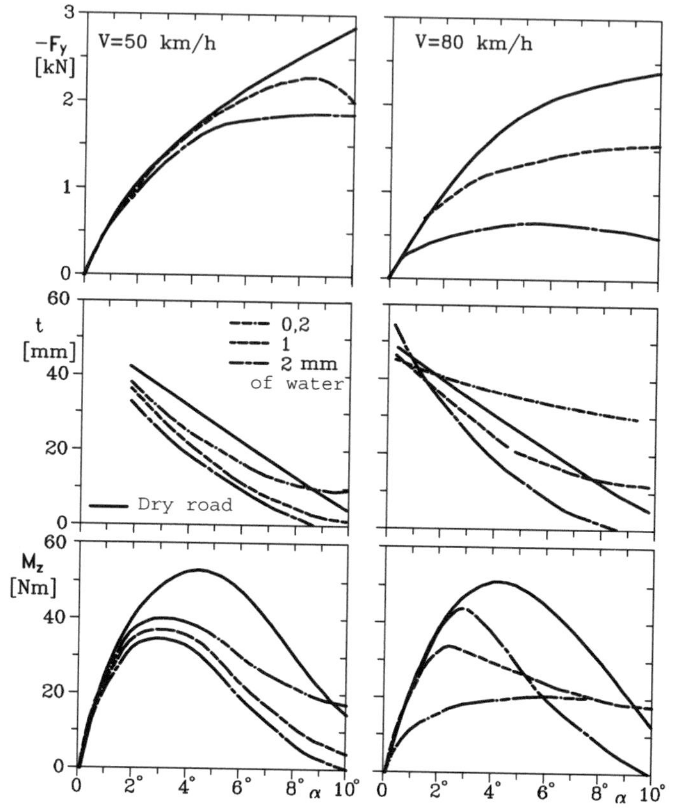

Fig. 2.48 Curves of $F_y(\alpha)$, $M_z(\alpha)$ and $t(\alpha)$ at two different speeds (50 and 80 km/h) on dry and wet road. Tire 155 R 15, $F_z = 3$ kN, $p = 150$ kPa

shown in Fig. 2.52. The camber thrust is more noticeable at low sideslip than when the sideslip is large, particularly when the wheel is less loaded. The aligning torque under the combined effect of sideslip and camber for a radial tire is shown in Fig. 2.53.

In theory both side slip angle and camber angle should be zero when cornering force and self aligning torque are zero. This is not always true in practice, as it shown on carpet diagrams on Figs. 2.54 and 2.55, respectively for the cornering force and the self aligning torque.

The reasons for this dissymmetry are the following. Firstly, the lateral behavior of tires exhibits a hysteresis, in such a way that when the zero sideslip angle condition is reached from a condition in which a force was exerted on the right, a small residual force in the same direction remains and the same holds when the wheel is centered from the opposite direction. This can give a feeling of lack of precision of the steering system and compel the driver to continuous corrections.

Moreover, the centre of the hysteresis cycle is not at the point in which both angle and force are equal to zero: Owing to lack of geometrical symmetry, a tire working

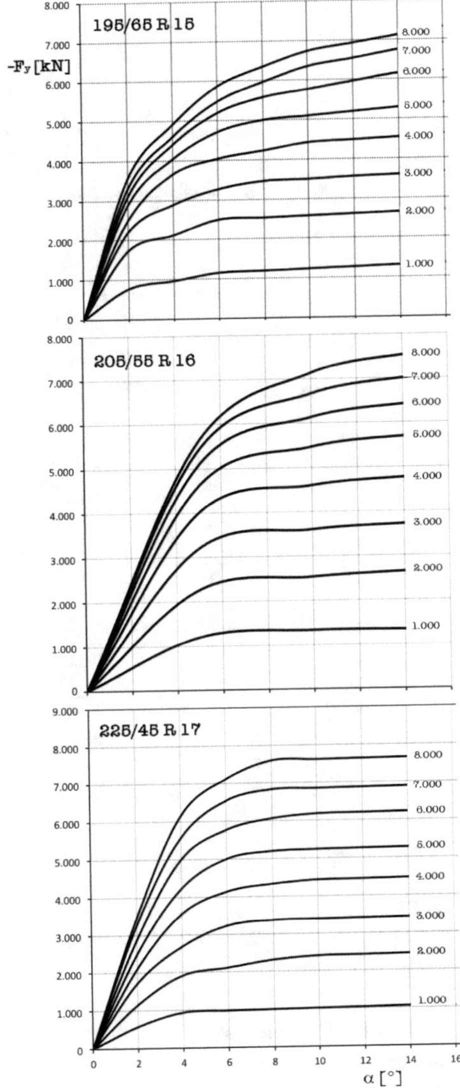

Fig. 2.49 Carpet diagrams of cornering force for three tires of same diameter but different aspect ratio: 195/65 R 15, 205/55 R 16 and 225/45 R 17. The parameter used is vertical load

in symmetrical conditions produces a side force. A first effect is due to a possible *conicity* of the outer surface of the tire: A conical drum would roll on a circular path whose centre coincides with the apex of the cone. Conicity is due to lack of precision during the manufacturing process and hence is linked with manufacturing quality control; its direction is random and its amount changes from tire to tire of the same model. If a tire is turned on the rim, the direction of the conicity is reversed, as is the force it causes when the tire is rolled along a straight path.

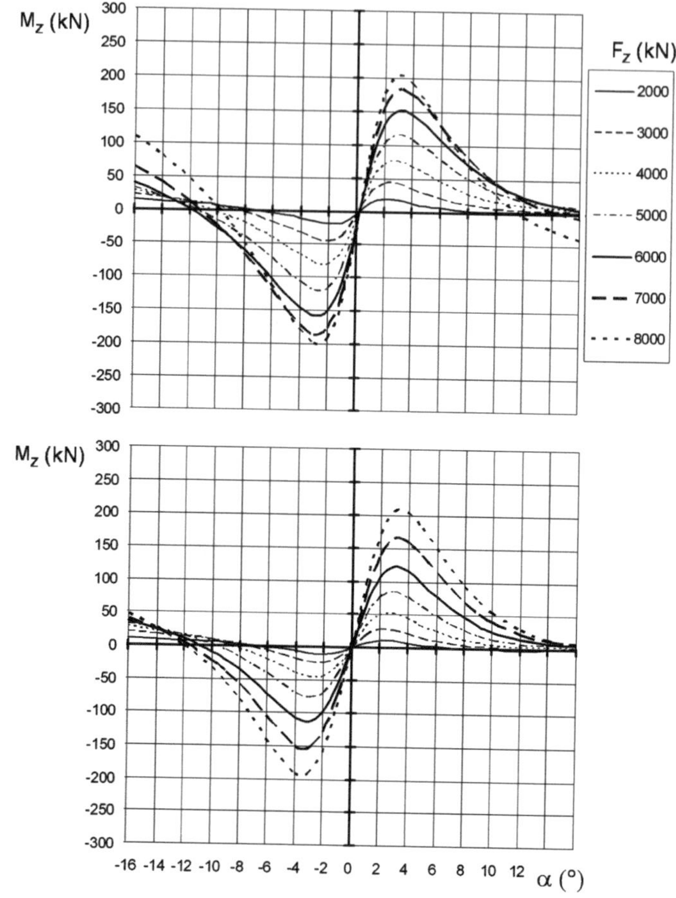

Fig. 2.50 Carpet diagrams of self aligning torque for two tires of same diameter but different aspect ratio: 195/65 R 15 and 225/45 R 17. The parameter used is vertical load

Another unavoidable lack of symmetry is linked with the angles of the various plies and their stacking order; the effect it causes is called *ply steer*. If the wheel is rolled free ply steer causes it to roll along a straight line angled with respect to the plane of symmetry; if the wheel rolls with no sideslip angle the generation of a side force results. If a tire is turned on the rim the direction of the force due to ply steer is not reversed. As it is caused by a factor included in the tire design, unlikely the effect of conicity, that of ply steer is very consistent between tires of the same model.

While conicity can be included into the models of the tire only in a statistical way, ply steer is one of the peculiarities of each tire and can be accounted for with precision. Note that while these effects are usually considered as a nuisance, opposite ply steer of the wheels of a given axle can even be used as a substitute of toe-in; while the latter increases the rolling resistance the first one has no effect on it.

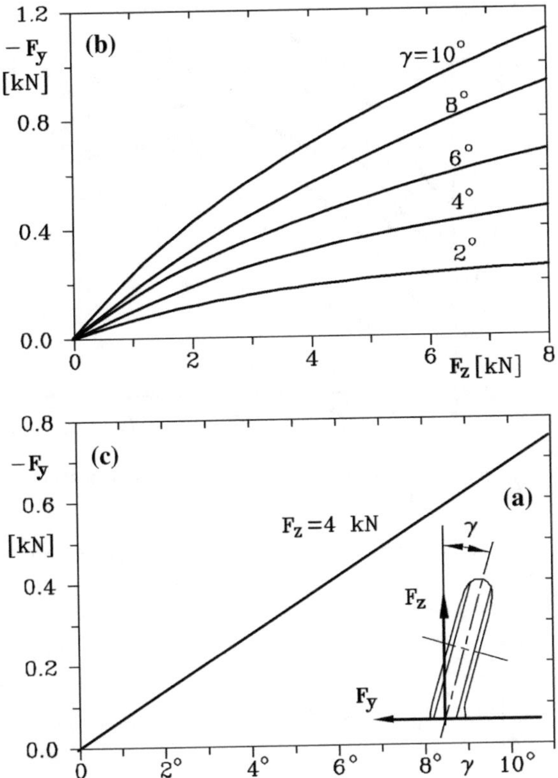

Fig. 2.51 Camber force. **a** Scheme; note that F_y is negative and directed to the opposite as shown. Camber force as function of vertical load (**b**) and camber angle (**c**). Tire 165 13, $p = 200\,\text{kPa}$

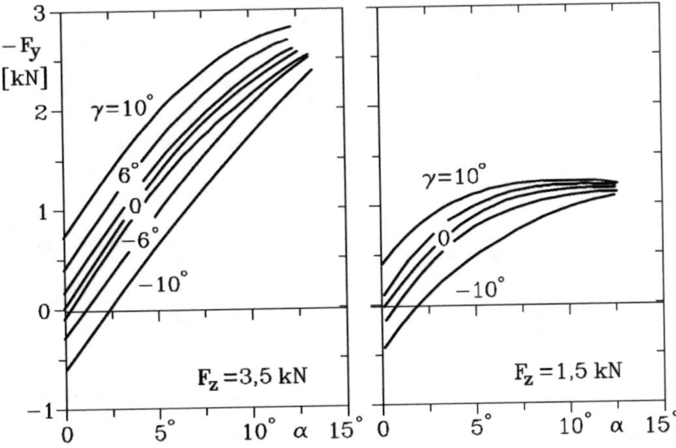

Fig. 2.52 Cornering force as function of the side slip angle, for different values of camber angle and for two values of the vertical force. Tire 135 13, $p = 160\,\text{kPa}$, $V = 40\,\text{km/h}$

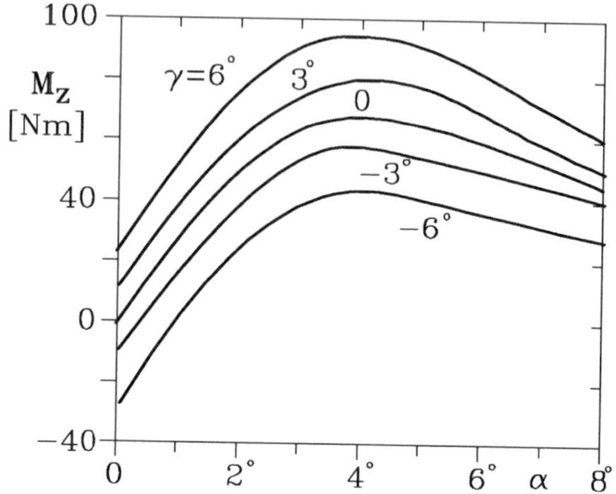

Fig. 2.53 Self aligning torque as function of sideslip angle, for different camber angles. Radial tire 185 R 14, $p = 230\,\text{kPa}$, $F_z = 3.5\,\text{kN}$

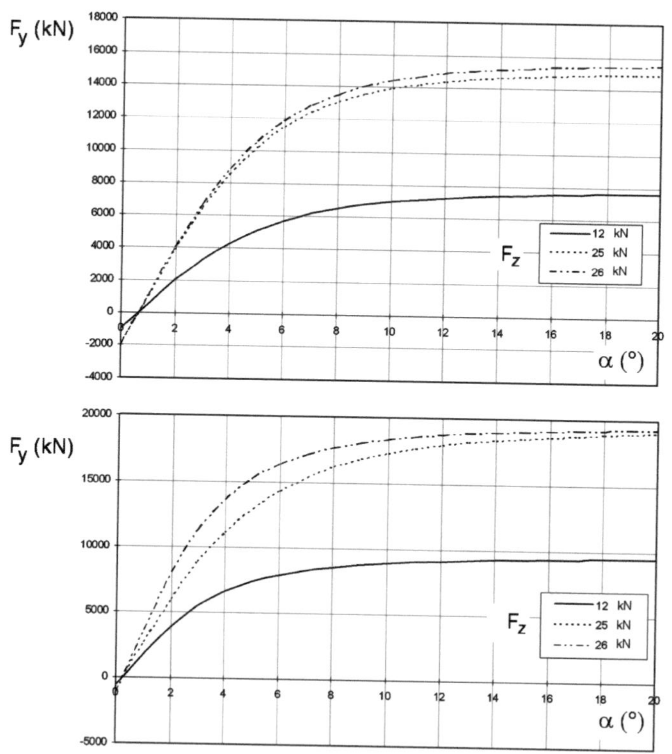

Fig. 2.54 Carpet diagrams for cornering force as function of side slip angle, for two industrial vehicle tires 315/80 R 22.5; the stiffer one has a reinforced belt for steering wheels (single wheel), while the other is for driving wheels (twin wheels). The parameter is the vertical force

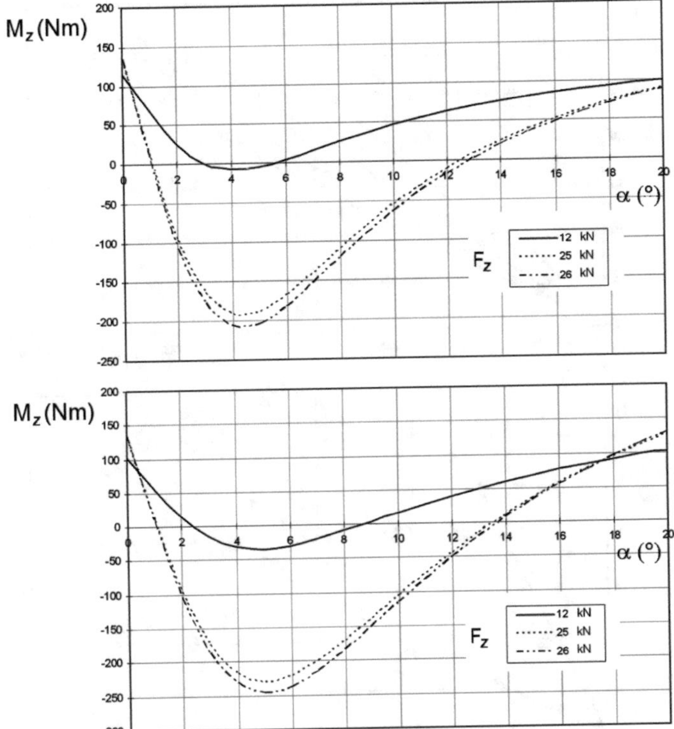

Fig. 2.55 Carpet diagram of self aligning torque as function of side slip angle for the same industrial vehicle tires 315/80 R 22.5 of previous figure

Generally speaking, the lateral force offset is subdivided into two parts: The part which does not change sign when the direction of rotation is reversed is said to be ply-steer force, while the part that changes sign is said to be conicity force.

As already stated, for low values of the sideslip angle the cornering force increases linearly with α

$$F_y = C_\alpha \alpha \ . \tag{2.26}$$

Coefficient

$$C_\alpha = \left(\frac{\partial F_x}{\partial \alpha}\right)_{\alpha=0}$$

is always negative, since a positive sideslip angle causes a negative side force. The *cornering stiffness* or *cornering power* C is usually expressed as a positive number, so it is arbitrarily defined and written as the derivative $\partial F_y/\partial \alpha$ changed in sign, and

$$F_y = -C\alpha \ . \tag{2.27}$$

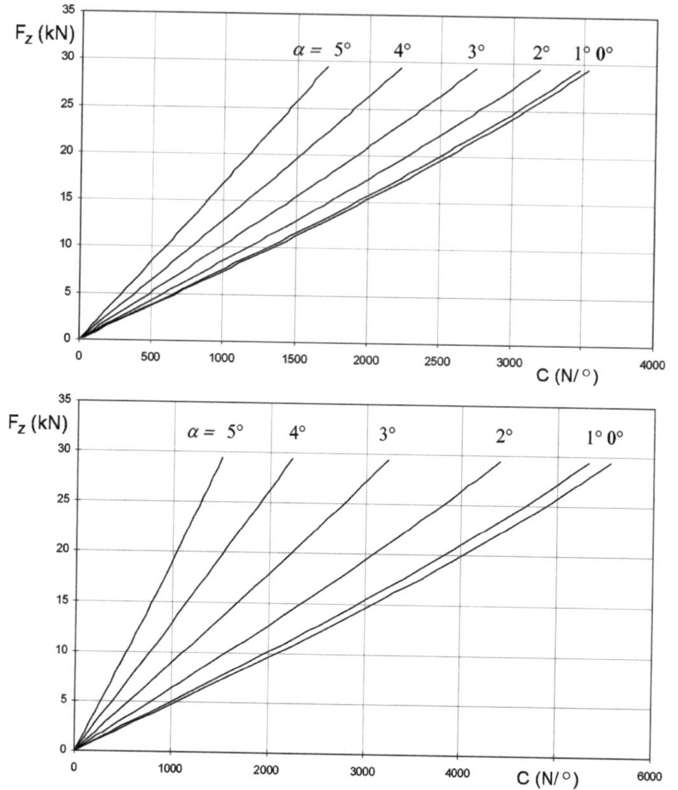

Fig. 2.56 Carpet plot of vertical force as function of the cornering stiffness for the two tires of previous figures. The parameter is the side slip angle

This relationship obviously holds only for small enough values of α.

Expression (2.27) is quite useful to study the dynamic behavior of vehicles under the assumption of small sideslip angles, as it actually occurs in normal driving conditions. In particular, it is essential in the study of the stability of linearized models.

Figure 2.56 shows a diagram of this parameter, as function of vertical load and side slip angle for same tires 315/80 R 22.5, already described on carpet plots.

The ratio between the cornering stiffness and the normal force is usually referred to as cornering stiffness coefficient (the term cornering coefficient is also used but SAE recommendation J670 suggests to avoid it for clarity).

Figure 2.57 reports the cornering stiffness value as a function of the vertical load of two of the tires (205/55 R16 and 225/45 R 17) whose carpet was already shown in Fig. 2.50.

For bias-ply tires it is of the order of $0.12\ \text{deg}^{-1} = 6.9\,\text{rad}^{-1}$ and for radial tires of the order of $0.15\ \text{deg}^{-1} = 8.6\,\text{rad}^{-1}$. As the cornering coefficient changes much from tire to tire, it is interesting to note the distribution of its value for tires of different types (Fig. 2.58a).

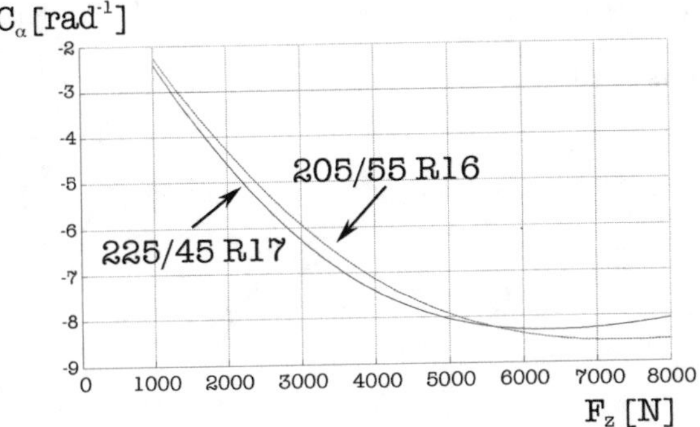

Fig. 2.57 Cornering stiffness as a function of the vertical load of a 205/55 R16 and a 225/45 R 17 tire

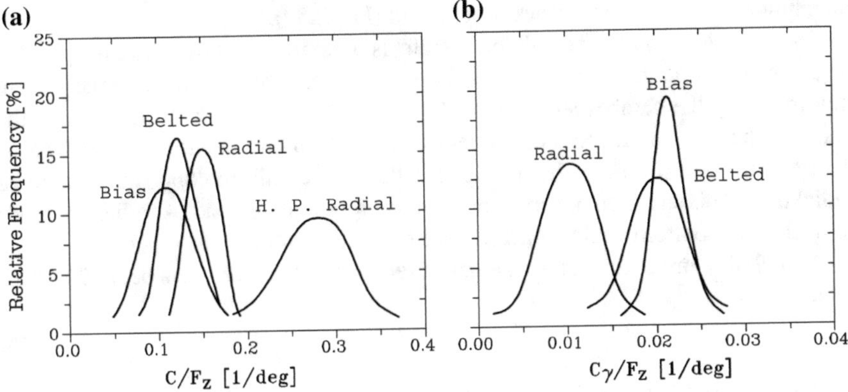

Fig. 2.58 Frequency distribution of the cornering stiffness (**a**) and camber stiffness (**b**) for different car tire types

In the same way the camber stiffness can be defined as the slope of the curve $F_y(\gamma)$ for $\gamma = 0$:

$$C_\gamma = \left(\frac{\partial F_y}{\partial \gamma}\right)_{\gamma=0} .$$

Note that the camber thrust produced by a positive camber angle is negative and hence the camber stiffness is negative. The ratio between the camber stiffness and the normal force is usually referred to as camber stiffness coefficient. This coefficient is higher for bias-ply tires than for radial tires: In the first case an average value is of the order of 0.021 deg^{-1} = 1.2 rad^{-1} and in the second is of the order of 0.01 deg^{-1} = 0.6 rad^{-1}. Also the distribution of the camber coefficient for tires of different types is of some interest (Fig. 2.58b).

The value of the camber stiffness is important in the case a wheel rolls on a road with a transversal slope with its mid plane remaining vertical: in this case there is a

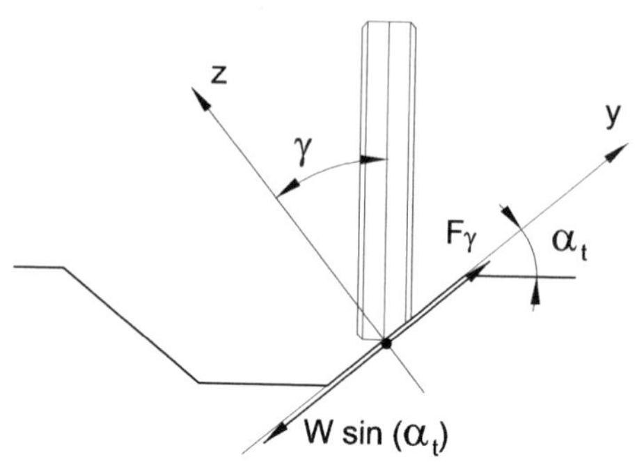

Fig. 2.59 Wheel on a road with lateral slope α_t: camber force and weight component along the slope

component of the weight which is directed downhill and the camber thrust which is directed uphill. The net effect can be in one direction or the other depending on the magnitude of the camber stiffness coefficient (Fig. 2.59).

The downhill component of the weight is $W \sin(\alpha_t) \approx W\alpha_t$, where α_t is the transversal inclination of the road while the camber thrust is equal to the weight multiplied by the camber stiffness coefficient and the angle. It is clear that if the value of the camber stiffness coefficient is larger than one (measured in rad^{-1}), as it occurs for bias ply tires, the net force is directed uphill; the opposite occurs for radial tires. This situation occurs when a rut is present in the road: A radial tire tends to track in the bottom while a bias-ply tire tends to climb out of the rut.

To include camber thrust into the linearized model, Eq. (2.27) can be modified as:

$$F_y = -C\alpha + C_\gamma \gamma \ . \tag{2.28}$$

It can be used with confidence for values of α up to about 4° and of γ up to 10°. Also the self aligning moment can be expressed by a linear law:

$$M_z = (M_z)_{,\alpha}\alpha \ , \tag{2.29}$$

where $(M_z)_\alpha$ is the derivative $\partial M_z/\partial\alpha$ computed for vanishingly small α and γ and is defined as aligning stiffness coefficient or simply aligning coefficient.

Here again the effect of the camber angle can be included by modifying Eq. (2.29) as:

$$M_z = (M_z)_{,\alpha}\alpha + (M_z)_{,\gamma}\gamma \ , \tag{2.30}$$

where $(M_z)_{,\gamma}$ is the derivative $\partial M_z/\partial\gamma$ computed for vanishingly small α and γ, but the second effect is so small that it is usually neglected.

Equation (2.30) supplies a good approximation of the aligning torque for a range of α far more limited than that in which Eq. (2.27) holds. It must however be noted that the importance of the aligning torque in the study of the behavior of the vehicle is limited and consequently a precision lower than that required for side forces can

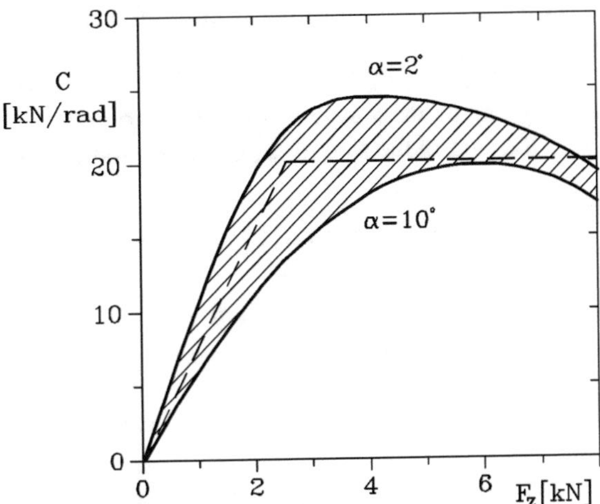

Fig. 2.60 Cornering stiffness in function of load F_z (curve shown as $\alpha = 10°$ is a sort of "secant" stiffness)

be accepted. Practically, a good approximation of the aligning torque is important only when studying the steering mechanism.

The aligning stiffness coefficient due to sideslip angle is of about 0.01 m/deg (Nm/N deg) for bias ply tires and of 0.013 m/deg for radial tires while that due to camber (aligning camber stiffness coefficient) is approximately of 0.001 m/deg for the first ones and of 0.0003 m/deg for the latter.

A small aligning moment is due to the curvature of the path even if the sideslip angle is equal to zero; however this effect is not negligible only if the radius of the trajectory is very small, of the order of a few meters, and consequently it is present only in low speed manoeuvres. It can be important for the dimensioning the steering system for the mentioned conditions.

The definition of the cornering coefficient implies that the cornering stiffness is linear with the normal load F_z; actually the cornering stiffness behaves in this way only for low values of force F_z and then increases to a lesser extent (Fig. 2.60). When the limit value has been reached it remains constant or slightly decreases. It is often expedient to approximate the cornering stiffness as a function of the load with two straight lines, the second of which is horizontal. Note that in the figure the line corresponding to a sideslip angle of 2° refers to the true cornering stiffness while the other curve ($\alpha = 10°$) is related to a sort of secant stiffness.

When the need for a more detailed numerical description of the lateral behavior of a tire arises, there is no difficulty, at least in theory, to approximate the experimental law $F_y(\alpha, \gamma, F_z, p, V, ...)$ and the similar relationship for the aligning torque, using the algorithms which are common in numerical analysis. This approach can be used with success in the numerical simulations of the behavior of the vehicle, even if it is often quite expensive in terms of time needed for data preparation and computation. A problem which is common to many numerical approaches like this is that of requiring a great amount of experimental data, which are often difficult, or costly, to obtain.

Polynomial approximations, with terms including the third power of the slip angle α, can be used.

As already stated, Eq. (2.25) can also be used to express the cornering force and the aligning moment as function of the various parameters.

In the case of the side force, the magic formula is:

$$F_y = D \sin\left(C \arctan\left\{B(1-E)(\alpha + S_h) + E \arctan\left[B(\alpha + S_h)\right]\right\}\right) + S_v \ , \tag{2.31}$$

where the product of coefficients B, C and D yields directly the cornering stiffness. The values of the other coefficients are:

$$C = a_0 \qquad D = \mu_{y_p} F_z \ ,$$

where a value of 1.30 is suggested for a_0 and $\mu_{y_p} = a_1 F_z + a_2$,

$$E = a_6 F_z + a_7 \ ,$$

$$BCD = a_3 \sin\left[2 \arctan\left(\frac{F_z}{a_4}\right)\right](1 - a_5|\gamma|) \ ,$$

$$S_h = a_8\gamma + a_9 F_z + a_{10} \ ,$$

$$S_v = a_{11}\gamma F_z + a_{12} F_z + a_{13} \ .$$

Product BCD is the cornering stiffness changed in sign.

To obtain a better description of the camber thrust, the constant a_{11} is often substituted by the linear law:

$$a_{11} = a_{111} F_z + a_{112} \ .$$

Coefficients S_h and S_v account for ply steer and conicity forces.

Similarly, in the case of the aligning torque the formula is:

$$M_z = D \sin\left(C \arctan\left\{B(1-E)(\alpha + S_h) + E \arctan\left[B(\alpha + S_h)\right]\right\}\right) + S_v \ , \tag{2.32}$$

$$C = c_0 \ , \qquad D = c_1 F_z^2 + c_2 F_z \ ,$$

where a value of 2.40 is suggested for c_0,

$$E = (c_7 F_z^2 + c_8 F_z + c_9)(1 - c_{10}|\gamma|) \ ,$$

$$BCD = (c_3 F_z^2 + c_4 F_z)(1 - c_6|\gamma|)e^{-c_5 F_z} \ ,$$

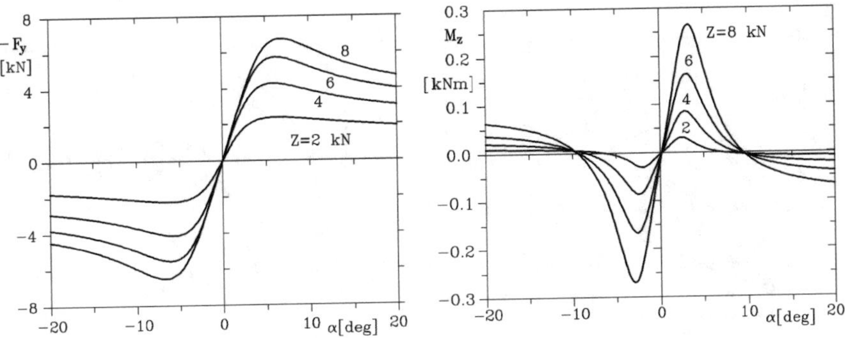

Fig. 2.61 Curves $F_y(\alpha)$ and $M_z(\alpha)$ obtained by using the magic formula (2.31) and (2.32). Radial tire 205/60 R 15 V

$$S_h = c_{11}\gamma + c_{12}F_z + c_{13} \, ,$$

$$S_v = (c_{14}F_z^2 + c_{15}F_z)\gamma + c_{16}F_z + c_{17} \, .$$

Also in this case the units introduced into the magic formula (2.31) and (2.32) are usually not consistent: The load F_z is expressed in kN, angles α and γ are in degrees, F_y and M_z are obtained in N and Nm respectively.

The curves $Y(\alpha)$ and $N_a(\alpha)$ for values of the vertical load F_z equal to 2, 4, 6 and 8 kN for a radial tire 205/60 VR 15 6J are shown in Fig. 2.61.

The magic formula can be used to model the behavior of the tire in a more complete way than to express the direct relationship of the side force and aligning moment with the sideslip angle. As an example, the carpet plots of the side force, the side force coefficient, the aligning moment, the camber thrust and the camber moment are reported in Fig. 2.62, together with the Gough diagram, for the same tire of Fig. 2.61.

It is also possible to build structural models of the tire to express the forces it exerts by taking into account the deformations and stresses their structure is subjected to. Apart from very complex numerical models, mainly based on the finite element method, which allow one to compute the required characteristics but are so complex that they are of little use in vehicle dynamics computations, it is possible to resort to simplified models, dealing with the tread band as a beam or as a string on elastic foundations.[9] These models allow one to obtain interesting results, particularly from a qualitative viewpoint, as they link the performance of the tire with its structural parameters, but their quantitative precision is usually smaller than that of empirical models, in particular of those based on the magic formula which is increasingly becoming a standard in tire modelling.

[9] See, for instance, J. R. Ellis, *Vehicle Dynamics*, Business books Ltd., London, 1969; G. Genta, *Meccanica dell'autoveicolo*, Levrotto & Bella, Torino, 1993.

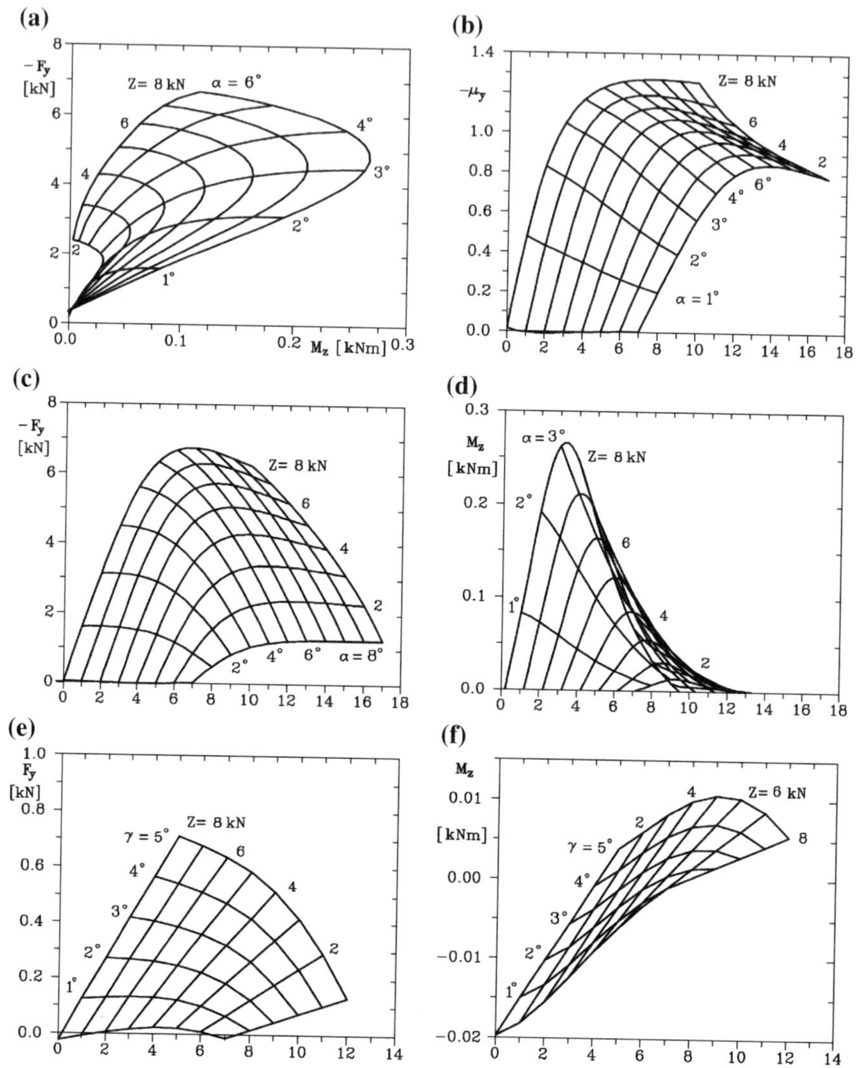

Fig. 2.62 **a** Gough diagram and carpet plot **b** of sideslip coefficient, **c** cornering force, **d** self aligning torque, **e** camber force and **f** camber moment for a radial tire 205/60 R 15 V, obtained using the magic formula

2.8 Interaction Between Longitudinal and Side Forces

The considerations seen in the preceding sections apply only in the case in which longitudinal and side forces are generated separately. If the tire produces simultaneously forces in X' and Y' directions the situation can be different as the traction used in one direction limits that available in the other.

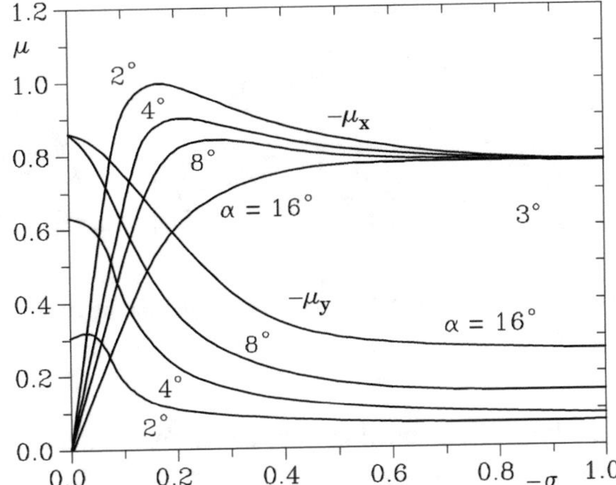

Fig. 2.63 Lateral and longitudinal force coefficients as function of longitudinal slip and side slip angle

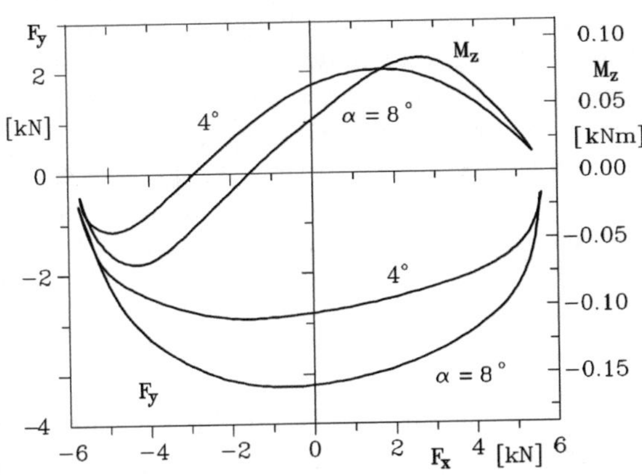

Fig. 2.64 Polar diagrams of lateral force on the wheel at constant side slip angle and self aligning torque as function of longitudinal force

By applying a driving or braking force to a tire which has a certain sideslip angle, the cornering force reduces and the same applies to the longitudinal force a tire can exert if it is called to exert also a lateral force.

An example is shown in Fig. 2.63: By setting the tire at a given sideslip angle and applying a braking torque, the shape of the curve $\mu_x(\sigma)$ is deeply changed. While the value in slipping conditions μ_s is almost unchanged, the peak value μ_p decreases to a greater extent.

It is then possible to obtain a polar diagram of the type shown in Fig. 2.64 in which the force in Y' direction is plotted versus the force in X' direction for any given value of the sideslip angle α. Each point of the curves is characterized by a different value of the longitudinal slip σ. In a similar way it is possible to plot a curve $F_y(F_x)$ at constant σ.

Strictly speaking, the curves are not exactly symmetrical with respect to F_y axis: Usually tires develop the maximum value of the force F_y when they exert a very

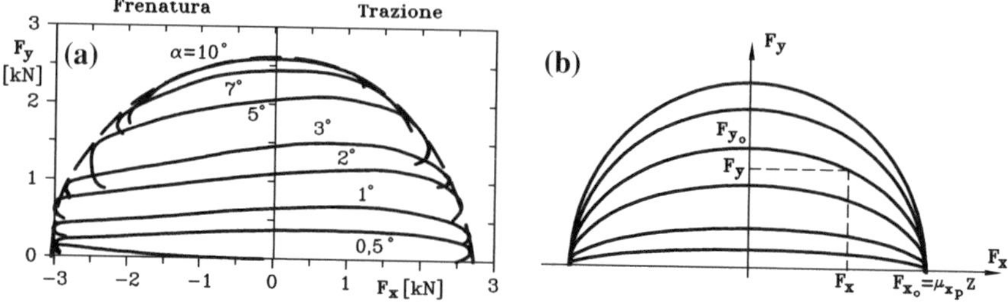

Fig. 2.65 Polar diagrams of force at constant side slip angle. **a** Experimental diagrams; **b** elliptic approximation

light longitudinal braking force and a slight braking force can actually increase the lateral force exerted at a moderate sideslip angle.

Together with the curves $F_y(F_x)$ two other curves $M_z(F_x)$ are also reported in Fig. 2.64. An application of driving forces results in an increase of the aligning torque, while a braking force causes the aligning moment to decrease to the point that it changes its sign when the braking limit is approached. This effect is destabilizing as it tends to increase the sideslip angle.

A set of experimental curves $F_y(F_x)$ at constant α is shown in Fig. 2.65a.

If F is the total force exerted on the wheel by the road while F_x and F_y are its components, the resultant force coefficient can be expressed as:

$$\mu = \frac{F}{F_z} = \sqrt{\mu_x^2 + \mu_y^2} \,. \tag{2.33}$$

The various curves plotted for different values of α are enveloped by the polar diagram of the maximum force the tire can exert. If it were a circle, the so-called *friction circle*, as in simple models it can be assumed to be, the maximum force coefficient would be independent of the direction.

Actually, not only the value of μ_x is greater than that of μ_y but, as already stated, there is some difference in longitudinal direction between driving and braking conditions. The envelope, as well as the whole diagram, is a function of many parameters. Apart from the already mentioned dependence on the type of tire and road conditions, there is a strong reduction of the maximum value of force F with the speed, which is particularly strong on wet road.

The three dimensional diagram of Fig. 2.66, limited to driving force F_x, gives an idea of this reduction. The phenomenon is emphasized by water, as on diagrams of Fig. 2.67.

A model allowing to approximate the curves $F_y(F_x)$ at constant α with simple functions can be quite useful. This can be obtained by using the elliptical approximation (Fig. 2.65b):

$$\left(\frac{F_y}{F_{y0}}\right)^2 + \left(\frac{F_x}{F_{x0}}\right)^2 = 1 \,, \tag{2.34}$$

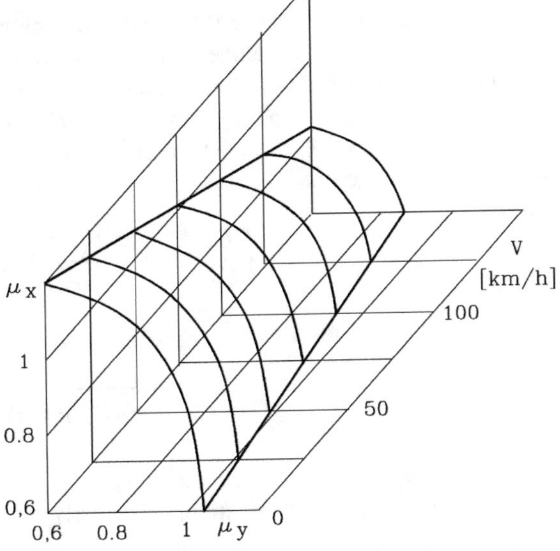

Fig. 2.66 Variation of the envelope curve of polar diagrams of force between tire and road at different speeds

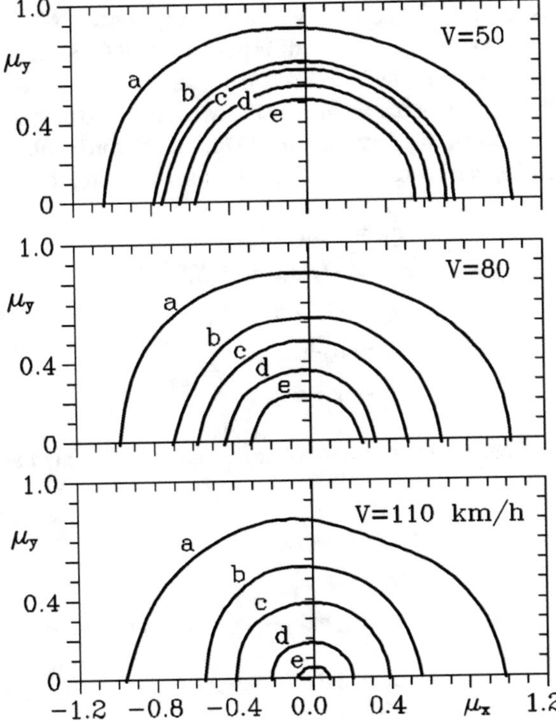

Fig. 2.67 Envelope curve of $F_y(F_x)$ curves, at different road conditions. **a** Dry road and **b** wet with 0.2 mm; **c** 0.5 mm; **d** 1 mm and **e** 2 mm of water

where forces F_{y_0} and F_{x_0} are respectively the force F_y exerted, at the given sideslip angle, when no force F_x is exerted and the maximum longitudinal force exerted at zero sideslip angle. The envelope curve is then elliptical, the friction ellipse.

If Eq. (2.34) is used in order to express function $F_y(F_x)$, the cornering stiffness of a tire which is exerting a longitudinal force F_x can be expressed as a function of the cornering stiffness C_0 (i.e. the cornering stiffness when no longitudinal force is produced) by the expression:

$$\left(\frac{C\alpha}{C_0\alpha}\right)^2 + \left(\frac{F_x}{F_{x_0}}\right)^2 = 1\,. \tag{2.35}$$

The cornering stiffness can be expressed by the function:

$$C = C_0\sqrt{1-(F_x/\mu_p F_z)^2}\,, \tag{2.36}$$

where force F_{x_0} has been substituted by $\mu_p F_z$.

Although a rough approximation, particularly for the case in which the longitudinal force approaches its maximum value (the differences between the curves of Fig. 2.65a and those of Fig. 2.65b are evident), the elliptical approximation is often used for all the cases in which the very concept of cornering stiffness is useful.

The empirical model of Eqs. (2.25) and (2.31) has been modified by Pacejka , to allow for the interaction between longitudinal and lateral forces in a better way than that of computing separately the two forces and then using the elliptic approximation.

First of all, formulae for longitudinal force calculation, in absence of side slip angle, has been redefined; they refer, therefore, to pure traction or braking:

$$\begin{aligned}
F_x &= F_{x0}(\sigma, F_z), \\
F_{x0} &= D_x \sin\Theta_x + S_{Vx}, \\
\Theta_x &= C_x \arctan\left\{B_x\sigma_x - E_x\left(B_x\sigma_x - \arctan\left(B_x\sigma_x\right)\right)\right\}, \\
\sigma_x &= \sigma + S_{Hx}, \\
\gamma_x &= \gamma\lambda_{\gamma x}
\end{aligned} \tag{2.37}$$

Coefficients of previous formulae are calculated according to following expressions:

$$\begin{aligned}
C_x &= p_{Cx1}\lambda_{Cx}, \\
D_x &= \mu_x F_z, \\
\mu_x &= (p_{Dx1} + p_{Dx2} df_z)\left\{1 - p_{Dx3}\gamma_x^2\right\}\lambda_{\mu x}, \\
E_x &= \left(p_{Ex1} + p_{Ex2} df_z + p_{Ex3} df_z^2\right)\left[1 - p_{Ex4}\frac{\sigma_x}{|\sigma_x|}\right]\lambda_{Ex}\,, \\
B_x C_x D_x &= F_z\left(p_{Kx1} + p_{Kx2} df_z\right)\lambda_{Kx} e^{p_{Kx3} df_z}, \\
S_{Hx} &= (p_{Hx1} + p_{Hx2} df_z)\lambda_{Hx}, \\
S_{Vx} &= F_z(p_{Vx1} + p_{Vx2} df_z)\lambda_{Vx}\lambda_{\mu x}.
\end{aligned} \tag{2.38}$$

Formulae are similar to the previous Eq. (2.25), except the difference on coefficients p and the introduction of scale coefficient λ, in most cases set to 1.

Parameter df_z is given by the following equation:

$$df_z = \frac{F_z - F_{z0}}{F_{z0}},$$

where F_{z0} is a reference value for the vertical load.

Also cornering force calculation formulae have been modified, to be used in case of absence of longitudinal slip, that is in case of pure side slip:

$$\begin{aligned}
F_y &= F_{y0}(\alpha, \gamma, F_z), \\
F_{y0} &= D_y \sin \Theta_y + S_{Vy}, \\
\Theta_y &= C_y \arctan \left\{ B_y \alpha_y - E_y \left(B_y \alpha_y - \arctan \left(B_y \alpha_y \right) \right) \right\}, \\
\alpha_y &= \alpha + S_{Hy}, \\
\gamma_y &= \gamma \lambda_{\gamma y}
\end{aligned} \tag{2.39}$$

Similarly, coefficients of previous equations are given by:

$$\begin{aligned}
C_y &= p_{Cy1} \lambda_{Cy}, \\
D_y &= \mu_y F_z, \\
\mu_y &= \left(p_{Dy1} + p_{Dy2} df_z \right) \left\{ 1 - p_{Dy3} \gamma_y^2 \right\} \lambda_{\mu y}, \\
E_y &= \left(p_{Ey1} + p_{Ey2} df_z \right) \left\{ 1 - \left(p_{y3} + p_{Ey4} \gamma_y \right) \frac{\alpha_y}{\left| \alpha_y \right|} \right\} \lambda_{Ey}, \\
B_y C_y D_y &= p_{Ky1} F_{z0} \sin \left[2 \arctan \left\{ F_z / (p_{Ky2} F_{z0} \lambda_{Fz0}) \right\} \right] \cdot \\
&\quad \cdot (1 - p_{Ky3} \left| \gamma_y \right|) \lambda_{Fz0} \lambda_{Ky}, \\
S_{Hy} &= (p_{Hy1} + p_{Hy2} df_z) \lambda_{Hy} + p_{Hy3} \gamma_y, \\
S_{Vy} &= F_z [\left(p_{Vy1} + p_{Vy2} df_z \right) \lambda_{Vy}] + \\
&\quad + F_z [(p_{Vy3} + p_{Vy4} df_z) \gamma_y \lambda_{\mu y}].
\end{aligned} \tag{2.40}$$

Also here, formulae are similar to the previous ones (2.31), made an exception for coefficients p and for scale factors λ, which, again, are seldom different as 1. The parameter df_z is given as previously.

In contemporary presence of both forces the longitudinal force has to be modified by multiplying the result of Eqs. (2.39) and (2.40), by the correction factor:

$$F_x = F_{x0} G_{x\alpha}(\alpha, \sigma, F_z)$$

The calculation of the correction factor is made by:

$$
\begin{aligned}
B_{x\alpha} &= r_{Bx1}\cos\{\arctan(r_{Bx2}\sigma)\}\lambda_{x\alpha}, \\
C_{x\alpha} &= r_{Cx1}, \\
D_{x\alpha} &= \frac{F_{x0}}{\cos\Phi_x}, \\
E_{x\alpha} &= r_{Ex1} + r_{Ex2}df_z, \\
S_{Hx\alpha} &= r_{Hx1}, \\
G_{x\alpha} &= \frac{\cos\Psi_x}{\cos\Phi_x}, \\
\Phi_x &= C_{x\alpha}\arctan\{B_{x\alpha}S_{Hx\alpha} - E_{x\alpha}(B_{x\alpha}S_{Hx\alpha}\arctan(B_{x\alpha}S_{Hx\alpha}))\}, \\
\Psi_x &= C_{x\alpha}\arctan\{B_{x\alpha}\alpha_s - E_{x\alpha}(B_{x\alpha}\alpha_s - \arctan(B_{x\alpha}\alpha_s))\}.
\end{aligned}
\tag{2.41}
$$

Also in this case we have the coefficients r and the scale factor λ.
In the same way:

$$
F_y = F_{y0}G_{y\sigma}(\alpha, \sigma, \gamma, F_z) + S_{Vy\sigma}.
$$

The correction factor is given by:

$$
\begin{aligned}
B_{y\sigma} &= r_{By1}\cos\left\{\arctan\left(r_{By2}(\alpha - r_{By3})\right)\right\}\lambda_{x\alpha}, \\
C_{y\sigma} &= r_{Cy1}, \\
D_{y\sigma} &= \frac{F_{y0}}{\cos\Phi_y}, \\
E_{y\sigma} &= r_{Ey1} + r_{Ey2}df_z, \\
S_{Hy\sigma} &= r_{Hy1} + r_{Hy2}df_z, \\
S_{Vy\sigma} &= D_{Vy\sigma}\sin\left[r_{Vy5}\arctan(r_{Vy6}\sigma)\right]\lambda_{Vy\sigma}, \\
D_{Vy\sigma} &= \mu_y F_z(r_{Vy1} + r_{Vy2}df_z + r_{Vy3}\gamma)\cos\left[\arctan(r_{Vy4}\alpha)\right] \\
G_{x\alpha} &= \frac{\cos\Psi_y}{\cos\Phi_y}, \\
\Phi_y &= C_{y\sigma}\arctan\left\{B_{y\sigma}S_{Hy\sigma} - E_{y\sigma}(B_{y\sigma}S_{Hy\sigma}\arctan(B_{y\sigma}S_{Hy\sigma}))\right\}, \\
\Psi_y &= C_{y\sigma}\arctan\left\{B_{y\sigma}\sigma_s - E_{y\sigma}\left(B_{y\sigma}\sigma_s - \arctan(B_{y\sigma}\sigma_s)\right)\right\}.
\end{aligned}
\tag{2.42}
$$

This way of modelling the interaction between F_x and F_y is quite complicated and not very well justified by theory.

Results are reasonable, better than those obtained by the elliptic approximation only; nevertheless there are cases where curves $F_y(F_x)$ have non realistic shapes and the above formulae must be carefully applied.

Research activities on this field are still going on, with the target of obtaining a simpler and more realistic set of formulae.

A graph similar to that of Fig. 2.65 and calculated with this method is shown on Fig. 2.68.

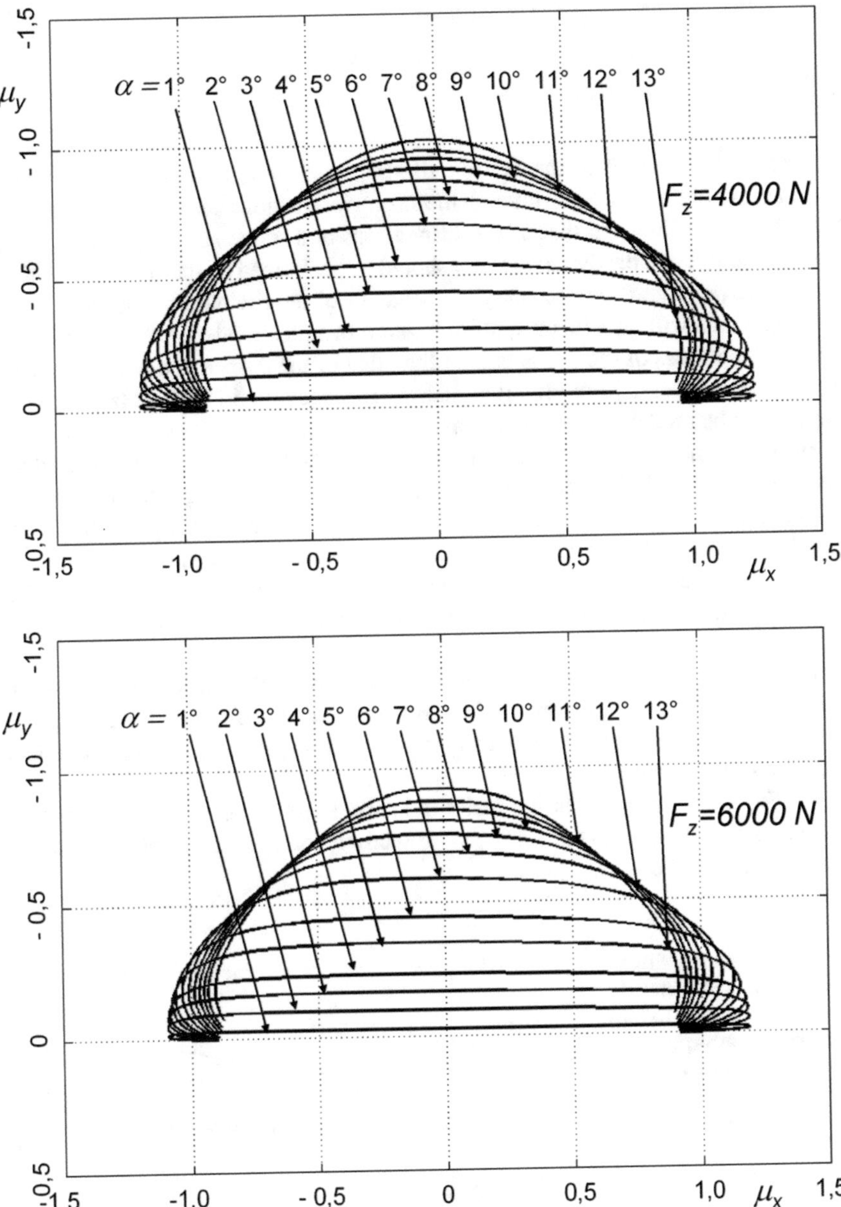

Fig. 2.68 Ellipses calculated with modified Pacejika formulae for the same 195/65 R 15, tire, used to calculate carpet plots previously shown

2.9 Outline on Dynamic Behavior

2.9.1 Vibration Modes

The dynamic behavior of the tire is quite important in determining both comfort and stability of the vehicle. Although the strong interactions between the vibrational behavior of the tires and that of the suspensions and of the suspended mass of the vehicle suggest that the dynamic study should be performed on the complete vehicle, it is at any rate interesting to study the behavior of the tire alone, at least to obtain the data which will later be introduced in more complex models.

As the stiffness of the tire is greater than that of the suspension, the behavior of the former is generally not important in low frequency motions (1–3 Hz), for which the tire can be modelled as a rigid body.

In an intermediate field of frequencies (10–20 Hz) the tire can be considered as a massless deformable element introduced into the system and it is possible to define a dynamic stiffness, in both vertical and transversal direction. It is possible to show that the tire should be characterized by a low vertical stiffness, in order to minimize vertical displacements of the suspended mass on irregular road surface, and high transversal stiffness, to react with small displacements to side forces applied to the vehicle.

It is evident that there is a notable interaction between the vertical behavior of the tire and that of the rest of the vehicle, between transversal stiffness and cornering stiffness and, finally, between vertical and lateral stiffness, because of elasto-kinematic behavior of the suspension.

On Fig. 2.69 the dynamic vertical and lateral stiffness of two 155 15 tires are shown, as function of speed. As it can be seen, dynamic stiffness of radial tires is less as bias ply tires.

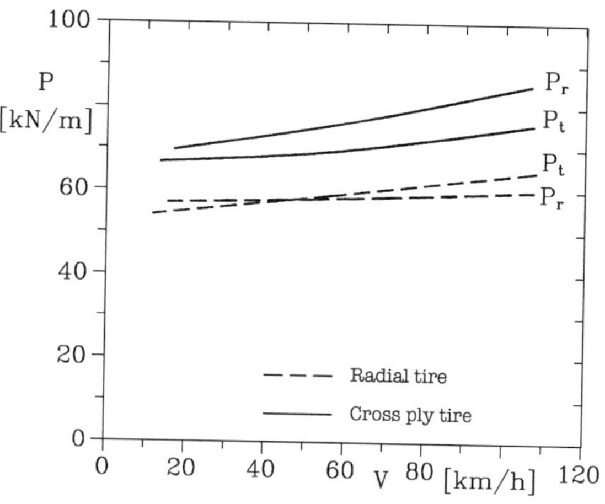

Fig. 2.69 Dynamic stiffness in vertical P_r and lateral P_t direction of two 155 15 tires of radial and cross ply type, in function of speed

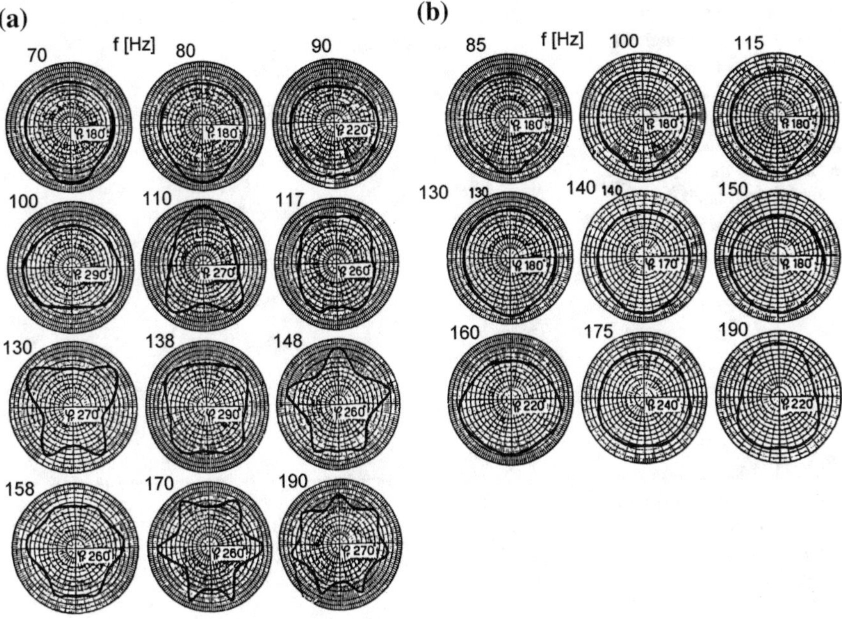

Fig. 2.70 Deformation of a 155 15 radial tire (**a**) and same size cross ply tire (**b**) at different excitation frequencies

At higher frequencies (over 50 Hz) the tire is vibrating according to its natural frequencies. On Fig. 2.70 deformation profiles are shown in correspondence of forced vibrations of 155 15 tires of both radial and cross ply type. Angles ϕ_r, shown in the picture are the phase retards between tire and exciting platform vibration. Radial tires present a peak between 60 and 90 Hz, according to tire dimension and other less relevant peaks are present at higher frequencies, with a reduced amplification factor.

Transmissibility, defined as the ratio between response and excitation amplitudes for a radial and a cross ply tire are shown on Fig. 2.71.

The figure shows the vertical response to a contact point vertical motion: in the same way it is possible to measure horizontal transmissibility.

The resonance of the unsprung mass because of the tire vertical elasticity takes place at a much lower frequency; the first resonance peak of the radial tire shown on the figure is bound to a mode where the tread is moving like a rigid body on the flanks. Higher frequency modes where the tread vibrates according to lobate shapes are less excited and more dampened.

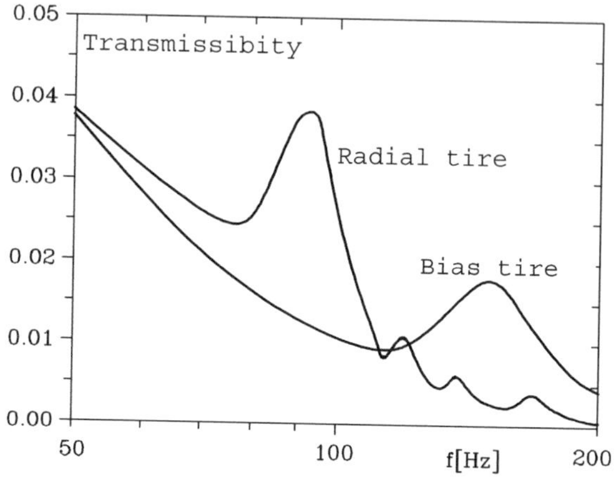

Fig. 2.71 Transmissibility for a radial and a bias tire 155 R 15 and 155 15 as function of frequency. An interval of frequency between 50 and 200 Hz is shown; the resonance of the unsprung mass for tire elasticity takes place at a lower frequency

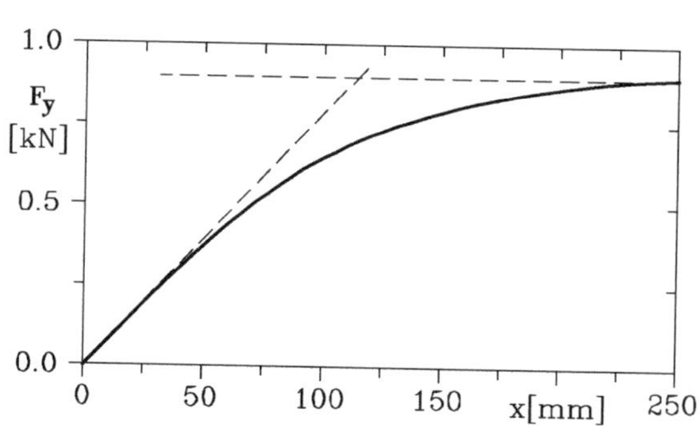

Fig. 2.72 Lateral forces applied to a standing tire then put in rotation for different time or distance. Tire 140 12, $F_z = 3.5$ kN, $p = 138$ kPa

2.9.2 *Cornering Dynamic Forces*

If the geometrical parameters (slip and camber angles) or the forces in X' and Z' directions are variable during motion, the values of the side force and of the aligning moment are at any instant different from the ones which would characterize a stationary condition with the same values of all parameters. As an example, if a tire is tilted about the vertical axis at standstill and then it is allowed to roll, the side force reaches the steady-state value only after a certain time, after rolling for a certain distance (Fig. 2.72), usually referred to as *relaxation length*.

This effect is usually not noticeable in normal driving as the time delay is very small, but the fact that there is a delay between the setting of the sideslip angle and the force generation is very important in dynamic conditions.

If the sideslip angle is changed with harmonic law in time, the side force and the aligning torque follow the sideslip angle with a certain delay, function of the frequency, and their value is lower than that obtained in quasi-static conditions, i.e. with very low frequency.

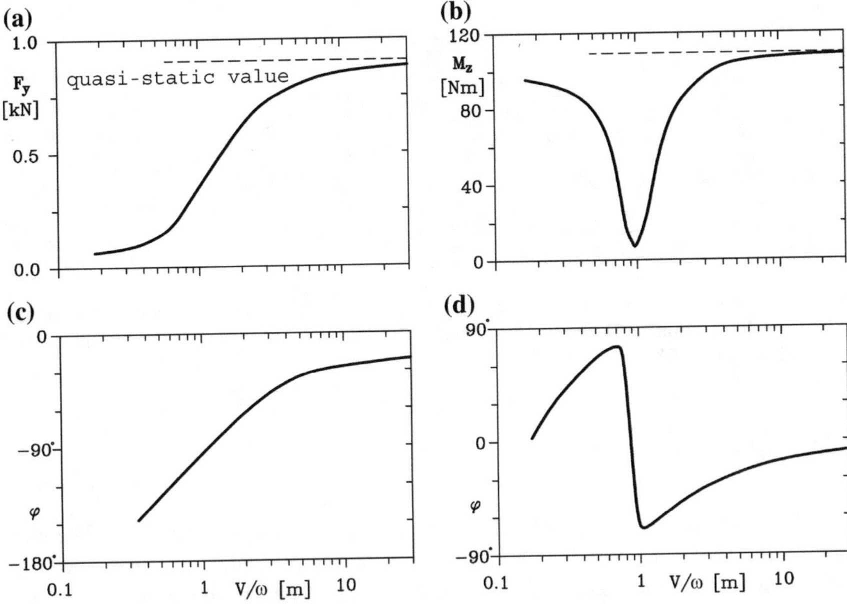

Fig. 2.73 Lateral force and self aligning torque on a tire having a side slip angle changing with harmonic law, between $-4°$ and $4°$, as function of the ratio V/ω between speed and frequency of the low $\alpha(t)$, the wave length. Peak value (**a**), (**c**) and phase difference (**b**), (**d**) of force and moment are reported. Tire 190 14, $F_z = 4.8\,\text{kN}$, $p = 165\,\text{kPa}$

If the frequency is not very high, at the speeds encountered in normal driving, the average values are not much lower than those characterizing static conditions, but a certain phase lag between the sideslip angle and the F_y force remains (Fig. 2.73).

The plots reported in the figure have been obtained for a given tire and are not easily generalized, particularly at high frequency. If a resonance of the tire occurs (as the phase of the aligning torque seems to suggest in the plot) the response is no longer dependent on the ratio V/ω but on the frequency and, to a lesser extent, on the speed separately.

More important for what practical applications are concerned is the case in which the load F_z applied by the wheel on the ground is variable, as is the case of rolling on uneven road (Fig. 2.74). The frequency can be very high, if the speed is high enough, and the decrease of lateral force due to dynamic effects can be large. In the figure the law $z(t)$ of the vertical displacement of the hub of the wheel is harmonic with a frequency of about 7 Hz while the response $F_y(t)$ is more complicated, with even an inversion of sign occurring at each cycle. The decrease of the average value of the lateral force at increasing frequency is shown in 2.74b.

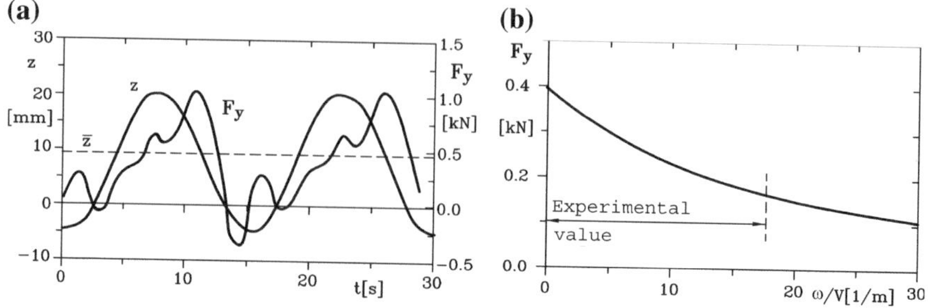

Fig. 2.74 **a** Lateral force on a tire working with constant side slip angle, but with a rim moving vertically with harmonic low $z(t)$. **b** Mean value of the force F_y as function of the ratio between ω (referred to $z(t)$) and V

2.10 Testing

The characteristics of tires can be measured using road or laboratory tests. The most common testing machines for tires are based on a drum which simulates the road; the tire can roll on the outer surface or on the inner surface of the drum (Fig. 2.75a).

The actual conditions are in a way intermediate between those encountered on the two types of test machines. To avoid the differences between the contact condition occurring in drum machines and the actual ones, more modern machines use a steel belt. The belt is kept flat by a hydrostatic bearing, which can be connected to a shaker to simulate the motion on uneven road. The simulation of the road surface is easier on drum than on belt machines, on which the surface is usually a plain metal surface.

The belt is kept flat by a hydrodynamic bearing, which can be excited with vibrations to simulate motion on a rough road. The ground simulation is easier in drum machine as in belt machine, where the surface is usually a steel band.

Other machines use a flat disc: In this case the contact surface is flat but the tire works with a ground moving along a circular path, generating a side force even with no sideslip. For low speed and low duration tests the wheel can roll also against

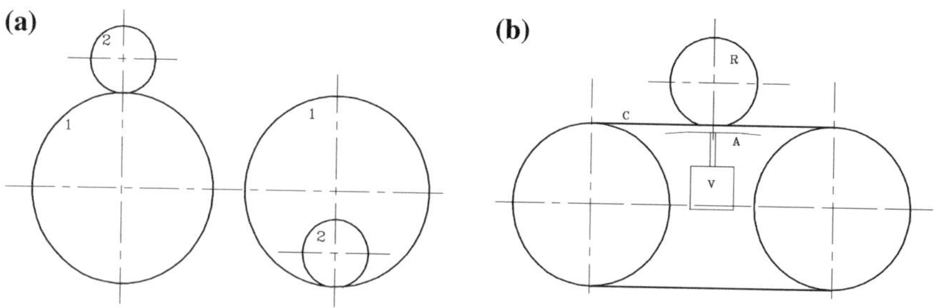

Fig. 2.75 Tire testing machines. **a** Drum machines with positive and negative drum curvature. 1: drum; 2: tire to be tested. **b** Belt machine. A: belt bearing; C: belt; R: tire to be tested; V: excitation

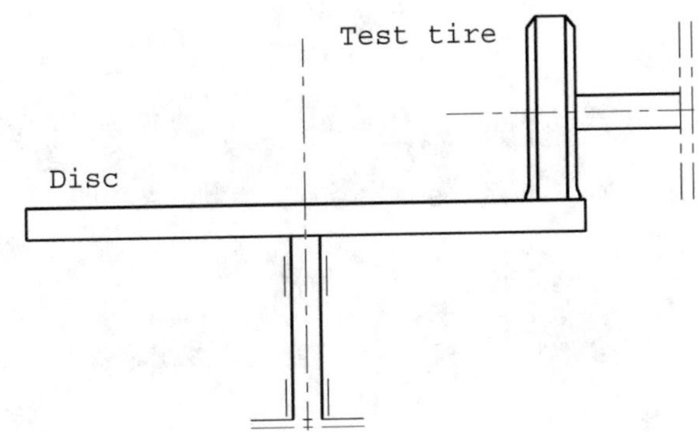

Fig. 2.76 Testing machine where road surface is simulated through a flat disc

a moving platform. With platforms it is possible to simulate easily different road surfaces (Fig. 2.76).

For low speed test of short duration the wheel can roll on a flat moving platform; different road surfaces can, in this case, be easily simulated.

In all test machines the tire is fit on a hub able to measure the three force and three moment components. Different driving and braking conditions can be simulated having two different motors to move the wheel and the road. Wheels can be maintained at given camber and side slip angles.

Laboratory test and, particularly those with roller machines, must be compared with those made on the road by means of particular test vehicles.

To test tires on the road wheel mounted on a dynamometric hub and provided with an independent motor is installed on a vehicle or a trailer. Figure 2.77 shows a picture of such kind of vehicle, used by the laboratory for vehicle testing of the Delft University.

Such vehicle feature all the necessary equipment for tire force and moment measurement in every possible condition. A second equal wheel is installed with an equal but opposite side slip angle, in order to allow the vehicle to drive on a straight line. A suitable device can wet the road according to a determined water layer thickness.

To measure the rolling resistance only simpler machines can be used.

On a wheel drum machine the wheel can be launched to a given speed, and then let to coast down by its inertia.

The equation for the motion of the wheel and drum assembly is:

$$\frac{d\Omega_R}{dt}\frac{J_R + J_S}{R_e} = -f F_z - \frac{|M_S|}{R_S}, \tag{2.43}$$

where R_S, J_S and M_S are the drum radius, its moment of inertia, reduced to the wheel axis and the total resistant moment of the wheel. M_S can be determined the deceleration of the wheel, under the vertical force F_z, which can be obtained with a simple vertical load.

The $\Omega_R(t)$ law can be recorded very easily and the roll resistance f can be calculated using the Eq. (2.43). This method has the inconvenience, common to all methods

Fig. 2.77 Vehicle for on road tire testing used by the vehicle research laboratory of the University of Delft. The test wheel is in the middle of the trailer

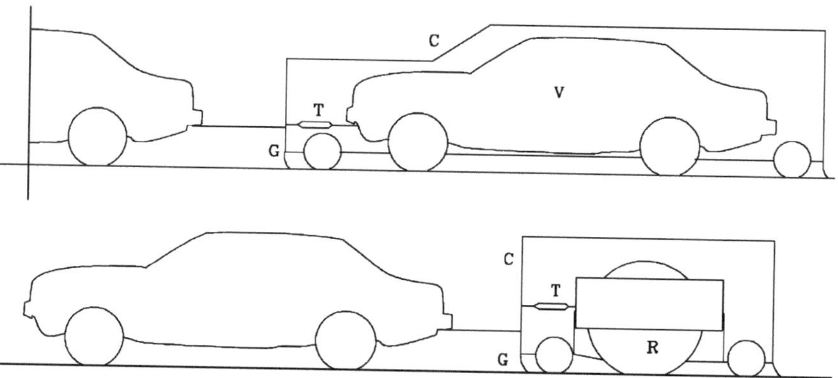

Fig. 2.78 Fixture to measure rolling resistance of a vehicle or a wheel on the road. C: trailer; G: trailer skirt; R: wheel under test; T: dynamometric pull bar; V: vehicle under test

starting from the acquisition of a coast down curve, of requiring the measurement of a magnitude, where the derivative is desired. Because the derivative of a function is very sensible to errors of the function measurement, this method requires particular precautions.

Other inconveniences are common to all tests on the drum and, particularly the difference of the surface, usually of steel and the different shape of the contact patch.

Moreover, as in all coast dawn tests, the measurement is made during a transient, where many magnitudes are changing during the test, the temperature in particular.

If the acceleration phase is long enough, the tire is reaching the steady state temperature at a speed lower as that of test start. If the temperature is higher as the appropriate one, rolling resistance is slightly lower. If the acceleration is higher, on the opposite, tire temperature may be too low, bringing to a higher value of rolling resistance.

On road tests can be performed pulling a vehicle or a single road, put inside a suitable trailer able to eliminate aerodynamic resistance in its interior (Fig. 2.78). A dynamometric pull bar connects the trailer with the test vehicle, allowing the measurement of the rolling resistance.

This method has the inconvenience of being sensitive to road slope: a slope of only 0.1% can bring to errors on 10% on rolling resistance coefficient f. The measurement can be made more difficult by possible longitudinal vibrations of the system.

Chapter 3
Suspensions

3.1 Introduction

For vehicle suspensions we mean a mechanism which links the wheel to the body directly or to a frame fixed to the body.

Since a rigid vehicle with more then three wheels is an hyperstatic system, it is necessary that the vehicle structure is flexible enough to allow the contemporary contact of the wheels with the ground or that the wheel are connected to a rigid body through a deformable system, the suspension. This second solution is adopted by most of the vehicles, while the first, widely applied to horse carriages in the past, is now used on particularly slow vehicles.

In many cases, to suspension deformation must be added the structure deformation, which plays an important role on handling and comfort characteristics of a vehicle.

To accomplish its task, suspensions must:

- allow a break down of forces, exchanged by the wheels with the ground, complying with design specifications in every load condition;
- determine the vehicle *trim* under the action of static and quasi static forces.

We shouldn't, in fact, forget that by introducing on a vehicle a deformable linkage, geometric variations of the body position are introduced as a function of payload and payload position; these variations are described through the three coordinates of the center of gravity and the three angles of the body reference system (yaw, roll, pitch angle). They are included under the name of vehicle static *trim*. A complete definition of the body reference system is given in the fourth part, in the second volume.

In addition to this function, which is accomplished basically with an elastic system, there is another one, not less important:

- to absorb and smooth down shocks that are received by the wheel from road asperities and transmitted to the body.

G. Genta and L. Morello, *The Automotive Chassis*, Mechanical Engineering Series,
https://doi.org/10.1007/978-3-030-35635-4_3

It should be remembered that this task requires the application of a suitable damping system; this function is so important that suspensions are applied also to two or three wheels vehicle which are not hyperstatic bodies.

In theory, tires alone could isolate the vehicle body from forces coming from the road, but their elastic and damping properties are not sufficient to reach suitable handling and comfort targets, unless at very low speed and on smooth roads.

Suspensions are therefore peculiar to reach an adequate road holding behavior (*handling*) and comfort and determine for each vehicle their own distinctive characteristics.

Wheels must be, therefore, free to move in a direction almost perpendicular to ground, in addition to their rotation and steering motion.

This vertical motion must be guided through the suspension linkages in order to guarantee a correct position of the tire with reference to the ground. The capacity of a tire to react with suitable forces is, in fact, determined by the angles between the equator plane of the wheel with the ground and with the hub speed.

If the suspension is a kind of filter between the road and the body, designed to limit the amount of forces coming from road asperities and maneuvers, this filter must not impair vehicle controllability, in possible driving situations. Road holding depends not only on mass properties of the vehicle (mass and moments of inertia), on its geometric properties (kind of traction, center of gravity position, wheelbase, track), on its tires but on suspensions too.

Suspensions are usually classified in three classes: *independent*, *dependent* and *semi-independent* suspensions.

The first class presents no mechanical linkage between the two hubs of the same axle; a force acting on a single wheel doesn't affect the other; while writing this sentence, steering linkages, anti roll bars or auxiliary frames are not taken into account.

Dependent wheel suspensions or rigid axles provide for a rigid linkage between the two wheels of the same axle; each motion of a wheel, caused by road asperities is affecting also the coupled wheel.

Semi-rigid suspensions have intermediate characteristics between the other two families. In these suspensions wheel hubs can't be considered as independent, because they are not linked with an articulated structure; on the contrary this structure has such mechanical characteristics that flexibility cannot be neglected. This category is practically including the so-called *twist axles*.

Another important characteristics, within these families, separates steering from non steering suspension; while independent suspensions can be, in principle, designed to become steering suspensions, dependent suspensions are no more applied to steering axles, industrial vehicles and off-road vehicles being an exception. The same applies with no exception to semi-rigid suspensions.

Most suspensions can be applied to driving and idling axles; this aspect has no significant impact on suspension design, except for joint stiffness.

Considering elastic and damping systems only, suspensions can be classified as *passive* or *active*. In the first family the elastic reaction of the suspension system is determined only by its deformation and the damping system can only waste part of the received energy; in the second family the suspension system can receive energy from other sources (the engine, or some intermediate storage of the engine energy) to affect body motion with the objective to limit this motion very close to its static equilibrium condition.

The objective of this kind of suspension is to limit body displacement to a minimum, necessary to transfer to the driver a sufficient perception of vehicle stability. We will comment on this kind of suspension in the chapter dedicated to chassis control systems and we will see that different level of activity are identified, according to the fact that the outside energy source is used to correct the static trim only, or the dynamic trim too. Most vehicle suspensions are passive.

Finally, two additional definitions.

We define as *sprung mass*, that part of the vehicle mass that is free to move with reference to the ground, because of the suspension application; the part of the mass that is not changing its position is called *unsprung mass*.

Some suspension components contribute part to sprung mass, part to unsprung mass. To evaluate the two contributions, the mass of these elements must be divided in two parts, concentrated ideally in the suspension joints, in such a way as to conserve the moment of inertia and the center of gravity position.

3.1.1 Suspension Components

To accomplish the above described functions, suspensions are built with different categories of components.

Bearing Components or Linkages

They are part of the mechanism linking the wheel to the body and guarantee the degrees of freedom of wheels and their correct positions with reference to the ground. They determine the relative motion of the wheel with reference to the vehicle body; they also transfer to the body part of the loads coming from the tire contact patch.

Primary Elastic Members

They include springs (coil, bar and leaf springs), anti roll bars and stop springs. These members connect elastically the wheel to the body and they store the energy coming from uneven road profile. They not only store this energy but determine body position as function of payload entity and position.

Secondary Elastic Members

They are elastic bushings of linkages joints. To some of these joints is given a certain elastic compliance.

At the origin this property was taken as a drawback of the possibility of avoiding joint lubrication, by using elastomeric joints.

More recently it was understood that his property could be exploited to better design the *elasto-kinematic* behavior of the suspension and its comport properties.

The deformation of these joints plays a very important role in determining vehicle handling.

Damping Members

They are basically shock absorbers, but we should remember that also primary and secondary elastic members have a non negligible capacity of wasting energy.

Shock absorbers are provided to waste the elastic energy store by elastic members and allow the oscillation damping of the vehicle body, avoiding stationary vibration or resonance.

3.1.2 Suspension Influence on Body Motion

Ideally, the suspensions should allow the wheels to move with respect to the body of the vehicle in a direction perpendicular to the road, maintaining the plane of the wheel parallel to itself and constraining all motions in x and y directions[1]: A suspension of a single wheel should be a system with a single degree of freedom, the displacement in z direction or suspension stroke.

It should be remembered that if the wheel is allowed to move along the Z' axis only, linkages should not transmit to the wheel hub body rotation, particularly roll angle.

However none of the linkages we will examine is able to satisfy this condition completely and every kind of car suspension has its own characteristic; the approximation to the ideal characteristic is part of the character of a car.

Wheel position is also affected by joints stiffness which even if considered spherical joints or cylindrical bushings are rather deformable; sometimes linkages too have to be considered as flexible members.

Wheel position is not only affected by body motion, but also by external forces.

---

[1] Usually the reference system adopted for suspensions is the same as the reference system used on body drawings; this is the vehicle reference system, explained in the second volume, but the origin of this system is set in the symmetry plane of the body, through the front wheel centers in the static load position.

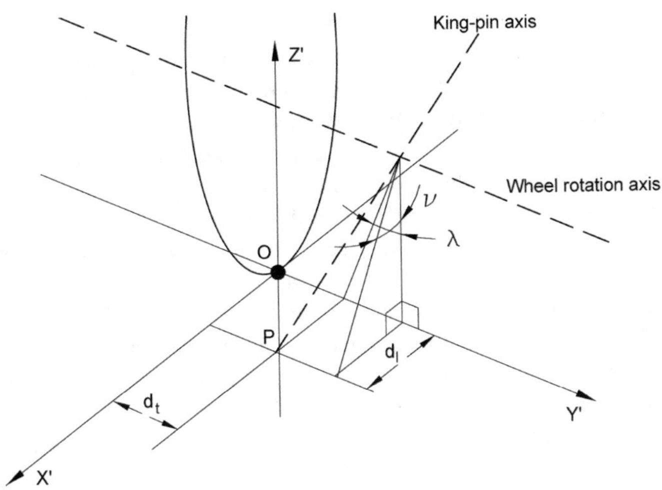

Fig. 3.1 Steering geometry

On the following lines we will describe angles and linear dimensions applied to describe suspensions behavior; these characteristics are described with reference to a front steering axle.

In a steering suspension, the *steering axis* or *king-pin axis* is the axis around which the hub holder or strut, and therefore the wheel is able to rotate or steer (Fig. 3.1). Also linkages of non steering suspensions can leave to the wheel a limited freedom to steer around a king-pin axis, according to body roll angle or applied forces; the effect of this additional steering angle will be discussed later.

The angle between king-pin axis and the Z' axis, projected on the $X'Z'$ plane is called *caster* angle (ν on figure). As it will be explained later, this angle is relevant to determine steering wheel return capabilities.

The distance between O, center of the contact patch of the tire, and P, intersection of the steering axis with the ground, projected on the $X'Z'$ plane is the *longitudinal trail* (d_l on figure).

The angle between the wheel equator plane and the $X'Z'$ plane is the inclination or *camber* angle, while the angle between the king-pin axis and the $Y'Z'$ plane is the *king-pin* angle (λ on figure).

We should remember that a suspension linking the wheel in such a way as to remain parallel to the car body (for instance as in longitudinal trailing arms) is applying to the wheel an inclination angle, with reference to the ground, equal to the body roll. This is particularly noxious, because the roll angle is such as to decrease the cornering stiffness of the tire. On two-wheeled vehicles, the behavior is opposite and brings advantage; the same could happen on variable trim vehicles, that will be described in the second volume.

The distance between O and P projected on a $Y'Z'$ plane is the king-pin offset (d_t on figure).

But an ideal suspension shouldn't allow the wheel to change its camber with reference to the ground; we will define as *camber recovery* angle the difference between the actual camber, with reference to the ground, and body roll. An ideal suspension should have a camber recovery equal to body roll.

With reference to the vehicle body, we define as *toe* angle the angle between the intersection of the equator plane of the wheel with the ground and the car x axis. It can be measured either in degrees or in millimeter of difference in transversal distance of two point on the wheel rim. It is called *toe-in* when wheel equator planes are crossing in front of the vehicle and *toe-out* in the opposite case.

Other dimensions to be considered are the *wheelbase*, the distance projected on a xz plane between center of tire contact area of the wheels of the same side and *track*, the distance projected on a yz plane between centers of the contact area of the wheels of the same axle. Quite often, referring to suspension geometrical properties *semi-wheelbase* and *semi-track* are distances of the center of the contact area from the xz and from the yz planes for the center of gravity of the vehicle.

When the vehicle body is moved in vertical direction (displacement z) or rotates about its x axis (roll angle ϕ) the position of the wheel changes, it is possible to plot camber angle γ, track t, characteristic angles of the steering system, steering angle δ, etc. as functions of z and ϕ. These functions are generally strongly nonlinear but they can be linearized about any equilibrium position and the derivatives $\partial t/\partial z$, $\partial t/\partial \phi$, $\partial \gamma/\partial z$, $\partial \gamma/\partial \phi$, $\partial \delta/\partial z$, $\partial \delta/\partial \phi$, etc.[2] can be easily defined. They can be considered as constant in the small motions about an equilibrium position and define the behavior of the suspension.

While outlining the design of a new suspension, mathematical models have today a fundamental importance.

With a simple approach, (see Fig. 3.2) starting from suspension drawings, a suspension scheme can be built up, where linkages are represented with rigid beams connected with kinematic couples (spherical joints, bushings, etc.) able to simulate the kinematic behavior by suspension stroke an wheel steering., taking into account the contribution of linkages and main elastic members only.

On a second approach, kinematic couples introduced in the model can be changed with elastic elements; it is now possible to calculate characteristics variations of the suspension as function of applied forces too. This second way of studying the suspension means analyzing the elasto-kinematic behavior.

This simple simulation applied iteratively have a fundamental importance in designing suspension with respect to assigned targets.

On following paragraphs we will analyze the most diffused suspension schemes existing on present cars and their main components.

[2] In the following the notation $(t)_{,z}$, $(t)_{,\phi}$, etc. will be used.

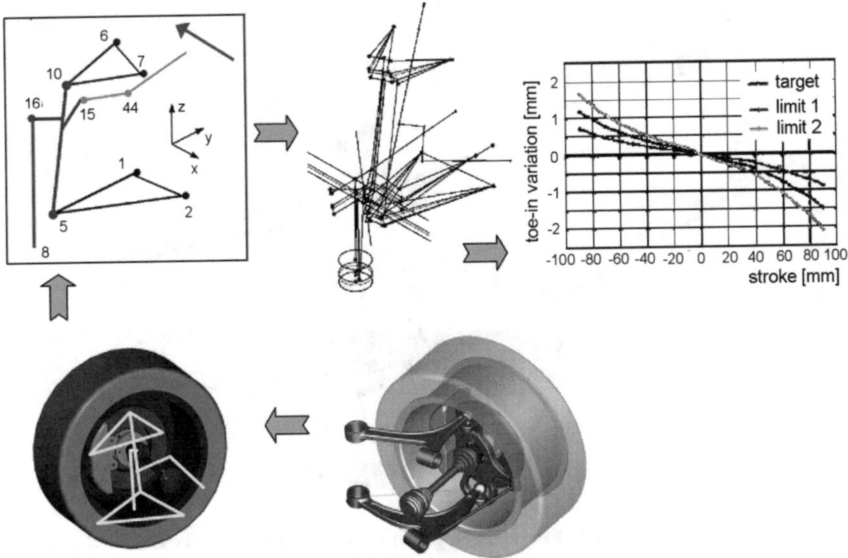

Fig. 3.2 Starting from suspension drawings a suspension scheme can be built up, where linkages are schematized with rigid beams connected with kinematic couples (spherical joints, bushings, etc.) able to simulate the kinematic behavior by suspension stroke an wheel steering

3.2 Independent Suspensions

If the wheels are suspended independently, linkages must constrain five out of the six degrees of freedom of the wheel (or better, of the wheel hub, because the wheel is then free to rotate about its axis). The unconstrained degree of freedom should be the translation in a direction perpendicular to the ground. None of the many devices which are currently used fulfills exactly this requirement.

As suspensions must restrain five degrees of freedom, they can be materialized as a system made of five bars with spherical hinges at the ends (Fig. 3.3), This layout, often referred to as multilink suspensions, which has the advantage of allowing a very large freedom of adjustment by changing the length of the bars by screwing in or out the joints, has little application outside the field of luxury cars for its complexity, even if simpler multilink suspensions are now more widespread. From five bars multilink suspensions almost all configurations can be obtained by grouping these bars in different ways.

Note that in general the motion of the wheel is not planar and as a consequence the study of the kinematic behavior is not easy. Nowadays it is however easy to obtain the exact kinematics of any suspension by using computer generated trajectories.

If points 1 and 2 and points 3 and 4 coincide with each other the corresponding bars become triangular elements: The suspension obtained is a transversal quadrilaterals suspension, often referred to as SLA (short-long arm) or A-arm suspension (Fig. 3.4).

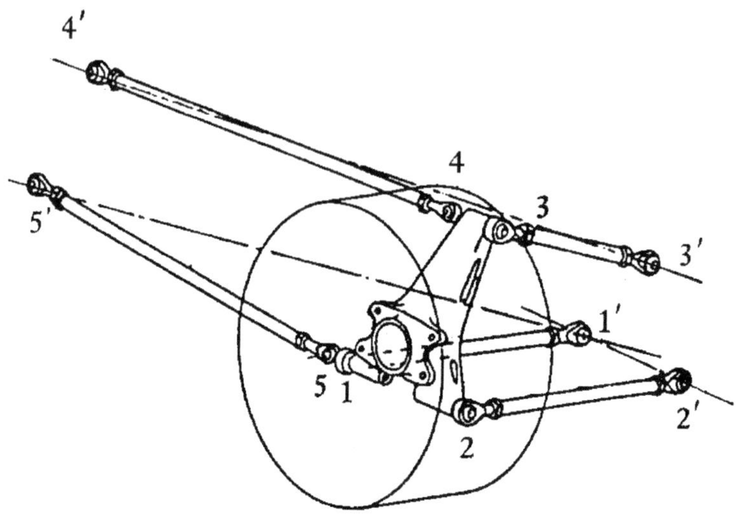

Fig. 3.3 Suspension based upon five linkages which cut five of the six degrees of freedom of the wheel

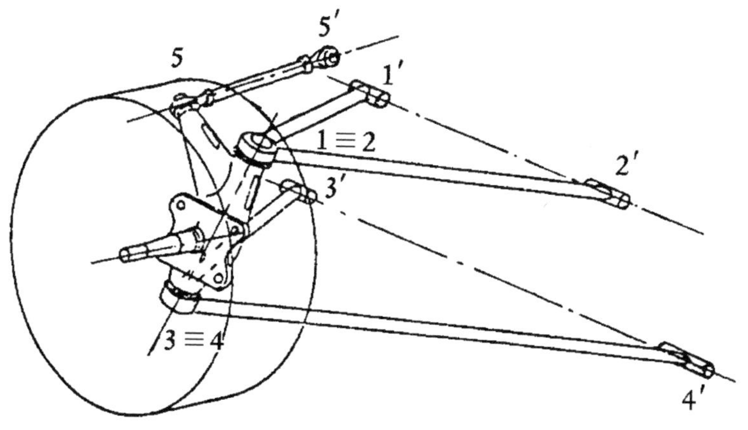

Fig. 3.4 Low transversal quadrilaterals suspension

If lines $1'2'$ and $3'4'$ are parallel, the motion of the wheel is contained in a plane perpendicular to the line $1'2'$ and the projection of the mechanism in such plane is an articulated quadrilateral, whose side $1'3'$ is made by the vehicle body. A front suspension of this type is shown in Fig. 3.80, in which the engine is also sketched to show that this solution allows to locate the mechanical parts of the vehicle with a greater freedom than solutions based on rigid axles.

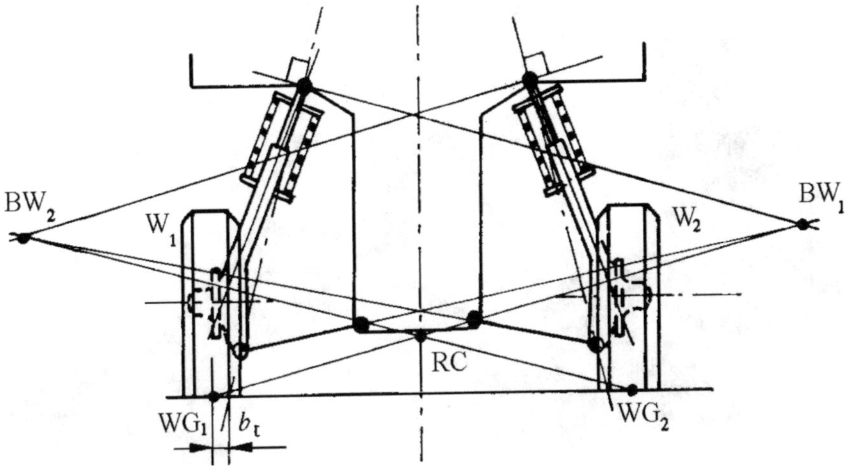

Fig. 3.5 Mc Pherson suspension. The figure shows the geometric construction to determine the roll center RC, not taking into account tire deformation

3.2.1 McPherson Suspension

If the upper triangle is substituted by a prismatic guide a McPherson suspension is obtained (Fig. 3.5). Its simplicity and the fact that it leaves much free space for the engine has made it a very common solution for cars front axles, particularly small ones.

The description of this suspension which is by large the most diffused configuration is divided in three parts.

The first part is addressed to define suspension geometry and its kinematic behavior; a second will follow on advantages and disadvantages implied by this suspension, as compared with the remaining types.

This explanation scheme will be adopted also for the other suspension types.

A third part is dedicated to the suspension and its most important components detail design.

Similar components with adaptations that can be easily imagined are also adopted on the remaining types. Therefore this section will not be repeated, unless for commenting relevant differences with the Mc Pherson suspension.

Description

The McPherson suspension is the most diffused one for front axles and is applied to all small and medium size cars. Some manufacturers adopt this solution also on large cars an sometimes for sport cars too; it is also applied sometimes to rear axles.

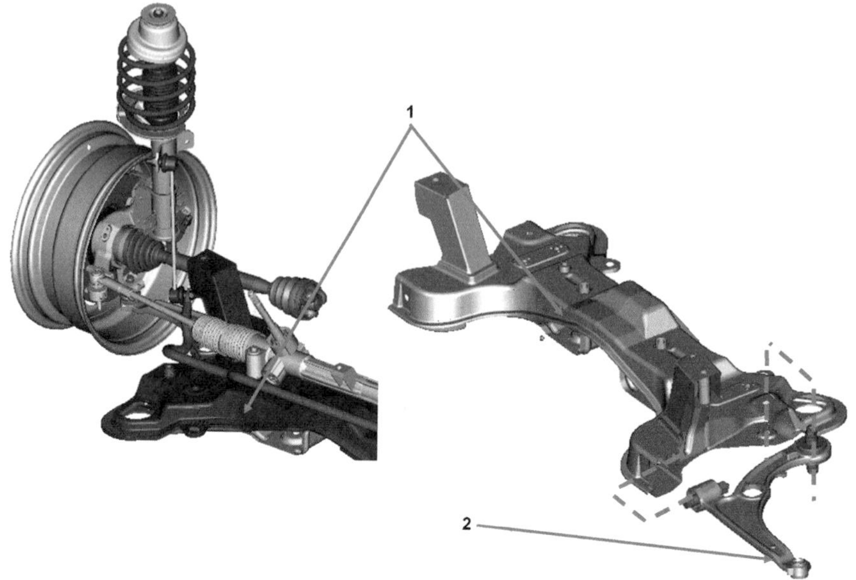

Fig. 3.6 McPherson suspension for the front axle of medium size front driven car; the lower arm (2) is articulated to the body through an auxiliary frame called also subframe (FCA)

We are going to describe first of all wheel linkages. The wheel is guided, during its vertical motion, by means of a lower arm and of a sliding guide, integral to the shock absorber. This device is also used as spring installation and an upper pivot links it to the body.

We refer to a solution applied to the front axle of a medium size car, represented on Fig. 3.6 and proceed describing the most relevant details.

The lower arm 2 is connect to the body through an auxiliary frame, also called *subframe* 1, in two different points, through elastic bushings 3. The lower arm is also linked to the strut 5 through a spherical joint 4, as shown on Fig. 3.7.

The strut is the element where is machined the seat used to install the outer ring of the wheel roll bearing. The inner ring is fit to a *hub*, which is flanged to the brake disc and wheel. On driving axles the hub is fixed to the drive shaft through a spline, allowing the torque transmission from the differential to the wheel. On the strut two flanges are used to fix the brake caliper.

At the base of the body of the shock absorbers 6 (Fig. 3.8) two brackets 7 are welded, which are bolted to the strut in a rigid way. The spring rests on two seats, a lower one 8 fixed to the shock absorber and an upper one fixed to a thrust bearing 10; the upper ring of this bearing rests on an elastic mount, fixed to the body, in wheel case.

The elastic mount is therefore the interface between the shock absorber piston and the spring, from one side, and the body, from the other side. The stop spring 13 is fit to the shock absorber piston and acts when the shock absorber tube is touching

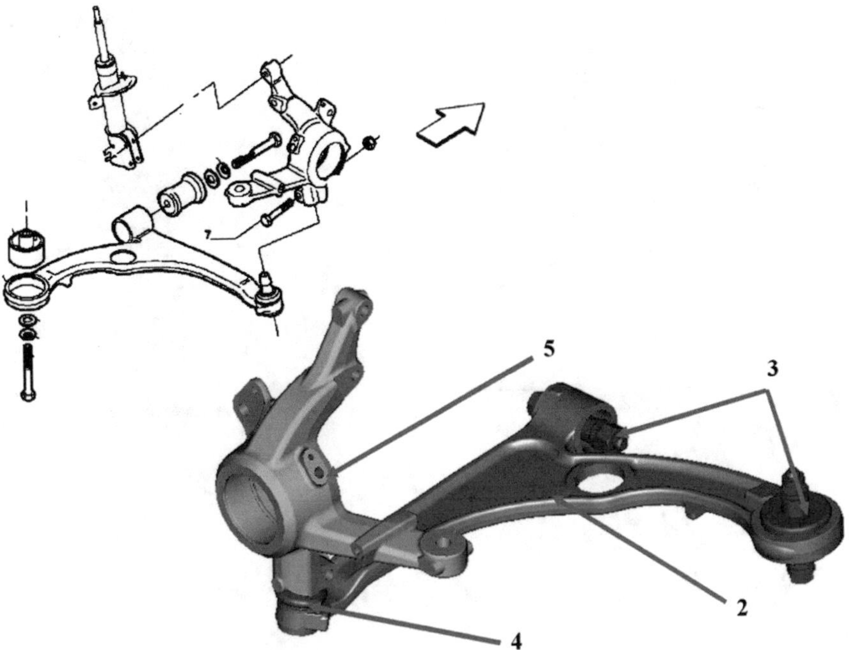

Fig. 3.7 The lower arm 2 is connected to the subframe 1 in two different points, through elastic bushings 3; a spherical joint 4 links the arm to the strut 5

its tip; this stop spring avoids direct metal to metal contacts in connection with large strokes and makes the elastic characteristic of springing more progressive.

The rack and pinion steering box 14 (Fig. 3.9), is bolted to the subframe and shows two steering tie rods 15, articulated to the rack through two spherical joints 16.

The anti roll bar 17 is fixed on the subframe, but is left free to rotate; it is connected to the shock absorber with two rods 18 (Fig. 3.10).

McPherson suspensions are using the shock absorber piston to guide the wheel along the suspension stroke.

This detail is causing that external forces applied to the wheel in the contact point with the ground determine, in function of the suspension geometry a lateral force F_L and a momentum M applied to the piston itself.

Reactions R_A and R_B and the piston rod flexion have a major influence on shock absorber characteristics.

As it will be explained, the shock absorber should exert a force ideally proportional to the suspension extension and compression speed; forces and flexion are, instead, causing a friction almost constant and independent of speed. This friction together with other kinds of friction is causing suspension *hysteresis*, that could be interpreted as the minimum force, we must apply to the suspension to have it moving.

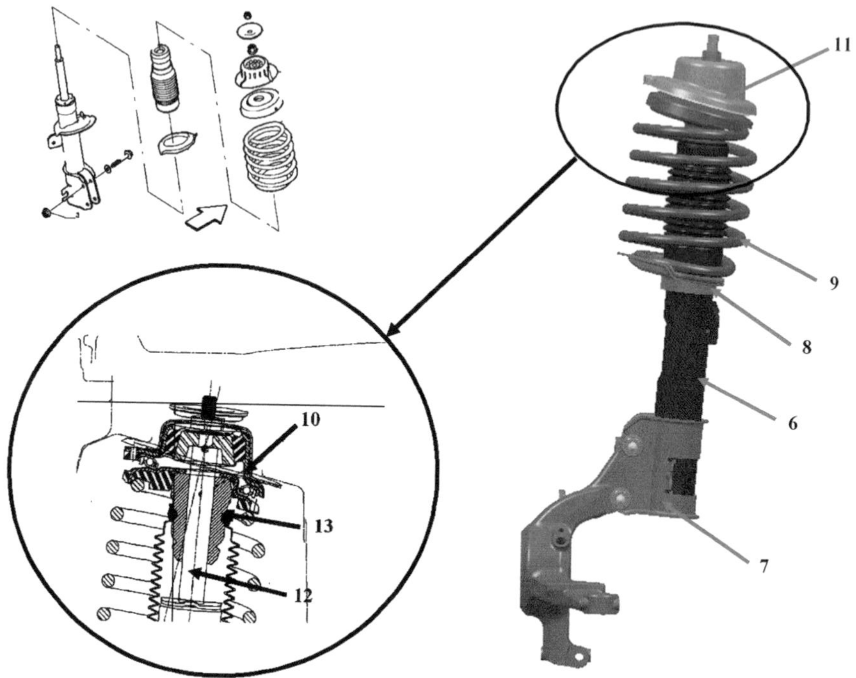

Fig. 3.8 At the base of the shock absorber 6 two welded brackets 7 are used to fix the strut in a rigid way; the spring 8 is contained between two seats, one connected to the shock absorber, the other to the thrust bearing 10; the upper ring of this bearing rests on the elastic element 11, fixed to the body

Hysteresis is not desired because is causing the suspension to block, when forces applied are lower than a minimum value. This characteristic of the Mc Pherson suspension must be interpreted as an inconvenient.[3]

With reference to Fig. 3.11, reactions R_A, R_B and deflections of the piston rod of the shock absorber ϑ_A and ϑ_B can be calculated using the following formulae.

It should be noted that the fictitious reaction R_T is the sum of the absolute values of reactions between piston rod and its bushing (R_A) and between piston and cylinder (R_B); this force multiplied by the friction coefficient (considered to be the same in the two couples) will show, neglecting the contribution of rod deformations, the friction force that reacts to rod motion, that is the hysteresis.

We define:

E the elastic modulus of the piston rod;

I the flexural moment of inertia of the rod cross section.

We can write:

[3]In reality, this inconvenient is not only present on Mc Pherson suspensions, but on all those suspensions where shock absorbers are partially loaded by wheel forces.

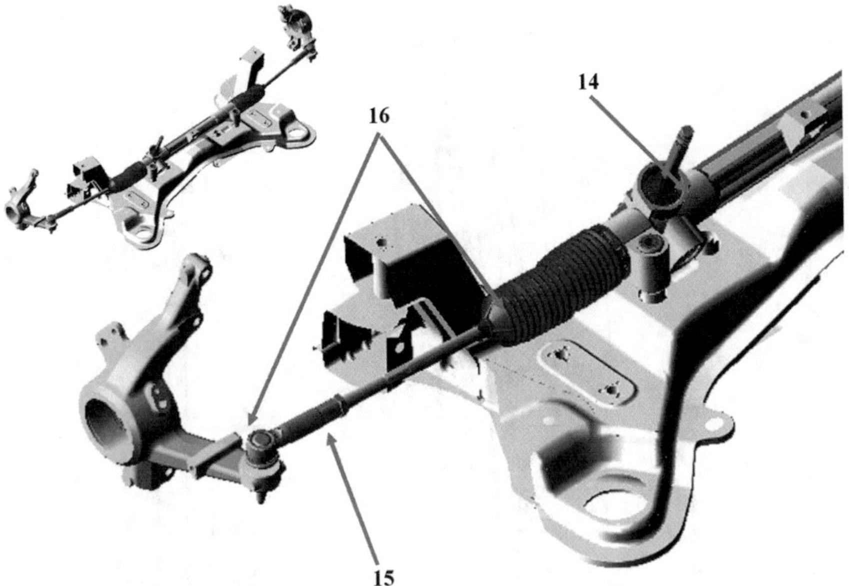

Fig. 3.9 Rack and pinion steering box mounted on the subframe of a front Mc Pherson suspension

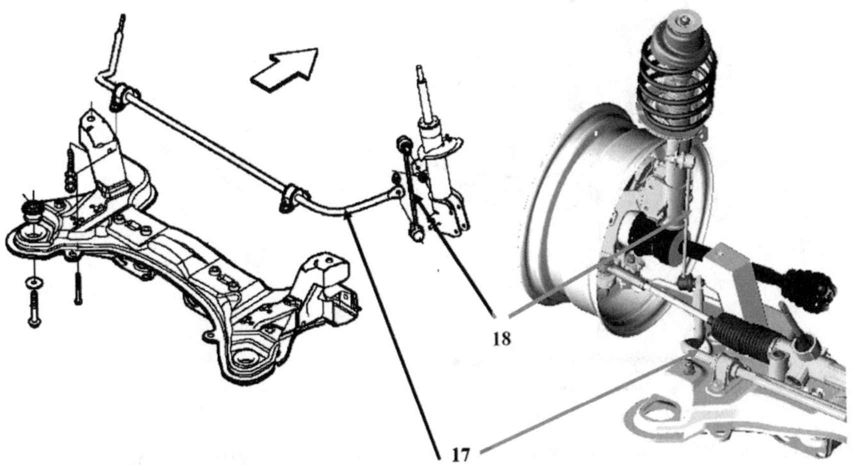

Fig. 3.10 The anti roll bar 17 is fixed on the subframe, but is left free to rotate; it is connected to the shock absorber with two rods 18

$$\begin{aligned}
R_A &= \frac{1}{b}\left[F_L\left(a+b\right)-M\right] \ , \\
R_B &= \frac{1}{b}\left(F_L a - M\right) \ , \\
R_T &= |R_A| + |R_B| \ ,
\end{aligned} \tag{3.1}$$

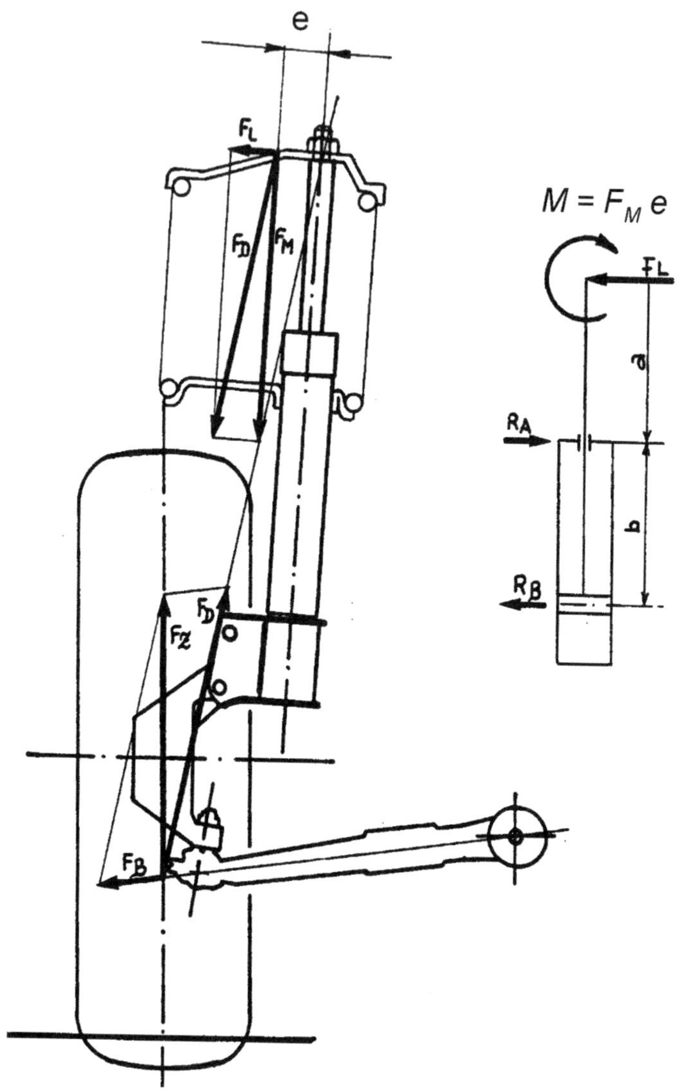

Fig. 3.11 Geometric scheme useful to calculate forces on sliding bushings of a shock absorber of a Mc Pherson suspension

$$
\begin{aligned}
\vartheta_A &= \frac{b}{3EI}\left(F_L a - M\right) , \\
\vartheta_B &= \frac{b}{6EI}\left(F_L a - M\right) .
\end{aligned}
\tag{3.2}
$$

Let us consider now the possibility of introducing an offset e between the rod and the spring symmetry axis. This offset will imply the application of a moment M,

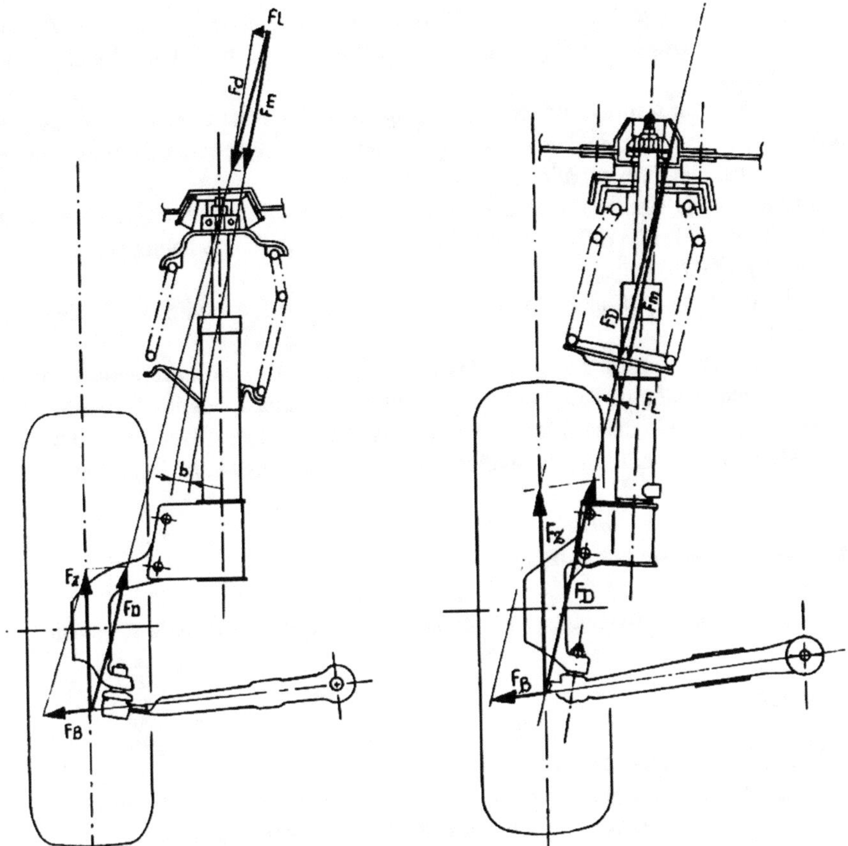

Fig. 3.12 Scheme of McPherson suspension with integral (left) and double (right) thrust bearing

exerted on the rod by the spring, as shown on the figure, that will reduce reactions acting on bushings. If:

$$M = F_L a \ , \tag{3.3}$$

we obtain the minimum value for R_T, while, in these conditions:

$$\begin{aligned} R_A &= R_T = F_L \ , \\ R_B &= 0 \ , \end{aligned} \tag{3.4}$$

$$\vartheta_A = \vartheta_B = 0 \ . \tag{3.5}$$

Magnitudes M and F_L change according to thrust bearing architecture and spring inclination axis.

With reference to Fig. 3.12, we identify two different families of Mc Pherson suspensions: those with integral thrust bearing (left) and those with double thrust bearing (right).

In the integral solution, where both spring and shock absorbers loads are exerted on the body through a single rubber element, reactions on the rod are represented by the formulae, we have explained.

In the double solution, spring load and shock absorber load are transferred to the body through two different rubber elements and the force acting on the rod is represented by F_L only.

In both cases is possible to reduce the hysteresis force, applying a suitable offset to the spring.

The zero condition for hysteresis is bound to the value we have chosen for the vertical load; changing this condition or applying other loads, as for example a cornering force, will cause a zero condition to change. Spring offset can be designed for a single load condition only, which is judged to be statistically important.

Advantages and Disadvantages

The comparison of advantages and disadvantages refers to front axle suspensions.

Advantages

- Design simplicity and reduced cost.
- Thanks to the relevant distance of body joints, forces exerted to the body are low, in comparison, for instance with a low double wishbone suspension.
- Higher suspension stroke, as in other suspensions, for instance a high double wishbone one, because on limitation on upper arm length.
- Contained transversal dimension, due to the absence of the upper arm; this fact is quite beneficial for transversal engine installation.
- Possibility of designing a good longitudinal flexibility, without affecting the caster angle too much.
- Good freedom in designing elasto-kinematic properties; camber recovery only is limited by viable positions for upper pivot and lower arm fixed joint.
- The ratio between suspension and shock absorber stroke is near to one. Shock absorbers are therefore working well with limited loads, low oil heating and valves wear.

Disadvantages

- Lower performance as far as camber recovery is concerned. See, for instance, the comparison between camber angle variation for a Mc Pherson and a double wishbone suspension, shown on Fig. 3.13.

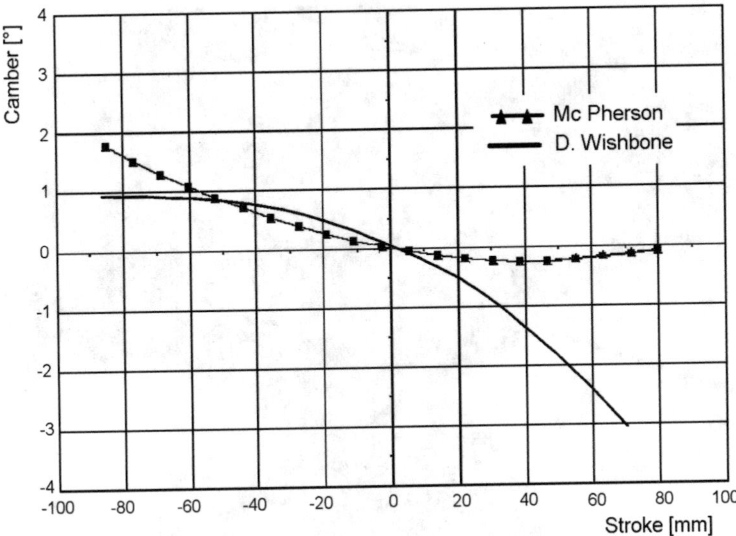

Fig. 3.13 Comparison between camber angle variation, as function of suspension stroke in a Mc Pherson and a double wishbone suspension; the difference motivates a better camber recovery in the double wishbone suspension

- Suspension characteristic geometry is causing a position for the upper pivot interface with the body, usually called *dome*, which usually far away of stiffest structures of the body, the side beams, as shown on Fig. 3.14. This is causing significant problems with suppression of vibrations and noise, coming from the road.
- Shock absorber piston rod deformation can increase friction and hysteresis.
- Notable height for the upper pivot, being spring and shock absorber set over the wheel; this fact could be a penalty for aerodynamic shape and for a sporty style for the body.

Design Details

Let us examine now some of the most relevant components of Mc Pherson suspensions, which can feature many different alternative versions.

Lower Arm

Lower swinging arms connect the moving parts as strut and wheel to the body directly or to the subframe.

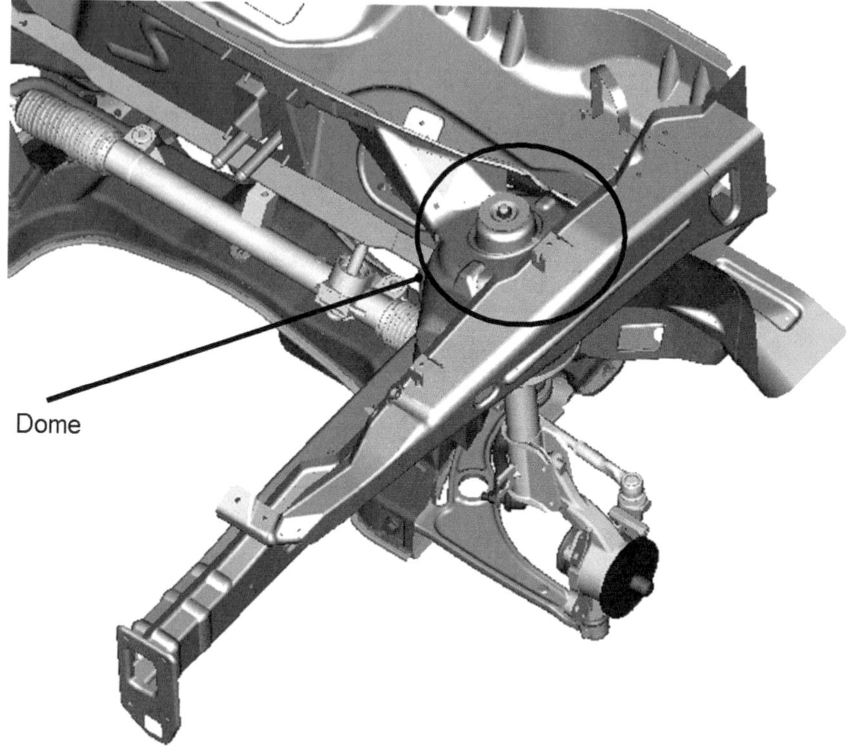

Fig. 3.14 The position of the dome imposed by Mc Pherson suspensions is usually far away of stiffest body structures, which are lower, close to side beams (FCA)

Lower arms have a particular shape able to allow both suspension and steering motions to the wheel, as shown on Fig. 3.15 and are connected to the subframe through elastic bushings and with spherical joints to the strut.

The choice of the shape of the lower arm is part of the concept design of the front suspension and must be compatible with the handling and comfort targets.

Once the suspension scheme has been decided, using a CAE tool, the envelope of the external shape of the tire can be designed, taking into account steering motion and suspension stroke, as shown on the right side of Fig. 3.15. The lower arm shape must be obviously outside of this envelope and must take into account of assembly and disassembly conditions (some times holes must be provided for screwdrivers and wrenches passage) and of limitations imposed by other parts of the car as powertrain or body.

The lower arm absorbs a relevant part of forces coming from the wheel during braking or cornering maneuvers. Because this components is fitted to the body through articulations, reaction forces must be contained in the plane through their centers.

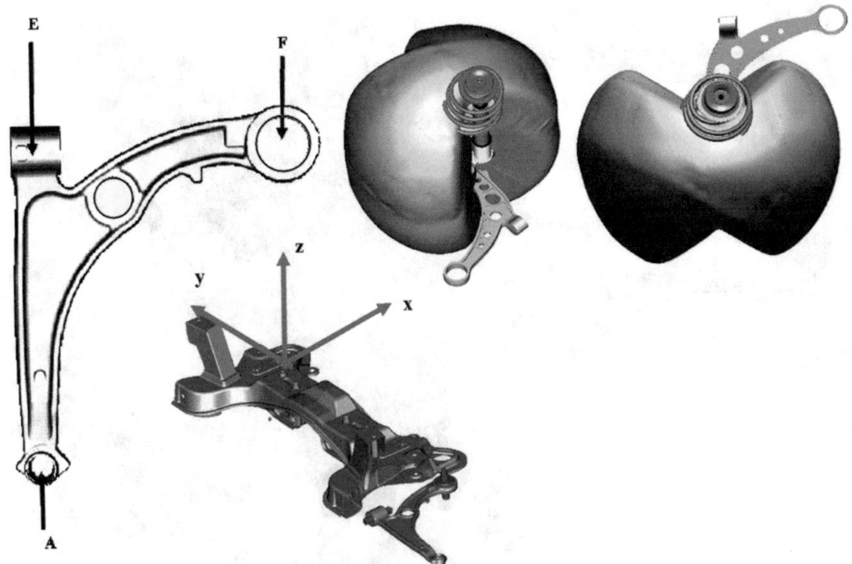

Fig. 3.15 Lower arm must allow both stroke and steering motions of the wheel, as shown by the envelope shape on the right; articulations are made by spheric joints or elastic bushings according to their function

The lower arm oscillates around the axis through the two bushing centers (points E and F of Fig. 3.15; remember the position of this axis with reference to the body). From a kinematic point of view the A point trajectory is not affected by the position of points E and F along their axis.

The axial position of bushing is, instead, relevant for the magnitude of the reaction forces in points E and F. Bushings are not rigid spherical joints but are made with elastomeric material and are deformed according to the amount of the force they receive.

The shape of the bushing is governing its flexibility: as we have already said suspension displacement under the applied loads are relevant for handling and comfort behavior of the vehicle.

This kind of considerations have led car manufacturers to design lower arms in a banana like shape (Fig. 3.16).

Because of this shape (Fig. 3.17, at left) the most important part of the lateral force applied to the A point is loading bushing E. This last must be designed very stiff as to limit camber variation (opposite to recovery of camber) under the effect of such forces.

For the same reason the moment of longitudinal forces, around point E, is loading transversely the bushing F.

Bushing F must show a high level of flexibility. The lower arm can therefore rotate around the rigid bushing E. As a consequence, when the wheel is affording an obstacle it can have a compliant back displacement, with beneficial effect on comfort.

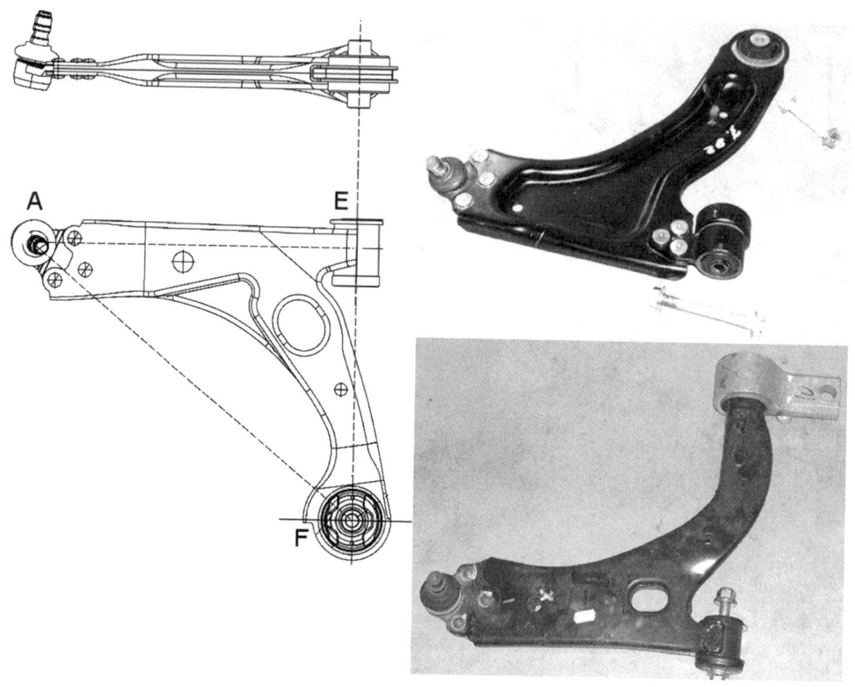

Fig. 3.16 The banana like shape of the lower arm of the Mc Pherson suspension is the result of a compromise between handling and comfort

But point A determines also the king-pin axis position, together with the upper pivot. To limit the toe angle variation coming from the wheel longitudinal motion, steering arms 7 should be positioned in a suitable way, to shape a parallelogram like on Fig. 3.17, at right.

On some cars lower arms are shaped like a isosceles triangle, where point A is the vertex; in this solution bushings E and F are almost equal as far as dimensions and stiffness are concerned. Also in this case the target is to obtain a good longitudinal flexibility, without affecting toe angle variations negatively. This scheme must be adopted with a different shape of subframe, that is positioned in a more advanced position as for banana lower arms.

Lower arms can be made by cast iron, by hot stamped steel or by cold stamped steel sheets; if weight reduction is a priority also stamped aluminum can be applied.

On Fig. 3.18 two examples are shown: one on cast iron and one on stamped steel. The iron technology allows to reach minimum cost, but these components feature reduced mechanical characteristics as far the rupture percentage elongation is concerned.

Lower arms made by stamped steel sheet are built welding together two semi shells. In this case the spherical joint housing has to be riveted, while bushings hous-

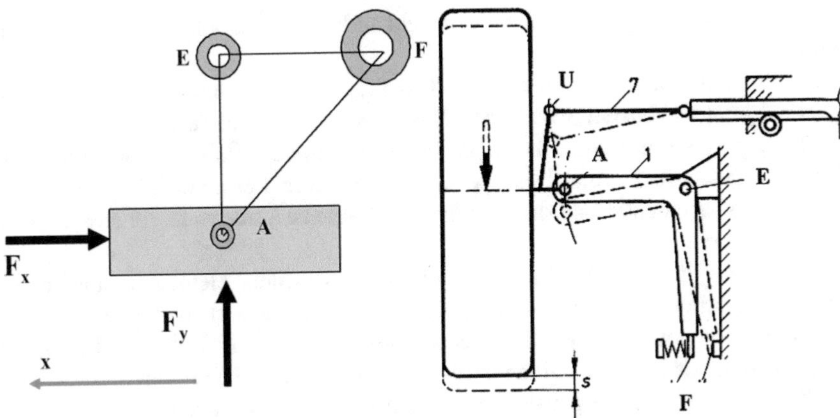

Fig. 3.17 Most part of lateral force through A is applied to point E; this bushing is designed as very stiff, to limit camber variation, because of the external load. Bushing F is designed to be flexible to allow the arm to have a compliant displacement when wheels are running on obstacles

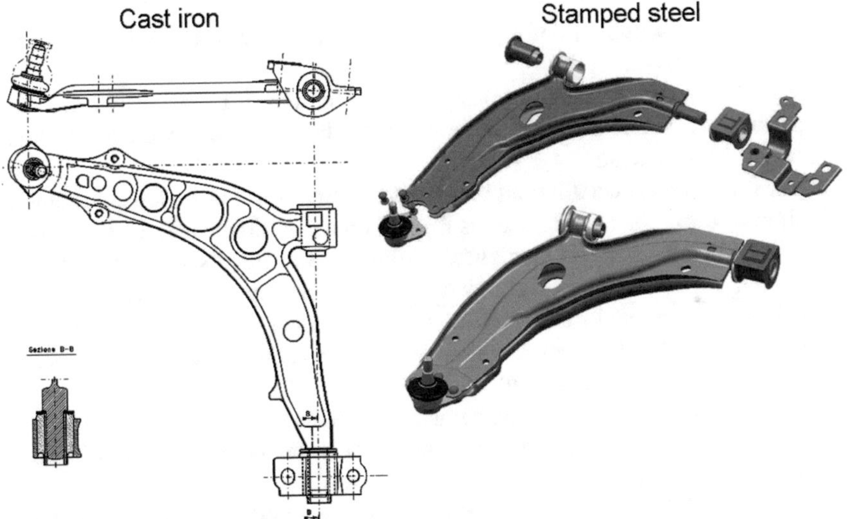

Fig. 3.18 Examples of lower arms made by cast iron or stamped steel sheet, in two shells to be welded together

ings show local reinforcements. This solution is characterized by reduced weight, better rupture elongation, but slightly increased cost.

Figure 3.15 shows an example of hot stamped aluminum arms; the shape of a hot stamp steel arm wouldn't show a very different shape, but for the notable reduction in cross section. With this technology very good mechanical properties can be obtained at a higher cost; they allow notable weight reduction, the aluminum solution in particular.

All this solutions must be machined to prepare the mounting interfaces with spherical joint and bushings with the necessary dimension tolerance.

The importance attributed to the percentage rupture elongation comes from considering accident shocks that the suspension might afford. Let us consider for example the case of an accidental side shock of the wheel against a side walk, for an excessive speed in a curve.

In this case the lower arm has the function of a sacrificial element, defined by its rupture properties, which include load, deformation and absorbed energy.

It should be noticed that in some cases the anti roll bar is directly linked to the lower arm; in this case a suitable interface on the arm must be provided.

During asymmetric suspension stroke, the lower arm is loaded also by bending moments, coming from anti roll bar, that intrust also bushing and spherical joint.

Elastic Bushings

Elastic bushings with their deformation allow motion of arm with reference to the car body or subframe.

They are made by an outside body 3 and inside body 1, respectively fixed with the suspension arm and the body or the subframe; between these two bodies a layer of rubber 2 is vulcanized.

Figure 3.19 shows two different installation alternatives. On the left side solution A is shown, where the outside body is flanged on the subframe and the inside body is press fitted on the arm. On the right side solution B feature the outside body press fitted into the arm and the inside body flanged to a fork into the subframe.

Rubber thickness and mix composition are designed to be able to differentiate stiffness according to the load direction. Vulcanized rubbers used in these applications must have a very high fatigue limit.

We have seen in the previous paragraph that bushings must be designed to grant the suspension a certain elasto-kinematic behavior, as far as toe angle and wheelbase (longitudinal displacement) variations are concerned.

Bushings with high stiffness will be finalized to a good road holding or handling behavior. In this case the layer of rubber between the two steel bodies will have a few millimeters thickness and the bushing outside diameter will be in the range between 35–45 mm.

The need for a high flexibility in bushings finalized to comfort performance requires a higher rubber thickness; therefore the outside diameter of the bushing could reach the value of about 60 mm.

Figure 3.20 shows, as an example, the elastic characteristics of the more flexible bushing of a Mc Pherson suspension. As diagrams can show, the stiffness in the x direction and in the y direction are very different; to this purpose the rubber body is not made like a solid ring, but presents a connection, between the inside and outside

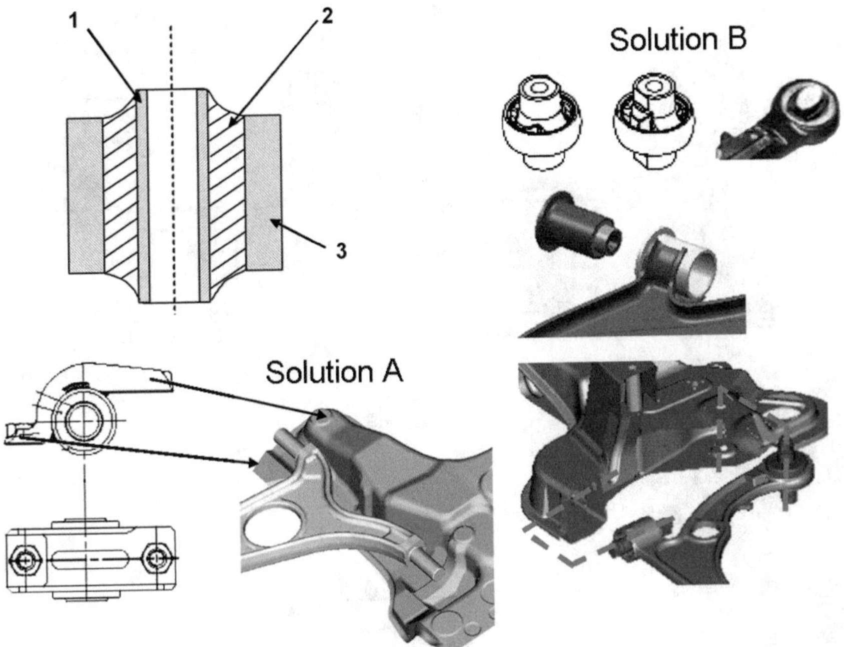

Fig. 3.19 Elastic bushings are made by an outside body 3 and inside body 1, respectively fixed with the suspension arm and the body or the subframe; between these two bodies a layer of rubber 2 is vulcanized. Note the two different installation solutions

bodies, made like a beam, bent in the y direction and compressed in the x direction; two stop springs made of rubber limit the bending deformation.

In a bushing have to be defined the following values of stiffness:

- radial stiffness along the two x and y directions of the body reference system, when the bushing internal diameter axis is, as in this case, directed to the z axis;
- axial stiffness, along the z axis;
- conical stiffness, measured by imposing to the bushing axis assigned angular deviations, with reference to the zero load condition;
- torsional stiffness, measured by imposing relative angular rotation of the steel body around their axis.

This values of radial stiffness must be recorded on drawings, in addition to geometric and material informations, to allow production quality control; for this purpose is necessary to show these elastic characteristics with an allowed tolerance band.

Conical and torsional stiffness control is not as critic as the previous one, because their impact on elasto-kinematics is limited; these values must be small, in order to allow a free arm motion. It is, instead, important to verify that the arm motion, caused by the suspension stroke, are not affecting fatigue life or are not changing the elastic behavior of bushings at high deformation angles.

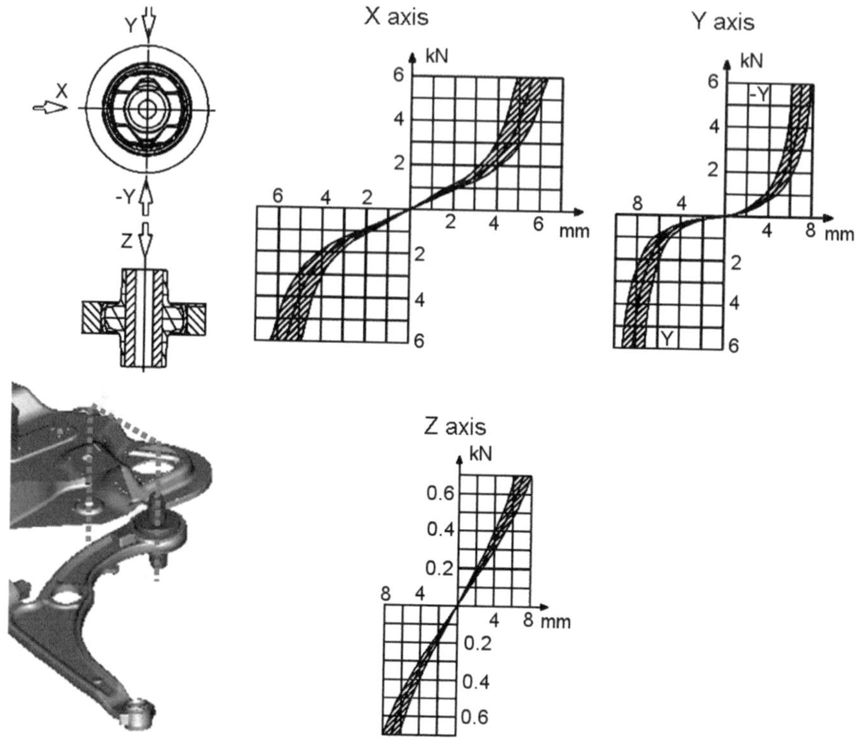

Fig. 3.20 The stiffness of the more flexible bushing are very different in the x and y directions; for this purpose the internal part of rubber is designed as a beam, being bent in one direction and compressed in the other, in order to have an asymmetric stiffness

Figure 3.21 shows, instead, the elastic characteristics of a stiff bushing, finalized to road holding and handling performance.

Many cars feature also hydraulic type bushings. These can add to the described elastic characteristics also some damping characteristic, by squeezing some quantity of oil through internal orifices.

This damping properties can have a beneficial effect in the field of comfort as far as harshness is concerned.

Spherical Joints

The articulation of suspension strut with lower arm is made by a spherical joint also called *steering knuckle*. Other spherical joints are used and the end of the steering bars, connecting steering rods with steering rack.

This kind of joint is very stiff in every direction but the radial one and allows very high rotations of the connected elements.

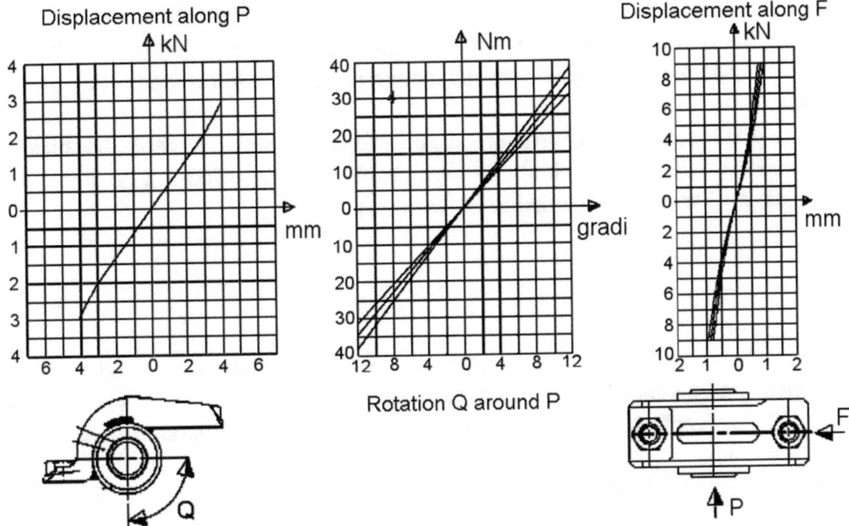

Fig. 3.21 These diagrams show the elastic characteristics of a stiff bushing, finalized to road holding behavior

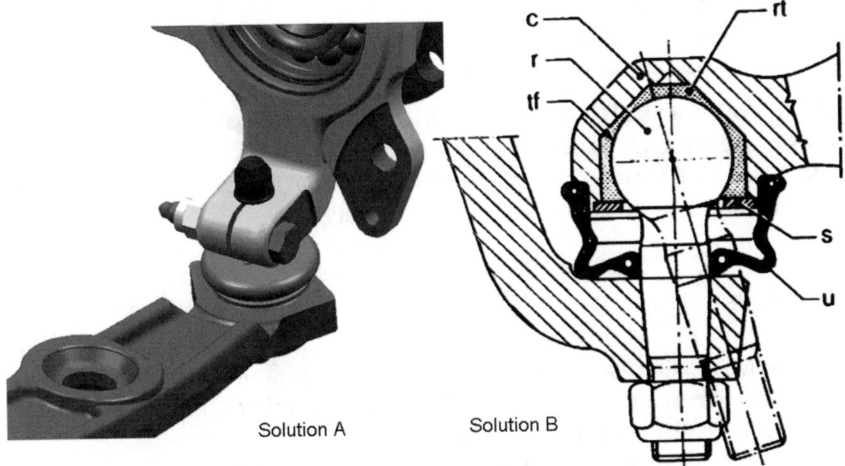

Fig. 3.22 The spherical head of the joint has a fixing pin. In the solution A a screw is used to press a cut hole in the strut around a cylindrical pin; in the solution B the pin is tapered and pressed into the strut hole by a traction nut

Figure 3.22 shows the spherical joint with its spherical head and a fixing pin. In the solution A a screw is used to press a cut hole in the strut around a cylindrical pin; in the solution B the pin is tapered and pressed into the strut hole by a traction nut.

The choice between the two alternative solutions has no precise rules, but is more due to tradition or existing tools for machining and assembling.

In both cases the material of the spherical cavity sliding surface is fixed into the arm directly; it is fixed inside the arm by rolling-in. A layer of pre-lubricated Teflon is set between the spherical head and its mounting cavity to lower the friction during rotation.

It is a good practice to prescribe on the drawings, in addition to geometric dimensions and materials, the maximum working angles to be allowed and the minimum radial stiffness which is assumed to be necessary.

Strut

The strut is connected with the majority of suspension components and transfers to them any force coming from the wheel. Its tasks condition its geometry and the severity of the applied loads characterize the strut by its squat shape with very thick cross sections.

We can identify the following functions of the strut (with reference to Fig. 3.9):

- to hold the wheel bearing in both radial and axial directions;
- to connect with the suspension arm through the spherical joint;
- to offer the fixing flange for brake calipers;
- to fix the brake cover;
- to fix the lower end of the spring and shock absorber unit;
- to connect with the steering arm with a spherical joint.

If the car is supplied by an ABS system or by a vehicle dynamics control system, the strut is also use for fixing the wheel speed sensor.

Materials used for this components include hot stamped steel and cast iron; some times also hot stamped aluminum is used on luxury cars.

Hub and Bearing

To allow hub rotation and to transfer the wheel forces to the strut, the hub is mounted with a double range ball bearing or a double tapered roller bearing; tapered roller bearings are applied with opposed conicity.

Figure 3.23 shows the hub assembly for a front wheel driven vehicle with ball or tapered roller bearings. This bearing configuration, specific to cars, is characterized by bearings with a single outside ring and double inside rings.

This is commonly called O configuration; the pre-load of the two bearings is determined by the play allowed between the two inside rings, that is brought to zero after mounting.

The solution used on most small and medium size cars provides a double range ball bearing, as we have seen until now. The hub pre load is obtained by tightening the nut of the hub at a certain torque. This kind of bearing is also called first generation.

Second generation bearings (Fig. 3.24) provide still two inside rings, but the outside ring is directly integrated into a flange to be screwed to the strut. This solution

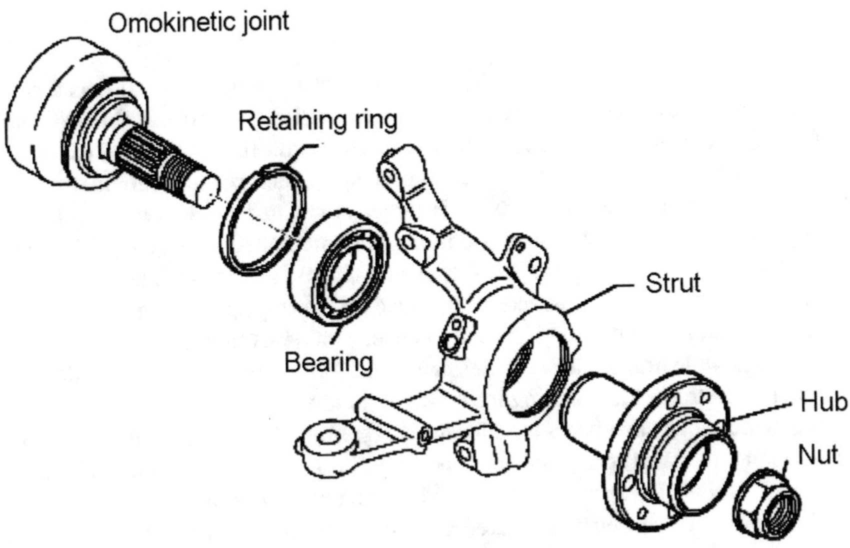

Fig. 3.23 Hub assembly of the front wheel for a front wheel driven vehicle

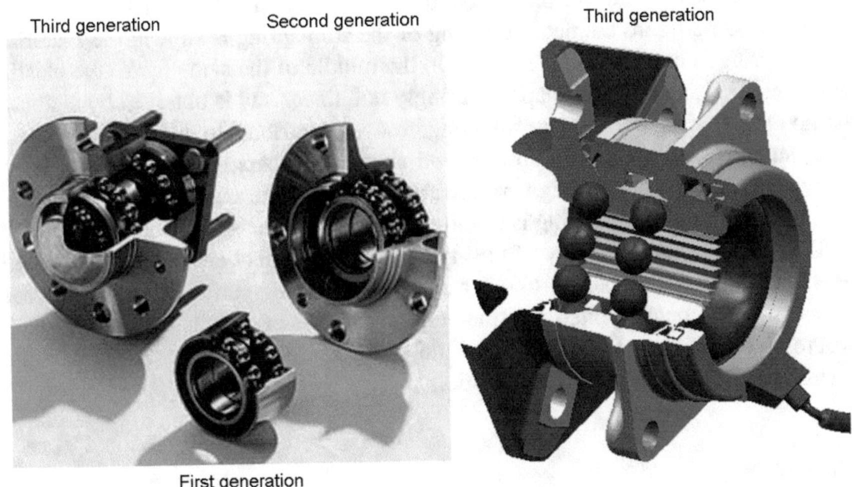

Fig. 3.24 Wheel bearings of first, second and third generation

simplifies strut machining operation to cut the seat for the bearing and reduces the assembly cost; on the contrary these bearing have a higher cost.

Third generation bearings feature a single inside ring, including the wheel hub, in addition to the mounting flange on the outside ring; this solution has the advantage of maximum reduction in space and weight and of having a sealed bearing; on the contrary the cost of the bearing is increased very much.

Coil Spring and Stop Spring

Coil spring alone would give the wheel an elastic characteristic almost linear, considering that the relationship between wheel stroke and spring stroke is almost linear.

As it will be explained in the chapter dedicated to comfort, in the second volume, a linear characteristic should be avoided, because spring deformation should decrease in proportion to the load increase. This target is reached by applying additional springs made of rubber, which in a limiting way are called *stop springs*, because their initial purpose was to limit the suspension stroke to a certain value.

The progressive spring characteristic, added by the stop spring, can keep the first natural frequency of the sprung mass at a constant value, even if the vehicle payload is changing, while limiting the compression stroke too; another stop spring is limiting the extension stroke and affecting the spring ratio too.

When the shock absorber is used for the coil spring installation, as in Mc Pherson suspensions, the stop spring too must be fitted into the shock absorber assembly. Figure 3.25 shows the example of a rear Mc Pherson suspension.

In case of front suspension differences will be minimal but for the lower mount. The compression stroke stop spring is coaxial with the shock absorber piston rod in the outside of the hydraulic cylinder; the extension stroke stop spring is much smaller and is again coaxial with the piston rod, but inside the hydraulic cylinder.

On the same figure the detail drawing of the stop spring is shown; the material of this spring is usually polyurethane. On the middle of the same figure the elastic characteristic is shown; the shape is strongly non linear and is obtained by shaping the tip of the spring like a cone. Also ring grooves contribute to this characteristic.

In suspensions where coil springs and shock absorbers are not unitized, stop springs can be separated; design criteria are in any case the same.

Extension stroke stop spring is usually always inside the shock absorber.

The main spring is usually of coil type; spring retainers are made by stamped steel sheet and are shaped as to offer a wide supporting surface to the coil; free coils can have constant pitch or variable pitch. In this case coils coming in touch progressively increase the total coil stiffness. The coil spring alone can feature the desired progressive elastic characteristic.

Anti Roll Bar

The roll angle resulting in a curve, because of suspension elastic characteristic given by normal springs, could be too much. In this case an additional anti roll bar is used.

This is an additional spring that increases suspension rigidity only by antisymmetric strokes, as in a curve, where one suspension is compressed and the other is extended. If both suspensions are moved for the same stroke, in the same direction, this additional sprig has no effect.

On our example, the anti roll bar (Fig. 3.26) is made by a bent tube connected to the subframe in two points. The tube ends are flattened and drilled, to be connected to a pendulum bar, linked to the shock absorber at the other end.

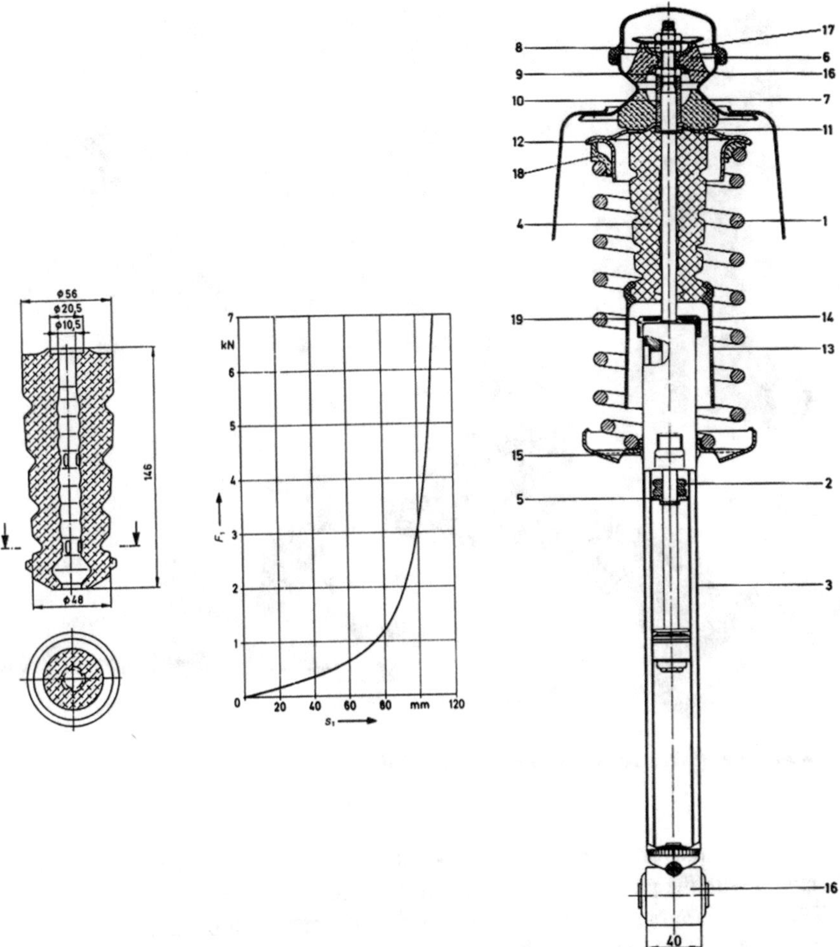

Fig. 3.25 Stop spring 4 interesting compression stroke is coaxial with the shock absorber piston rod, outside of the hydraulic cylinder; the extension stroke stop spring 2 is much smaller and is inside the hydraulic cylinder

During body roll the wheel outside the curve is moving to compression, that inside is moving to extension. The body motion torques the linear part of the bar which tends to limit the difference in strokes, limiting in this way the angle of roll (Fig. 3.27).

The pendulum bar limits the force exchange with the suspension to the anti roll torque and avoids bar stressing during symmetric strokes; this happens if the bar is directly flanged to the suspension arms, as on some small car.

In this last case, in fact, as showed by Fig. 3.28, the interface between bar and arm is asked to follow the circular trajectory of the arm and this fact introduces into the roll bar an additional bending stress; this arrangement has the further disadvantage

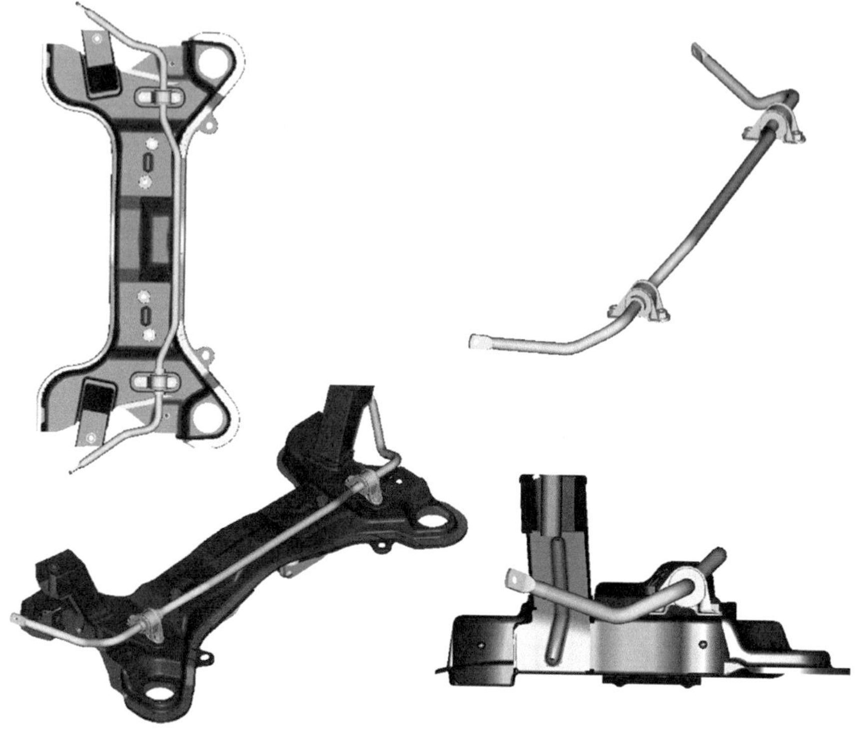

Fig. 3.26 Anti roll bar mounted on the suspension subframe

of having an unfavorable ratio between bar torsion angle and suspension stroke, as compared with the pendulum bar application, therefore reducing bar effect.

The pendulum bar can also be connected to the suspension arm, usually with a shorter length; in this case arm swinging should not introduce additional deformation to the anti roll bar but the pure torsion, in order not to affect comfort by symmetric strokes.

Figure 3.28 shows on section A-A also the elastic bushing mounts of the anti roll bar, to the sub frame; this mounts are common to any kind of bar.

Shock Absorber

Shock absorbers used presently on vehicles are only hydraulic; the damping force is obtained by squeezing with a piston a certain amount of oil through small orifices controlled by valves. The flow of the oil through the orifice occurs, at a given area of the flow cross section, with a certain pressure drop, almost proportional to flow rate.

Fig. 3.27 Anti roll bar connected to the shock absorber through a pendulum bar

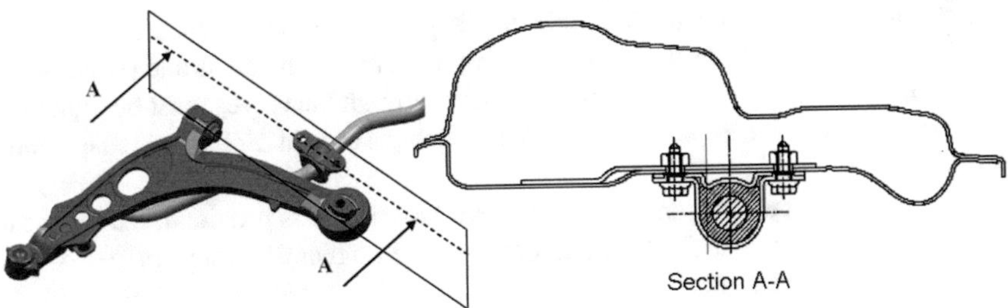

Fig. 3.28 Anti roll bar directly fixed to the suspension arm

Shock absorber can be classified as *structural* type or *conventional* non structural type, according to their capability of supporting loads perpendicular to the piston rod.

The first type is used on Mc Pherson suspensions.

The shock absorber is fit to the suspension strut and is carrying on it part of the loads coming from the contact point of the tire; these loads generate on the piston rod shear forces and bending moments as we have seen on Fig. 3.11. They have a consequence on piston bar diameter and sliding bushings.

The second case can be found on almost all remaining types of suspension, where shock absorbers receive forces only along its piston rod.

Shock absorbers, finally, can be unitized with the coil spring or can be separated; there are Mc Pherson suspension where the two elements are separated, as there are other type of suspension where they are unitized, with benefit for installation space and assembling easiness.

On conventional shock absorbers the only load is coming from the internal pressure that loads, by definition, the piston rod.

It should be noticed, however, that also in this case transversal forces could be applied, causing hysteresis increase and additional stress to the piston bar; this situation may occur when, because of the stroke of the suspension, the shock absorber is changing its inclination. In this case, the torsional or conical stiffness of elastic bushings, connecting the shock absorber to the suspension and the car body can introduce transversal forces; this can also happen on sliding bushings, because of their internal friction.

Figure 3.29 shows on the left a scheme, useful to understand the operation of a hydraulic shock absorber. The upper part of the shock absorber is fixed to the sprung mass, while the lower one to the unsprung mass; if the shock absorber has no structural function the joints should be equivalent to a spherical one.

During its stroke the suspension is moving the piston rod 1 and the piston 2, inside the cylinder 3. The volumes outlined with broken lines A, B and C are filled with oil; the piston motion, because of the incompressibility of the oil, is squeezing a certain flow rate from chamber A, to chamber B, through the valves V_1 and V_2.

From this flow a damping force is generated; it should be noted that valves V_1 and V_2 are outlined as check valves; in this way is possible to obtain different damping coefficients during compression stroke (valve V_1) and extension stroke (valve V_2).

The oil pressure generated by friction losses through the valves must be tightened by the seal 4, to avoid oil spillage; this seal is causing a certain friction and suspension hysteresis.

To improve seal effectiveness an additional labyrinth 5 is provided, reducing the pressure acting on the seal to a value close to the ambient pressure of a chamber connected to the holes 6.

The penetration and the extraction of the piston rod must be accompanied by a displacement of an oil volume equal to what was occupied or freed by the rod.

In our scheme, volume C, with free surface, contained in the cylinder 7 provides to this purpose, through the check valves V_3 and V_4, communicating with the volumes A and B. The protection tube 9, avoids water or dust contamination on the rod; this protection can be improved with a rubber bellow, not shown on this figure.

This type of shock absorber is also called *double tube* because of the double outside cylinder. Two chambers are therefore available (inner and outer) containing

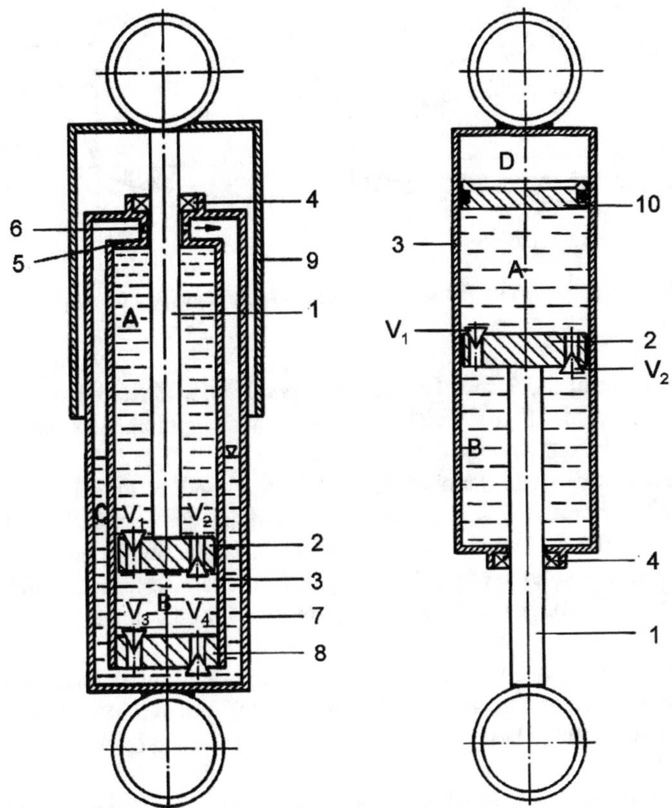

Fig. 3.29 Hydraulic scheme for understanding the operation of a double tube (left) and mono tube (right) shock absorber

oil. The piston with the valves assembly enter the inner chamber, a second valves assembly separates the inner from the outer chamber.

In the lower part of the outer chamber, a quantity of pressurized gas compensate for the oil volume variations because of the piston motion. The inside pressure of the compressed gas avoids cavitation near the valves, where pressure drops are present.

Valves assemblies have different functions, according to the direction of flow. During suspension extension the piston valves control the damping force, while lower valves allow the compensation flow between inner and outer chambers. During suspension compression the lower valves have the damping function, while those on the piston allow an almost free flow.

Some of the orifices are closed by pre loaded valves which open when pressure drop goes over a certain limit. In this way the damping force can be limited for high compression or extension speeds.

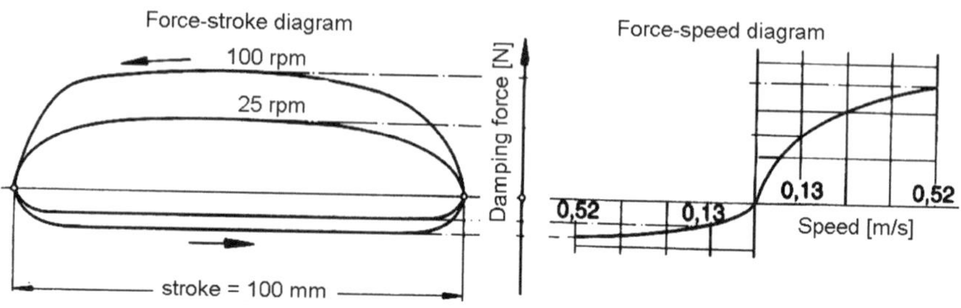

Fig. 3.30 Diagram force-stroke of a shock absorber and conversion to a force-speed diagram

A second kind of shock absorber is called *mono tube*, with reference to the scheme on the right of the same figure. This kind of shock absorber is longer, but with a smaller diameter as compared with the double tube one.

In this case only two oil chambers are available, A and B on the scheme. The valve assembly on piston 2, with rod 1, controls the damping force in both directions.

Particularly the check valve V_1 controls the compression force, while V_2 the extension force. A volume of pressurized gas D is separated from the oil through a floating piston 10. This gas volume is compensating for oil volume variations due to piston rod motion.

The advantage is a simplification of the design, while the disadvantage is an increase of pressure on the seal 4, with potential leakages and increased hysteresis. The pressure in chamber D applies to the suspension a non negligible force, in parallel to coil springs.

The force-speed characteristic of a vehicle shock absorber can look like Fig. 3.30 at right; on the same figure, on the right, also a force-stroke diagram is shown, for two different speeds. This last diagrams refers to a simple test machine normally used to evaluate shock absorber; this machine moves the shock absorber through a crank, rotating at constant speed, from one side and holds it on the other side, through a load cell.

In this machine the stroke is constant (usually 100 mm) and the speed can be set at different values; the shock absorber extension and compression speeds are, therefore, not constant. The cycle shown in the diagram represents the extension force in its upper side and the compression force in the lower one.

The diagram on the right of the figure is derived by the first, calculating for each point of the first the real compression or extension speed. The line is interpolated in the clouds of points that are obtained in this way. The extension of this cloud in the force direction is determined by the shock absorbed friction, mainly due to seals. This machine doesn't show correctly the effect of side forces on the rod.

The asymmetry of the diagram can be controlled by adjusting the valves.

The internal gas is giving to the shock absorber a certain pre load, obtained by multiplication of the internal pressure by the rod cross area; the polytropic low of gas justifies also a variation of this force with the stroke

The following figures regard other details of shock absorbers.

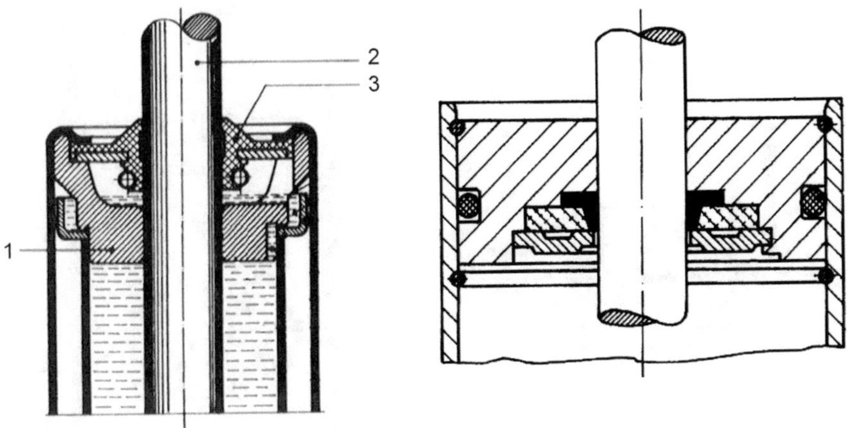

Fig. 3.31 Seals on a double tube (left) and a mono tube (right) shock absorbers

Figure 3.31 shows on the left the detail of the seal on a double tube shock absorber.

The seal is made by two parts: the body 1, which guides the rod 2, and the double lip seal 3. The body 1 shapes a labyrinth between the pressure and the compensation chambers.

The oil spilling from the labyrinth is conveyed to the bowl on the body 1 and doesn't leak to the outside because of the elastic double lip ring; this seal is not very stressed and the lib elasticity can be reduced to have a consequently reduced friction force.

The oil reservoir in the bowl can take few days do empty into the lower chamber, because of the unavoidable spills of valves V_3 and V_4; consequently the danger of having air bubbles in the pressure chamber is reduced. Air bubbles can introduce clicking during the first suspension strokes after a long stop.

The same figure on the right shows the seal of a mono tube shock absorber.

The seal is in this case single and is made by a ring of particular shape made by elastomers, which surrounds the rod tightly; seal is effective but hysteresis is increased significantly.

Figure 3.32 shows two different valves assemblies: one on the bottom (left) and one on the piston (right) for a double tube shock absorber; in a mono tube one differences are not substantial.

The valve on the bottom is working during the compression stroke. The check valve features a bevel disc spring, on the lower side, which opens the holes 2, cut on the bottom; spring rigidity and hole diameters control the damping coefficient at low and high speed respectively.

The valve 3, on the upper part of the bottom has a very flexible spring and hole 4 very large. The necessary pressure for oil return must be, in fact, low; this valve is working during extension stroke, to return to the A chamber the oil necessary to replace the rod volume.

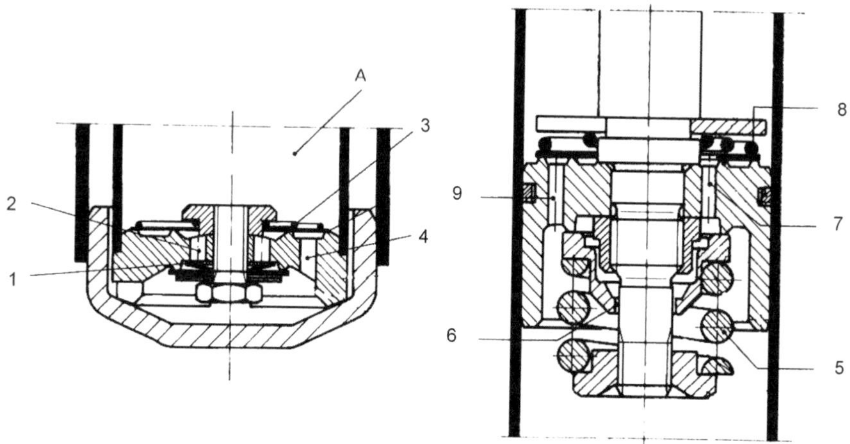

Fig. 3.32 Valves assemblies of a double tube shock absorber

The valve on the piston (at right) is controlled, during the extension stroke by the coil spring 5, which allows the lift of the disc 6, opening holes 7 cut on the piston; also in this case the spring is controlling low speed force, while the high speed value is controlled by holes diameter.

The upper valve, closed by spring 8 is working during oil transfer only, through holes 9, to compensate for the oil moved by the rod.

Springs and valves bodies are easily interchangeable, to build shock absorbers for different applications on the same production line; during final stages of a new car development the choice of dimensions of valves assemblies is usually made empirically, according to a trial and error process, starting from first approximation values, obtained by mathematical models.

The characteristic curve of the damping force versus speed can be classified, with reference to its shape, as:

- *constant*, when the derivative of force, with respect to extension or compression speed is almost constant but, usually, with different values in compression and extension;
- *progressive*, when the derivative is increasing with speed;
- *degressive*, in the opposite case, when the derivative is decreasing with speed.

The derivative of the force with respect to speed is also called *damping coefficient*; a constant damping coefficient appear ideal according to the Boursier De Carbon theory, we will explain in the second volume; a progressive characteristic could be suitable to improve handling quality of a sport car, with some sacrifice on comfort; a degressive characteristic can, instead, improve comfort on bumpy roads, where suspension stroke speed can be high.

The rubber bushing between shock absorber and car body is used to filter dynamic forces over 30–40 Hz of frequency, where shock absorbers are loosing their capacity

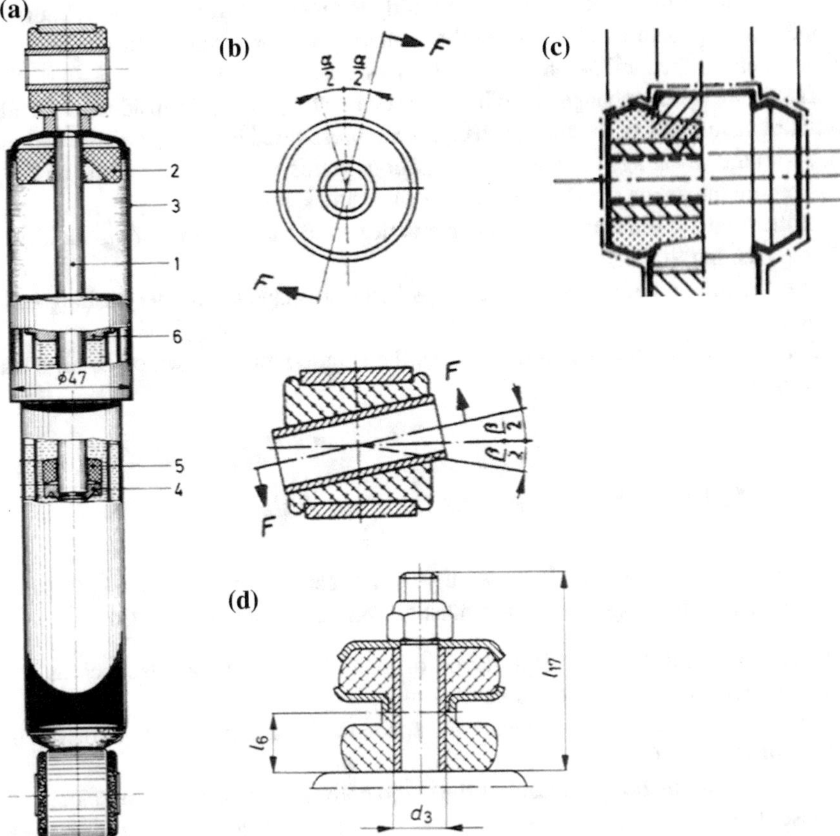

Fig. 3.33 Kinds of body and suspension mounts for a non structural shock absorber; the conical stiffness of these mounts can affect shock absorber friction

significantly; these vibrations cause noise which is diffusing to the car interior through the body structure.

On Mc Pherson suspensions, the design of this bushing integrated within the upper pivot can be made in two ways: simple or double.

In the simple configuration the shock absorber is loading the body together with the coil spring and the stop spring; being the load significant rubber must be well sized and is usually stiff.

With the double configuration, a dedicated rubber bushing is carrying the shock absorber load only. Static load is negligible, except for the gas pressure effect, if any. In this case the rubber can be lighter, with improvements in flexibility and a better filtering capacity.

For non structural shock absorbers the body mount is made by a simple cylindrical rubber bushing (detail a of Fig. 3.33); the symmetry axis of bushings must be oriented to make easier the inclination changes due to suspension stroke.

If the value of this change of inclination is contained ($<10°$), the inside and outside cylindrical surfaces of the rubber bushings are press fitted between the hole and the pin, with the advantage of avoiding lubrication; in this case the shear of the bushing is stressing the piston and the piston rod of the shock absorber.

If the angle is greater, it is a better practice to use a self lubricating additional sliding bushing to reduce hysteresis.

If other rotation angles are present it is better to shape the rubber bushing as in detail c.

For very limited rotation angles ($<5°$), the simple mount shown on detail d can be adopted.

3.2.2 *Mc Pherson Suspensions for Rear Axle*

As we have already said Mc Pherson suspensions can be applied also to rear axles.

In general, a rear axle suspension must perform the following functions:

- to be compatible with the installation of many bulky components, such as fuel tank, spare wheel, silencer and exhaust pipe;
- for rear wheel drive, must provide the space for differential and final drive and transmission shafts;
- must allow to the body a space for the cargo compartment as big as possible;
- must allow a suspension stroke longer as for the front axle, because the rear axle load variations are larger.

Fortunately the wheel positions envelope is smaller because steering angles are limited to the much smaller toe angles variations.

The Mc Pherson configuration allows a better elasto-kinematic performance, as compared with the other typical solution for rear trains we will see later on, with some sacrifice on cargo compartment width and with component cost.

Figure 3.34 shows the rear suspension of a medium size passenger car.

The kinematic scheme of this suspension can be derived from that of the front axle suspension by changing the steering link with a bar 1 connected between the strut 2 and the subframe 3. The cross bar 4 is for cornering forces, while the bar 5 is for longitudinal forces; together they work as the lower arm of the front axle suspension.

Also in this case the application of a subframe doesn't affect, in first approximation the elasto-kinematics of the suspension.

The spring-shock absorber unit 6 is articulated to the body to a dome in the wheel case. The anti roll bar 7 is supported by the subframe and is linked to the shock absorber with a pendulum rod 8.

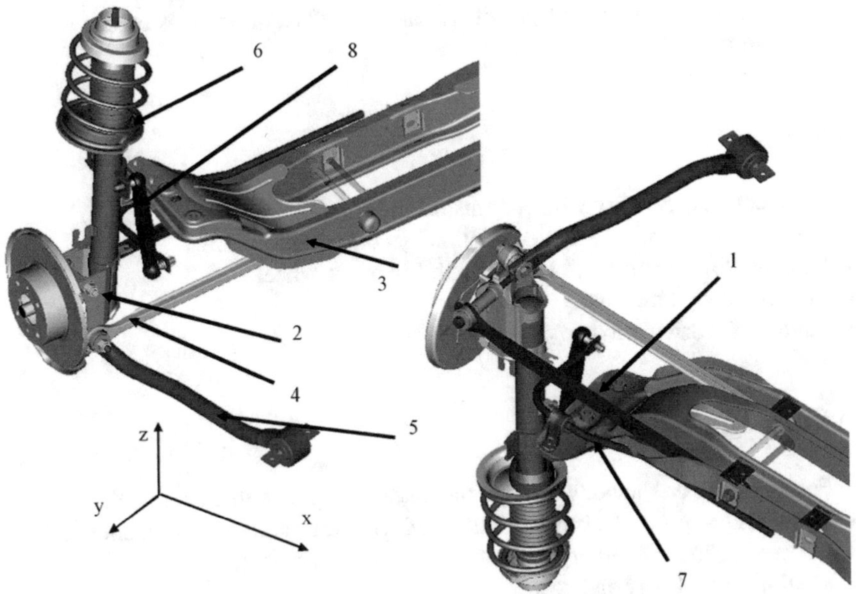

Fig. 3.34 Rear Mc Pherson suspension for a medium size car with front wheel drive. Can be derived from the same type of suspension for the front axle, by imagining to change the steering rod with the lever 4 (FCA)

The appearance is quite different from that of the front wheel suspension even if the kinematic scheme is similar and the number of links is the same.

The three links 1, 4 and 5 lie on the same plane; links 4 and 5 have the same function of an arm with virtual center; in the same way link 1 is like a steering bar.

Either because of the mutual position with the wheel and for the higher stiffness of their bushings, the bar 4 will absorb a high percentage of cornering forces.

Because of its orientation, the bar 5 will absorb a high percentage of the braking forces and other longitudinal loads; its bushing connecting with the body could be conveniently flexible for a better filtration of the longitudinal shocks, originated by obstacles, without affecting toe angles, because the steering mechanism is again an articulated parallelogram.

Finally, the flexibility of the bushing of the articulation with the body of bar 1 can be finalized to correct the understeering behavior of the car, under the action of cornering forces. In conclusion, each of the elastic bushings can be designed for a single performance, without negative consequences on the remaining ones.

There also examples of rear Mc Pherson suspensions that are more similar to the front axle type; in this case, the strut rotation, causing the wheel steering is avoided with a steering bar fixed, in the inner end, to the body, called usually *false steering bar*; the articulation position of the fixed end can be designed to obtained the desired toe angle variation.

Advantages and disadvantages of rear axle Mc Pherson suspension can be summarized in the following list.

Advantages

- Good camber recovery with suspension stroke.
- Potential for proper toe angle variation by cornering forces.
- Potential for proper longitudinal flexibility of the wheel.
- Reduced unsprung mass.
- Suitable for driving axles.

Disadvantages

- Forces acting on the body through the dome are applied to a flexible point.
- The coil spring-shock absorber unit is reducing the cargo compartment width.
- Structural shock absorber with potential hysteresis problems.
- Medium complexity and cost.

For its characteristic is used on medium size high segment cars or on sport cars.

3.2.3 Double Wishbone Suspension

Double wishbone suspensions are applied to luxury sedans and sport cars, because they allow a design of the elasto-kinematic parameters that is very suitable for an optimum compromise between handling and comfort performance.

Since the upper arm is in general shorter as the lower one, to allow a given camber recovery, they are also called short and long arm suspensions, with the acronym of SLA suspensions.

Description

Figure 3.35 shows two different kinds of double wishbone suspension; the first one, at the left, is easily adaptable to front wheel driven cars with transversal powertrain but, as the second at the right, can also be applied to longitudinal powertrains.

If we compare these suspensions with the Mc Pherson ones, we can notice the upper arm, which carries on the function of the shock absorber sliding bearing. The shock absorber has no structural functions, because static and dynamic forces don't load it, but the damping force.

Hysteresis and the consequent penalty on comfort is therefore limited.

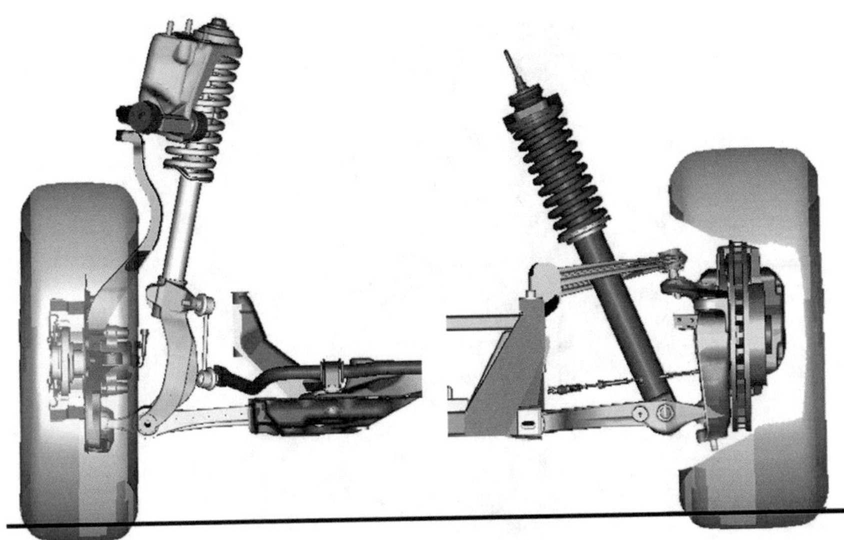

Fig. 3.35 Front wheels double wishbone suspensions, of the high type, on the left and of the low type, on the right; only the high double wishbone suspension can be easily applied to transversal powertrains

According to the upper arm position with respect to the wheel, suspensions of this type are classified as high (on the left) or low (on the right).

The difference between these two solutions is imposed by the transversal bulk of the engine or differential, in case of rear wheel driven cars, and by the track that must be obtained for the car; as a matter of fact, the high double wishbone suspension is the only one that can be adopted on front wheel driven car with transversal engine and is applied to other cars also, because, sometime, it suits the structure of a unitized body very well. Low double wishbone suspensions, featuring a very good elasto-kinematic performance are very intrusive for the engine compartment and are practically applied only to luxury touring cars.

Figure 3.36 shows a high double wishbone suspension in detail; we refer to what we have said about the Mc Pherson suspension for many common details, as steering bars, subframe, steering mechanism, anti roll bar and elastic bushings.

The upper arm 1 and the lower one 2 are linked to the body through elastic bushings. The same arms are linked to the strut 3 through spherical joints 4 and 5 which allow the steering rotation of the strut. The line connecting the two spherical joints is the king-pin axis.

The shock absorber and the coil spring are unitized as in the Mc Pherson suspension; this assembly is connected to the lower arm through the elastic bushings 7 and is shaped like a fork, to clear the bulk of the half shaft.

The shock absorber piston rod is connected to the body through an elastic bearing; in this case there is no pivot bearing, because the steering motion interests the wheel

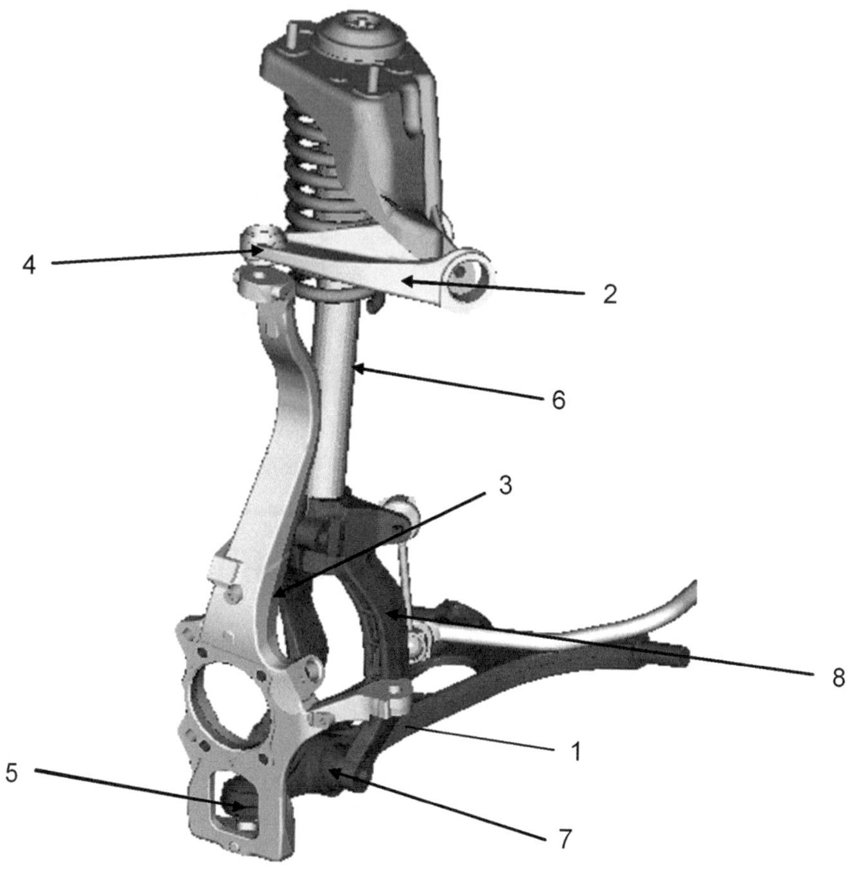

Fig. 3.36 The lower arm 1 and the upper one 2 are linked to the car body with rubber bushings. These arms are also linked, at the other end, to the strut 3, through spherical joints 4 and 5. The fork arm 8 connects the lower arm with the shock absorber (FCA)

strut only. A rest includes upper arm bushings and shock absorber bearing and is bolted to the body.

On other applications this rest can have a different shape or can be avoided, being bushings and bearings directly linked to the car body; this solution is justified by standardization reasons of the body, that can be produced in the same plant together with other bodies equipped with Mc Pherson suspension.

The lower arm is connected to a subframe. The torsion bar is connected to the shock absorber through a pendulum rod. As in previous suspension, the subframe is used for the steering rack installation too.

Advantages and Disadvantages

Advantages

- Optimum design of elasto-kinematic parameters, particularly as far as camber recovery is concerned.
- Shock absorbers have no structural function; comfort can be improved, because of hysteresis reduction.
- Possibility of lowering the hood profile, particularly for the low version.

Disadvantages

- Production costs are higher, because of the increased number of parts; an arm an its bushings are added, as compared with the Mc Pherson suspension.
- Additional parts for upper suspension attachment.
- The space taken by the upper arm is notable. Transversal engines oblige to adopt the high version: the reduced length of the upper arm of this version compromises the possibility of reaching best elasto-kinematic performance.
- The increased number of joints and bearings can affect wheel angles because of bushing rubber permanent deformations, with negative consequences on tires wear.
- The high value of braking loads can have a negative influence on longitudinal flexibility.

Design Details

On high wishbone solution the strut is characterized by a gooseneck shape, if observed on an yz plane. This shape is conditioned by the tire bulk and by the upper joint position, with reference to the engine bulk. This thin and highly stressed component is made usually by hot stamped steel.

The example on Fig. 3.36 shows an arm inclination in the xy view; this is only due, again, by the transversal position of the powertrain.

Figure 3.37 show alternative design of a low double wishbone suspension, as far as spring and shock absorber position is concerned.

Solution A is characterized by having shock absorbers and spring applied to the upper arm.

Solution B shows a spring made by a torsion bar and the shock absorber applied to the lower arm: this solution allow a very limited hood height and introduces limited values of stress in upper body parts. The same result, on height only, can be obtained by applying a unitized coil shock absorber, if the space is sufficient.

A last variant of this suspension can have, as springing element, a transversal left spring, which can integrate the function of one of the arms too; the distance between the bearing of the leaf spring can be designed as to obtain the desired symmetrical

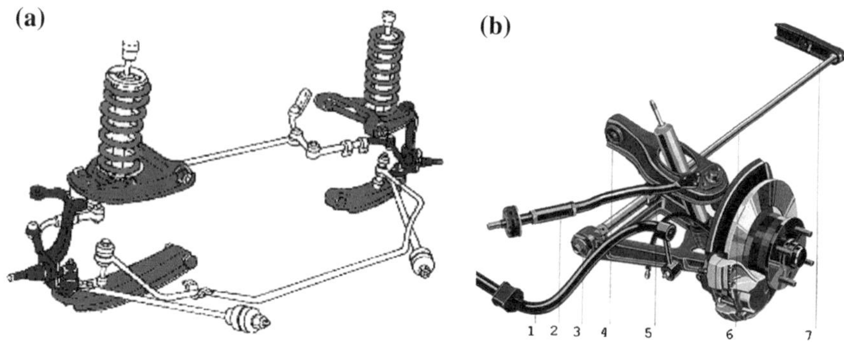

Fig. 3.37 Different alternative positions for the coil spring and of the shock absorber for a low double wishbone suspension

and asymmetrical flexibility without applying any anti roll bar. This solution is now rarely applied because of the hysteresis penalties of the leaf spring.

The high wishbone solution can be easily adapted to pneumatic springs, particularly suitable to high class sedans and SUVs. With this kind of suspension the vehicle trim can be easily changed according to the kind of the road; for instance the clearance between the body and the ground can be set high in off-road driving or low in motorway driving, to reduce the drag coefficient; in addition, the clearance can be kept constant independently of the vehicle load.

An example of this application is shown in Fig. 3.38; we can note the very simple shape of the spring assembly connecting to the lower arm, in this case goose neck like, to clear the bulk of the half shaft.

The diameter of the pneumatic spring is contained to comply with the space available in the upper arm; as can be seen in Fig. 3.39, the lower part of the pneumatic chamber is tapered, to provide an increasing stiffness with the compression stroke.

We should remember that the stiffness k of an air spring is not constant as in coil springs or torsion bars; the last two springs feature a constant stiffness with the spring stroke s, while the first features a stiffness that can change with the stroke. The reasons are two:

- The shape of the piston area A_p can change with the stroke as in the example in Fig. 3.39;
- The pressure p is changing with the inside volume of the pneumatic chamber; this change is dependent on the heat that is transferred from the chamber to the environment. Fast changes can be supposed to be adiabatic, as in case of springing motion, while slow changes can be supposed to be isothermal, as in case of imposed trim adjustments.

The stiffness of a pneumatic spring at a given position in its stroke can be written as the derivative of the spring force F with respect to the stroke.

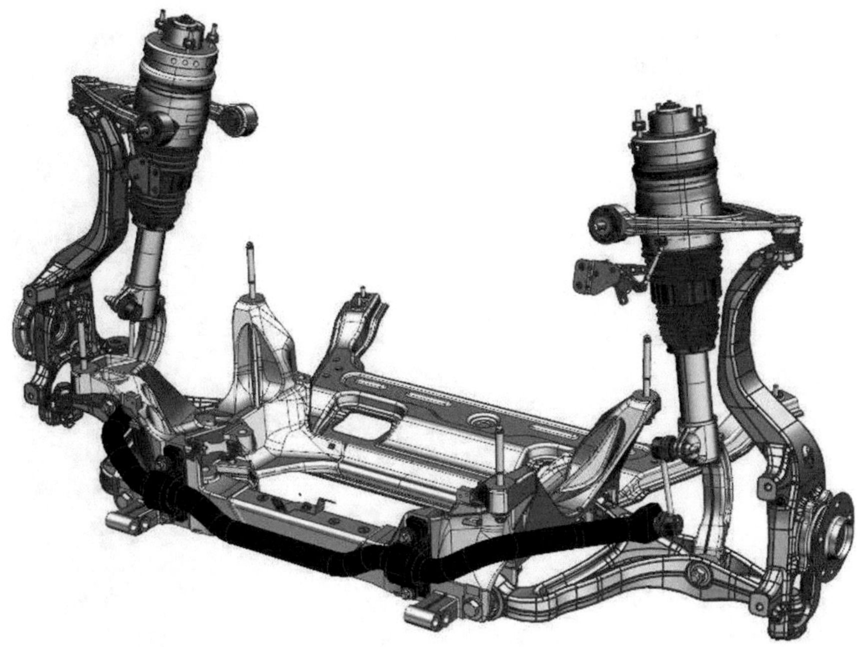

Fig. 3.38 Example of double wishbone front suspension with pneumatic springs suitable for a high class SUV (FCA)

$$k = \frac{dF}{ds} = \frac{d(A_p p)}{ds} = p\frac{dA_p}{ds} + A_p\frac{dp}{ds} = p\frac{dA_p}{ds} + A_p^2\frac{p}{V}n,$$

where V is the volume of the air chamber and n is the exponent of the volume in the gas equation ($n = 1$ for isothermal transformations, $n = 1$ for adiabatic transformations). The first part of the last equation is depending from the piston shape only, while the second part in increasing as the volume decreases because of the spring compression stroke.

Designing a particular shape for the lower part of the pneumatic chamber can make a pneumatic spring constant (stiffness constant with stroke), progressive (stiffness increasing with stroke) or degressive (stiffness decreasing with stroke), as shown in Fig. 3.40; changing inflation pressure is a tool for a further adjustment of the stiffness.

To contain the bulk of the spring assembly, pneumatic chambers are usually made with a thin layer of elastomeric material, protected by a steel or aluminum capsule, in car suspensions, while a kevlar reinforced elastomeric bellows without protection is used in truck suspensions where there are no problems of space.

Fig. 3.39 The air spring contains inside the shock absorber and the stop spring to comply with the space available in the upper arm

3.2.4 *Virtual Centres Suspensions*

On suspensions we have examined up to this point, the strut is linked to two spherical joints, which identify king-pin axis. In many cases king-pin axis can't be set in a position suitable to obtain a desired value (limited or negative) of the king-pin offset, without setting the wheel with an undesired camber angle.

Some manufactures solve this problem, by adopting, particularly on high class cars, single or double *virtual center* links for the suspension strut; in this case one or two arms are substituted by a double number of linkages, each of them with its spherical joint. The king-pin axis is no more identified by the physical position of a joint, but by a virtual point given by the intersection of the two lines connecting the articulation points of a linkage.

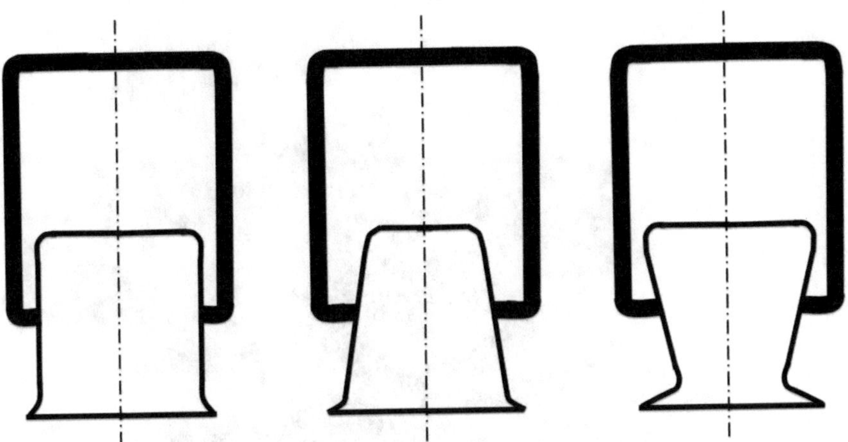

Fig. 3.40 Schemes of three types of pneumatic springs; from left: constant stiffness, progressive stiffness, degressive stiffness

This scheme can be applied to both arms or for the lower arm only, crucial to obtain a negative king-pin offset. In this second case also the lower arm of a Mc Pherson suspension can receive a virtual centre articulation.

In the case where only the lower arm is adopting virtual centers, this last will be substituted by two arms, as in Fig. 3.41, while the upper arm will receive a conventional shape.

The king-pin axis can be determined by the already known geometric construction, by intersecting the two planes through the lines connecting the four articulation points of the arms and the upper physical center, as shown on Fig. 3.41.

On Mc Pherson suspensions the upper physical center is the upper pivot.

The virtual center configuration allows a negative king-pin offset also by large tires and bulky brake discs. As known a negative king-pin offset can improve vehicle stability during braking, when one of the braking circuits has failed or when different friction coefficients are applied to the two wheels of the front axle.

The disadvantage of this solution is the introduction of additional spherical joints. Arms must be shaped in a particular way as to avoid the wheel envelope space.

With this design feature king-pin axis is changing with steering angle; as a consequence king-pin offset must be verified for each steering angle.

As outlined, this solution can be applied also to the upper arm of a double wishbone suspension, as shown on Fig. 3.42. In the same way, as we have explained in the previous case, the king-pin axis can be obtained by intersecting the two planes through the two lines for the four articulation points of the two rear arms and of the two front arms.

It is possible to reduce the value of the king-pin offset, even by large disc brakes.

King-pin offset reduction can also have benefits on steering wheel return, which can be affected by the value of the traction force.

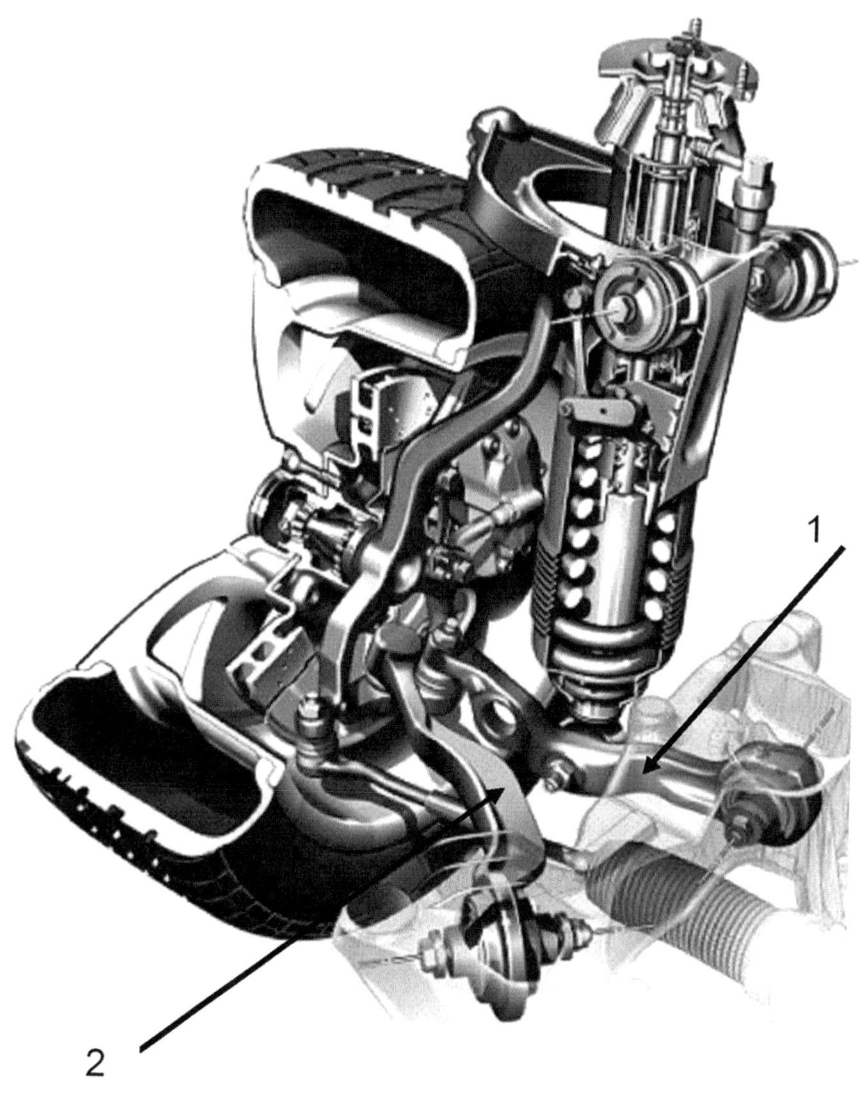

Fig. 3.41 High double wishbone suspension with lower virtual center articulation; king-pin axis is determined by the intersection of the two lines through articulations of arms 1 and 2 (Mercedes)

3.2.5 *Trailing Arm Suspensions*

This architecture is widely applied on small and medium cars on rear axles only.

Both wheels are fixes to a longitudinal arm free to rotate with reference to the body: the rotation axis of the two arms is usually the same and is parallel to the y vehicle axis.

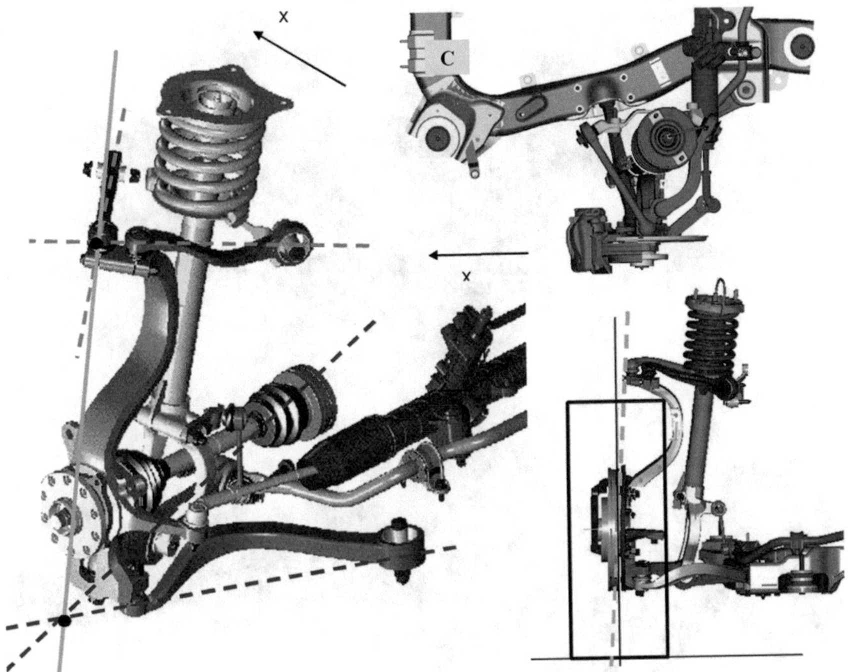

Fig. 3.42 High double wishbone suspension with both virtual center suspension arms; the figure shows the geometrical construction to obtain king pin axis position (FCA)

During suspension strokes wheelbase is affected but toe angles remain unchanged. Wheel comber with reference to the ground is equal to the car body roll; there is no camber recovery.

Description

Figure 3.43 shows two perspective views of a trailing arm suspension for a front wheel driven medium class car.

The axle features a couple of oscillating arms connected to a subframe through a transversal rotation axis; the subframe is connected to the body by four elastic bearings. This subframe is made by two shell of stamped steel sheet, welded to a tubular cross beam. Subframe can be not applied and in this case arms bearing are directly fixed to the body.

On each shell is shaped the upper spring rest. A bracket offers the upper shock absorber mount.

The suspension arm is, in this case, made of an iron casting, integrating the lower, moving, coil spring rest, shock absorber bushing case and the wheel bearing case or pin.

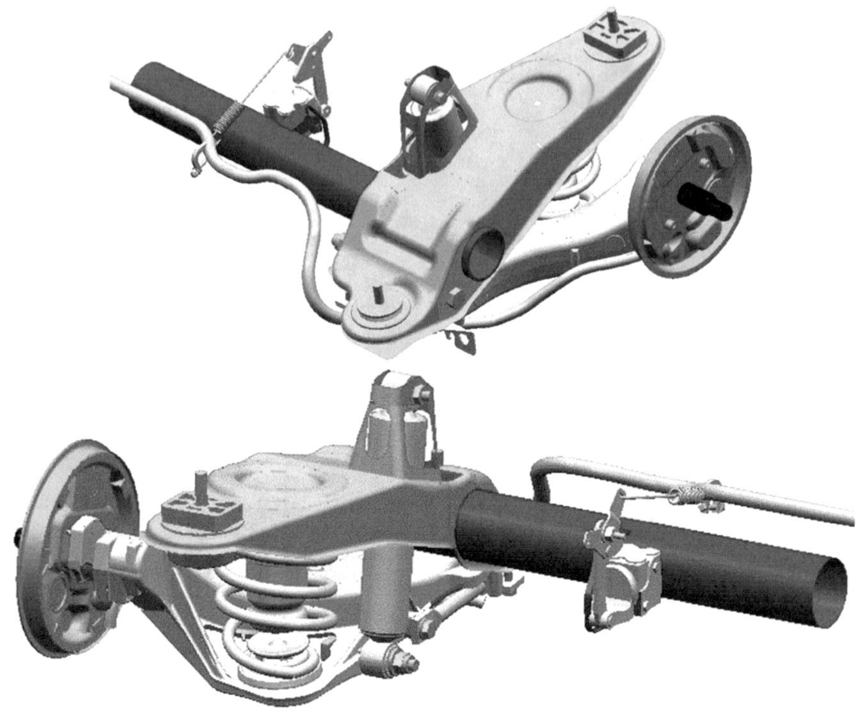

Fig. 3.43 Rear suspension with trailing arms (FCA)

The anti roll bar is fixed to the arms directly in two different points on each arm; there are no fixation points on the body; for this reason this is also called floating bar.

Figure 3.44 shows a detail of the arm of the above suspension. It shows an opposed conicity roller bearing used for the articulation of the arm to the subframe. Being this bearing quite stiff, there won't be any toe angle, or camber variation, or other displacements caused by arms deformations because of forces applied to the wheel.

As far as the camber angle is concerned we must take into account the arm torsional deformations too, caused by vertical and lateral forces.

The choice of roller bearings instead of the more traditional rubber bushings is caused by the high value of the reaction forces, due to the reduced distance a between the bearings; a higher value could allow the use of rubber bushing but would reduce the space available for fuel reservoir and spare wheel.

Figure 3.45 shows a cross section of the wheel assembly. The wheel bearing inside ring is press fitted to a pin, which is again press fitted into the suspension arm; the bearing is a second generation one. The arm shows a flange for fixing a drum brake or the caliper of a disc brake; notice that the braking torque is applying to the arms a reaction torque, which rends to reduce the body rear up-lift, due to inertia force.

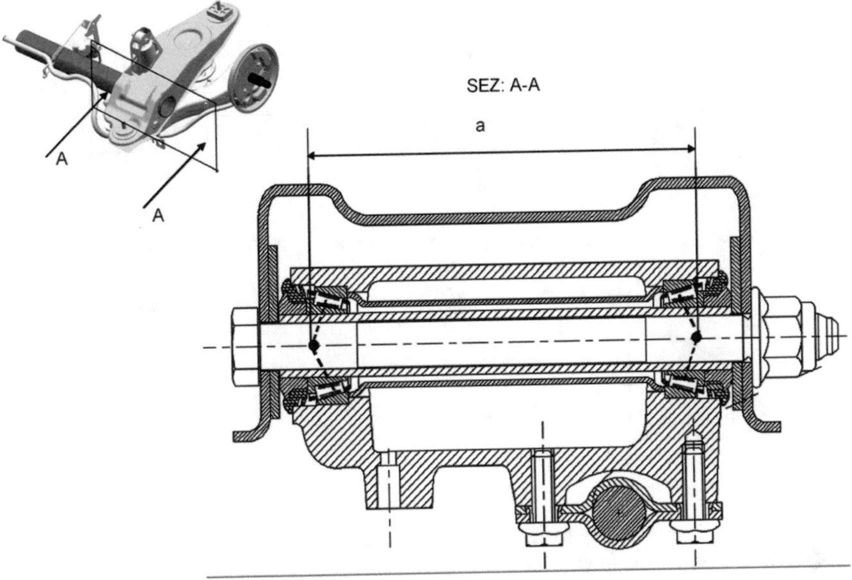

Fig. 3.44 Detail of tapered roll bearings of the arm of the trailing arm suspension of the previous figure; one of the fixation points of the floating anti roll bar can be also seen

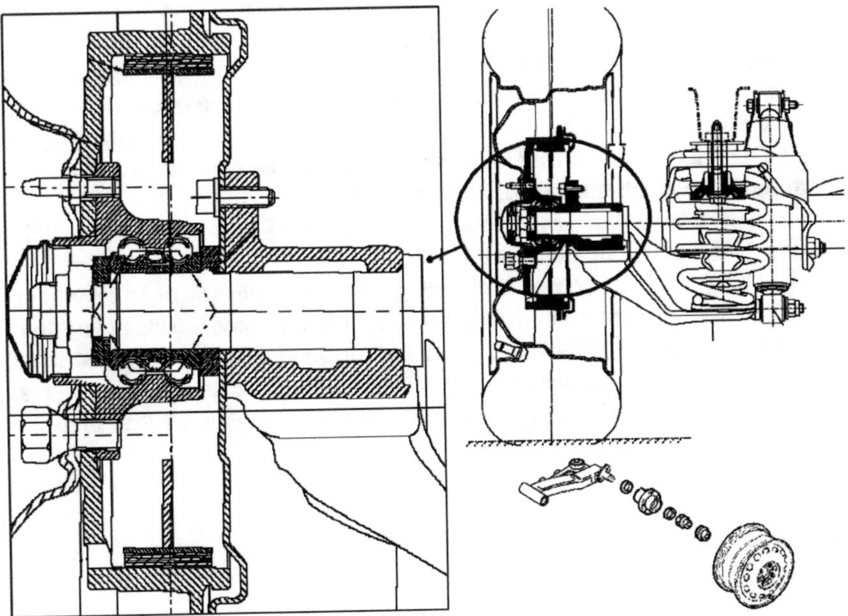

Fig. 3.45 Detail of the fixed wheel pin of the second generation bearing of the wheel hub of a trailing arms suspension

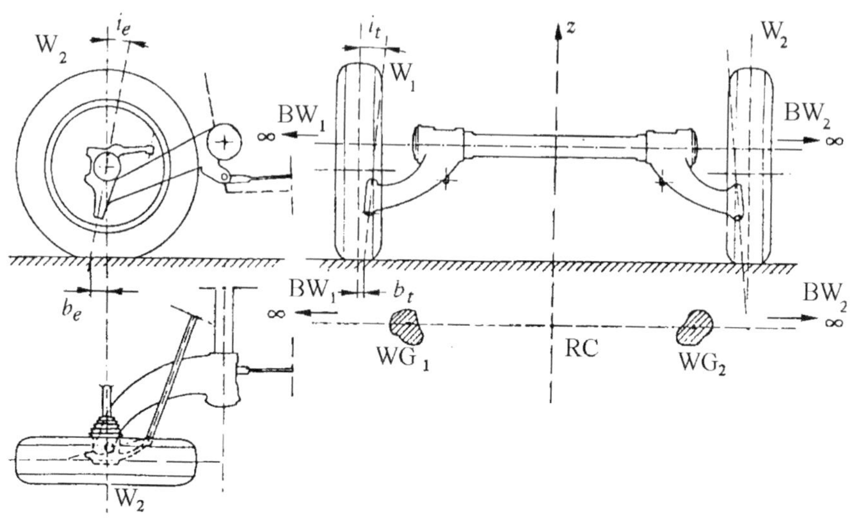

Fig. 3.46 Front axle with longitudinal trailing arms; springs working through pull bars are not represented. Roll center position has been calculated taking into account tire deformation. different rotation centers are represented (see on this topic the dedicated paragraph)

This example applies two coil springs as elastic elements; torsion bars can also be applied to reduce the vertical height of the cargo floor.

In this case, usually three torsion bars are applied: one for each suspension are dedicated to symmetrical springing and are fixed at the arm, on one side and to the subframe to the other; a third bar is fixed at each side on the two suspension arms and act as anti roll bar in asymmetric springing. Sometimes this kind of arrangement is causing different wheelbases for each side of the car.

In the past this kind of suspension has also been applied to steering front axles, as in Fig. 3.46, but this solution is no more used.

In fact king-pin axis direction is changing either by symmetric or by asymmetric strokes and the return torque of the steering wheel can increase too much on curves, because of the increase of the longitudinal trail of the external, more loaded wheel. In addition to that the longitudinal motion of the wheel, because of vertical stroke, is opposite to the direction along which obstacles are met, with negative impact on comfort.

Advantages and Disadvantages

Advantages

- Hysteresis is very low with roller bearing versions.
- The suspension intrusion into the baggage compartment is very low.

- High simplicity and, therefore, low production cost.
- Easy to be assembled.
- Suitable also for driving axles.
- Reduced value for unsprung mass.

Disadvantages

- Transversal deformations of trailing arms, because of cornering forces, have an oversteering effect.
- There is no camber recovery.
- Low longitudinal flexibility because of the rigidity of highly loaded bearings.
- There are no independent parameters to be tuned for improving the elasto-kinematic behavior.
- High vibration transmittance from the wheel because of bearing stiffness, again due to the value of acting loads.

This solution is used predominantly on low segments.

3.2.6 Semi Trailing Arms Suspension

As a difference with trailing arms suspensions, which feature arms rotating around the same transversal axis, semi trailing arms or triangles are rotating around two different symmetric axis, with reference to the car zx symmetry plane.

This axis show an inclination angle either on the zy plane and on the xy plane; a particular case of this kind of suspension is the cross arms suspension, where arms rotate around a longitudinal axis.

Also this suspension is applied to rear axles only.

Figure 3.47 is shown a comparison between a trailing arm scheme (a) and a semi trailing arm one (b).

Arm inclination angles allow to obtain a modest camber recovery and a certain toe angle variation with understeering effect, with a slight improvement of the elasto-kinematic behavior.

Description

Figure 3.48 shows the semi trailing rear suspension of a small front wheel driven car.

The axle is made by a couple of swinging triangular arms (1 on scheme b) articulated to a subframe (1 on scheme a) being fixed to the car body in points A and B.

The subframe is made by a cylindrical tube with two stamped steel shells at the two ends; arms too are made by two stamped semi shells spot welded together. This

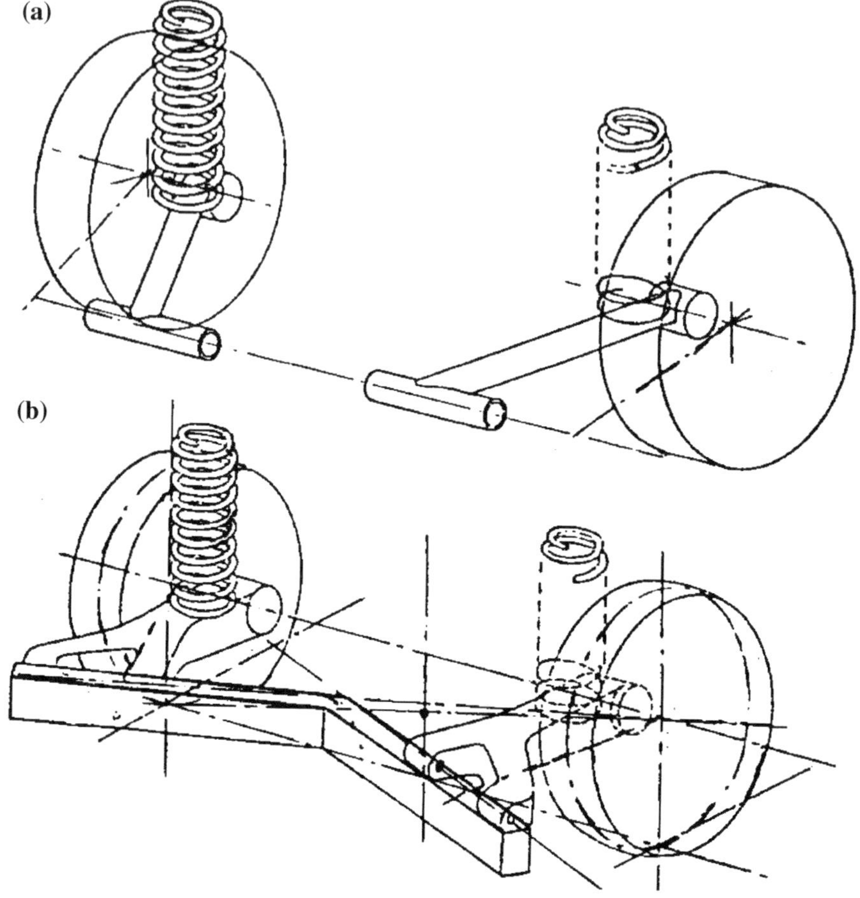

Fig. 3.47 Trailing arm suspension (**a**) and semi trailing arm suspension (**b**)

figure shows also the position os the coil spring and the stop spring (2 and 10, on scheme a) and of the shock absorber (2, on scheme b).

As a difference with trailing arm suspensions it has been possible to use elastic bushings for arm connections to the body, thanks also to the increased distance between the two articulation points.

We can say that this distance is not caused by the suspension kind, but rather by the dimension and installation of the underbody components; on this car the increased space is justified by the installation of the spare wheel under the hood, possible because of the small diameter.

Also if this application has only historic value, because of its extreme roughness, we show on Fig. 3.49 a semi trailing arm suspension suitable for rear engine cars; this example explains also the main reason for using this kind of suspension, because it simplifies the design of drive shafts.

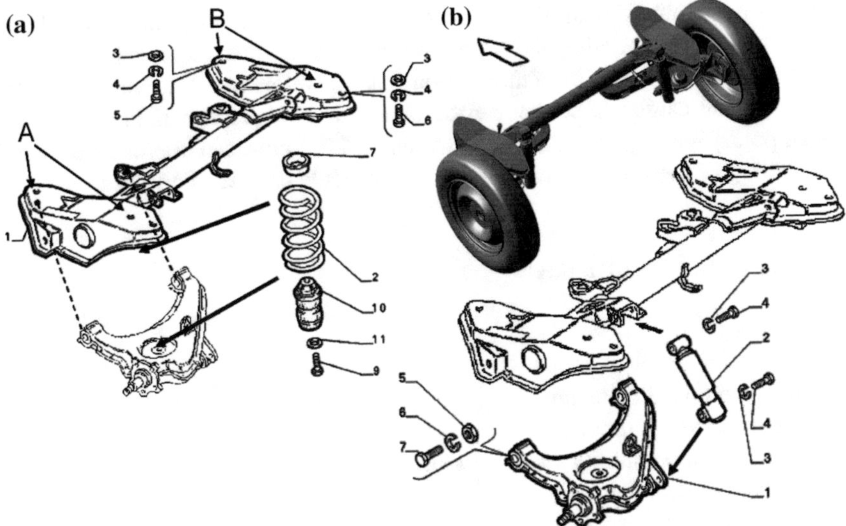

Fig. 3.48 Assembly view and details of a semi trailing arms suspension for a small front wheel driven car (FCA)

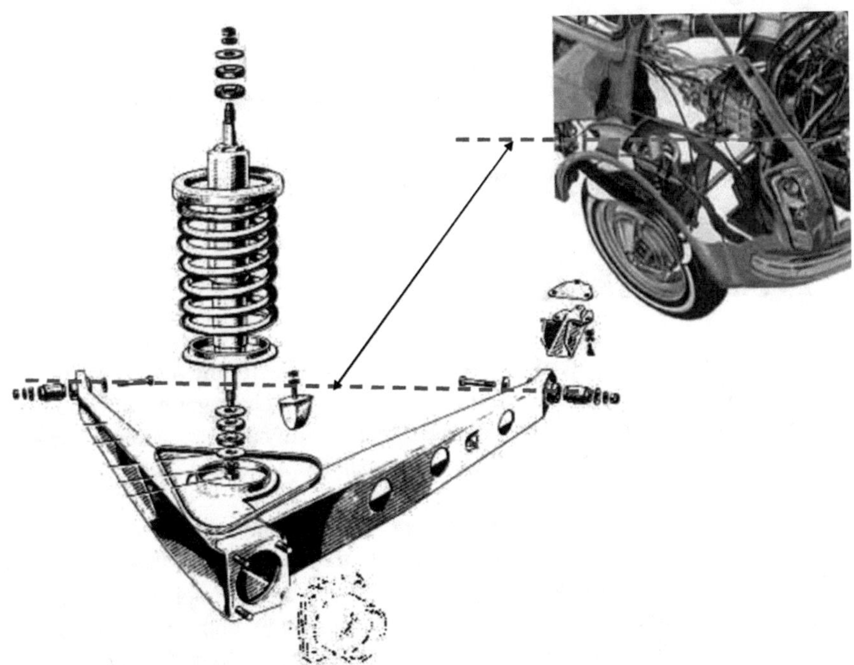

Fig. 3.49 Semi trailing arms suspension in a small car with rear wheel drive and rear engine; the arm rotation axis crosses the center of the constant speed joint (FCA)

The arm rotation axis can cross the center of the constant speed joint; this condition avoids shaft length variation and allows the application of very simple joints.

Because of the inclination of the rotation axis with reference to the body symmetry plane a partial camber recovery can be obtained. This practice is limited by the contemporary toe angle variation that, within certain limits, can improve the natural oversteering of a rear engine car but can be not viable over certain values.

Advantages and Disadvantages

Advantages

- Limited vertical dimension.
- Limited unsprung mass.
- Decent design possibility on elasto-kinematic properties.
- Construction simplicity.
- Suitable for rear wheel drive

Disadvantages

- Transversal dimension penalizing underbody components lay out.
- The excessive track variation because of the suspension stroke can wear tires prematurely.

This solution is presently used only on microcars or quadricycles.

3.2.7 Guided Trailing Arm Suspensions

Guided trailing arms suspensions have been conceived as improvement to trailing arms suspensions; with respect to this last suspension more linkages have been added and the number of degrees of freedom of the trailing arm has been increased, trying to conserve the natural advantage of this suspension of a reduced intrusion into the cargo compartment.

There are not, at this time, names for this suspensions which are shared universally by chassis engineers: we have selected this one to point out that there is a more sophisticated connection of the trailing arm to the body than a simple rotary bearing.

Many car manufactures, for commercial reasons, have improperly called this suspension as multilink suspension; in the opinion of the Authors this name should be reserved to the five linkages suspensions only.

On guided trailing arms suspensions, two or three additional arms are linked to the trailing arm, in order to improve elasto-kinematic performance of the suspension.

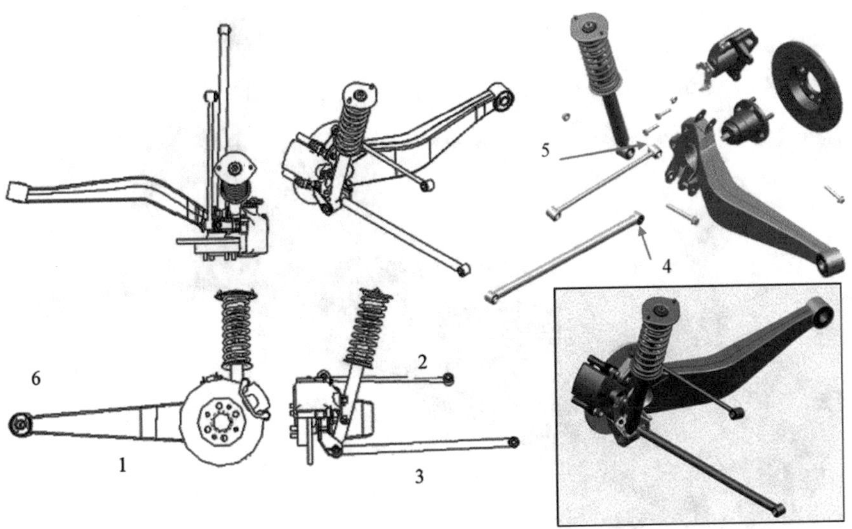

Fig. 3.50 Scheme for a guided trailing arms suspension; the articulation point 6 must allow two rotational degrees of freedom to feature the desired camber angle recovery

To restore the correct number of degrees of freedom of the mechanism the connection between trailing arm and body or subframe is made by a sufficiently flexible rubber bushing.

Figure 3.50 shows an example of a guided trailing arms suspension; the trailing arm 1 is guided, in this case, by two additional cross arms 2 and 3.

Cross arms identify a steering axis through the two elastic bushings 4 and 5. This axis is designed to create a given understeering toe angle variation, under the action of the cornering force or braking force. The elastic bushing 6 of the trailing arm, as dimension suggest, can grant a good longitudinal flexibility for a good comfort.

The position of bushings allows longitudinal motions without undesirable steering rotation of the wheel.

Figure 3.51 shows a different example. The application of three cross arms can be noticed on this suspension. The introduction of the third arm improves the elasto-kinematic behavior further.

With reference with the lower figure, arms 1 and 2 shape a transversely articulated parallelogram, which can grant a correct camber recovery. The linkage 2 is made by a shell structure of stamped steel sheet, relatively stiff, because spring loads are applied to it; linkage 1 is, on the contrary very light, because only transversal forces are applied to it; this linkage is curved to avoid the volume of the side beam of the underbody.

The linkage 3 can react to longitudinal loads with controlled deformation to stabilize steering with an understeering effect, as we have seen in other suspensions.

The trailing arm 4 is also made by steel sheet and shows a notable flexibility; if it would be rigid there will be too few degrees of freedom and the suspension would be

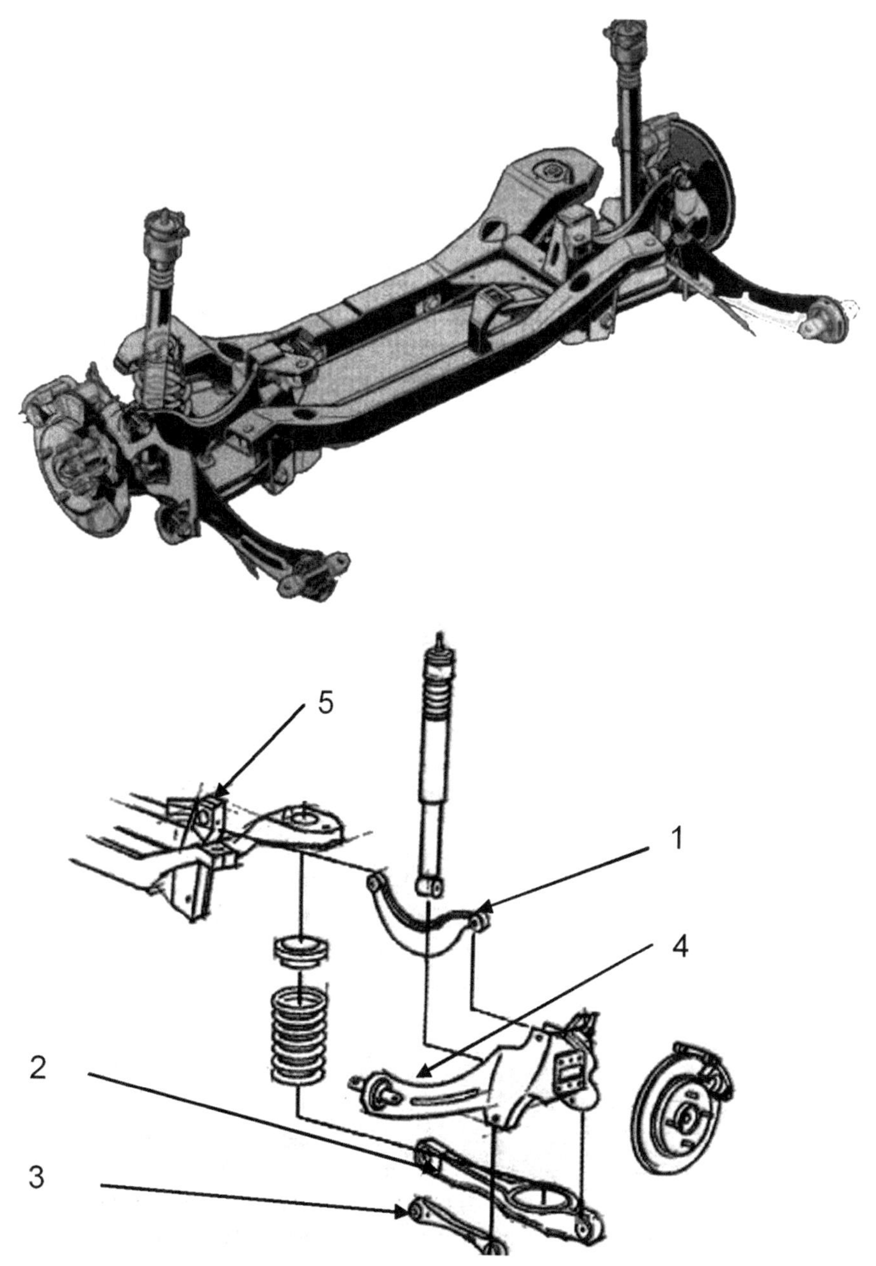

Fig. 3.51 Outline drawing of a guided trailing arm suspension; to understand the way of operation also the torsional deformation of arm 4 must be taken into account (Ford)

blocked. The arm can react only to forces acting through the articulation points, in longitudinal direction and to bending moments as those given by the braking force; camber variations can take place thanks to its torsional deformations. With a suitable elastic bushing the wheel can move longitudinally on obstacles, with benefits on comfort, without penalty on toe angle variation.

Transversal arms are connected to the subframe 5, suspended to the body through elastic rubber bearings.

Advantages and Disadvantages

Advantages

- Toe angle variation under cornering and longitudinal forces with stabilizing effect.
- Suitable camber recovery.
- Good longitudinal flexibility.
- Limited under floor volume similar to simpler suspensions.

Disadvantages

- Wheel case restriction if shock absorbers are applied to trailing arms.
- Many adjustment points for correct assembly on the subframe.
- Higher cost complexity as compared to previous solutions.

For its characteristics is used on medium cars, where it represents a good compromise between sophisticated multilink suspensions and lighter trailing arms or Mc Pherson suspensions; this architecture could be widely applied in the future.

3.2.8 *Multilink Suspensions*

The solution shown on Fig. 3.52 obtains, with penalties on weight and cost, the best result on comfort and handling; in this case, referring to a high class SUV, also air springs are applied.

By this suspension the strut is connected to the body with five linkages, as many as the degrees of freedom to be subtracted to the strut, to leave the suspension stroke motion only.

This or similar schemes are applied on most large and luxury cars rear axles, for front and rear wheel drive.

This suspension can be also considered as a double wishbone suspension with both lower and upper steering virtual points (4 and 5 for the upper arm and 2 and 3 for the lower one). The fifth linkage 1, or false steering linkage, is used to avoid or control the wheel steering motion.

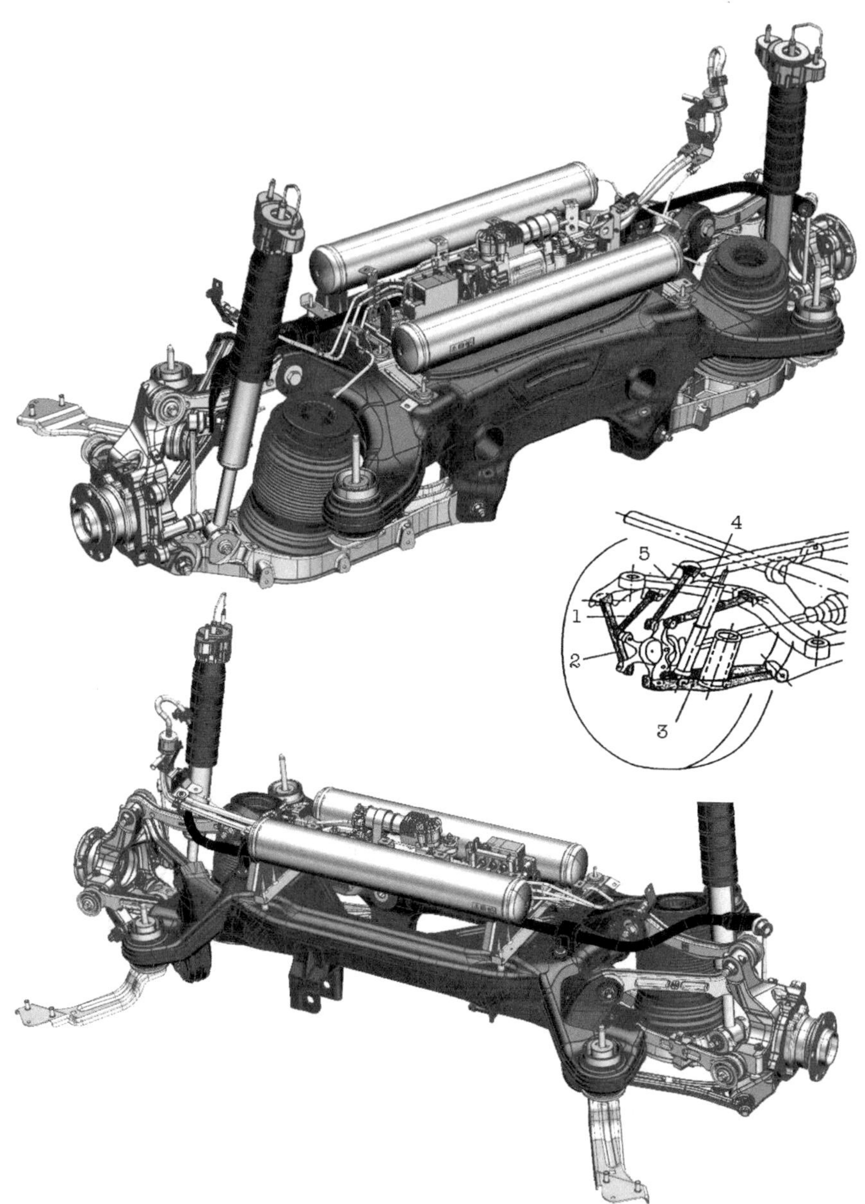

Fig. 3.52 Classic multilink suspension for the rear axle of a rear wheel driven car with air springs; it features five linkages, one for each degree of freedom subtracted to the wheel, to allow the suspension stroke only (FCA)

The difference with a front wheel double wishbone suspension are limited; nevertheless elasto-kinematic behavior will be different, considering the different role played by the two axles.

Every fixed articulation point is set on a subframe, because of the already explained benefits on pre-assembly capability and comfort.

When adopted on a rear driving axle, toe angle variations depend on the traction force, in addition to cornering and braking force; this fact must be taken into account while designing rubber bushings.

Any function can be obtained in the field of:

- toe angle variations stabilizing turning and braking;
- longitudinal flexibility, without undesirable toe angle variations;
- camber recovery by roll angle.

In addition to that, for rear wheel driven cars, wheels can increase their toe-in angle as function of the traction force, with reduction of the phenomenon of steering angle reduction at increasing traction force.

This phenomenon, called *torque steering*, is justified if we recall that the application of a longitudinal force is increasing the side slip angle of the rear wheels, at a given cornering force; to avoid a path change front wheel steering angle must be decreased.

In summary the design freedom allowed by this suspension is notably increased, particularly for a driving axle.

Similar considerations can be also made for the multilink suspension of a large front wheel drives sedan (Fig. 3.53).

This solution is characterized by the triangular arm 1, a cross beam 2 and a false steering linkage 3; the remaining linkage 4 connects the triangle to the suspension strut.

In this case we can again assimilate the scheme to a double wishbone suspension with the arms 1 and 2. The linkage 4 without fixed connection, increases wheelbase variation as function of the wheel stroke, with benefits on comfort. In this case torque steering is not to be taken into account.

Advantages and Disadvantages

Advantages

- Stabilizing toe angle variations as function of cornering and braking forces.
- Camber recovery.
- Wheelbase increase at compression stroke.
- Suitable to avoid torque steering effect on rear wheel driven cars.

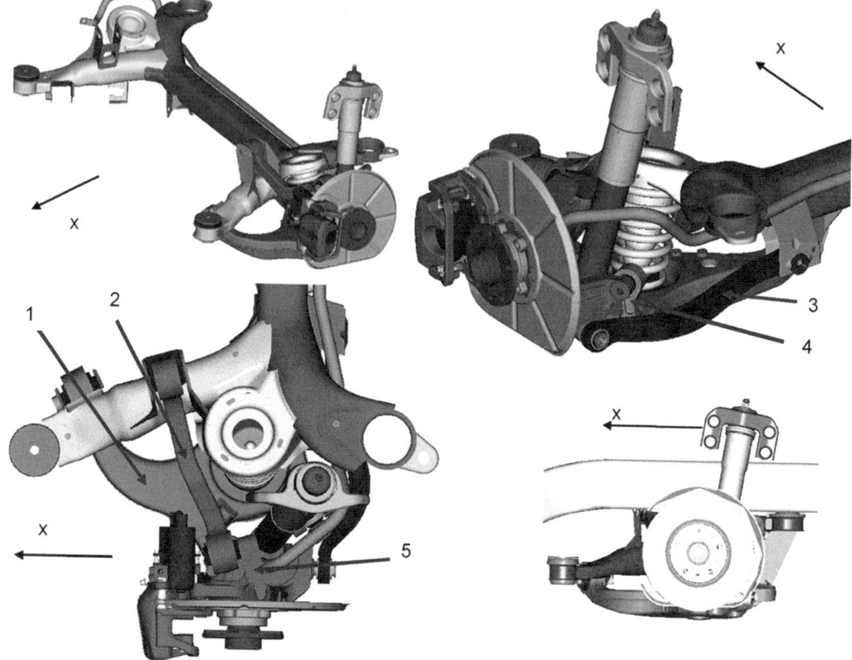

Fig. 3.53 Particular kind of multilink suspension for front wheel driven cars; the linkage 4 is used to increase wheelbase variations during compression stroke with benefits on comfort (FCA)

Disadvantages

- High mechanical complexity.
- Long development time.
- High production cost.
- High volume and weight.
- High sensitivity to variation of bushings elastic behavior.

3.3 Semi Independent Suspensions

3.3.1 Twist Beam Suspension

Description

Twist beam suspension (Fig. 3.54) can be imagined as two trailing arms 4, fixed to the body with elastic bushing 2; the intrinsic instability of the resulting structure is corrected by a cross beam 3.

The spring shock absorber unit is fixed between arms and body.

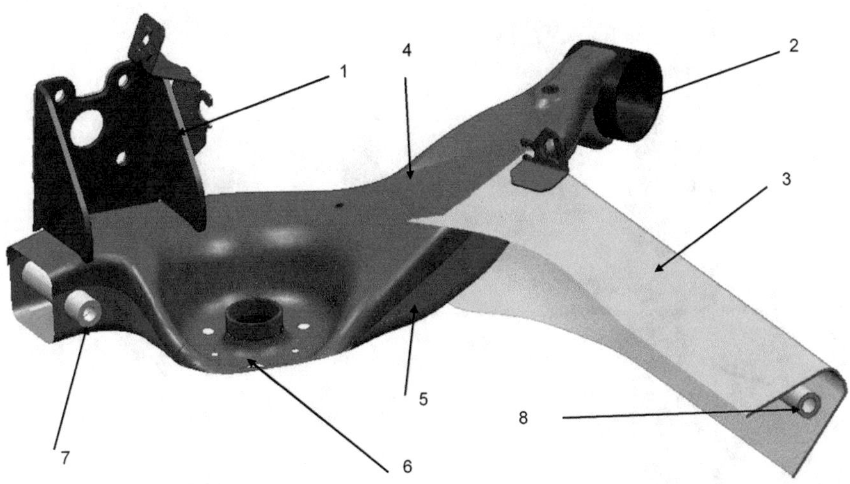

Fig. 3.54 The longitudinal arm 4 of this twist beam suspension is carrying the wheel hub welded on it. The torsion cross beam 3 has a U shaped cross section and is welded to the longitudinal arms. The anti roll bar is welded too

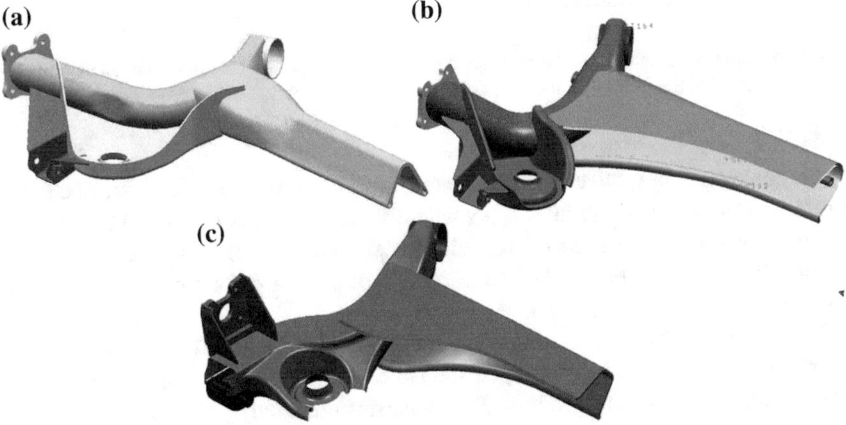

Fig. 3.55 Design alternatives for a twist beam axle: **a** tubular type; **b** hybrid type; **c** type made by stamped parts only

Figure 3.54 shows further details. The arm is made by two stamped and welded steel shells. The strut will be flanged to the plate 1, welded on the arm. The spring seat 6 is shaped on one of the stamped shells, building up the arm; the lower shock absorber mount 7 will be screwed to a tube, also welded on the shell 5.

The cross beam 3 has a U shaped cross section and is welded to the arms at its end. Also the anti roll bar is welded to the arms.

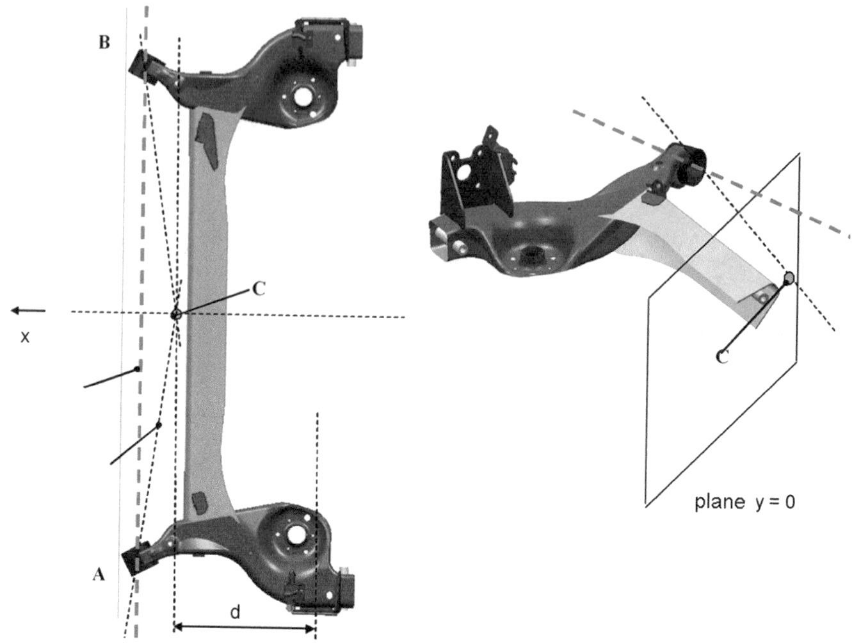

Fig. 3.56 Scheme to determine the rotation axis of the arms of a twist beam axle in an asymmetric suspension stroke; the point C is the shear center of the cross section of the cross beam

Some design alternatives for twist beam axles are shown on Fig. 3.55. The detail a refers to the so-called tubular type, where either the cross beam and the arm are made by cuts of steel, stamped and welded each to the other; also the twisting cross beam is made by a tube crushed to the shape of a V, in order to obtain the elastic properties, we will discuss later.

The detail b refers to the hybrid type, where the arm is made by a cut of tube and the twist beam is made by a stamped steel sheet. A possible variant of this solution can include arms made by cast iron. The choice is motivated by process optimization, in regard to the existing production tooling.

By symmetric suspension stroke, arms are rotating around the axis AB (Fig. 3.56), given by joining the two center of the arms elastic bushings; there are no changes of toe and camber angles because of the suspension stroke, but for the structure deformation caused by external forces.

By asymmetric stroke, the cross beam is twisted by the difference of torque applied to the arms and, only because of this deformation, arms can have different angles, around non coincident axis.

To determine axes BC and AC, around which the arms are rotating, it is necessary to join the bushing center of each arm with the shear center of the cross beam, in the symmetry plane of the car (point C). It should be remembered that the shape of the cross beam is such as to obtain a high bending stiffness (regard the cornering forces),

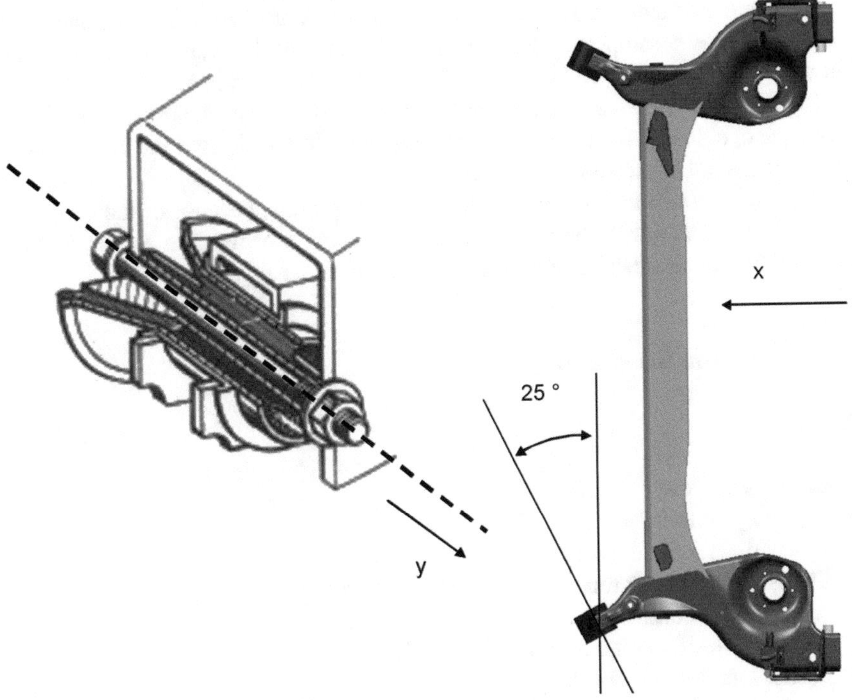

Fig. 3.57 Twist beam axles have a natural tendency to steer because of cornering force application. This tendency can be hindered with particular rubber bushing (left) or by inclining bushing axis (right)

with a modest, but suitable torsion stiffness. For this reason camber angle variations will be very small by symmetric strokes.

It is therefore possible to design toe angle and camber angle variations, by designing the section of the cross beam and the dimension d, relative to the wheel offset with reference with the shear center.

If the cross beam is made by a crushed tube, the V cross section of the tube can be designed to avoid the need of an anti roll bar.

On Fig. 3.57. at left, the details of an elastic bushing are represented. The bushings axis is parallel to the car y axis; to limit toe angle variation caused by cornering forces, the inside metal body of the bushing has a tapered shape which limit this rotation.

The same effect, suitable to obtain an understeering behavior on the car, can be obtained by inclining bushing axis on the xy plane, as shown on the right scheme.

Different arrangement are possible for coil springs and shock absorbers. The choice between the different alternative is a compromise between car comfort and the width of the cargo compartment between the wheel cases.

As a matter of fact the optimum position for shock absorbers is perpendicular to the arm as far as possible from the articulation point to the body; in this case the ratio between shock absorber and wheel stroke ratio is the highest one.

On the contrary this same positions brings the tip of the shock absorber inside the wheel case, reducing the cargo compartment width, particularly if the coil spring is installed coaxial to the shock absorber.

A further system to limit lateral deformation of arms, because of the cornering forces, is using a Panhard bar; this system is suitable for heavy cars, for instant high class MPVs, when other solution can affect comfort negatively.

Advantages and Disadvantages

Advantages

- Design simplicity.
- Assembly simplicity.
- Reduced vertical dimension.
- Almost total camber recovery by asymmetric strokes.
- Possibility of controlling toe angle by body roll.
- Unsprung mass smaller as with a rigid axle.
- Decent longitudinal elasticity.

Disadvantages

- High wheel case width, because of expected camber variation.
- Low roll stiffness.
- Highly stressed parts (twist beam and its welding).
- Unsuitable to driving axles.
- Toe angles too sensitive to loads.
- Remarkable different behavior by empty and full load condition.

Twist beam axles are used on small and medium cars extensively.

3.4 Rigid Axle Suspensions

This solution virtually abandoned on cars, finds applications on commercial and industrial vehicles and on many off road vehicles.

By driving axles the suspension cross structure (axle) can integrate final drive, differential and driving shafts.

In this case unsprung mass is certainly heavier as by independent suspensions.

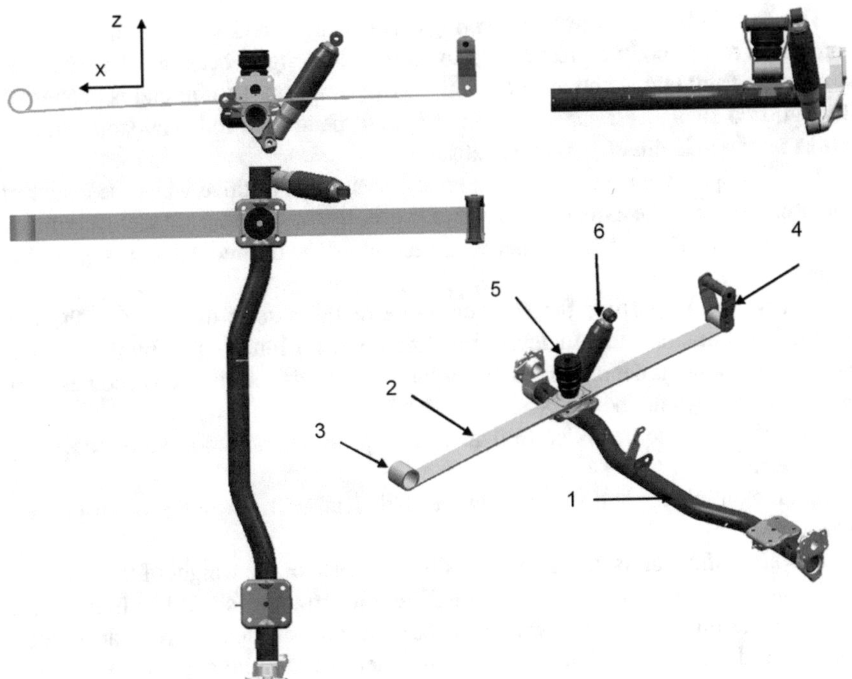

Fig. 3.58 Simple rigid axle suspension with single leaf springs for a small commercial vehicle (FCA)

By front wheel driven cars, rear idle axles have a simple structure connecting the two wheels; the unsprung mass value is not far away of the range of independent suspensions.

Considering the kinematic point of view, this solution features no track variation by roll and parallel springing motion. Wheel camber referred to the ground doesn't change because of body roll.

We will consider two different design alternative: rigid axles with leaf springs, where also the kinematic function is performed by springs and the guided rigid axle, where the kinematic function is performed by dedicated linkages, as by independent suspensions. In this case elastic elements can be different as leaf springs.

Rigid Axles with Leaf Springs

Figure 3.58 shows the drawing of a very simple solution suitable to front wheel driven vehicles; in this case it is a small commercial vehicle.

The two wheel are connected by the axle 1, whose shape is caused by the bulk of the other under body components, in this case, the spare wheel.

The axle is linked to two leaf springs with flanges and screws. The spring is connected to the body as to be able to change its length, because of suspension stroke; the front end is connected to a fixed eye 3, where the rear end is connected through two *swinging shackles*, a kind of small pendulum rod. The stop spring 5 works on the axle directly, as shock absorbers 6.

The leaf spring is bent on the zx plane; curvature is positive when the center of curvature is above the axle; this curvature is peculiar to axle that are self-steering by body roll; this curvature is also necessary to obtain the desired value of suspension stroke.

If curvature is positive the steering angle of the axle is oversteering the car; by negative curvature it is understeering and gives no contribution by flat springs. Because curvature is decreasing as load is increasing, this contribution becomes more understeering as load increases.

This effect has to be added to that of leaf spring lateral deformation, normally oversteering.

In our example the leaf spring shows a single leaf, with advantage on suspension hysteresis.

This simplification is not always possible, because of the weight of the vehicle.

When leaves are many they feature different length; they should be free to slide one over the other, in such a way as to build a uniform flexural resistance body. The stack of leaves is, in fact, by semi elliptic leaf springs, bound to have the same curvature and the resulting structure can be considered as a different arrangement of a beam with rhomboid shape, supported at the ends and loaded in the middle, as shown on the lower part of Fig. 3.59.

In some cases single leaf spring are used, as in our example or on Fig. 3.59, at right; in this case the leaf is tapered according to a theoretical parabolic shape, to obtain again a uniform flexural strength body. In other cases tapered leaves can be joined in twin assemblies as in the upper right. If the fabrication process is able to dominate the leave thickness leaves stacks can provide contact at the end only, with benefits on the hysteresis.

The central part of the stack is fixed to the axle through U shaped brackets (see Fig. 3.64 in the following paragraph); other brackets keep leaves in position.

The longest leave has two eyes, sometimes copied by the second leave, for fixation to the body. On semi elliptical leaf springs, layers of self lubricating plastic strips are interposed between the leaves. They decrease friction and avoid leaf sticking after long stops, because of corrosion.

Advantages and Disadvantages

Advantages

- Extreme design simplicity.
- The only suspension with full camber recovery by roll.
- Robustness suitable to off road and heavy duty applications.

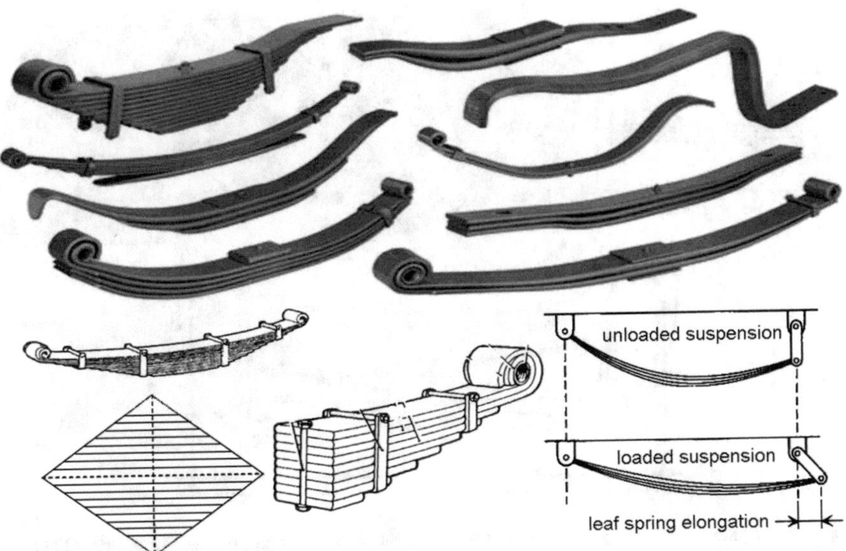

Fig. 3.59 Different kind of leaf springs for rigid axles (upper part); the scheme shown on the left of the bottom part, explains how a semi elliptic leaf spring can be seen as a uniform flexural strength body; the scheme on the left shows the motion of the leaf spring end because of suspension stroke

- Wheel stroke is greater as spring stoke, by roll motion; this property is useful on off-road vehicle to clear transversal trenches.

Disadvantages

- High elevation of the body floor on the cargo compartment.
- High restriction on wheel cases reducing cargo compartment width.
- Heavy unsprung mass.
- Wheels working angles can't be adjusted.
- Poor elasto-kinematic characteristic.
- Low longitudinal elasticity.
- Low roll stiffness.
- Vibrations caused by S deformations of the leaves.

Rigid Guided Axles

To improve the transversal and longitudinal guidance of the axle and to obtain a better elasto-kinematic behavior, the axle can be complicated by adding suitable linkages.

This additional linkage are essential if springs are only capable to absorb vertical loads as coil springs and pneumatic springs.

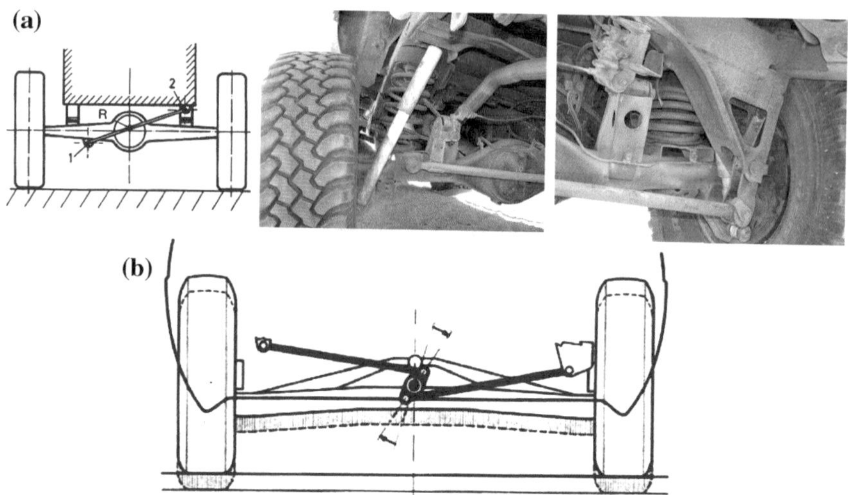

Fig. 3.60 Systems for axle guidance; on the upper part a Panhard bar, on the lower side a De Dion mechanism

Figure 3.60 (scheme A) shows a solution where the reaction to cornering forces is assigned to a Panhard bar, linked to the car body on one end (point 2) and to the axle, at the other end (point 1). If coil springs are adopted other linkages are necessary, not visible on this figure, as longitudinal parallelograms, to build up the necessary longitudinal guidance.

Panhard bars can also reduce the natural oversteering behavior of a leaf spring rigid axle, setting limits to lateral springs elasticity.

Scheme B shows a solution where the vertical guidance of the axle is performed by a Watt quadrangle. In this case the axle is usually shaped as a rigid triangle, where the basis is connecting wheel hubs and the vertex is set on the symmetry plane of the body in front of the wheels and is connected through a spherical joint; this schemed is named De Dion axle.

The Watt quadrangle, by its shape, is obliging the middle of the axle to move on the symmetry plane of the body. The front spherical joint can receive a certain flexibility as to reach a given value of understeering.

Figure 3.61 shows possible linkages for a rigid axle in connection with the application of coil springs.

Scheme A shows two longitudinal arms and an anti roll. Arms react to longitudinal loads, while their inclination can absorb also cornering forces; the anti roll bar can take longitudinal loads, contributing to react to the torques applied to the axle.

Scheme B provides four longitudinal arms and a Panhard bar. Longitudinal bars react to longitudinal forces and driving and braking torques, while cornering forces are absorbed by the Panhard bar. In a possible variant to this version Panhard bar can be avoided by a suitable inclination of the longitudinal beams. In both cases elastic bushings rigidity can be designed with an assigned elasto-kinematic behavior.

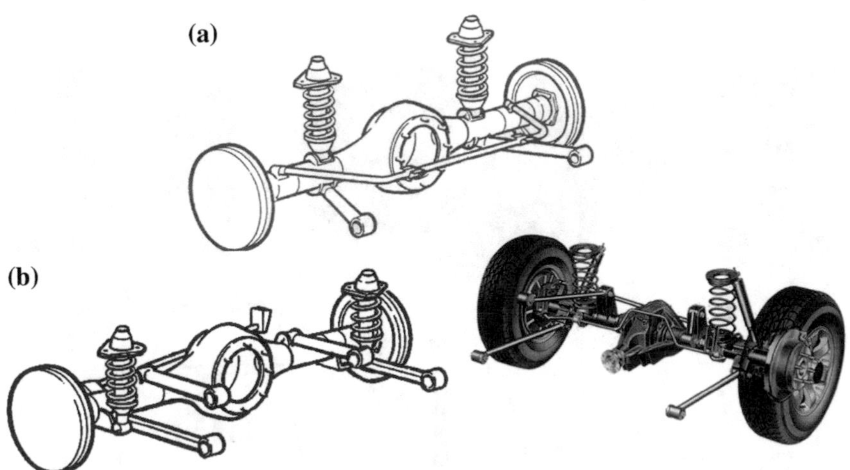

Fig. 3.61 Different rigid axle linkages suitable for coil springs; the anti roll bar itself (upper scheme) can work as a linkage, with function

3.5 Industrial Vehicles Suspensions

Industrial vehicle suspensions must allow notable load variations particularly on rear axles; we should remember the ratio between curb weight and full load on a truck. This ratio is lower of an order of magnitude, in comparison with cars.

Industrial vehicles have a structure including a chassis frame made by side beams and cross beams, bearing chassis components, cabin and pay load.

Front and rear suspensions on trucks are usually rigid axle type, similar to those introduced in previous paragraphs. Buses and commercial vehicles adopt also independent suspensions.

Many vehicles of this kind adopt pneumatic springs instead of leaf springs. The advantage of this choice is the possibility to control vehicle trim and the springing characteristic as function of the actual load.

Their application implies an increase of cost and a more complicated axle mechanism, because the guidance function of leaf spring is missing; the necessity of compressor and storage pressure bottles is partially overcome, because they are already justified by the braking system.

On light industrial vehicles, there is still a large diffusion of leaf springs.

Sometimes to increase progressively the suspension stiffness at higher loads, more leaf springs set in parallel are applied, as shown by the example on the second left raw of Fig. 3.59. The spring set in parallel works when pay load overcomes a certain limits; this reduces vehicle trim variations and approximates the constant natural frequency condition in a better way.

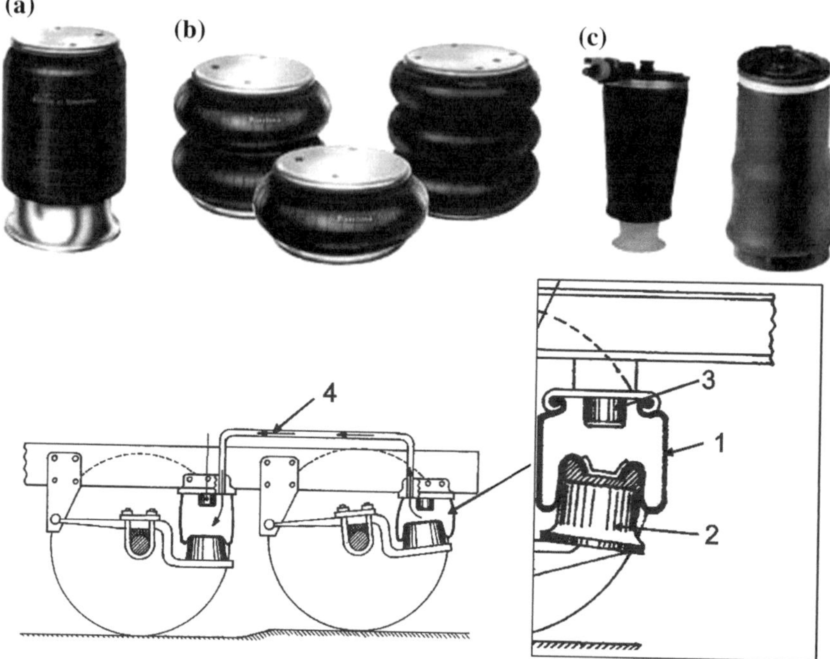

Fig. 3.62 Different kind of pneumatic springs for the suspension main spring; if the elastic elements of tandem axles are pneumatically connected the two axles built up an isostatic system in an easy way (lower scheme at left); stop spring are usually included inside the rubber element (lower scheme at right)

3.5.1 *Pneumatic Springs*

Air spring allow to change and control vehicle trim and to reduce the load excursion effect on suspension geometry, with benefits on comfort and, on trucks, on transported goods integrity.

As it will be explained later, the condition to minimize sprung mass acceleration implies a spring stiffness proportional to the sprung mass (constant natural frequency).

Because of the value of pay load variations by industrial vehicles, springs rigidity should be highly progressive. But springs are designed for maximum pay load and prove to be too stiff at partial pay load or at empty vehicle.

Air spring overcome this inconvenient, by changing their inflation pressure, as previously explained for pneumatic double wishbone suspensions.

Figure 3.62 shows some examples of pneumatic springs and a simplified scheme of their installation on the vehicle. The spring is a rubber bellow reenforced with textile fibers, like a tire, containing pressurized air.

Types a and c show a bellow like a tube 1 (lower detail), which can roll up on a metal surface 2, to adapt for different lengths; this configuration is suitable for light vehicles where the spring diameter is reduced with reference to the stroke; the bellow diameter, at the same inflation pressure, must be proportional to the square root of the sprung mass.

For types b the bellow is reinforced with metallic rings; this shape is suitable (notice that this picture is not scaled to the others) for heavier vehicles where the diameter has the same magnitude of the useful stroke.

Air, with its characteristic to change the volume as function of pressure, is the elastic element of the suspension. With a compressor and controllable valves is possible to adjust the pressure to have the same vehicle trim at any load.

Pressure variations can also adjust vehicle height, when needed (for instance to lower the saddle of a tractor, when carrying the semitrailer on).

The air physical transformation can be assimilated to an isotherm, during slow trim variations; it must be assimilated to a polytropic or adiabatic during springing motion, where the time allowed to thermal exchange is very low.

In this last case, spring characteristic provides no constant flexibility, not considering possible effective diameter changes.

The detail shows also how the rubber stop spring can be integrated into the pressure bellow.

The lower scheme shows also how is possible to obtain an interconnected suspension on a tandem axle. Tandem axles are used to limit road loads by heavy vehicles. The interconnection eliminates the intrinsic hyperstaticity of the system.

Similar springing elements can also be used by light vehicle and cars when the vehicle trim must be adjustable, on off road vehicle, for instance, where the ground clearance is increased by off road driving only.

In a dedicated chapter we will discuss also the control techniques applied to these systems.

3.5.2 *Front Suspension*

Rigid Axles

On European industrial vehicles the steering wheel is usually set over the front axle or in the front overhang.

This kind of architecture is, therefore, characterized by the application of a longitudinal steering rod connecting the steering box with the steering knuckle, as can be seen on Fig. 3.63.

As the figure shows, the steering box of rack and sector type with ball recirculation, rotates the drop arm 1, articulated to the longitudinal steering rod 2. This last is connected to the steering arm 3, part of the steering knuckle, or stub axle. A cross steering bar 4 connects the right steering knuckle, as to obtain correct steering.

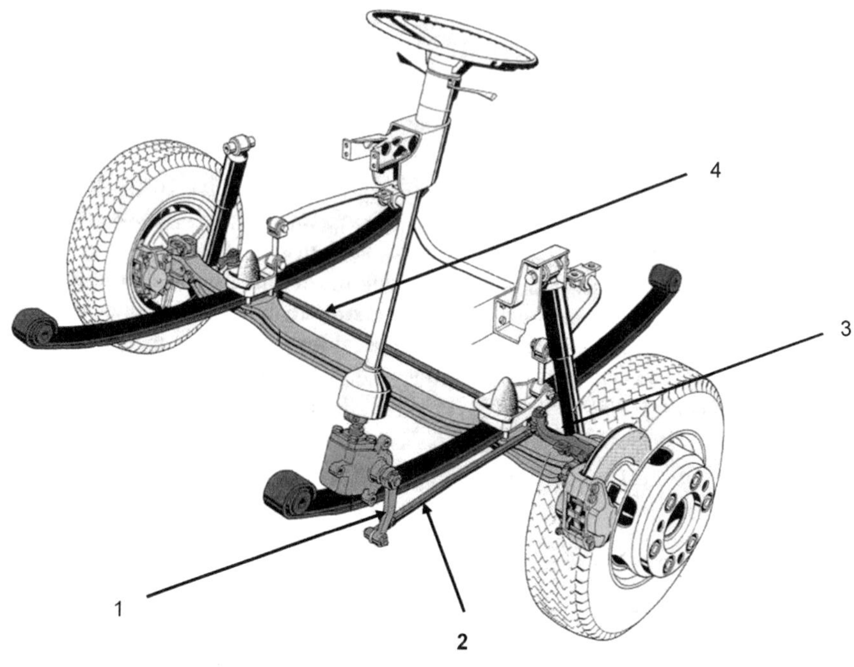

Fig. 3.63 Steering mechanism for a front rigid axle of an industrial vehicle (Iveco)

To limit steering angle consequent to suspension stroke, the articulation between drop arm and longitudinal steering rod must be set at the instantaneous rotation center of the axle, or leaf spring, relative to the body, close to the front eye of the spring.

Figure 3.64 shows a rigid steering axle with its main components. The leaf spring features two equal length tapered leaves fixed to the chassis frame in the known way. Leafs are contacting each other at the ends at in the middle only; in this way the behavior of a uniform stress body is obtained, with a very limited friction and hysteresis.

A third very short leaf can increase the flexural stiffness of the spring assembly for high loads and decrease S deformation during braking.

The S deformation is called in this way because of the shape of elastic deformation of the leaf spring, caused by the application to the axle of a pure torque.

It is particularly annoying, because the moment of inertia of the axle, around the y axis, can lead to periodic deformations, where the axle accepts a torque higher as the friction limit for a short while and then slips when the inertia force is disappearing. The result is vibrations and loss of braking efficiency or traction.

Leaves are fixed to the axle 3 through U bolts. The anti roll bar 6 is moving with the axle (therefore must be accounted as unsprung mass), while its ends are connected to the chassis frame, through pendulum bars 7; rubber stop springs 4 work on the axle directly.

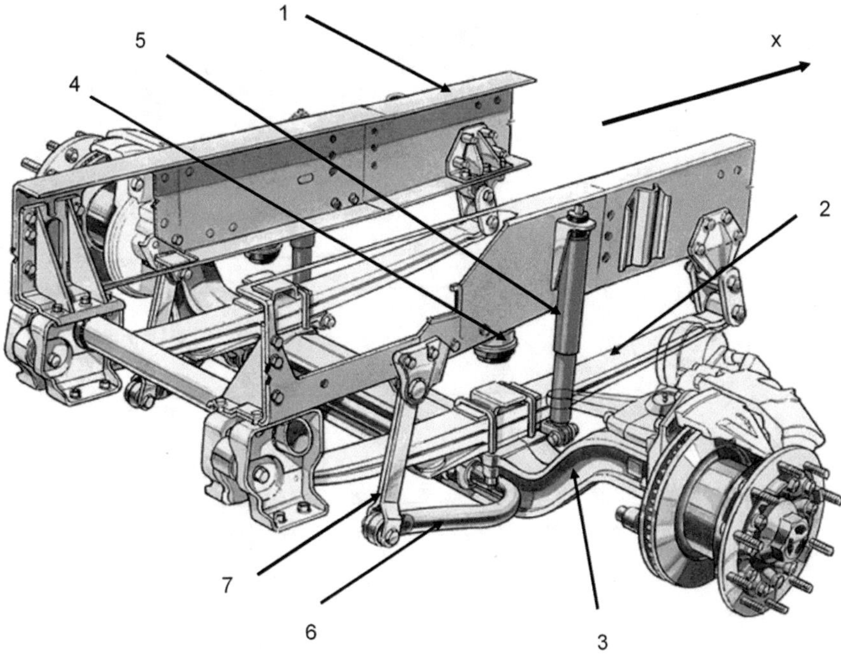

Fig. 3.64 Design details of a front rigid axle suspension for an industrial vehicle (Iveco)

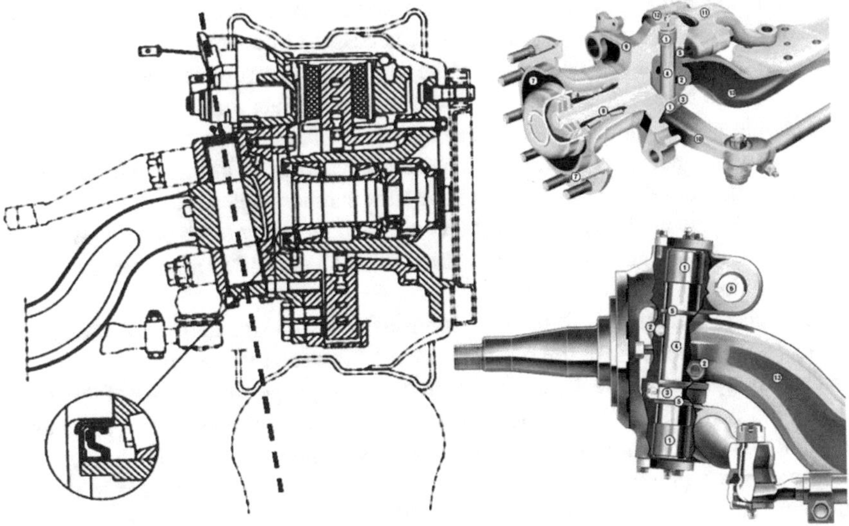

Fig. 3.65 Stub axle of a rigid steering axle, showing king pin axis position

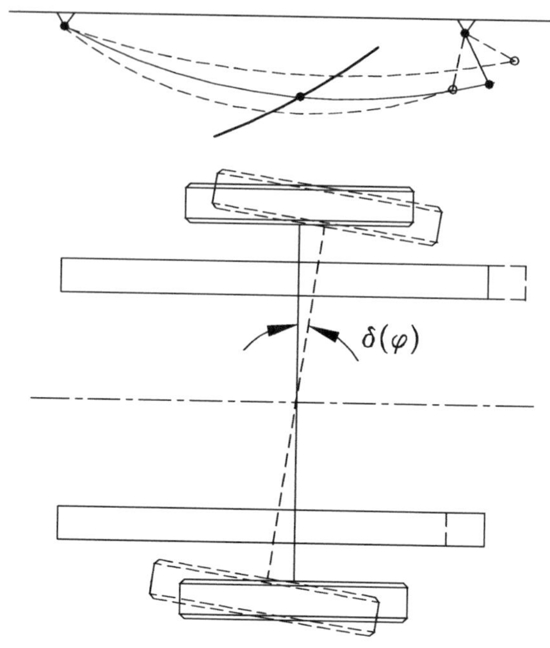

Fig. 3.66 Roll steering of a rigid axle

The drawing on Fig. 3.65 shows a cross section through the front wheel axes. King-pin axis and king-pin offset are identified by the rotation axis of the steering knuckle. The wheel disc has a bell shape, necessary to contain king-pin offset, in consideration of the space necessary for the large disc brake; this shape is standardized for all vehicle wheels and must take into account the fact that by rear twin wheels tires should not contact together or trap rubble.

As we explained for cars this axle has the advantage of keeping wheels always perpendicular to the ground; on such large tires camber angles are not acceptable.

A further advantage is the possibility of keeping every suspension component below the chassis frame with simplification of its shape and more space for the pay load or cabin.

As a disadvantage it is difficult to control the elasto-kinematic characteristic of the suspension, because of the rigid connection between the wheels.

On bumpy roads wheels can be steered by the suspension stroke. Figure 3.66 shows the path of the leaf spring connection point to the axle, during suspension motion; this point is moving forwards during suspension extension and back, during suspension compression. By parallel springing the axle shouldn't steer in the longitudinal steering rod articulation is set in the centre of curvature of this path. By asymmetric springing (body roll) a certain steering angle will be present.

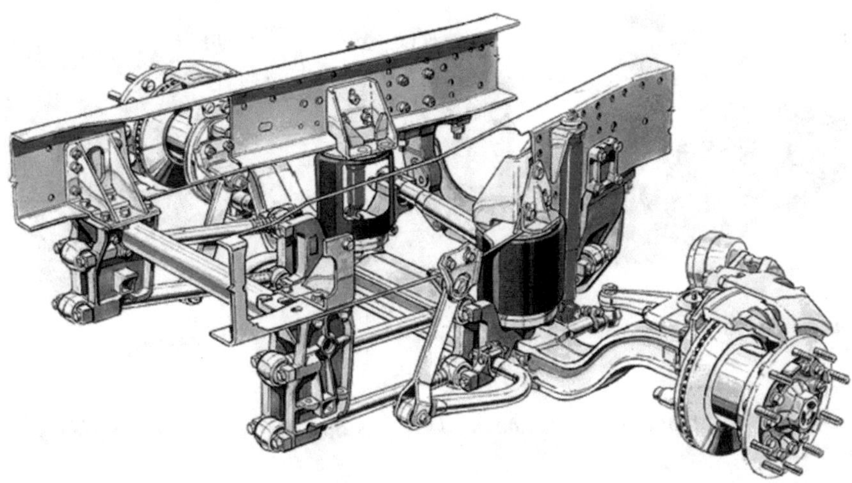

Fig. 3.67 Rigid axle suspension for an industrial vehicle; the application of pneumatic elastic elements is made in conjunction with longitudinal and transversal guiding linkages (Iveco)

Guided Rigid Axles

Figure 3.67 shows a rigid axle suspension where the axle is guided by two longitudinal parallelograms, similar to those we have seen for small vehicles.

There are two couples of longitudinal rods dedicated to the reaction to longitudinal forces and torques. A Panhard bar is dedicated, as usual, to cornering forces. Springing is obtained through pneumatic bellows.

The advantage of this arrangement is a better control of the axle steering by roll and the possibility of application of pneumatic springs. This kind of mechanism can be applied also with leaf springs, which will be free to move longitudinally at both ends; this mechanism avoids the S deformation completely.

Double Wishbone Suspensions

The need for a better ride comfort for passengers and for a better handling performance due to the increased speed has led industrial vehicle designers to apply to buses a front wheel double wishbone suspension, similar to those applied to cars.

A further need for these industrial vehicles is to have the passenger compartment floor as low as possible in order to make boarding easier; this need is particularly important by urban buses, where there is a frequent exchange of passengers. Rigid axle cause a too high position for the chassis frame and this is a further advantage for double wishbone suspensions.

Figure 3.68 shows two examples of this suspension. Air springs are applied on both suspensions; the suspension mechanism scheme is the same as by cars. Air

Fig. 3.68 Front double wishbone suspensions with pneumatic spring for busses

springs allow urban buses to incline the body by stops, in order to lower the entrance door further.

3.5.3 Rear Suspensions

The need to have twin tires always perpendicular to the ground doesn't allow suspension configurations different from rigid axles.

The high value of the load spread on heavy duty vehicles has practically standardized the application of pneumatic springs; by tractors the vehicle can be lowered when the semi trailer must be hooked.

Figure 3.69 shows a rigid axle suspension architecture widely applied to heavy duty industrial vehicles. The axle is guided by two lower longitudinal bars 2, which react to longitudinal loads; the upper triangular arm is connected to the axle through a spherical joint 1 and reacts to cornering forces and to torques.

Pneumatic spring bellows 4 are double for each side to allow standardization with the front ones.

On highly loaded vehicles tandem axles must be sometimes applied in order to comply with the legislative limits of load on a single axle.

Tandem axles, or multiple axles must distribute the load in a uniform way to the ground and adapt to uneven roads or bumps.

An isostatic condition is also necessary to better distribute the reaction to payload when this is not uniformly distributed.

Isostatic condition is easy to be satisfied by air springs by a suitable interconnection of pressure air pipes.

Tandem axles using leaf springs obtain this condition using two cantilever springs free to rotate in their middle section, as Fig. 3.70 shows.

The axles are suspended by symmetric longitudinal triangles and bars, as in the previous figure; this mechanism causes the two axles to carry on the same vertical load.

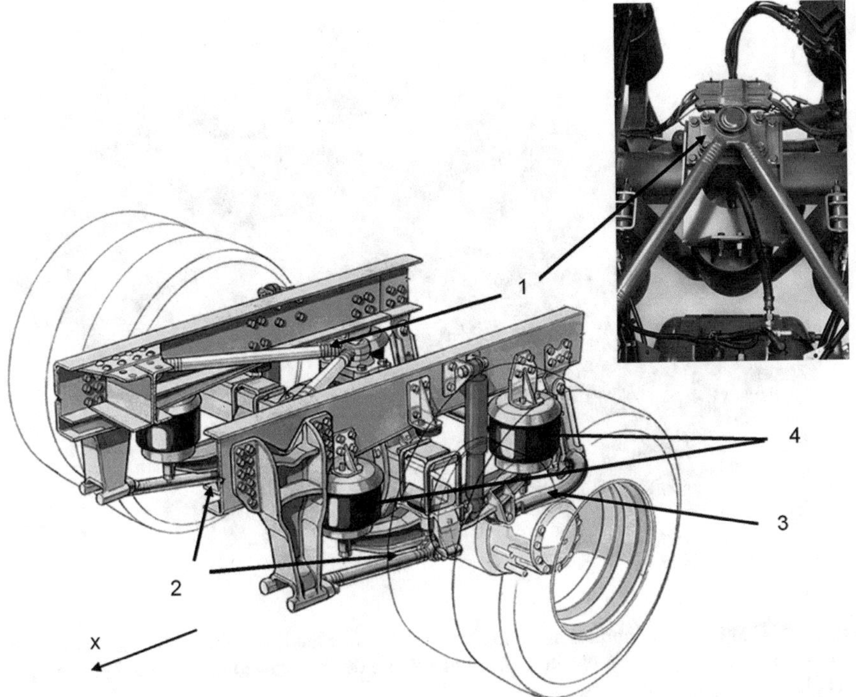

Fig. 3.69 Rear suspension for a heavy duty industrial vehicle, featuring pneumatic springs; the axle is guided by an upper triangular arm and two lower longitudinal arms (Iveco)

Leaf springs cannot react to pitch; front shock absorber have to limit by suitable stop rubber spring any excessive leaf spring rotation due to driving on bumps.

3.6 Design and Testing

3.6.1 Design Preliminary Outline

The design preliminary outline of a suspension scheme consists essentially of the determination of the geometric coordinates of the articulation points and of the position of the main axis as king-pin axis, shock absorber axis, etc., in order to reach the assigned target for elasto-kinematic performance.

The choice of the suspension scheme has been made previously, taking into account the volumes of the mechanical components of the chassis, the minimum clearance from the ground, the body structure and the other conditions set on costs and available production technology.

By automobiles, suspensions architecture is also one of the selling points.

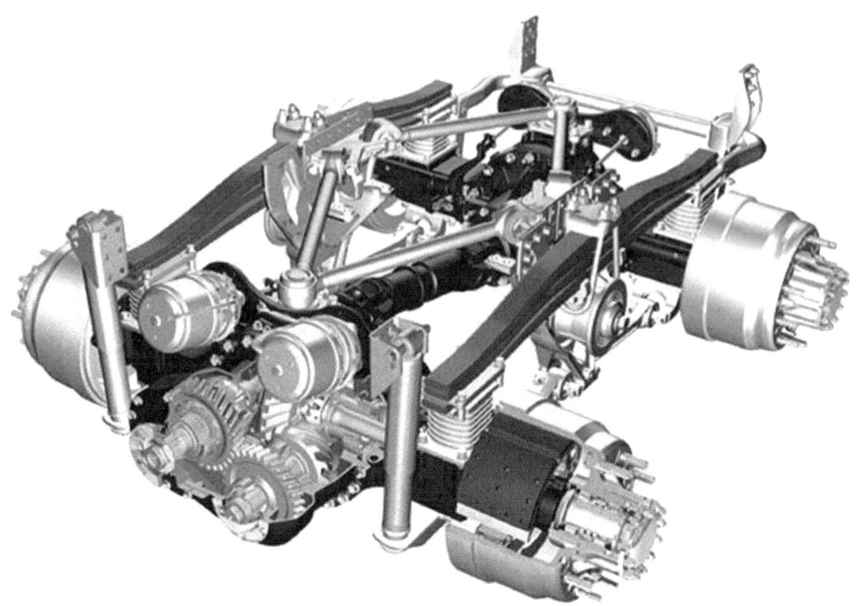

Fig. 3.70 Cantilever leaf springs allow to obtain an isostatic tandem suspension. Leaf springs are connected to the frame by a pin in their middle and axles are guided by longitudinal bars and triangles (Iveco)

On this paragraph a step by step logic flow of the design outline operations is explained for a Mc Pherson suspension; the extrapolation to other suspension is left to the reader.

We refer to the case of a front steering axle by a transversal powertrain.

The first step consist of the identification of the king-pin axis position. This should be consistent with what has been decided for king-pin offset and trail in relation to steering wheel feeling and return and braking stability.

By Mc Pherson suspension the king-pin axis is determined by the center of the spherical joint A and the center of the upper pivot B (Fig. 3.71). The point A is set along the king-pin axis in such a way as the spherical joint could find its place between the constant velocity joint and the wheel rim.

To reduce as much as possible the reaction to lateral forces on the lower arm, the point A is set as low as possible. The lower limit is done by wheel rim and braking disc positions.

Point B position must allow the assigned suspension stroke and the correct operation of coil spring and stop spring.

Point B elevation must be as low as possible to comply with a low hood profile; hood profile has influence on style and aerodynamics performance and the clearance between hood profile and suspension upper bearing is depending on requirements of passive safety of impacts against pedestrians.

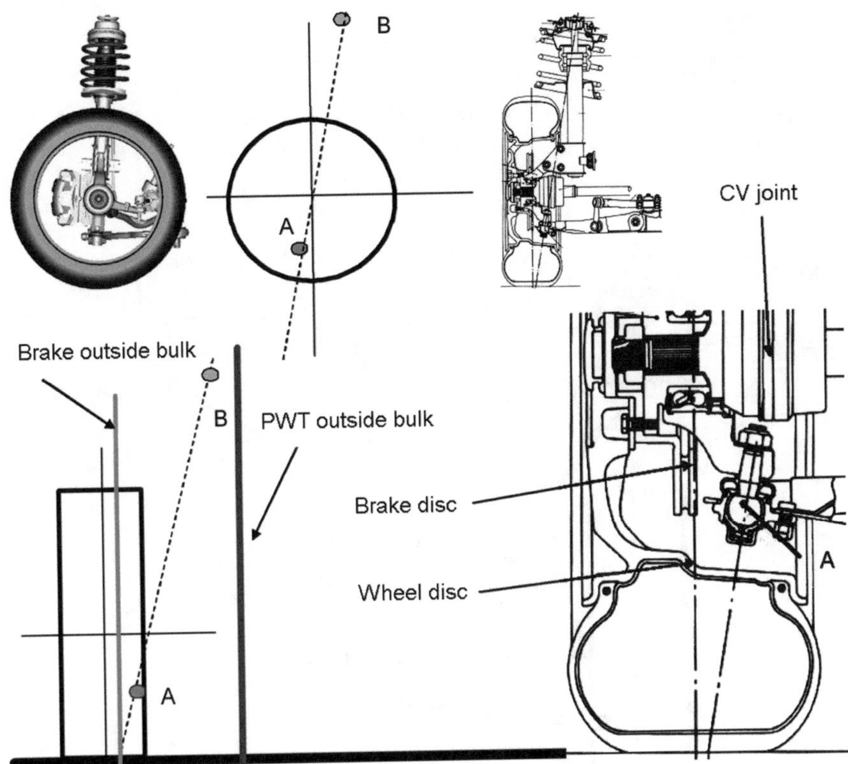

Fig. 3.71 Points A and B positions of king pin axis is conditioned by powertrain bulk, on one side, and brake disc bulk, on the other side, and by constant velocity joint position

When the king-pin axle is set, the shock absorber axis can be set. This is joining points B and C, on the piston rod of the shock absorber, as shown on Fig. 3.72.

To avoid mechanical complication on the strut design, the point C, as seen from the y direction, should be set on the king-pin axis.

The position of point D, setting the rotation axis of the lower arm, is a consequence of the desired camber recovery and track variation, as function of suspension stroke (see Fig. 3.72). D is the intersection of the arm rotation axis with a zy plane determined by point A position.

Because of kinematics of this mechanism, to increase camber recovery, point B must be closer to the car inside or the shock absorber inclination increased; this positions must be compromised with possible mechanical interferences and with king-pin axis position, which practically limit camber recovery.

As an alternative, camber recovery can be improved by rising the point D. This modification implies, as we will see later, the elevation of the suspension roll center, with impact with other vehicle performances (Fig. 3.73).

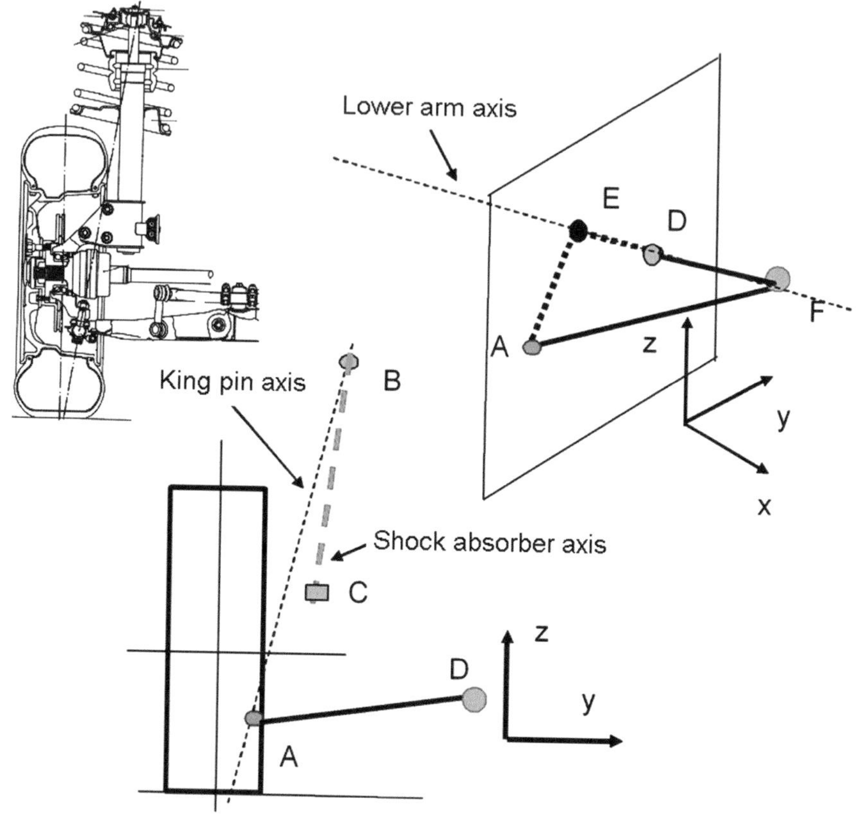

Fig. 3.72 The arm rotation axis EF is determined to obtain the desired camber recovery, once A and B are determined

Now the rotation axis of the arm can be determined by setting points E and F, centers of the elastic bushings.

Most solutions provide the same y coordinate for these two points; we will adopt this assumptions for sake of simplicity.

The z coordinates of these points is decided to limit pitch angles during acceleration or braking; we will explain later about anti dive properties of their position.

The x coordinates of these points determines the shape of the arm and are bound to the space available for the articulation points and the kind of subframe that will be adopted.

Their position has also direct consequences on the value of the forces that are applied to the points E and F, by longitudinal shocks against obstacles and by cornering forces.

Combining an a suitable way the position of these points and the radial and axial stiffness of the elastic bushings is possible to control longitudinal wheel flexibility,

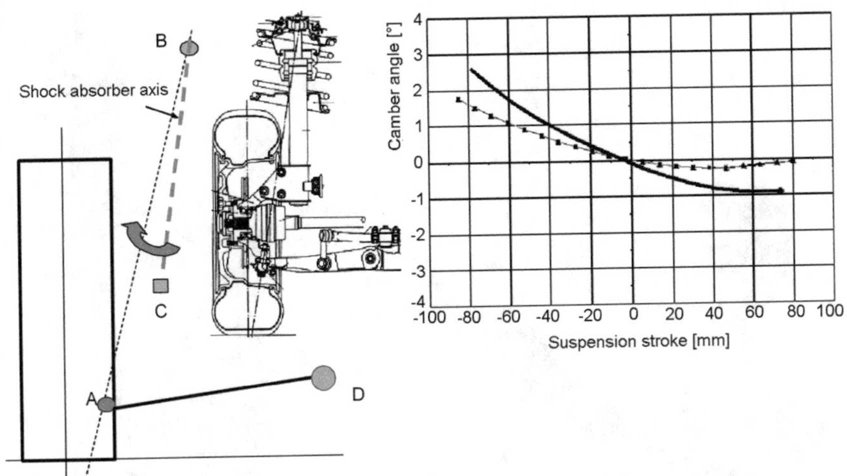

Fig. 3.73 Correction on the point D position can change the shape of the camber angle variation diagram as function of the suspension stroke

impacting comfort, and camber angle variations by cornering forces, impacting tire cornering stiffness and, therefore, handling.

At the end, the positions of the articulation of the steering mechanism will be set, as shown on Fig. 3.74; additional explanations on this subject will be given on the chapter dedicated to the steering system.

The choice of point H position (spheric joint between strut and steering rod) is determined considering the Ackermann correct steering condition, the torque permissible on the steering wheel and the position available for the steering rack.

Point I (spheric joint between steering rod and rack) is determined in order to limit toe angle variations because of the suspension stroke. On a view from the z axis, points A, E, I, H must shape a parallelogram in order to limit toe angle variation caused by longitudinal motions of the wheel.

After the preliminary design outline of the suspension, the detail design of suspension components can start, taking into account the structural mission of this part of the chassis. This mission can be considered according to two different points of view:

- mechanical integrity, defined by the fatigue and shock resistance of the suspension components;
- deformations under the applied loads or elasto-kinematic behavior.

We are going to analyze these two missions briefly.

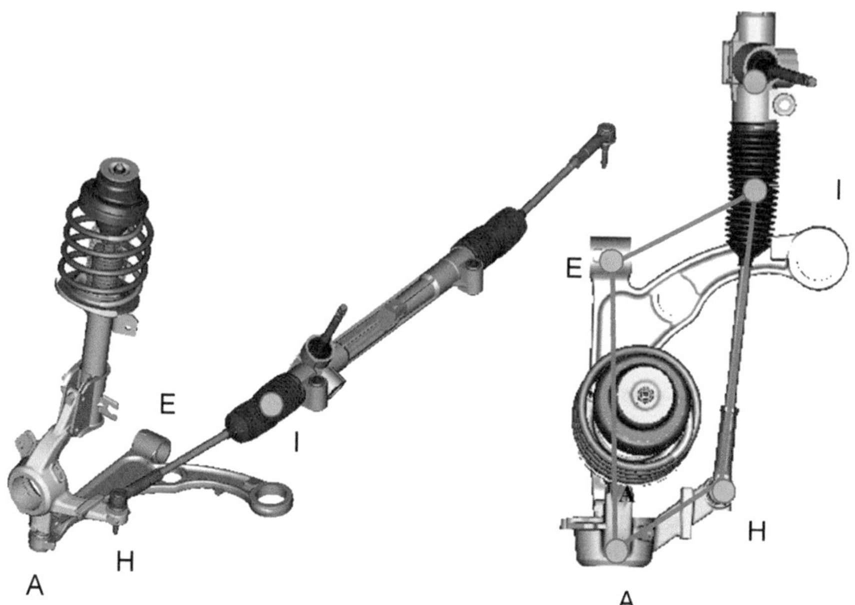

Fig. 3.74 The point E position should allow the wheel to move back against an obstacle, without sensible variations of the toe angle and camber angle

3.6.2 *Structural Integrity*

Fatigue Behavior

The evaluation of the fatigue behavior may be accomplished, through calculations or tests; in this last case dedicated facilities are used, as benches and measurements systems. In both cases, reference is made to load specifications that have the function to simulate in a synthetic way all the operational life of the component under investigation.

The load synthesis is made applying scientific methods to real load histories on road circuit chosen according to the manufacturer experience and the purpose of the vehicle.

Structural analysis can be performed on the complete suspension or on single components. In the first case it is a good practice to include in the test also part of the body or subframe structure, where the suspension is connected.

The suspension under test is usually installed on a bench including a real part of the car body; this dummy proves to be the less expensive installation fixture to be used. The wheel hub is linked to a three axis actuator, applying loads in the three directions of the car reference system.

Load and displacements, which are applied, are extracted from on road measurements, significant of the average life of the vehicle; at regular intervals of time the general integrity of the suspension is controlled (cracks, bushings or bearings deterioration, etc.).

The mission loads definition is made, as we have said, by recording forces F_x, F_y and F_z and moments M_x, M_y and M_z applied to the wheel hub on an assigned road circuit, where a similar vehicle equipped with measurement and recording devices is driven.

It is obviously very important do define the minimum life, in term of distance travelled, necessary on a given course; different courses can be travelled in sequence.

When our suspension under test has reached the target life, the following evaluations are made:

- bolts and screws thrust decay;
- main characteristics of disassembled components, to be compared with those at the beginning of the test, with particular reference to bushings, bearings, springs, stop springs, spherical joints and shock absorbers.

Usually the suspension is assembled again and the test is continued until at least three failures are reached or the life is doubled.

The suspension target life must be at least the same of the entire vehicle, usually 200,000 km for a car and in the range 300,000–800,000 km for industrial vehicle, according to the kind of application.

Broken components or not accomplishing their design specifications are analyzed and modified until they reach the desired life.

When single components are tested, forces and moments are applied to their mounting interfaces, which are calculated with mathematical models starting from forces and moments applied to the wheels. This test procedure is particularly cost effective when only single components are modified on a suspension or when tests must be repeated to gather a sufficient confidence level.

In this case simplified test are defined, which are normally more severe and which the experience has demonstrated to be equivalent to the baseline test.

As an example of this approach, we can summarize an example for a suspension arm; by tests or by calculations the component is linked to a testing block and is loaded with the wheel forces.

The test cycle can include:

- longitudinal loads, due to braking forces (at least 3,000 load blocks representing a stop brake with the highest friction coefficient);
- cornering forces, due to curves (at least 300,000 load blocks each of them simulating a bend with the highest friction coefficient).

By this example are not taken into account loads due to road asperities, because they are not so relevant to the suspension arm.

If the suspension is applied to a driving axle, also traction load at maximum friction coefficient must be taken into account.

If the part under test is interested by vertical loads also maximum loads applied by stop springs are considered; usually they are assumed to correspond to a vertical acceleration of 35 m/s^2.

For a more accurate evaluation of a component life stresses of the most loaded parts are calculated with analysis codes and their results are interpreted using a fatigue damage criterion.

The scheme of this method is the following. Let us consider that this component undergoes n different load histories ($P_k(t)$, with $k = 1, n$). P_k is a generic load applied to a point of our structure.

Using the stress analysis code stresses are calculated, at first, generated by the forces applied to the same points ($\sigma_{ij,k}(x)$, with $k = 1, n$) where loads P_k are applied.

The fatigue code can calculate the time history of stresses by a linear superimpositions of the effects, as:

$$\sigma_{ij}(x, t) = \sum \sigma_{ij,k}(x) P_k(t).$$

Suitable algorithms allow to summarize any three axial stress status in a single axis stress having the same danger; on metal components the distortion energy criterion is usually applied.

Afterwords the fatigue life of each point of the structure can be evaluated, with reference to the material characteristics.

A method frequently used is the Miner hypothesis, saying that a body which undergoes different stress levels is reaching the end of its life (rupture) when the sum of the partial damages of each of these levels reaches the value of 1. The partial damage is, by definition, the ratio between the number of cycles actually performed and the number of cycles necessary to reach a rupture at a given stress.

Also defects of the material and shape oddities must be taken into account.

A simple way to evaluate this effects is to evaluate the materials fatigue by testing samples extracted by parts of the same components or by similar components produced by the same process.

In this way many Wöhler curves can be achieved, suitable to evaluate the influence of the main factors on the material fatigue life.

Misuse

It is a common practice to take into account a reasonable *misuse*: misuse is a not correct, but possible use of the component; the idea behind is that, after one of this misuses, the driver understands to have made a mistake, that probably learned to never repeat again, but it would be unreasonable to have the mission interrupted for this mistake only.

A minimum number of misuses must be allowed without rupture; in the case of suspensions misuse usually implies the applications of shocks.

Certain obstacles must be afforded without damages or permanent deformations; a typical test example consists of driving at 50 km/h on an obstacle 100 mm high and 100 mm long.

A different kind of misuse test is hitting a wheel against a sidewalk, as it could happen driving in a curve at an excessive speed.

In this case the structural integrity of the suspension cannot be guaranteed, particularly for the lower arm; occupant safety must be, instead, protected within certain limits of acceptable severity.

After a maneuver of this kind, the following results must be obtained:

- arms mustn't brake as to maintain their guidance to the wheel until the car is stopped;
- permanent deformation on the arm must be visible and such as to discourage any prosecution of the trip.

Crash Test

By automotive engineering crashes against external obstacles or other vehicles make reference to a standard test specified by law.

Suspensions must be designed as to not intrude into the passenger compartment as a result of a standard crash test.

These structures must prove to collapse in a non aggressive way; on the other side subframes, wheels and other elements have to play an important role into the entire car structure collapse, after a crash test, to conserve the passenger compartment geometrical integrity.

3.6.3 Elasto-kinematic Behavior

After the explanation on different suspension kinds and after having analyzed their advantages and the disadvantages, it is clear that a certain technical hierarchy could be established, ranking suspension types by increasing function, but also by increasing cost.

Front suspension could be ranked starting from Mc Pherson to double wishbones with virtual centers, while rear suspension from trailing arms and twist axles to multilink.

If the designer decision is easy for a high performance car with no problems on price, or for a small economy car, where costumers are keen to sacrifice for a contained price, the choice is rather difficult for most medium cars.

Many times very good results can be achieved by a cheaper suspensions if the design is sufficiently refined, particularly as far as tire sizing and elastic bushings are concerned.

Design Rules

Some design rules can be set to address suspension design to obtain good handling and comport performance.

Tires Working Angles

Let us start from the tire working angles and their impact on handling performance.

If it is true that the working angles variations, in consequence of the body roll, can correct significantly handling properties by making the vehicle more or less understeering, it is also true that such corrections are delayed as compared with the maneuver starting time; the best rule is to design suspension mechanism without any variation of toe angles and camber angles with the suspension stroke.

The desired level of understeering may be achieved by a convenient tire sizing and by optimizing weight distribution also in consequence of transfers due to inertia forces.

This condition on toe angles is relatively easy to be reached; as far as camber is concerned the condition can be reached only if camber angles are recovering the roll angle effect, but for rigid axle where the problem is not existing.

This condition should not be assumed without reasonable criticism; the camber stiffness of selected tires is fundamental.

What must be reduced is the additional cornering force rising from camber angles; if on a given tire camber stiffness is about 1/10 of the cornering stiffness (that means the cornering force rising from a side slip angle of 1°, with no camber, is the same rising from a camber angle of 10°, with no side slip angle), tires with a lower camber stiffness can be manufactured making a simpler Mc Pherson suspension performing like a more sophisticated double wishbone one.

On prevailing use conditions toe angles and camber angles must be set to zero to minimize wear and rolling resistance.

For the same reason is better to minimize track variations because of suspension stroke.

Comfort

Considering comfort, a first rule can be set on total suspension flexibility determined by coil springs and stop springs; a convenient value for the first natural frequency of the sprung mass should be set around 1 Hz, optimum for a human body in sitting position.

This condition implies that flexibility should be proportional to the vertical suspended mass or that suspension must become stiffer as the pay load is increased.

Suspension total stroke must be limited for practical reasons: wheel case dimension and car lifting with a jack, in order to change a punctured wheel; it should be at least comparable with the extension of maximum road asperities to be afforded.

This condition has, as consequence, a maximum stroke of about 70 mm in both direction (extension and compression) starting from the reference condition of vehicle at minimum load.

In the compression direction, the stroke necessary to reach the maximum pay load condition must be also added. By wide load variations (station wagons, minivans, large sedans, etc.) suspension with trim control are necessary.

To think about vertical obstacles only is a rough approximation of the reality. Because of the vehicle motion each obstacle hits the wheel also in an horizontal direction. It is fundamental to assure to each wheel also a horizontal suspension; longitudinal stiffness should be no more then ten times of the vertical one.

This result may be obtained by designing a mechanism as to have a convenient wheel base variation with suspension stroke; for front wheels, wheel base must decrease by compression stroke, while the rear ones must do the opposite. In addition elastic bushings must assure an additional convenient longitudinal flexibility.

In the first case the suspension is again a mechanism with one degree of freedom, but the path of center of the tire patch has been conveniently inclined in the vertical xz plane. In the second case an additional degree of freedom has been added: the longitudinal motion.

The suspension mechanism must never comply horizontally with toe angles variations, but in a reduced way for stabilizing braking.

Steering

Further considerations can be made for the steering mechanism.

Contrary to what commonly thought reaching Ackermann condition is not vital until the difference between the effective and the theoretical angle (the so-called *steering error*) is lower as about 4 deg at maximum steering angle. In fact side slip angles is partially vanishing steering angle accuracy.

King-pin offset can affect steering wheel torque during braking; the rule for a negative king-pin offset is universally adopted, because is stabilizing the car path during braking (if the braking force increases on one wheel, the added steering action left by the late response of the driver, is able to counterbalance the yaw moment caused by braking forces unbalance).

The longitudinal trail, by cornering forces, can apply to the steering wheel an additional self aligning torque; this contribution is particularly important because can compensate tire self aligning torque decrease by large side slip angles, but still with good cornering force.

It is better not to exaggerate with front wheel driven cars, because of the aligning torque caused by traction forces.

Caster angle is bound to the longitudinal trail; an appropriate value can have a positive impact on turning radius, because the king-pin inclination is causing additional camber angles on wheels.

Elasto-kinematic Calculation

Independent Suspensions

Elasto-kinematic behavior predictions can be made by modelling an articulated system including all spherical or cylindrical couples included in the mechanism. If couples are considered to be stiff, the kinematic behavior can be predicted; if they are supposed to be compliant to the applied forces, the elasto-kinematic behavior will be predicted.

It is convenient to calculate the suspension configuration by kinematic contributions only ant then to add to this configuration the effect of elastic displacements. This procedure is approximate, because it doesn't take into account contributions of the modified position on the values of forces; this approach is nevertheless acceptable in practical cases.

By this kind of calculations, conventionally, all displacement and rotations of the wheel are referred to the car reference system and not to the wheel; in other words the body is supposed to be fixed at a test bench block and the actual position of the wheel is calculated by imposing to the hub a given displacement along the z axis.

At this displaced position a certain force F_z will be necessary to maintain the elastic suspension system in equilibrium; when necessary, to this new displaced position forces F_x and F_y will be applied to examine their additional effect.

It must be recalled that toe and camber angles that are so evaluated cannot be introduced into the formulae used to calculate tire forces, as they are, but they must be corrected with the effects of body and wheel rotations, as roll angle, vehicle side slip angle and wheel steering angle.

Usually the study of displacement and rotations caused by suspension stroke are separated from the study of the effect of applied forces.

In the first case, vertical loads (at the wheel center or at the contact patch center), toe, camber, track, wheelbase variations are calculated with reference to the static configuration, as function of the suspension stroke (referred to the body) and for different steering angles.

These characteristics may have a different behavior in the case of symmetric stroke (the wheels of the same axle have the same stroke) or asymmetric (when the displacements are opposite, with the same absolute value).

By elasto-kinematic calculations, suspension elements and articulations to the body have further displacements because of the applied loads.

By calculating the effect of additional forces but the vertical ones, the procedure is the same; it should be remembered that the path of the center of the contact patch, where forces are applied is not perfectly directed along the z axis; therefore the application of F_x and F_y forces causes additional suspension stroke.

At this new position, the effect of bearing and bushings deformations can be calculated.

On Fig. 3.75 some typical diagrams are shown that are obtained with this kind of analysis. The diagram at the upper left reports stroke as function of the vertical force. The stroke Δz is measured starting from the reference conditions of car with full

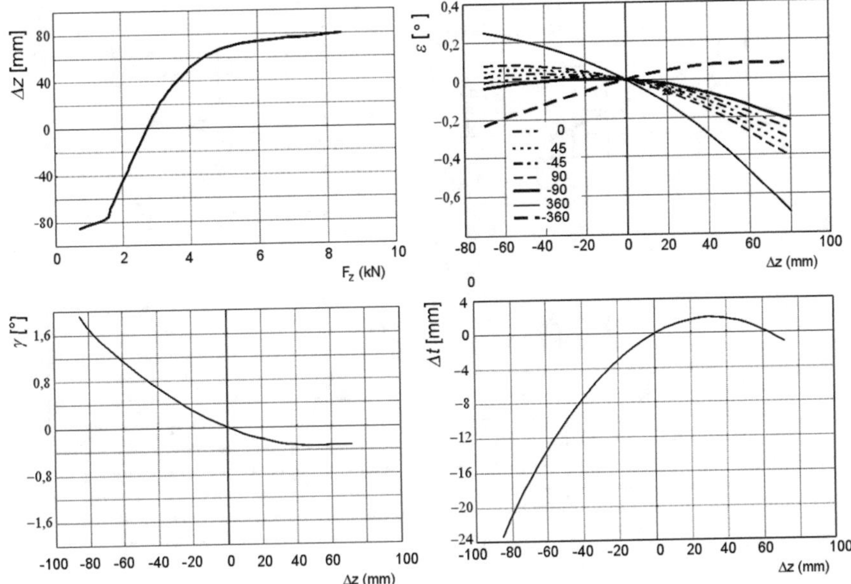

Fig. 3.75 Typical diagrams of suspension geometric parameters. From the top right, going clockwise, we can see suspension stroke diagram as function of vertical load, toe angle diagram as function of suspension stroke for different steering angles, camber angle diagram as function of suspension stroke and semi track diagram as function of suspension stroke

tanks and two people with their baggage; Δz is positive in compression and negative in extension.

There is a part of the line almost linear, when only the coil spring is working (the contribution of elastic bushings is very small), while at increasing loads the curvature of the diagram is increasing till a maximum value, as stop spring start to work. Suspension flexibility is defined as the partial derivative $\partial z/\partial F_z$, calculated for $\Delta z = 0$. The total suspension stroke is done by the difference between the maximum compression and the maximum extension.

The diagram at the upper right shows, instead, toe angle as function of the suspension stroke, for different steering angles, while the diagram at the lower left concerns camber angle as function of the suspension stroke.

Finally, the last diagram reports semi track variation as function of the suspension stroke.

These diagrams can be used to calculate also the same magnitudes as function of roll angle by reading the diagrams at symmetric values, which easily provide the roll angle. It should be noticed that each of these diagrams as function of roll are valid only for the value of the symmetric stroke, we have considered as starting point of the roll motion.

On the following Fig. 3.76 are represented the effects of the applied forces on the same characteristic parameters.

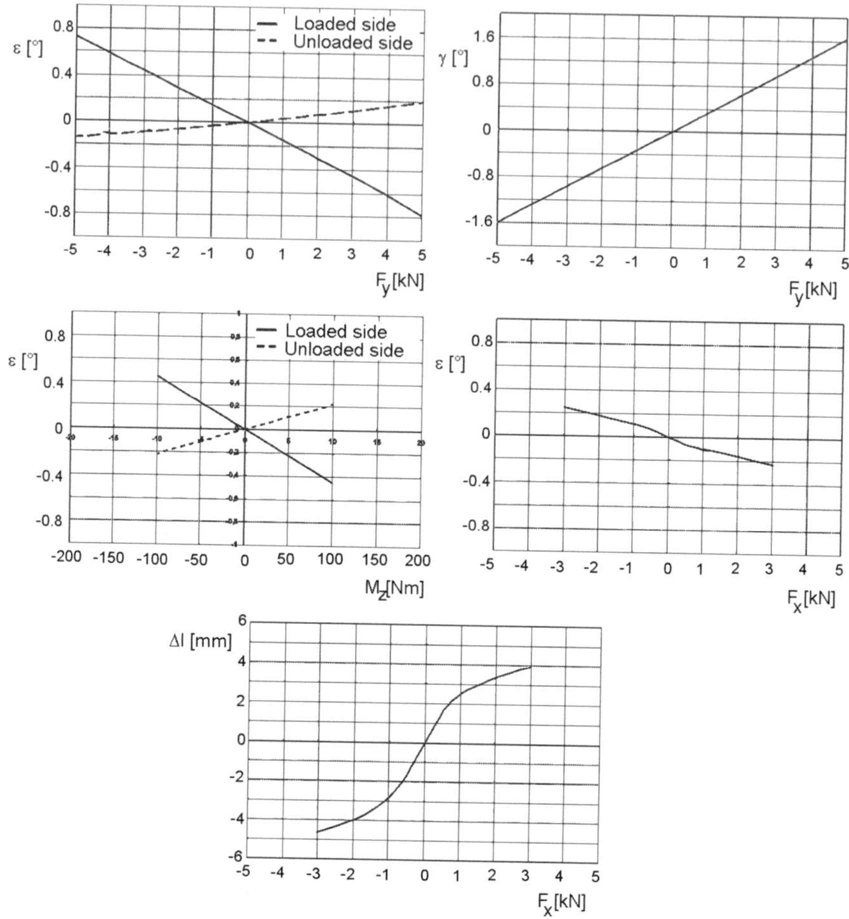

Fig. 3.76 Typical diagrams describing the elasto-kinematic behavior of a suspension; notice how steering rack can couple the two wheels of the same axle

Toe angles variations, as function of the cornering force F_y, are reported on the upper left diagram. In this diagrams the force is applied to a suspension only (called loaded side), but the effects are also measured on the other suspension of the same axle (called unloaded side): notice how the unloaded side wheel responds, because of the mechanical connection established by the steering rack.

The following diagrams show the camber angle variation, as function of the cornering force F_y.

Toe angle variations, as function of the self aligning torque M_z is shown on the lower left; in this case too the effect on the unloaded wheel can be observed.

The last two diagrams show respectively toe angles variation and wheelbase variations as function of the braking or driving force F_x. This last diagram is useful to evaluate the longitudinal absorption capacity of longitudinal shocks.

Very often these diagrams are linearized by substituting these curves with their tangent at the origin, in order to be input in simple mathematical models: in this case the derivatives of this parameter are considered.

The main ones are $(t)_{,z}$, $(t)_{,\phi}$, $(\gamma)_{,z}$, $(\gamma)_{,\phi}$ and $(\varepsilon)_{,z}$, $(\varepsilon)_{,\phi}$.

Rigid Axles

If wheels are connected by a rigid beam, this last must have two degrees of freedom with respect to the body, one translation in the z direction and one rotation around the roll axis. This rigid beam can be guided, as we have seen, by linkages or by the leafs springs or by both.

The elasto-kinematic behavior must, in this case, be described by two coordinates; a common choice is to have diagrams as function of roll angle, with suspension stroke as parameter.

The solution with leaf springs, with structural function, has the disadvantages of approximating the correct kinematic behavior in which the axle can move only along z coordinate and rotate about the roll axis, in a poor way. The stiffness in x and y directions, although far higher than that in z direction, is not high enough, as is the stiffness for rotations about the y axis and, above all, any rolling motion is linked to a steering of the whole axis (roll steer). In other words, the derivative $\partial\delta/\partial\phi$ can be quite high (Fig. 3.66). The latter characteristic is due to the fact that the motion of the points in which the axle is connected to the springs is not exactly vertical, although the deviation from the vertical direction of the trajectory shown in the figure is exaggerated, to show the phenomenon.

The solution with guiding linkages can avoid this poor characteristic and improve the behavior also by longitudinal flexibility.

In all cases track and camber angle are constant, this last if tire vertical deformation is neglected. Therefore $(t)_{,z} = (t)_{,\phi} = (\gamma)_{,z} = (\gamma)_{,\phi} = 0$.

Roll Axis

Roll axis is the instantaneous rotation axis of the vehicle, when the body is in symmetric condition (i.e.: $\phi = 0$). It is an instantaneous rotation axis because the roll motion is not a pure rotation around a physical articulation; therefore this axis can be defined for small rotations only, starting from a well defined initial condition.

The points where roll axis is crossing vertical planes through wheel centers of same axles are called suspension *roll centers* CR.

For symmetry reasons, the roll axis must lie in the symmetry plane of the vehicle (xz plane) and therefore also the roll centres of the suspensions must lie in it. Note that in case of a two-axle vehicle the roll centre of each suspension can be determined from the characteristics of the relevant suspension only and that the roll axis can be defined as a line connecting the roll centres of the two suspensions. If the vehicle has more than two axles, the roll centres of the suspensions need not be aligned: A roll axis still exist, but it does not pass through the roll centres of the single suspensions, considered as insulated.

The roll centre of each suspension can also be defined as the point on a plane perpendicular to the ground and to the symmetry plane in which the application of a lateral force F_y to the vehicle body does not cause any roll. The two definitions obviously coincide.

Other important points are:

- the center of rotation of the body with respect to the wheels CW and
- the center of rotation of the wheels with respect to the ground RS.

Rigid Axles

As in a rigid axle the two wheels are rigidly connected, the two points are located in the symmetry plane. Moreover, if the compliance of the tires is neglected, the wheels cannot rotate with respect to the ground: Points CW are located at infinity on the intersection between the ground and the plane parallel to yz plane passing through the centres of the wheels and the two points BW coincide with the roll centre RC.

If the compliance of the tires is accounted for, points RS lie on the symmetry plane slightly below the ground, but their positions are not exactly defined as they depend on the deflection of the tires and hence on the forces applied to them; it is however possible to define a zone in which they lie. Also points CW coincide and are located in the symmetry plane of the vehicle below the roll centre CR.

In some case points CW are physically defined as there is a material hinge between the axle and the body (Fig. 3.77a–d). If the axle is guided laterally by leaf springs, CW is on the symmetry plane, at the level of the attachment of the springs to the body (Fig. 3.77e). The lateral deflection of springs causes CW to be located at a lower level and the inflection of the tires causes the roll centre CR to be located below CW.

With lateral deformations of tires, the center will be at the intersection of the perpendiculars to the displacement vectors, caused the application of M_x and F_y, as in Fig. 3.77f.

A four-link suspension is shown in Fig. 3.78. To obtain the position of the roll centre, the intersections A and B of the axes of links 1-1′ and 2-2′ must be found first. They lie in the symmetry plane of the vehicle. The roll centre is found as the intersection of line AB with the plane perpendicular to the ground containing the centres of the wheels. If two links are parallel (say links 1 and 1′) the intersection is at infinity and line AB is parallel to the projection of the relevant links on the symmetry plane.

The situation for a three-link suspension is similar (Fig. 3.79). The only difference is that point B is the intersection of the axis of the transversal link with the plane of symmetry.

In the case of a rigid axle with leaf springs, the roll centre is located at the intersection between the projection on the symmetry plane of a line connecting the points in which the springs are connected to the body and the perpendicular to the

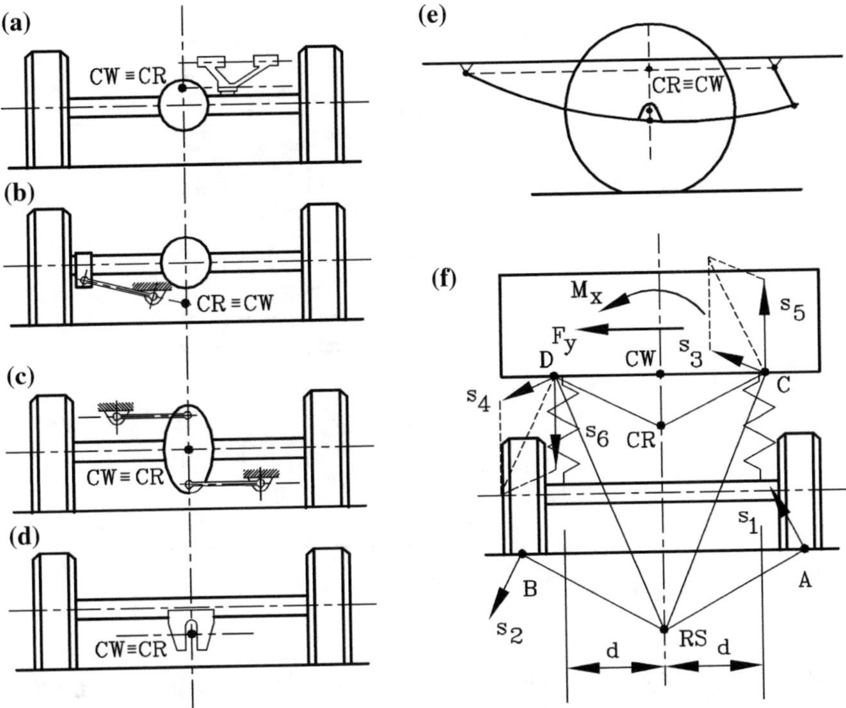

Fig. 3.77 From **a** to **e** position of point CW on some rigid axle suspensions; tires are stiff. Scheme **f** shows the position of RS, CR and CW, taking into account tires deformation

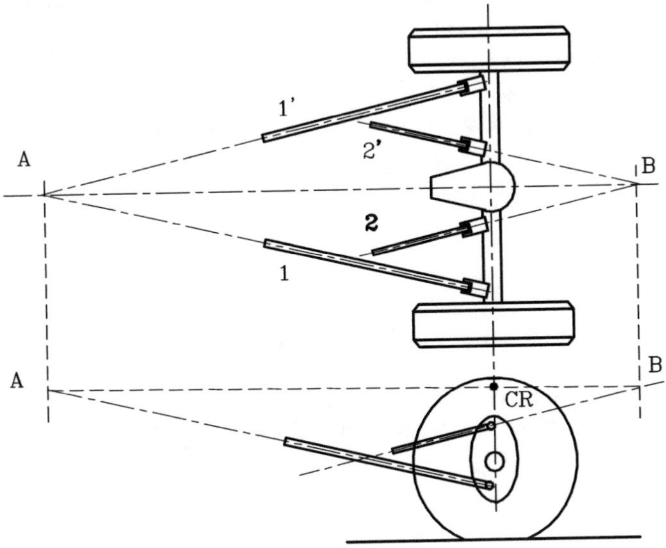

Fig. 3.78 Four link suspension. Position of point CR

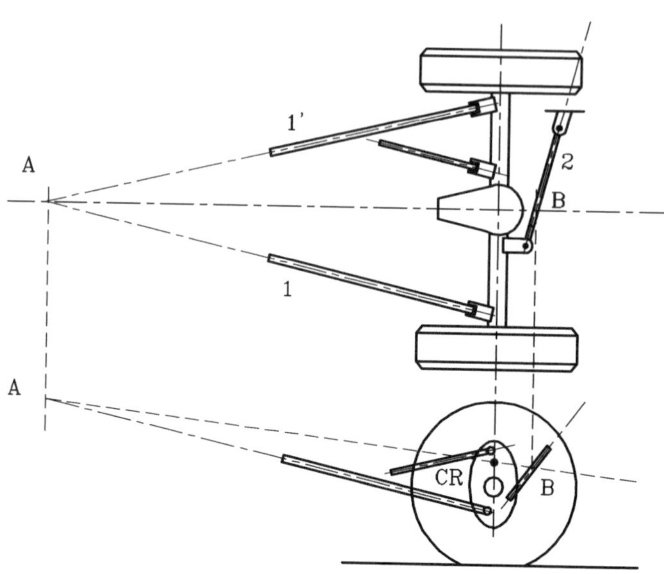

Fig. 3.79 Three-link suspension. Position of point CR

ground through the centre of the wheel (Fig. 3.77e). If the compliance of the tires is not neglected, to find an approximate location of CW, CR and RS it is possible to proceed as in Fig. 3.77f: A force F_y and a moment M_x are applied to the body: Displacements s_1, s_2, s_3 and s_4 of points A, B, C and D are due to the inflection of the tires, while the compliance of springs causes displacements s_5 and s_6 of points C and D. The intersection of the lines perpendicular to s_1 and s_2 locate point RS; CW is the intersection of the lines perpendicular to s_5 and s_6 and that of the perpendiculars to the resultants of s_3 and s_5 and of s_4 and s_6 locate point CR.

The derivative $(\gamma)_{,\phi}$ can be approximately computed as:

$$\frac{\partial \gamma}{\partial \phi} = \frac{\chi}{\chi + \Pi} , \tag{3.6}$$

where:

$$\chi = 2kd^2 + \chi_t ,$$

(k is the stiffness of the springs and χ_t that of the anti-roll bar) and Π are respectively the rolling stiffness of the suspension and of tires.

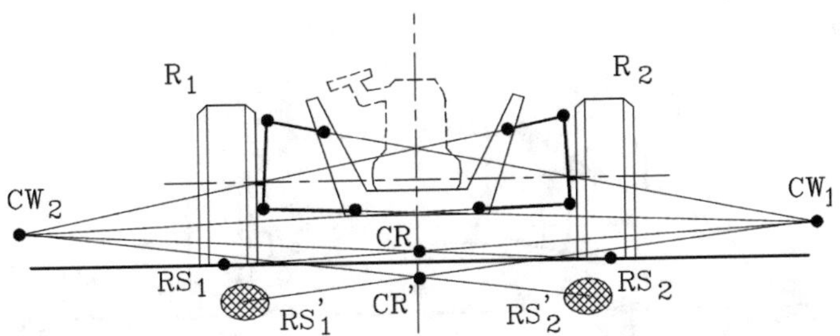

Fig. 3.80 Position of roll center CR for a low double wishbone suspension with $1'2'$ and $3'4'$ arms parallel to x axis. CR: position with rigid tires; CR$'$: position with deformable tires

Independent Suspensions

Note that in general the wheel motion is not planar and as a consequence the study of the kinematic behavior is not easy. Nowadays it is however easy to obtain the exact kinematics of any suspension by using computer generated trajectories.

With double wishbone suspensions, the points CW1 and CW2 of the two wheels are located at the intersection of the upper arms and lower arms axis that can cross outside of the vehicle (Fig. 3.81a,) or inside (Fig. 3.80). It is possible to obtain $\partial t/\partial z = 0$, by imposing that CW1 and CW2 are on the ground (Fig. 3.81b), but this condition can be obtained for a certain value of the pay load. If $\partial \gamma/\partial \phi$ must be set to zero, the points CW1, CW2 and CR must lie on the symmetry plane (Fig. 3.81c).

If the wheels were thin rigid discs, points RS1 and RS2 would coincide with the centres of the contact areas. If the compliance of tires is accounted for their approximated position can be located under the ground, slightly inboard of the centres of contact. By connecting points CW1 and RS1 and points CW2 and RS2 and intersecting such lines the roll centre CR, which lies in the symmetry plane, can be located. In the case of transversal articulated quadrilaterals, it is usually close to the ground or, if the deformation of the tires is considered, even below it. If the axes of the hinges of the two triangular linkages are not horizontal or not parallel (Fig. 3.82) the determination of the roll centre and of the motion of the latter is far more complicated.

The construction necessary to obtain the different centers of a Mc Pherson suspension is shown on Fig. 3.5.

A different approach is that of using trailing arms (Fig. 3.47). The arms can be hinged to an axis which is perpendicular to the symmetry plane of the vehicle but this is not always the case. In the first case the track remains constant,

$$\frac{\partial t}{\partial z} = \frac{\partial t}{\partial \phi} = 0$$

and the camber angle does not change in the vertical motions and is equal to the roll angle:

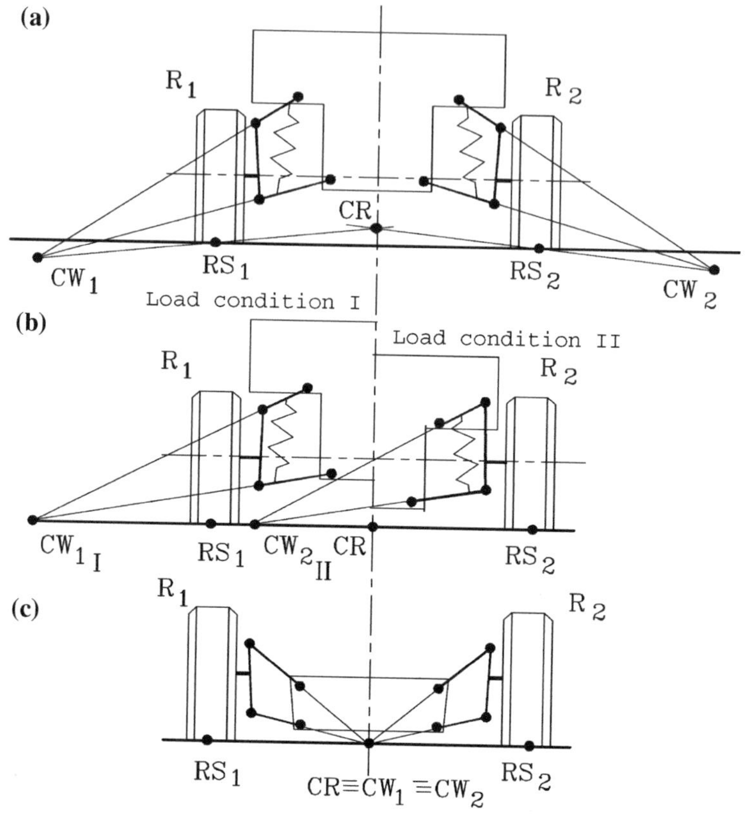

Fig. 3.81 Low double wishbone suspensions

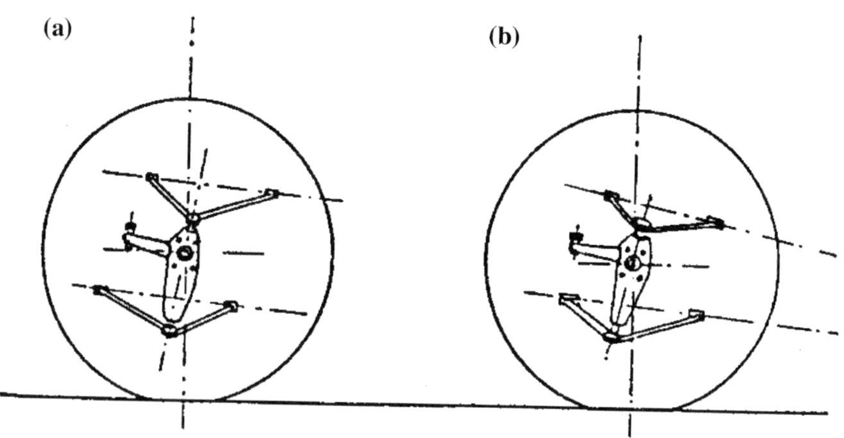

Fig. 3.82 Low double wishbone suspension with non horizontal articulation axis (**a**) and non parallel axis (**b**). The inclination of these axis determines the position of the pitch center

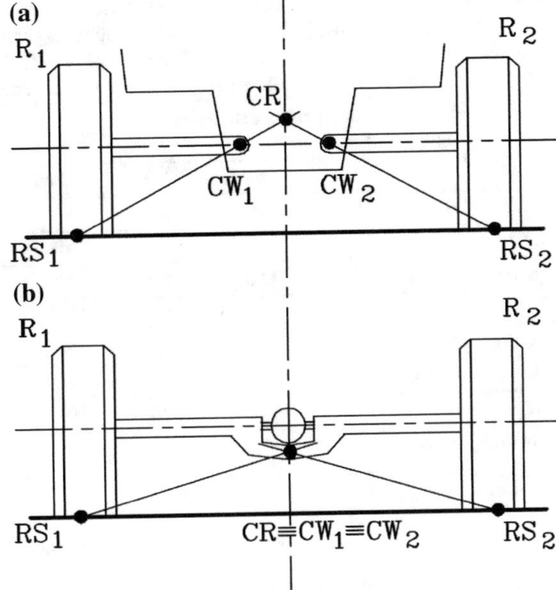

Fig. 3.83 Swing arms suspension with different (**a**) or coincident (**b**) hinges

$$\frac{\partial \gamma}{\partial z} = 0, \quad \frac{\partial \gamma}{\partial \phi} = 1 .$$

If the compliance of tires is neglected, the roll centre is on the ground (Fig. 3.47a) or slightly below (Fig. 3.47b).

Another solution is that based on swing arms. The hinges of the arms can be located at different points (Fig. 3.83a) or be coincident (Fig. 3.83b). The roll centre can be quite high on the road and the values of $\partial t/\partial z$ and of $\partial t/\partial \phi$ cannot be small. Swing arms can be connected to the powertrain block instead of being hinged to the body, as it was common in small cars with rear engine. The fact that the engine was suspended through rubber blocks decreased the precision requirements of the suspension.

Conclusions

The main kinematic characteristics of the various type of suspensions are summarized in Table 3.1. This overview does not cover all the solutions which have been used or have been suggested like, as an example, the multilink solution, which is common on luxury cars. It is similar to a suspension with transversal quadrilaterals but the greater number of distinct elements allows a more detailed adjustment.

Table 3.1 Main characteristics of some suspensions. TA: Trailing arms; SAL: Swinging arms (limit case); SA: Swinging arms; DWP: Double wishbone with parallel arms; DWL: Double wishbone (limit case); DW: Double wishbone; MP: Mc Pherson. A, CR high on the ground; B, below or on the ground, C, about at the wheel center D, over the wheel center; E, on the ground

	Rigid axle	Independent suspensions						
		TA	SAL	SA	DWP	DWL	DW	MP
$\partial t/\partial z$	0	0	#0	#0	#0[a]	0	#0[b]	#0
$\partial t/\partial \phi$	0	0	0	#0	#0	#0	#0	#0
$\partial \gamma/\partial z$	0	0	#0	#0	#0	#0	#0	#0
$\partial \gamma/\partial \phi$	0	1	0	#0	1	#0	$> 0 < 1$	$< 0 > 1$
CR Pos.	A	B	C	D	B	E	B	B
CR Var.	0 ($\approx$0)	0	0	0	#0	0	#0	#0
$\partial i/\partial z$[c]	0	#0	#0	#0	0	0	#0	#0

[a]0 in one configuration
[b]0 in two configurations
[c]i: kingpin inclination

Pitch Axis: Anti-dive and Anti-squat Configurations

When the vehicle accelerates or brakes a load transfer between front and rear wheels occurs. This causes the body to pitch up (lift or squat) or down (dive).

As for the roll, a pitch axis can be defined. The pitch axis is the axis of instantaneous rotation for pitch. It is an instantaneous rotation, because the pitch motion is not a pure rotation around a well defined axis, but this axis can be defined only for small rotation starting form a defined position, such as that where x axis is parallel to the X axis of the inertial reference system.

Apparently, as forces F_{z_1} and F_{z_2} can be approximated as:

$$\begin{cases} F_{z_1} = F^*_{z_1} - m\dfrac{h_G}{l}\dot{V} \\ F_{z_2} = F^*_{z_2} + m\dfrac{h_G}{l}\dot{V} \, , \end{cases} \tag{3.7}$$

where forces $F^*_{z_i}$ are those occurring when the vehicle does not accelerate, the lift of the front and the rear of the body are respectively:

$$\begin{cases} \Delta z_1 = m\dfrac{h_G}{lK_a}\dot{V} \\ \Delta z_2 = -m\dfrac{h_G}{lK_r}\dot{V} \, , \end{cases} \tag{3.8}$$

where K_f and K_r are the vertical stiffness of the front and rear suspensions. The pitch angle due to an acceleration is then:

$$\theta = \frac{1}{l}(-\Delta z_1 + \Delta z_2) = -m\frac{h_G}{l^2}\dot{V}\left(\frac{1}{K_a} + \frac{1}{K_r}\right) . \tag{3.9}$$

A positive value of θ occurs when the vehicle dives (pitches down) as it occurs with a negative acceleration, hence the minus sign in the formula.

This expression is however an oversimplification, for two reasons: Firstly the longitudinal forces due to driving or braking wheels can cause themselves a pitching moment due to the coupling of the suspensions and, secondly, driving and braking torque reactions can be applied, at least partly, to the suspensions instead of the body, inducing further pitching. Both effects cause pitching even in constant speed driving.

If suspensions allow wheels to move also in x direction, i.e. if the characteristic $\partial x/\partial z$ is not vanishingly small, a fraction $F_x \partial x/\partial z$ of force F_x acting between the road and the wheel acts on the suspension and causes pitching. Equation (3.7) then becomes:

$$\begin{cases} F_{z_1} = F^*_{z_1} - m\frac{h_G}{l}\dot{V} - \left(\frac{\partial x}{\partial z}\right)_1 F_{x1} \\ F_{z_2} = F^*_{z_2} + m\frac{h_G}{l}\dot{V} - \left(\frac{\partial x}{\partial z}\right)_2 F_{x2} \end{cases} \tag{3.10}$$

and Eq. (3.9) transforms into:

$$\theta = -m\frac{h_G}{l^2}\dot{V}\left(\frac{1}{K_a} + \frac{1}{K_r}\right) - \left(\frac{\partial x}{\partial z}\right)_1 \frac{F_{x1}}{lK_a} + \left(\frac{\partial x}{\partial z}\right)_2 \frac{F_{x2}}{lK_r} . \tag{3.11}$$

If only longitudinal forces needed to accelerate or to brake the vehicle are considered and the percentage of the longitudinal force assigned to the front axle is k_l, it follows:

$$F_{x_1} = k_l m\dot{V} , \qquad F_{x_2} = (1 - k_l)m\dot{V} .$$

Equation (3.9) yields:

$$\theta = -m\frac{\dot{V}}{l}\left[\frac{h_G}{lK_a} + \frac{h_G}{lK_r} + \frac{k_l}{K_a}\left(\frac{\partial x}{\partial z}\right)_1 - \frac{(1 - k_l)}{K_r}\left(\frac{\partial x}{\partial z}\right)_2\right] . \tag{3.12}$$

Obviously Eq. (3.9) holds in the case of acceleration and braking alike, provided that the sign of $\dot{V}$ is correct and a suitable value for k_l is used.

Consider for example the trailing arm suspension of Fig. 3.84a. With simple geometrical reasoning it is easy to assess that:

$$\left(\frac{\partial x}{\partial z}\right) = \frac{e}{d} . \tag{3.13}$$

A similar equation holds also for the suspension of Fig. 3.84b. If a torque M_y is applied to the sprung mass, it causes an increase of the force acting on the spring equal to M_y/d. As the torque linked to a generation of driving or braking forces is equal to $-F_x R_l$, the result is that the application of the braking torque to the suspension can be accounted for by substituting:

$$\left(\frac{\partial x}{\partial z}\right)_i + \left(\frac{R_l}{d}\right)_i \quad \text{a:} \quad \left(\frac{\partial x}{\partial z}\right)_i .$$

Note that d is positive when point A is in front of the wheel and negative otherwise.

The driving torque is applied to the unsprung mass in the case of live axles, while in De Dion axles and independent suspensions it is applied directly to the vehicle body and this correction does not apply. Braking torques are on the contrary applied usually to the unsprung masses, so the term in R_l/d must always be accounted for. However, if the torque transmission between the sprung and the unsprung masses is supplied by linkages which prevent any relative rotation about y axis, as d tends to infinity, these effects are minimized.

The above relationships allow one to design the suspensions to compensate, usually partially, for squat or dive. A total compensation occurs when Eq. (3.11) yields $\theta = 0$. If:

$$\frac{h_G}{l K_a} + \frac{k_l}{K_a}\left(\frac{\partial x}{\partial z}\right)_1 = 0 \tag{3.14}$$

the front of the car does not lift in acceleration or dive in braking, while if:

$$\frac{h_G}{l K_r} - \frac{(1-k_l)}{K_r}\left(\frac{\partial x}{\partial z}\right)_2 = 0 \tag{3.15}$$

the rear does not squat in acceleration or lift in braking.

Note that in case of a single driving axle either $k_l = 0$ or $k_l = 1$ and both front and rear compensations cannot be performed together. To obtain a complete compensation the term in square brackets in Eq. (3.11) must vanish and the front of the car must dive to compensate for the squat of the rear axle in front drives or the rear must lift in rear drives.

In case of braking a total compensation of the front axle leads to the condition:

$$\frac{k_l}{K_a}\left[\left(\frac{\partial x}{\partial z}\right)_1 + \left(\frac{R_l}{d}\right)_1\right] = -\frac{h_G}{l K_a}, \tag{3.16}$$

i.e.:

$$\frac{k_l}{K_a}\left(\frac{e + R_l}{d}\right)_1 = -\frac{h_G}{l K_a} \tag{3.17}$$

and that of the rear axle to:

$$\frac{(1-k_l)}{K_r}\left[\left(\frac{\partial x}{\partial z}\right)_2 + \left(\frac{R_l}{d}\right)_2\right] = \frac{h_G}{l K_r}, \tag{3.18}$$

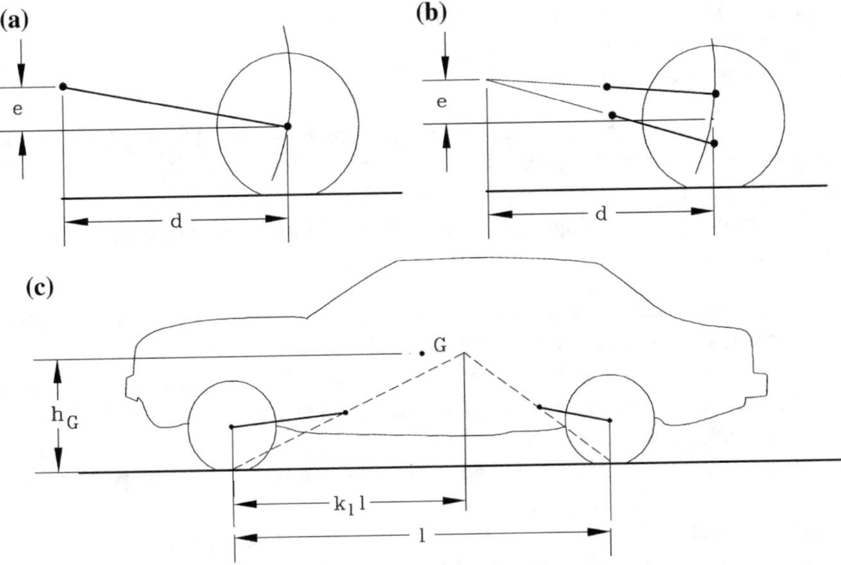

Fig. 3.84 Relationship between $\partial x/\partial z$ and suspension geometry, for trailing arms (**a**) and double wishbone (**b**) suspensions. If arms are parallel, $d \to \infty$. How to establish the anti-dive effect (**c**)

i.e.:

$$\frac{(1-k_l)}{K_r}\left(\frac{e+R_l}{d}\right)_2=\frac{h_G}{l K_r}\,. \tag{3.19}$$

A simple geometrical construction is shown in Fig. 3.84c; if pivots are on the dotted lines, a complete compensation in braking is obtained while if they lie below the lines dive compensation is only partial. If they are above the line, the front will lift and the rear will squat in braking.

A similar scheme could be draft for traction conditions; in this case the dotted line should start at the wheel center and not on the contact point.

Usually no complete anti-dive compensation is obtained for many reasons, psychological (a flat braking is not desirable) and objective (complete anti-dive compensation can lead to overcompensation of acceleration squat and to geometries which are poor both for comfort and performance).

3.6.4 Bench Testing Methods

The most relevant problems of testing a suspension are those that are common to the development of remaining vehicle components that are:

- reducing lead time to significant prototypes;
- improving velocity of performing tests;
- containing testing costs.

For these reasons is preferable to test suspensions as outside of the road as possible, using dedicated indoor benches; obviously the final approval will, in any case, be given after an on road test of the complete vehicle.

The advantage of this approach is bound to the simplification of the system to be analyzed.

- The test object can be simplified, by limiting prototypes to the really new components and by utilizing remaining ones taken from the shelf of the production shop. This approach can be used for both a suspension to be bench tested and to a vehicle, drivable on the road, with the new suspensions. Vehicle of this kind are usually called *mule cars*.
- Test outputs can be easily obtained in an objective way for prototype simplicity and because it is not tested on the road; these outputs can be elaborates through mathematical models, to predict the entirely new vehicle performance.
- Bench test can be easily automatized, with benefits on costs and on test results confidence level.

This approach is clearly advantaging only if test results have been anticipated by calculations; in this case the most relevant issues only are investigated, while the less relevant ones can be totally neglected.

Elasto-kinematic Bench

Bench tests, finalized to elasto-kinematic behavior measurement are performed essentially in order to:

- verify the accomplishment of the projected characteristics;
- determine the input data of vehicle dynamics simulation models, used to analyze and verify the performance of a prototype car or interpret the performance of a competitor's car (*reverse engineering*).

Generally speaking an elasto-kinematic bench is a system able to impose to suspensions of a real car or of a test fixture, which is fixed on the bench, an assigned displacement or set of forces.

The bench is composed by:

- actuators (can be hydraulic or electric);
- actuators control system;
- measurement and recording system;
- a suitable software package for data acquisition and elaboration and hardware control.

Fig. 3.85 Elasto-kinematic bench with hydraulic actuators and anthropomorphic arm measuring system. Hydraulic actuators

As an example, an elasto-kinematic bench with hydraulic actuators and anthropomorphic arm measuring system is briefly described.

Actuators are composed by a couple of moving platforms, shown on Fig. 3.85, able to apply assigned displacement and forces to the center of the contact patch of the tire or to the wheel hub.

Displacements can be parallel or opposed, along the three axes; also rotations can be imposed around the z axis; both displacement and rotations are applied according to sinusoidal laws at very low frequency (≈0.005 Hz) within the desired measuring range.

The possibility to apply load cycles is essential to evaluate frictions and hystereses.

Also assigned vertical or cornering forces and self aligning torque can be applied.

The anthropomorphic arm system of Fig. 3.86 allows to detect the spatial motion of the wheel hubs of the same axle, consequent to the application of the said displacements and forces; three displacements and three angles are monitored.

The body is fixed through the structure 1 and is considered as the reference system; measurements can be repeated with different steering angles.

While interpreting measurements body displacements must be also evaluated; when assessing the suspension only, their contribution should be eliminated; they

Fig. 3.86 Elasto-kinematic bench with hydraulic actuators and anthropomorphic arm measuring system. The anthropomorphic arm detects the spatial wheel motion, consequent to forces application or suspension stroke

are to be taken into account, in any case, when the entire car behavior is under investigation.

We complete this example by reporting some of the technical specifications of this bench:

vertical stroke:	± 150 mm
longitudinal displacement:	± 75 mm
lateral displacement:	± 75 mm
angle of rotation around Z':	$\pm 90°$
steering angle rotation:	$\pm 180°$
vertical force:	$\pm 20{,}000$ N
longitudinal force:	$\pm 10{,}000$ N
lateral force:	$\pm 10{,}000$ N
torque around Z':	$\pm 1{,}000$ Nm
steering wheel torque:	± 50 Nm.

Vibrations Characterization Bench

This bench is also called four axes bench, because of the number of available independent actuators; it allows measurements of the vibrational behavior of the entire vehicle and of its suspensions, because of the application of periodic forces at its wheels, which simulate, in a controlled environment, the road profile.

Each wheel rests on an independent vertical actuator. This actuator is characterized by the following limits:

- input forces can be only vertical; therefore horizontal components due to car motion are neglected;
- wheels are not rotating; dynamic contribution of tires given by rolling is, therefore, neglected.

The bench is, nevertheless, very useful to monitor the filtering capacity of the suspension, to compare different suspensions, to reproduce and solve problems detected on the road, in terms of vibration, noise and squeaks and to supply experimental correlations to comfort mathematical models.

The bench includes:

- four vertical actuators;
- actuation control system;
- measurement system;
- software package for data acquisition and control.

Different kinds of vertical inputs are available:

- sinusoidal input at constant frequency;
- sinusoidal input at increasing frequency (frequency *sweep*);
- *random* not correlated inputs on the four wheels;
- assigned road profiles reproduction.

The first three input types allow to calculate the transmissibility of the tire-suspension-body system, as ratio between the input point at the wheel and the measurement point. By changing excitation amplitude nonlinearity can be discovered. Sinusoidal inputs and frequency sweeps easily allow their detection.

Not correlated random inputs are useful for a quick characterization: with a single test a predefined frequency fields my be investigated. By data elaboration it is possible to separate transmissibility contributions of parallel springing, roll and pitch motions.

Road profiles reproduction is a direct evaluation of the suspension comfort performance.

To characterize suspension comfort, three axial accelerations are measured at wheel hubs, body spring seats and shock absorbers body connections.

The comparison with the accelerations of the body at foot rest, seat rails and steering wheels is a very common indicator for the comfort performance.

If the bench isn't noisy also induced noise can be evaluated.

The technical specifications of a bench of this kind may be:

suspension stroke:	±100 mm
maximum actuation speed:	2.8 m/s
maximum acceleration:	270 m/s^2
passing band:	up to 200 Hz.

Fatigue Bench

Fatigue testing can be applied to the entire suspension or to a suspension component (like a suspension arm or a spring, etc.), separated of the system.

The benches in use depend on the complexity of the system to be tested. Components are tested using hydraulic pulsator that can be combined with other, according to the system complexity; a single hydraulic actuator should have these specifications:

- three axial capability (can apply forces along x, y, z axes contemporarily);
- load programmer;
- complete body or universal fixation structure;
- signal recorder.

Fig. 3.87 Three axial bench: the suspensions are mounted on a body, fixed to the bench with the structure 1. Suspensions are tested by actuators that can transfer to the wheel longitudinal, vertical and cornering forces

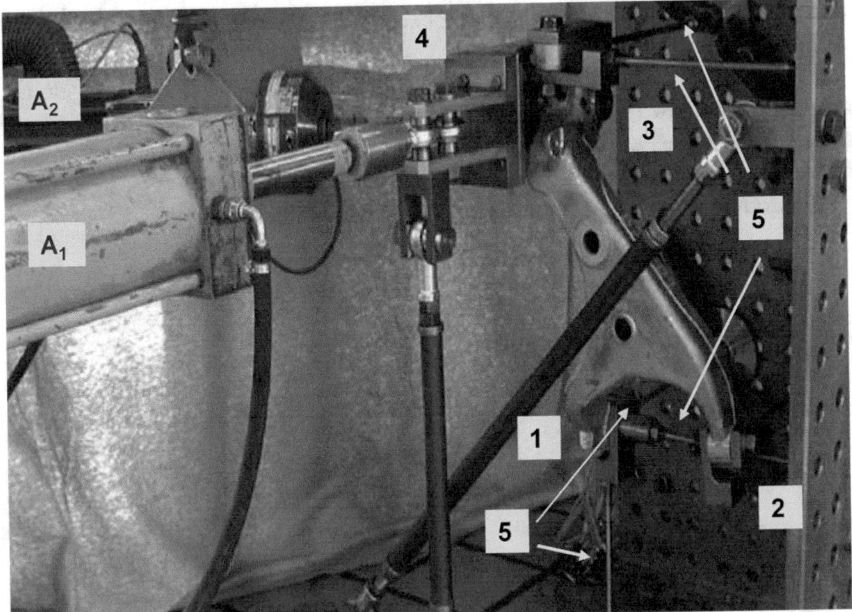

Fig. 3.88 Fatigue test of a single suspension arm

The suspension to be tested is mounted on the body or body dummy, set on the bench with a combination of actuators; the same bench can also fatigue test an entire car body.

Figure 3.87 show a bench of this kind: suspensions are installed on the body, which is fixed to the bench through the structure 1. Suspensions are loaded by a set of actuators that can apply vertical, longitudinal and cornering forces.

Longitudinal and cornering forces actuators are working through rods 3 and 4.

Actuators must be designed for an unlimited life.

These actuators apply load blocks which a load time history gathered on a real roads; some parts of this history are eliminated, if they do not contribute to structure damage, to shorten the test duration.

When a single component is tested, the bench is much smaller and can be used for different test: it consists of a general purpose fixture for assembling the test components and the actuators and some hydraulic actuators.

This benches can be easily retooled to perform test on different components.

Figure 3.88 represents the bench setup for testing a longitudinal arm of a guided arms suspension. The point 1 corresponds to the body articulation, while points 2 and 3 to the other arm articulations; the point 4 is the wheel hub flange.

The arm is installed on a test bed with thin rods 5, able to react with axial forces only and set as to reproduce the total mechanism, missing on the bench. Particularly, the point 1 is fixed by three rods, as to eliminate any linear displacement, the point

2 has the only vertical displacement blocked, while the point 3 has the lateral and vertical displacement constricted.

The point 4, corresponding to the wheel hub flange, is loaded by two hydraulic actuators; the vertical actuator A_1 simulates the vertical static load, while a second actuator A_2 applies a periodic load on the yz plane, with a given inclination as to reproduce the cornering force and the consequent load transfer.

A different setup could simulate traction and braking forces.

Chapter 4
Steering System

4.1 Introduction

As far as path control is concerned, vehicles may be classified according to two categories:

- *guided vehicles*, or better, *kinematically guided vehicles*, whose trajectory is fixed by a set of kinematic constraints;
- *piloted vehicles*, in which the trajectory, a tri-dimensional or a planar curve, is determined by a guidance system, controlled by a human pilot or by a device, usually electro-mechanical. The guidance system acts by exerting forces on the vehicle which are able to change its trajectory.

In the first case the kinematic constraint exerts all forces needed to modify the trajectory without any deformation, i.e. is assumed to be infinitely stiff and infinitely strong. A perfect kinematic guidance is therefore an abstraction, although it is well approximated in many actual cases.

In the second case the forces are due to the changes of the attitude of the vehicle which in turn are caused by forces and moments due to the guidance devices. These vehicles can be said to be dynamically guided.

Apart from the cases in which the forces needed to change the trajectory are directly exerted by thrusters (usually rockets), there can be two cases:

- the attitude changes can be quite large, large enough to be directly felt by the pilot or driver, or
- small enough to be unnoticed.

The first case is that of aerodynamically or hydrodynamically controlled vehicles, in which the pilot acts on a control surface, causing the changes of attitude needed to generate the forces which modify the trajectory. There is also usually a certain delay between the changes of attitude and the actual generation of forces and consequently the drivers feels clearly that a dynamic control, i.e. a control through the application of forces, takes place.

G. Genta and L. Morello, *The Automotive Chassis*, Mechanical Engineering Series,
https://doi.org/10.1007/978-3-030-35635-4_4

In the case of road vehicles the situation is similar but the driver has a completely different impression: The driver operates the steering wheel causing some wheels to work with a sideslip and to generate lateral forces. These forces cause a change of attitude of the vehicle (change of angle β) and then a sideslip of all wheels: The resulting forces bend the trajectory.

However the linearity of the behaviour of the tire and the very high value of the cornering stiffness give the driver the impression of a kinematic, not dynamic, driving. Wheels seem to be in pure rolling and the trajectory seems to be determined by the directions of the midplanes of the wheels.

This impression has influenced the study of the handling of motor vehicles for a long time, originating the very concept of *kinematic steering* and in a sense hiding the true meaning of the phenomena.

The impressions of the driver is in good accordance with this kinematic approach, at least for all the linear part of the behaviour of the tire. When high values of the sideslip angles are reached, the average driver has the impression of losing control of the vehicle, much more so if this occurs abruptly. This impression is confirmed by the fact that in normal road conditions, particularly if radial tires are used, the sideslip angles become large only when approaching the limit lateral forces.

These considerations are only an indication, as there are cases in between those considered here like kinematic guidance with deformable constraints or magnetic levitation vehicles. The difference turns out to be more quantitative than qualitative, and depends mostly on the greater or smaller stiffness with which the vehicle responds to the variations of attitude due to the guidance devices.

We should also notice that it is conceptually possible to generate the torque, which initially modifies the vehicle attitude angle, by different means as those we have explained. Instead of steering and applying a side slip to the steering wheels, an unequal traction or braking force can be applied to the wheels of the same axle or an aerodynamic force, outside of the vehicle symmetry plane.

In both cases, this torque is causing a side slip angle to the wheels which curves the vehicle path. We could imagine a vehicle where no axle is steering and which is controlled by means of differential wheel braking. Tracked vehicles are controlled in a similar way, but their dynamics is quite different, because of the intrinsic difference between tires and tracks.

As we will see later on, differential braking is used by some control system, applied to improve the vehicle dynamic behavior or handling: the path is set up by the driver, by steering the wheels, while an automatic control system applies corrections by differential braking or, sometimes, by additional steering angles.

In the following paragraphs we will comment on the main components which are part of the steering system, including:

- *steering mechanism* that is the linkages system steering the front wheels, in a particular way around the king-pin axis, connecting steering arms, moving with the suspension stroke, with the steering box;
- *steering box* transforming steering wheel rotation into a displacement of the steering tie rods;
- *steering column* connecting steering wheel with steering box.

4.2 Steering Mechanism

Before explaining the steering mechanism configuration, let us introduce some further consideration on steering piloted road vehicles.

Low speed or kinematic steering is defined as the motion of a wheeled vehicle determined by pure rolling of the wheels. The velocities of the centres of all the wheels lie in their midplane, i.e., the sideslip angles α are vanishingly small. In these conditions the wheels can exert no cornering force to balance the centrifugal force due to the curvature of the trajectory. Kinematic steering is possible only if the velocity is vanishingly small.

Consider a vehicle with 4 wheels, two of which can steer (Fig. 4.1). The relationship that must be verified to allow kinematic steering is easily found, by imposing that the perpendiculars to the midplanes of front wheels meet the ones of rear wheels at the same point:

$$\tan(\delta_1) = \frac{l}{R_1 - \frac{t}{2}}, \qquad \tan(\delta_2) = \frac{l}{R_1 + \frac{t}{2}}. \tag{4.1}$$

Instead of the track t, Eq. (4.1) should contain the distance between the king-pin axes of the wheels, or better, between their intersections with the ground.

By eliminating R_1 between the two equations, a direct relationship between δ_1 and δ_2 is readily found:

$$\cot(\delta_1) - \cot(\delta_2) = \frac{t}{l}. \tag{4.2}$$

A device allowing to steer wheels complying exactly with Eq. (4.2), is usually referred to as *Ackerman steering* or *Ackerman geometry*. No actual steering mechanism allows to follow exactly such law and a *steering error* $\Delta\delta_2$, defined as the difference between the actual value of δ_2 and that obtained from Eq. (4.2) can be obtained as a function of δ_1.

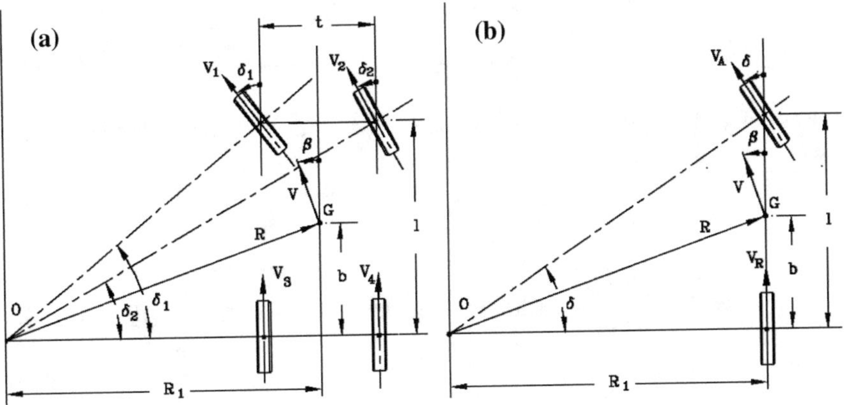

Fig. 4.1 Kinematic steering of a four-wheeled and a two-wheeled vehicle

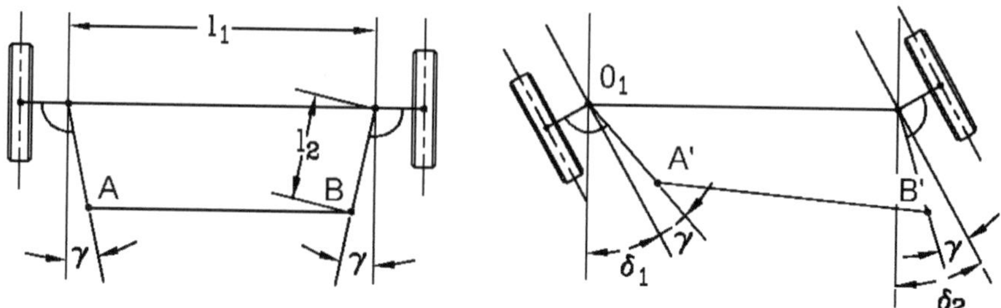

Fig. 4.2 Scheme of a steering mechanism made by an articulated quadrilateral

Consider for instance the device based on an articulated quadrilateral shown in Fig. 4.2, used by rigid axle suspensions with a screw and sector steering box.

By this simple mechanism, the track rod AB coordinates wheels steering.

To calculate the relationship between δ_1 and δ_2, is sufficient to develop some geometrical considerations. The track rod length AB is:

$$\left(\overline{\mathrm{A}-\mathrm{B}}\right) = l_1 - 2l_2 \sin(\gamma) \ . \tag{4.3}$$

When steering wheel is turned, the new positions of A and B, said A′ and B′, with reference to a reference system, having the origin in O_1 and the x axis in cross direction, are:

$$\left(\overline{\mathrm{A}'-\mathrm{O}_1}\right) = \left\{ \begin{array}{c} l_2 \sin(\gamma + \delta_1) \\ -l_2 \cos(\gamma + \delta_1) \end{array} \right\} , \tag{4.4}$$

$$\left(\overline{\mathrm{B}'-\mathrm{O}_1}\right) = \left\{ \begin{array}{c} l_1 - l_2 \sin(\gamma - \delta_2) \\ -l_2 \cos(\gamma - \delta_2) \end{array} \right\} , \tag{4.5}$$

The square of the distance between A′ and B′ is:

$$\begin{aligned} \left(\overline{\mathrm{A}'-\mathrm{B}'}\right)^2 = & [l_2 \sin(\gamma + \delta_1) + l_2 \sin(\gamma - \delta_2) - l_1]^2 + \\ & [l_2 \cos(\gamma + \delta_1) - l_2 \cos(\gamma - \delta_2)]^2 \ . \end{aligned} \tag{4.6}$$

But this distance is equal to the square of $\left(\overline{\mathrm{A}-\mathrm{B}}\right)$, and a relation between δ_1 and δ_2 can be obtained:

$$\begin{aligned} & \sin(\gamma - \delta_2) + \sin(\gamma + \delta_1) = \\ & = \frac{l_1}{l_2} - \sqrt{\left[\frac{l_1}{l_2} - 2\sin(\gamma)\right]^2 - [\cos(\gamma - \delta_2) - \cos(\gamma + \delta_1)]^2} \ , \end{aligned} \tag{4.7}$$

that is:

$$1+\sin(\gamma-\delta_2)\sin(\gamma+\delta_1)-\lambda\sin(\gamma-\delta_2)-\lambda\sin(\gamma+\delta_1)+ \tag{4.8}$$

$$+[\lambda-2\sin(\gamma)]\sin(\gamma)-\cos(\gamma-\delta_2)\cos(\gamma+\delta_1)=0\ ,$$

where $\lambda = l_1/l_2$.

Equation (4.7) may be solved as to obtain δ_2 as function of δ_1. With laborious calculations, the following relationship is obtained:

$$A\sin^2(\gamma-\delta_2)+B\sin(\gamma-\delta_2)+C=0\ ,$$

where:

$$A = 1+\lambda^2-2\sin(\gamma+\delta_1)$$

$$B = 2D\,[\sin(\gamma+\delta_1)-\alpha]$$

$$C = D^2-\cos^2(\gamma-\delta_1)$$

$$D = 1-\lambda\sin(\gamma+\delta_1)+[\alpha-2\sin(\gamma)]\sin(\gamma)\ ,$$

which allows to calculate δ_2.

If the two steering arms converge on the rear wheels axle, at the intersection with the vehicle symmetry plane, as proposed by Jeantaud, the angle γ can be calculated as:

$$\gamma = \text{artg}\left(\frac{l_1}{2l}\right).$$

The relationship $\delta_2\,(\delta_1)$ obtained with the Jeantaud quadrilateral is compared with that obtained using the correct kinematic relationship and three different values of γ in Fig. 4.3a. The steering error $\Delta\delta_2 = \delta_2 - \delta_{2_c}$ can be calculated by any value of δ_1, as shown on Fig. 4.3b.

The Jeantaud condition leads to a value of $\gamma = 13, 3°$ and fairly large steering errors. Three other curves are plotted for γ: 16°, 18° and 20° (in all 3 cases the steering arms converge behind the rear axle); the error is decreasing, for low values of the steering angle, as γ decreases. Nevertheless low errors at low steering angles are accompanied by large errors at large steering angles; a compromise is necessary: on our case a value of $\gamma = 18°$ can be a reasonable choice.

The complete calculation of the steering error for a rack and pinion steering box is omitted for sake of brevity; we can refer to Fig. 4.4.

If we suppose that rack and tie rods OA are lying on the same plane (this is not always true) or the triangle OAB is always on a horizontal plane, we can understand visually that the steering error depends on dimensions a and h, which are conditioned by the place allowable in the engine compartment.

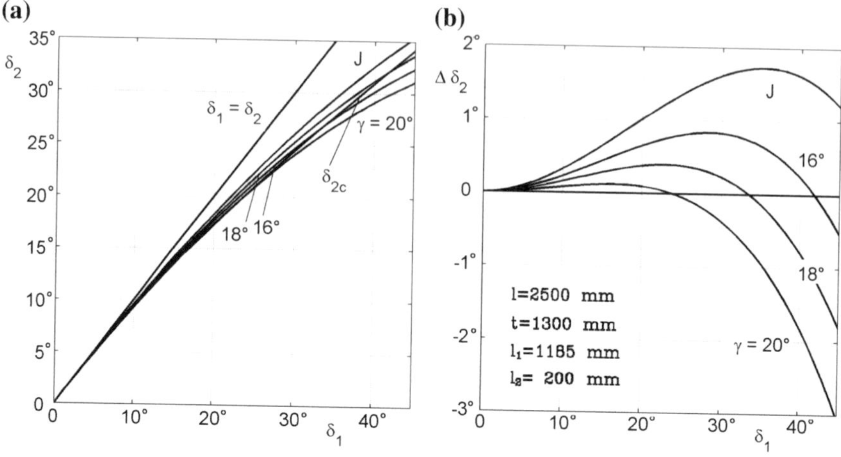

Fig. 4.3 **a** Relationship δ_2 (δ_1) obtained with the Jeantaud quadrilateral (curve J) compared with that obtained using the correct kinematic relationship and three different values of γ. **b** Steering error, $\Delta\delta_2 = \delta_2 - \delta_{2_c}$ as function of δ_1 for the same steering systems

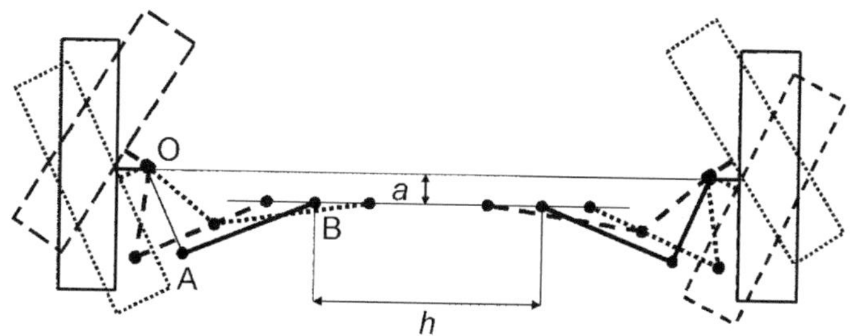

Fig. 4.4 Kinematic scheme to calculate steering error by a rack and pinion steering box

Let us consider, finally, the case of a screw and sector steering box in connection with an independent wheel suspension; in this case a track rod cannot be applied, because it could introduce undesirable toe angle variations by suspension asymmetric strokes.

Different steering mechanism can be applied; some of them are shown on Fig. 4.5.

On the upper side of the figure, the case of a light duty vehicle is represented, for instance an off-road vehicle. The drop arm, moved by the steering box, is part of an articulated quadrilateral, where pull arms are connected or the drop arm is directly doubled. The choice between the two mechanisms is imposed by the available space and by the elasto-kinematic behavior of the suspension.

On the lower side, at the right, the case is shown of a bus or, more in general, the case of a vehicle by which the distance between steering wheel and steering axle is relevant; in this case, a longitudinal pull rod, similar to that of a rigid axle (on the left), is moving an additional swing arm.

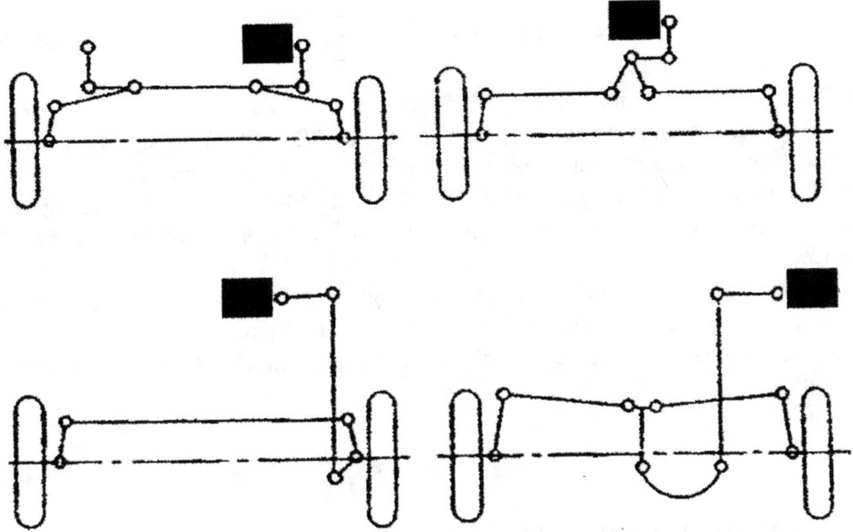

Fig. 4.5 Steering mechanisms for a screw and sector steering box, applied to an independent wheel suspension; on the upper side of the figure, the case of a car; on the lower side, the case of a bus with rigid axle (left) can be compared with the case of an independent suspension (right)

Many efforts have been made in the past to minimize the steering error; nevertheless the importance of a correct kinematic steering is often overestimated.

As a matter of fact, we should remember that:

- a wheel side slip angle is always present, by steering;
- most suspensions cause an additional steer angle with roll;
- in most cases steering wheels must have a slight toe-in angle;
- additional steering angles are caused by suspensions stroke and deformation.

All these facts reduce the importance of the steering error and suggest to consider this issue from an ampler point of view.

The steering error has, nevertheless, a significant effect on tire wear of the front wheels and on steering wheel self aligning, influencing the driver feeling of the steering wheel.

It is important that the reaction torque of the steering wheel increases progressively with steering angle, feature that can be obtained with a suitable geometry.

The radius of the trajectory of the centre of mass of the vehicle is:

$$R = \sqrt{b^2 + R_1^2} = \sqrt{b^2 + l^2 \cot^2(\delta)} \; , \tag{4.9}$$

where δ is the steering angle of the equivalent two-wheeled vehicle (Fig. 4.1b), also called *monotrace* model. Although it should be computed by averaging the cotangents of the angles of the two wheels,

$$\cot(\delta) = \frac{R_1}{l} = \frac{\cot(\delta_1) + \cot(\delta_2)}{2}, \tag{4.10}$$

it is very close to the direct average of the angles.

Consider for example the same vehicle of Fig. 4.2 with centre of mass at mid-wheelbase on a curve with a radius $R = 10$ m. The correct values of the steering angles are $\delta_1 = 15.090°$, $\delta_2 = 13.305°$ and $\delta = 14.142°$. By direct averaging the steering angles of the wheels, it would follow that $\delta = 14.197°$, with an error of only 0.36%.

Therefore, conventionally, the steering angle δ of a mono trace model is the average of the steering angles of the wheels of the same axle.

In case the radius of the trajectory is large if compared with the wheelbase of the vehicle, Eq. (4.9) reduces to:

$$R \approx l\cot(\delta) \approx \frac{l}{\delta}\,. \tag{4.11}$$

Equation (4.11) can be rewritten in the form:

$$\frac{1}{R\delta} \approx \frac{1}{l}\,. \tag{4.12}$$

The expression $1/R\delta$ has an important physical meaning: It is the ratio between the response of the vehicle, in terms of curvature $1/R$ of the trajectory, and the input which causes it. It is therefore a sort of transfer function for the directional control and can be referred to as *trajectory curvature gain*. In kinematic steering conditions it is equal to the reciprocal of the wheelbase.

In the case of a vehicle with independent suspension and rack and pinion steering box, the steering mechanism is quite different, as shown on Fig. 4.4.

The Jeantaud's condition could be only verified if tie rods and rack are aligned and steering arms are crossing in the middle of the rear axle center line.

The complete explicit calculation of the steering error is laborious.

By using a mathematical model, we can arrive to the diagram shown on Fig. 4.6, where errors are rather higher as compared with the previous case.

We can assume, in any case, bearing in mind what we have said, that a maximum error of 5° is still acceptable.

The same figure shows a second diagram, where the steering error is calculated at the maximum traction force, in this front wheel driven vehicle; this variation is the effect of the longitudinal displacement of the king-pin axis, by effect of the traction force.

The wheels steering angle must be also calculated, as function of steering wheel angle; the result of a calculation of this kind is reported on Fig. 4.7; this result is obtained by a simple mathematical model of the articulated system representing the steering mechanism.

On the right of the diagram the right steering angle is reported for a maximum steering wheel rotation of about 360°, in counterclockwise direction; on the other side, the same diagram is shown for the right wheel.

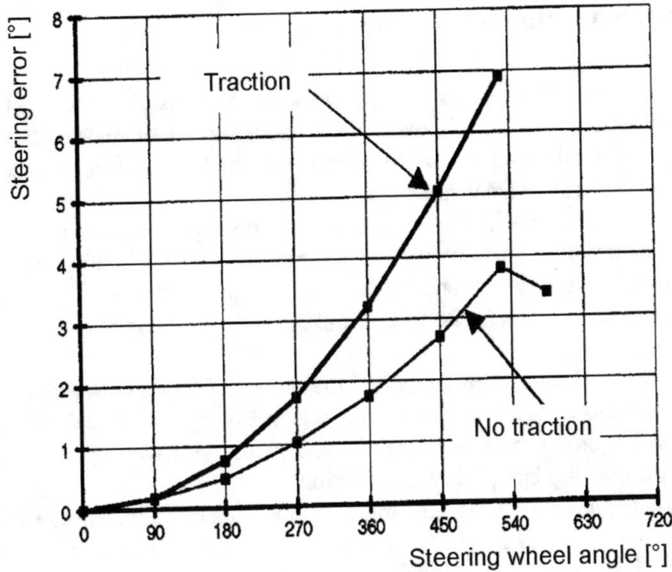

Fig. 4.6 Steering error as function of steering angle in a front wheel driven car, with rack and pinion steering box. Two diagrams are shown, with and without traction force

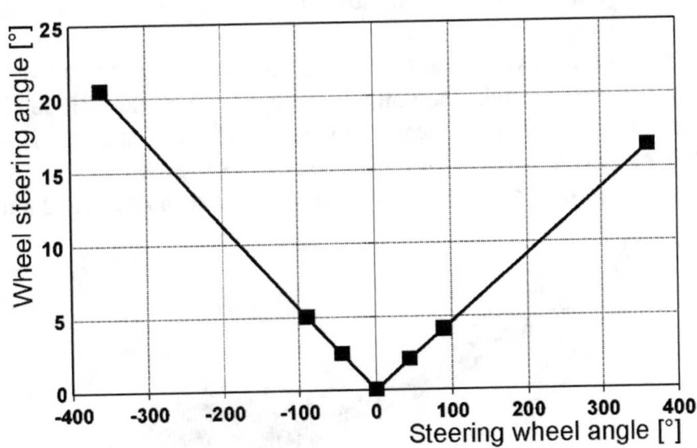

Fig. 4.7 Diagram of wheel steering angle as function of steering wheel angle; at the right the left wheel steering angle, at the left the right wheel steering angle

The overall transmission ratio of the steering mechanism is, therefore, about 20:1. This value is typical for a car without power steering assistance; for a power assisted steering system this value can be reduced with the benefit of a more immediate response.

The two curves are almost linear, at least on the investigated field; they can loose this property at the rack full stroke, at about 400° of steering wheel rotation.

4.3 Rack and Pinion Steering Box

A rack and pinion steering box is applied today to almost all cars and light duty industrial vehicles; the other alternative, the screw and sector steering box, with its variants, is practically reserved to heavy duty industrial vehicles or to those off-road vehicles still featuring a front rigid axle.

The device shown on in Fig. 4.8 transforms, through the geared couple of the pinion 3 and the rack 1, the rotary motion of the steering wheel, imposed by the driver, into a linear motion of the spheric heads 2, which operate the steering mechanism.

The rack accomplishes in the mean time the task of steering wheels and that of the track rod.

Due to the mechanism simplicity and the reduced friction between teeth flanks, mechanical efficiency is usually very good; this fact is appreciated because it reduces the reaction torque on the steering wheel and gives the driver a good and faithful feeling of the existing lateral tire-road friction.

As a disadvantage, steering transmission ratio cannot be increased over certain values, because is limited by tooth size, imposed by fatigue resistance of the material and by the minimum number of teeth that can be cut without interference.

The steering wheel is, therefore, always rather immediate; this fact, that is positive in general, doesn't allow the application of this mechanism to heavy vehicles. Therefore, when power steering was limited to luxury cars, rack and pinion was also limited to small cars.

Figure 4.9 shows a cross section of this steering box through the pinion axis; this version is not power assisted. The pinion 1 is supported by the ball bearing 2 and by the needle bearing 3; the ball bearing reacts to radial and axial loads, while the needle bearing reacts to radial loads only. The rack is supported by the sliding block 4, pushed by the spring 5 that controls the pressure between rack 6 and pinion 1.

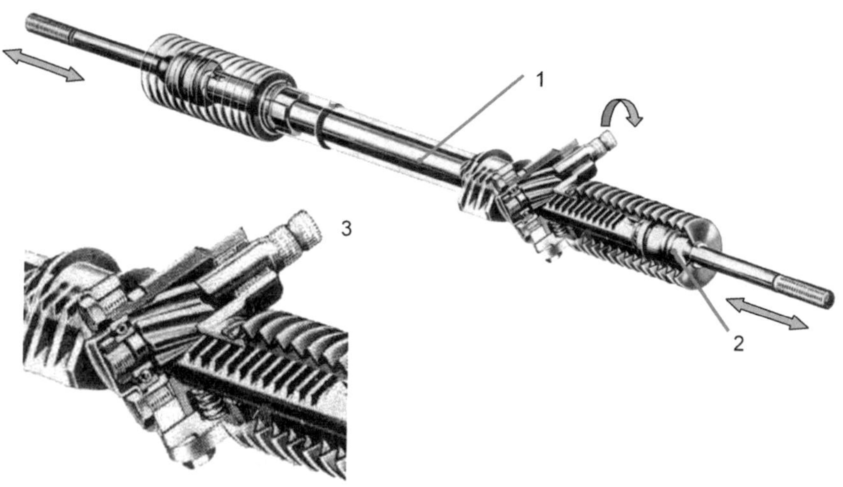

Fig. 4.8 Rack and pinion steering box; the gearing point between rack and pinion is enlarged in the lower view

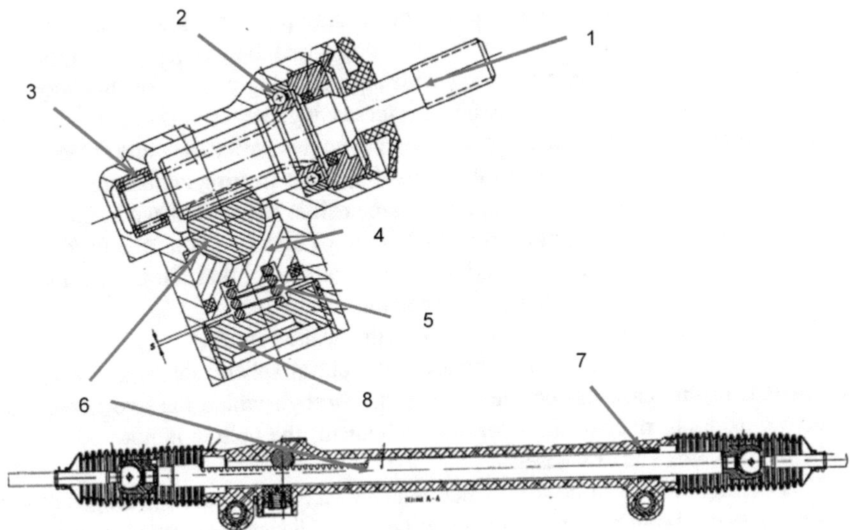

Fig. 4.9 Enlarged detail of a rack and pinion steering box

It should be noticed that gears are of the helical type in order to allow a larger transverse contact ratio.

A sliding bush 7, made by Teflon, offers the second bearing point, at the other end of the steering rack.

The pre-loaded spring 5 overrides backlash between pinion and rack and controls the internal friction between parts. Since the sliding block 4 is mounted with a determined clearance, the spring can also absorb dynamic loads at the rack stops; the threaded ring 8 determines the value of this clearance, adjusted at the assembly line.

Rack and pinion are splash lubricated inside the box of tubular shape; also the spheric heads connecting the rack with the tie rods are set in the same oil. Two flexible rubber bellows avoid lubricant spillage and dust contamination.

The steering box is fixed to the car body or to the subframe, through the two holes shown in the same figure; those mounts feature rubber bushing, to filter noise and vibrations.

Transmission ratio is in part determined by mechanism geometry, but a major role is played by the primitive radius d of the pinion.

Once the pinion is built, or a certain value for its base circle d_b is set, this simple relationship applies:

$$d = d_b / \cos\theta \; , \tag{4.13}$$

where θ represents the pressure angle of the cutting rack tool which generated the tooth wheel; by correcting tooth profile it is possible to have a correct contact with rack with different angle too.

This feature is exploited on racks with decreasing pressure angle on each tooth, starting from those teeth meshing with the pinion when the wheels are straight and going to those teeth that are meshing with high steering angles. In this way the pitch radius is increasing as steering angle is increasing; the result is that the steering control is immediate at high speed, when steering angles are small, while the steering reaction torque is reduced in parking maneuvers, when steering angles are large.

A critical point of the rack is its lateral dimension, which is determined by the necessary stroke (steering angle from stop to stop) and by a convenient distance between its bearing points; it should also be remembered that the pinion and the steering column must be positioned between the drivers feet.

The transversal bulk can sometimes affect the desired result.

To solve this problem, steering boxes with central spheric heads are designed (Fig. 4.10); in this case the box features a cut through which the two heads can emerge outside, to operate the steering mechanism; the bellow is now set inside, while the two rack end are closed.

This configuration can be used when is necessary to increase the length of the two tie rods to obtain a correct kinematic behavior during suspension stroke; this happens when steering arms are pointing forwards, instead of backwards, for reasons of installation, as shown on the picture or when the car is too narrow.

The kinematic behavior is very much conditioned by rack position, but the freedom allowed to position this component is very limited.

Engine dimensions, according to its position, are conditioning rack installation.

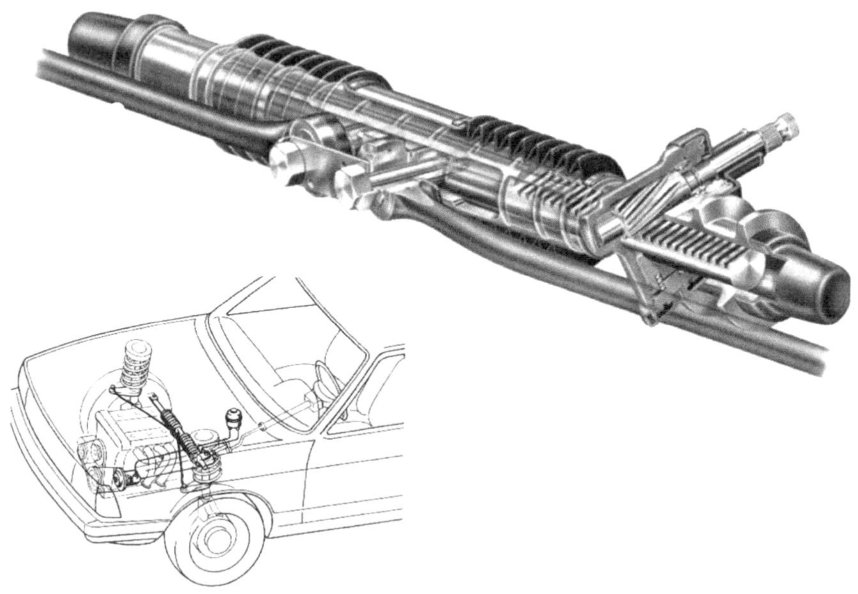

Fig. 4.10 Rack and pinion steering box with central spheric heads, particularly useful to solve installation problems; notice also the different configuration of the steering mechanism

By cars with longitudinal engine and rear wheel drive, the rack can be installed in front of the engine, when the front axle is also in front of the engine, with steering arms pointing forwards, or behind the wheel center, again with steering arms pointing forwards, in any case the rack must be installed below the oil sump. By front wheel driven cars with transversal engine, the rack must be installed behind the engine and steering arms are pointing usually backward.

Another condition on rack installation is posed by obtaining a correct position for the steering wheel.

4.4 Screw and Sector Steering Box

Screw and sector steering boxes are applied, as we have explained, mainly on industrial vehicles either with rigid or with independent suspensions (buses).

There are no conceptual limits to the application of rack and pinion steering boxes to rigid axles, but we should consider that the rack should be mounted on the axle and it is difficult, with the available space, to provide for a constant velocity transmission between the pinion and the steering wheel.

Figure 3.63, in the previous chapter, shows the installation of this kind of steering box on an industrial vehicle with driver compartment on the engine. The mechanical transmission, connecting the steering wheel with the steering arm has the configuration shown on Fig. 4.11.

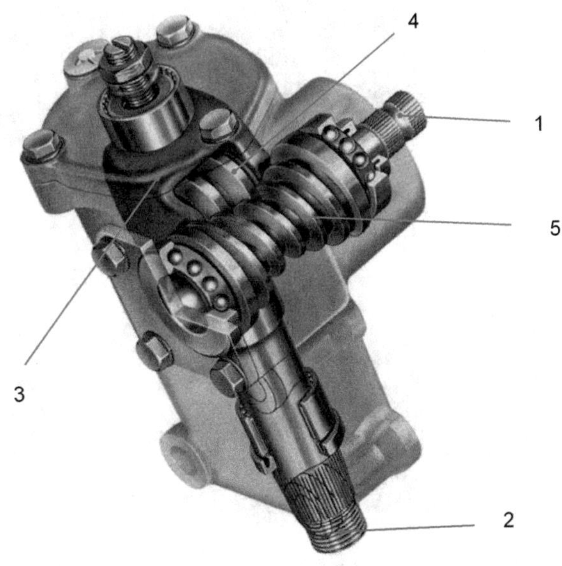

Fig. 4.11 Screw and sector steering box. The tooth wheel sector has been made with a roller, to decrease mechanical friction. The shape of the screw is copying the roller profile in order to increase the number of working threads

This box could include a very simple screw and a tooth wheel sector, moving the steering arm, as it has happened in the past; this solution would imply very high internal friction.

We remember that such kind of transmission is almost irreversible, that means that the torque can be transmitted from the screw to the sector and not vice versa. This fact would imply lack of steering wheel self alignment and of feeling on the existing friction between tires and ground.

Reversible kinematic couples, working according to the same principle, can be produced, but with much lower transmission ratios, adopting a multiple threads screw; but in this case, the advantage of reducing the steering wheel torque is almost disappearing.

To counteract the irreversibility, a solution has been developed where the screw 5, operated by the steering column 1, is of globoidal type and the sector shows a threaded roller 4.

The globoidal screw has a thin section in the center, to increase the number of working threads. The threaded roller, whose profile is matching the screw, works as a rack, capable of motion according to the screw rotation, with consistent reduction of mechanical friction.

The circumferential motion of the roller is transmitted to the spline shaft 2, through the fork 3; the shaft is connected to the steering arm.

Another way of reducing friction is shown on Fig. 4.12 with a recirculating balls screw.

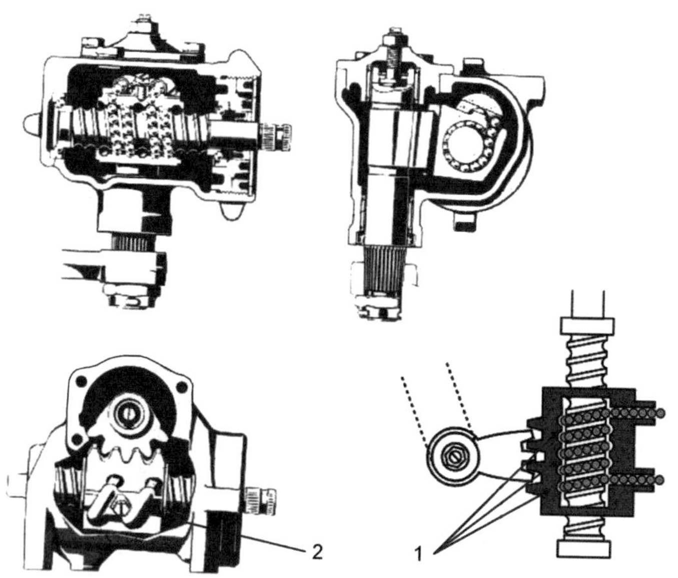

Fig. 4.12 Screw and sector steering box with recirculating balls; this kind of screw reduces the friction to a minimum

In this case the globoidal screw is limited to a small portion. This screw is also threaded in the inside bore, matching with a second screw fixed to the steering column. The shape of the thread is particular, building up an helical channel 1 (shown on the scheme at the lower right), inside which a number of balls can roll and move. The pipe 2 is used to close the circuit full of balls, to allow their recirculation.

In this case, again, the slip motion is changed into rolling motion, with friction reduction; since the contact between balls and channel walls is very limited, the friction is reduced to a minimum.

4.5 Steering Column

Figure 4.13 shows the assembly of a steering column for a medium size car; the miniature on the left can identify the steering column position inside the car.

The function of this component is to transfer the torque applied to the steering wheel by the driver to the steering box. Very seldom, steering box and steering wheel position allow, in modern cars, to have a straight steering column; for this reason the column is made by three sections, where the first and the last are, respectively, connected to the steering wheel and to the steering box.

The medium section is connected to the first and the third through universal joints; the three shaft lay-out must grand a constant speed transmission and therefore the shafts must lie on the same plane and the working angles must be equal.

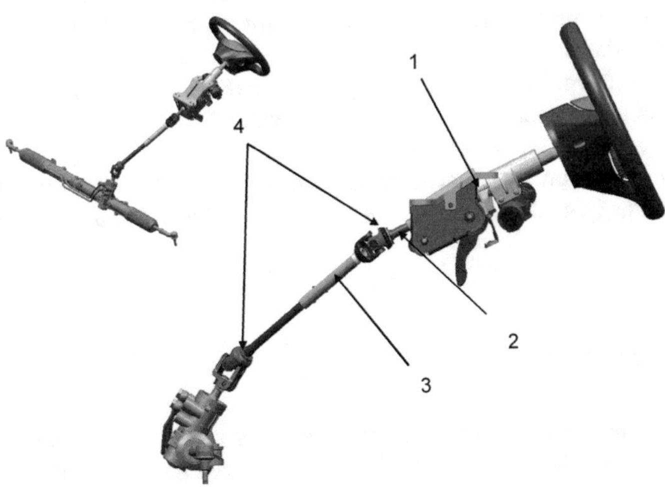

Fig. 4.13 Steering column assembly, capable of adapting steering wheel position on the zx vertical plane. The miniature allows to understand the inclined position of the medium part of the column, suitable to allow structural collapse with no intrusion of the steering wheel into the passengers compartment

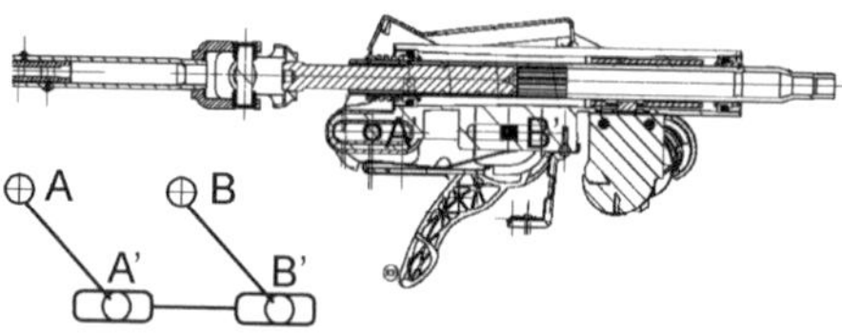

Fig. 4.14 Cross section of the adjustable structure bearing the first section of the steering column and steering wheel: the articulated quadrilateral AA′BB′ allows this adjustment

This kind of lay-out is also justified by steering wheel position adjustment and by passive safety.

The first section 2 of the column is fixed to the body through a structure 1, containing a couple of needle bearings or bushings. It rotates the mid section 3 through an universal joint. The vertical position of the structure 1 can be adjusted to adapt steering wheel position to driver size.

Figure 4.14 shows a cross section of the structure 1.

It is divided in two parts: one fixed to the body (not shown on this figure), one supporting the bearing of the first section of the steering column.

This second part is connected to the first through an articulated quadrilateral, where the joints between cranks (AA′ and BB′) and rod A′B′ allow rotation and linear displacement, as shown on the same figure. This part can be shifted in vertical and horizontal directions.

One of the A′ and B′ articulation is done by a screw that can fix the crank in a given position or leave the steering wheel free to be adjusted. Obviously the universal joint between the first and the second section must feature a sliding spline.

Major legislative regulations require the steering column to be collapsible, that means that in case of a reference crash against barrier the steering column can reduce its length, to allow steering box displacement without steering wheel intrusion into the drivers compartment over a certain homologation limit.

This homologation requirement can be satisfied if the central section of the steering column (3 on Fig. 4.13) can change its length and if the first and third sections are not aligned.

4.6 Power Steering

The constant increase of passive safety requirements, of comfort and convenience devices and of interior room have lead to a general mass increase of all cars, despite the efforts devoted to apply light weight materials and design; this mass increase has affected loads between tires and ground and, therefore, the torque to be applied to the steering wheel to rotate front wheels.

The reduction of the steering wheel torque to levels that are ergonomically acceptable has been obtained by applying power assistance to the existing mechanical systems; power steering has spread from heavy luxury cars and industrial vehicles to virtually all cars. The most diffused power assistance system used to be hydraulic, sometimes integrated by an electro-hydraulic device with electronic control, to adjust the assistance to the vehicle speed; such hydraulic systems show slight differences if they are applied to rack and pinion steering boxes or to screw and sector steering boxes.

Hydraulic systems affect car fuel consumption negatively, not only because of their low overall efficiency (the power necessary for assistance is draft from the engine, transformed from mechanical to hydraulic and again from hydraulic to mechanical and control is obtained by wasting the flow in excess) but also because the pump feeding the oil circuit is always in motion, even when the car is not steering. Many on-road measurements have shown that hydraulic power steering systems are in the average increasing the fuel consumption by about 8% in city driving and 4% in highway driving.

A partial solution to this problem was obtained by having the pump driven by an auxiliary electric motor, working with a duty cycle to maintain an oil reservoir at a preset pressure; this kind of system does not improve significantly the overall mechanical efficiency but eliminates any power absorption when the car is not steering.

In recent years, full electrical assistance systems were widely applied, reducing the field of application of hydraulic systems to very heavy vehicles only.

4.6.1 Hydraulic Rack and Pinion Steering Box

Figure 4.15 offers a general lay-out of the hydraulic power steering system for the rack and pinion box.

This system applies, as power source, the hydraulic pressure of a certain flow of oil, generated by a pump driven by the engine.

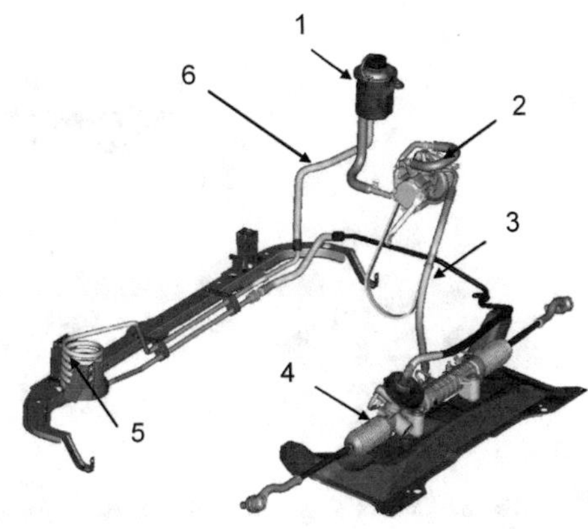

Fig. 4.15 Hydraulic power steering system for a small front wheel driven car with rack and pinion steering box

The system includes the following components:

1. oil reservoir;
2. pump, normally blade type, driven by the accessory belt of the engine;
3. high pressure tube;
4. steering box;
5. cooling serpentine;
6. return, low pressure tube.

This scheme applies to screw and sector steering boxes too.

The cooling serpentine is necessary to waste the heat generated by pressure control choke valves. In addition to that, the flow rate generated by the volumetric pump is depending on engine speed only; what is not necessary to power assistance is by passed from high to low pressure, with additional heat waste.

An inadequate cooling can allow the oil to work close to the boiling point, with lack of efficiency and danger of cavitation.

Thanks to the choking valves and to other circuit losses, this system grants a certain damping characteristic, with benefits oh the steering wheel return and vibrations. See also what was said about shimmy.

It appears that the relative simplicity of this system has a negative impact on fuel consumption; as a matter of fact the fuel consumption increase is about 2–3% on the average driving cycle. To limit this negative effect some system has been developed, where the oil pump is driven by an electric motor and feeds a pressure accumulator. In this case the electric pump is working only when the oil pressure in the accumulator drops below a certain threshold value, with potential savings on hydraulic losses.

This system has also the advantage of allowing a more flexible installation of the pump, on small cars engine compartments, with problems of space; this system had a short life and was abandoned in favour of totally electric systems.

As shown by the upper scheme on Fig. 4.16, a double effect hydraulic cylinder is integrated into the steering box, able to move the rack. M and R are the input and

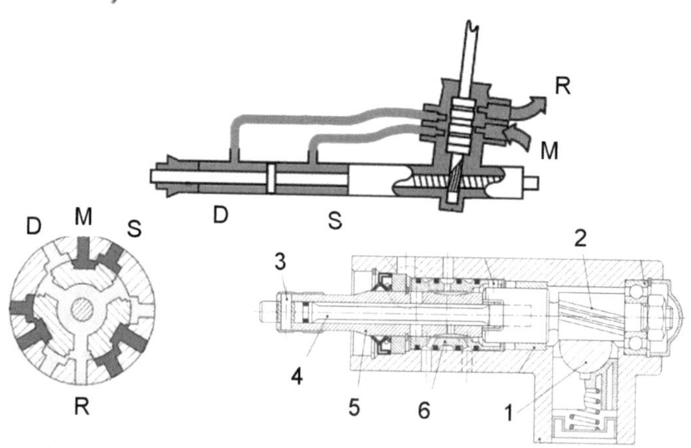

Fig. 4.16 Schema of a hydraulically assisted rack and pinion steering box (upper side); detail of the hydraulic control valve (lower side)

output nipples of the oil flow and chambers S and D on the power cylinder assist left and right steering.

The lower part of the same figure shows an enlargement of the distributor and control valve which puts in pressure chambers S and D of the power cylinder; the pressure force is added to that applied to the rack by the steering wheel.

The double effect power piston can have one of the faces at intake pressure or both at the exhaust pressure, according to the angular position of the distributor and control valve. This valve is made by two coaxial cylinders, with relative rotation motion, featuring a number of windows, which can put the S and D chambers into communication with ports M and R, as shown on the cross section on the left.

In the position shown on the figure, the oil pressure is sent to the S chamber, while the R chamber is to exhaust. The higher the pressure, the higher the clockwise rotation angle of the inside cylinder of the valve, until a maximum opening of the window is reached; by counterclockwise rotation the pressure will be reduced till the closure of the window.

If the counterclockwise rotation of the cylinder in continued the same functions are performed for the other face of the piston.

The inside cylinder 5, as can be seen on the right cross section, is connected through a pin 3 to one end of a torsion spring 4, loaded by the torque applied to the pinion 1 by the steering column. In the same way the external cylinder 6 is connected to the other end on the torsion spring.

The torsional stiffness of the spring 4 must be determined in such a way as windows are completely open at the maximum acceptable steering wheel torque. At lower torque, the spring deformation will be proportionally reduced; the steering wheel torque, even if reduced, will reproduce the aligning torque variation, thus rendering a correct feeling of the dynamic behavior of the car.

This trailing mechanism reduces the assistance torque to zero, as soon as the steering wheel torque is set to zero. It should be noticed that the operation is the same when the steering wheel is moving the wheels or when wheels are applying torque to the steering wheel, during steering wheel return or driving on an asymmetric obstacle.

We should draw our attention on the fact that pressure pulses or variations can generate vibrations and noise; therefore it is a good practice to apply rubber bushings to the steering box mounts and damping joints to tubes under pressure.

4.6.2 Hydraulic Screw and Sector Steering Box

Figure 4.17 shows the scheme of a hydraulic power assisted screw and sector steering box, with balls recirculation, suitable to industrial vehicles.

As can be noticed, by comparing this figure with the previous one, the valve scheme, here featuring two spool valves 9 and 10, moved by the fork 3, connected to the torsion spring 18, the working principle is substantially the same. In this steering box, the short rack 2 is also the double effect working piston, inside the cylinder 1.

The two valve types we have shown are not necessarily bound to the steering box with which they have been presented, but can be applied to both kinds.

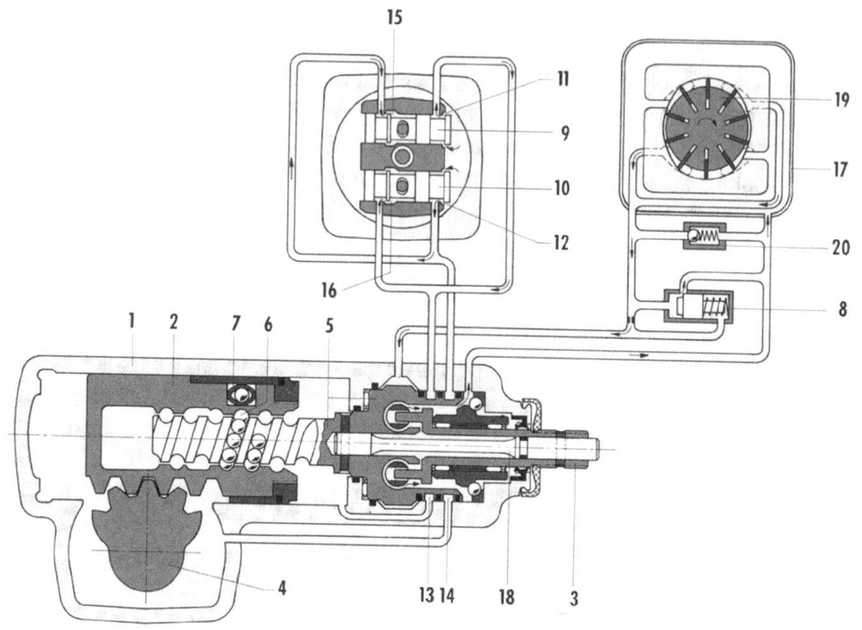

Fig. 4.17 Cross section of a recirculating balls screw and sector steering box, with hydraulic power assistance (below) and the detail of the control valve (above)

4.6.3 Electric Power Assistance

Starting from small and medium size cars, electric power steering (EPS) systems have obtained a wide diffusion. They are usually applied to rack and pinion steering boxes and are made substantially with an electric motor adding torque to the steering mechanism.

We should not forget that only EPS fits very well the needs of electric and hybrid cars where the prime mover is supposed to be stopped for significant periods of time.

In addition, an electric motor can be interfaced with an electronic control system more easily than a hydraulic pump: from this point of view, EPS can be considered as a cornerstone of an independent steering control, suitable to collision avoidance or even to autonomous driving systems.

Several EPS systems are now available and their choice is mainly bound to cost, performance (vehicle weight and speed of actuation) and the space available in the engine compartment; the following list is roughly ranked by the value of the force F_r they can exert on the rack and, therefore, by vehicle mass.

Column EPS

In column EPS (C-EPS), the electric motor is applied to the steering column, acting on that with a suitable gear and is located in the passenger compartment; this solution is particularly suitable for small cars where the engine bay is always crowded but has the disadvantage of increasing the torque on the steering column and featuring a reduced integration with the steering mechanism with negative effect on cost. The force on the rack F_r can range between 6 and 8 kN.

Pinion EPS

In pinion EPS (P-EPS), the electric motor is geared on the pinion directly, after the universal joint on the steering column; it grants a better road feedback in comparison with the C-EPS; it is probably the cheapest solution for its good degree of integration and the most suitable one for small cars newly designed form the ground up that take provision for the additional space it needs. The force on the rack F_r can range between 6 and 10 kN.

Dual Pinion EPS

Dual pinion EPS (D-EPS) feature about the same performance of the previous solution with more flexibility in the installation at a small disadvantage in cost; it can use the existing space usually allowed for the rack and pinion mechanism in other-side-drive versions. The force on the rack F_r can range between 6 and 10 kN.

Belt EPS

Increasing the electric motor rotation speed can reduce the motor cost and weight at an assigned performance; this is what is done in belt EPS (B-EPS). The motor is connected to a recirculating balls screw directly acting on the rack with a belt that enables the mechanism to be easily reversible (for prompt return to the aligned wheels position) for it good mechanical efficiency. The force on the rack F_r can range between 7 and 16 kN.

Rack EPS

In rack EPS (R-EPS), the electric motor is coaxial with the recirculating balls screw acting on the rack; this solution is suitable to large cars that can enjoy a good reversibility of the steering mechanism despite of the significant inertia of the electric motor. The force on the rack F_r can easily reach 20 kN.

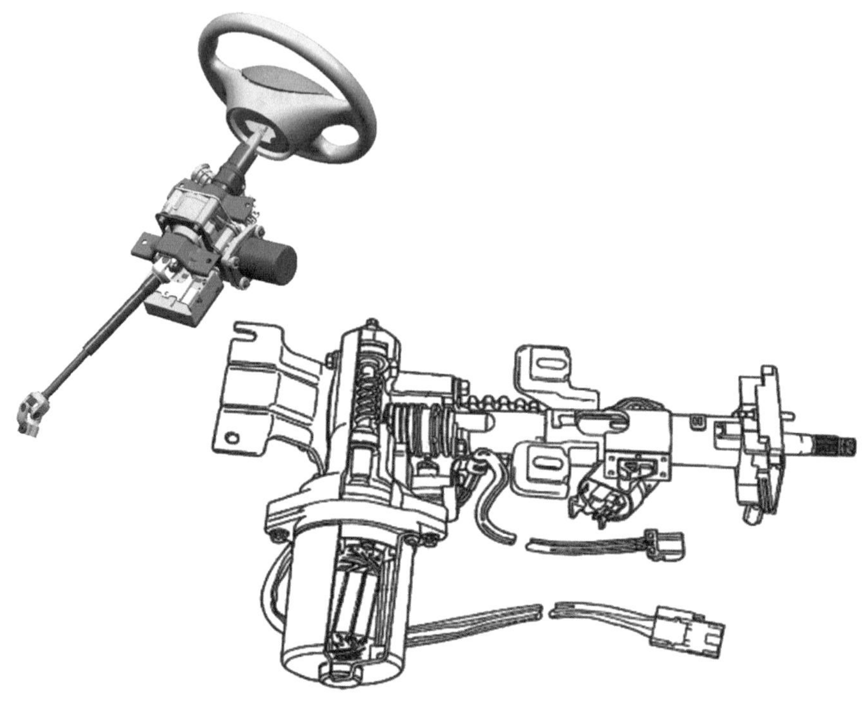

Fig. 4.18 Example of C-EPS. Above, a perspective view of the assembly and below a detail of the motor with reduction gearbox (FIAT)

Figure 4.18 shows a C-EPS system with an assembly including a modified steering column and an electric motor with reduction gearbox, in this case a worm gear.

An example of application of a DP-EPS is shown in Fig. 4.19, in this case the rack has two independent pinions, one connected with the steering column, one connected with the electric motor through a worm gear.

On a larger car, in Fig. 4.20, we see a very compact B-EPS; the two details above show the two-steps transmission connecting the motor with the steering rack: the first step is a toothed belt, while the second is a recirculating balls screw, chosen for its high mechanical efficiency and compactness.

In any case, the electric motor regulation is done by an electronic controller, which again has the function of generating an assistance torque proportional to the steering torque.

The control system includes sensors able to measure steering torque, vehicle speed, steering wheel speed and angle. These data are used to calculate the optimum torque to be applied by the motor; it is function of vehicle speed, to improve driver sensitivity to the smaller steering torque values. The assistance is increased at large steering angle, typical of parking maneuvers and of narrow turns, where actuation speed is more important then sensitivity.

Fig. 4.19 Example of application of a DP-EPS. The rack has two independent pinions, one connected with the steering column, one connected with the electric motor through a worm gear (FCA)

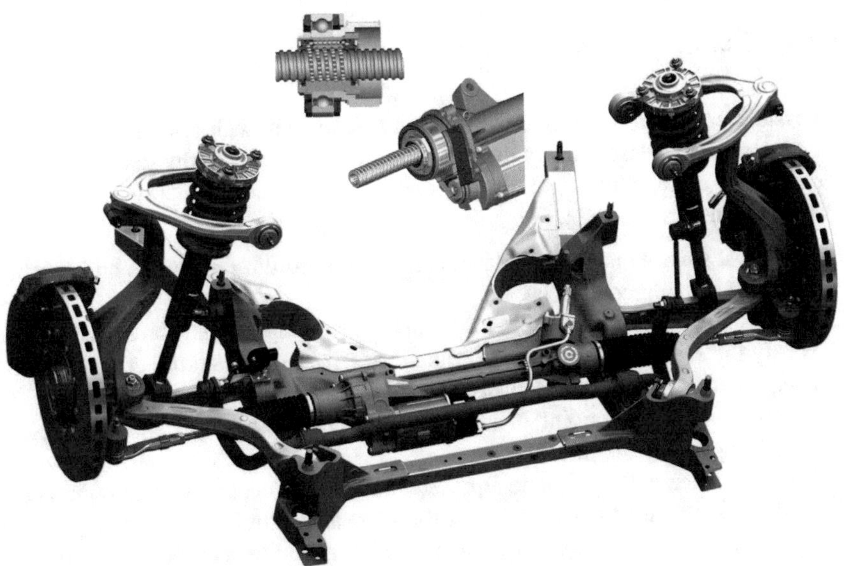

Fig. 4.20 Example of B-EPS. Above is shown the two-steps transmission connecting the motor with the steering rack (FCA)

This optimum torque is applied with the correct direction, given by steering angle acquisition. Sometimes a switch on the dashboard allows the driver to obtain a larger or smaller assistance, according to driving environment.

The motor is a simple direct current unit with permanent magnets.

A limited inertia of the motor (consider that the motor inertia must be added to the steering wheel inertia, with its value increased by the square of the transmission ratio to the steering column) is mandatory to have a good return of the steering wheel to the aligned position without oscillation; system damping can be enhanced by a suitable current control of the motor.

These systems have a number of advantages if compared with a conventional hydraulic unit:

- oil circuit and radiator are eliminated, with consequent simplification of the engine compartment lay-out;
- start-stop devices do not influence steering assistance;
- active safety is increased, because the assistance is available also when the engine stalls or is switched off;
- easier possibility of having a more sophisticated torque control as in hydraulic systems;
- low impact on mechanical components design.

4.7 Design and Testing

4.7.1 Outline Design

A possible procedure to outline the steering mechanism, with reference to the rack and pinion case, is described by the following steps; the case of the screw and sector can be easily extrapolated.

Let us consider Fig. 4.21.

1. The most suitable position for the rack is set, according to engine and gearbox bulk and steering column lay-out.
2. On the plant view, starting from king-pin trace steering arm may be draft, trying to satisfy the Jeantaud condition, to contain steering error. Steering arms can point backwards or forwards, but must comply with the wheel bulk. In this way x and y coordinates of the point A are decided, while those of point B are decided according to the rack position.
3. With reference to the yz plane represented on the figure, the trace of point A can be draft or calculated as function of the suspension stroke, with no variation of toe angle. The trace of this path can be approximated by a circle.
4. The center on this circle should be coincident with the B articulation of the steering arm, to avoid toe angle variation versus suspension stroke; if this condition isn't verified the z coordinate of the point A can be changed, until point B satisfies the rack installation requirements.

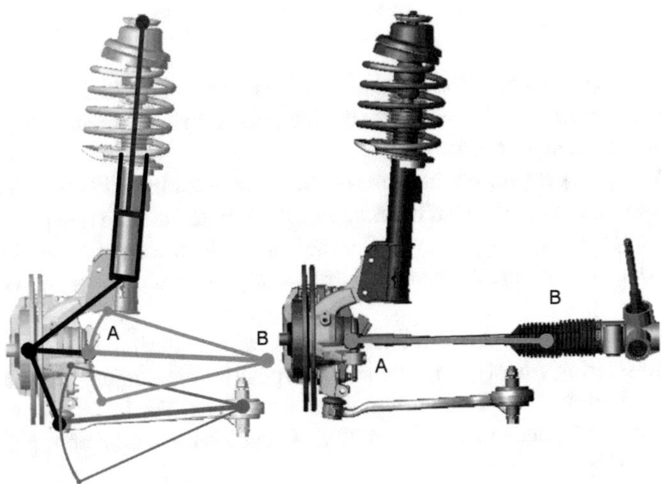

Fig. 4.21 Scheme useful for positioning rack and steering arm

- By draft or calculation the angle between steering arm and steering rod must be evaluated in order to verify that the mechanism doesn't block (this angle must, prudentially, be in the interval between 20° and 160°)
- The rack length must be sufficient to allow the necessary stroke; often the two B points (right and left side of the car) are too close. In this case a rack with central spheric heads must be adopted and this procedure repeated from the beginning.
- If car understeering should be changed the z elevation of point A can be changed, until the desired toe angle variation is obtained.

4.7.2 *Mission*

The steering system mission includes requirements on material fatigue, crash worthiness and elasto-kinematic behaviors.

Fatigue

The fatigue design of the steering system includes the experimental verification of all components and of the complete system. The target of this design is to predict the life of this system under significant test conditions. The mission can be derived from that introduced for the suspension and the entire body.

In this case it is necessary to establish not only the forces acting on the wheel, but also the appropriate steering angle.

Given the importance of this system it is suggested to verify all parts to unlimited life.

Crash

Usually reference is made to the standard test against barrier, with increased severity, as suggested by EuroNCAP, to evaluate the intrusion into the passenger compartment and the potential danger for the driver.

Other safety targets are set, because the steering column includes also the driver air bag system; the driver's head must impact the inflated air bag in a preestablished way, to be protected during its travel after the impact; in other words steering column and wheel deformation must not affect substantially the deployment geometry of the air bag.

A first approximation criterion consists in evaluating if the impact between driver's dummy and steering wheel can take place. In a positive case, injury criteria are used, limiting the dummy head acceleration after the impact. For instance, an optimum performance corresponds to a value of 72 g of acceleration and a poor performance to a value of 88 g.

Elasto-Kinematic Behavior

The elasto-kinematic study of the steering system includes the evaluation, as function of the suspension stroke, at different steering angles, of the following magnitudes:

- caster angle;
- king-pin angle;
- longitudinal trail;
- king-pin offset.

The diagrams of these magnitudes are reported on Fig. 4.22 for a front wheel driven medium size car, as a reference for a suspension of acceptable behavior.

King-pin angle variations can be small, as those of king-pin offset; on the contrary, longitudinal trail is affected by wheel steer angle mainly.

The design value of the caster angle will determine the design value of longitudinal trail as well, while its variation in function of the suspension stroke will affect not only the longitudinal trail behavior, but also comfort and braking capacity on bumpy roads, because of the dynamic response of sprung mass.

The design value of king-pin angle has effects on king-pin offset and on steering wheel return (due to the vertical load on the steering axle), for small lateral accelerations. The king-pin angle variation will influence king-pin offset variation and the torque felt on the steering wheel during braking.

Longitudinal trail, which is very sensitive to steering angles, has influence on vehicle stability on straight roads and determines the steering wheel torque in curves, in addition to the self aligning torque of the tire; it can, therefore, conceal the self aligning torque decrease for high tire side slip angles, giving a different perception (through the steering wheel torque) of the friction between tire and ground.

Steering wheel torque should never be too low or negative if the cornering force can still increase, when tire side slip angle is increasing. It should be remembered

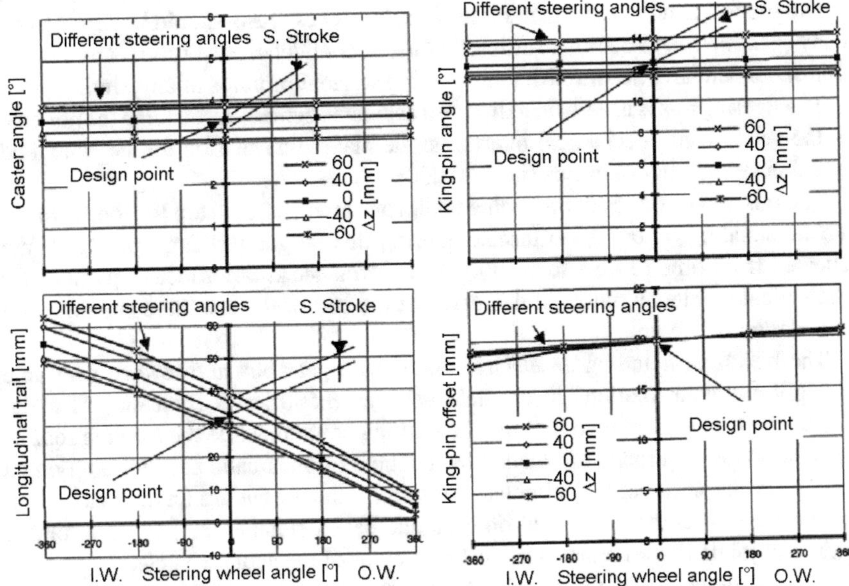

Fig. 4.22 Diagrams of main geometrical characteristics of the steering system as function of suspension stroke and steering angle. I.W. (inside wheel) and O.W. (outside wheel) refer to the position of the wheel on a curved path; outside wheel has a positive stroke

that the steering wheel torque is determined by the sum of torques acting on the two wheels; what is lost on the wheel inside the path, can be compensated by an increase on the outside wheel.

King-pin offset variation have influence on the steering wheel torque and on vehicle stability during braking on roads with different friction coefficient on the two wheels of the same axle. As a matter of fact, in this case, the torque generated on the steering wheel, different on the two wheels, is such as to direct steering to counteract the applied yaw moment.

4.7.3 *Bench Testing Methods*

The components that usually are evaluated separately are steering rods, steering box, steering column and power steering pump; with small adaptations what will be written on this last subject will apply also to electric power systems.

The steering rod is assembled to the steering box, fixed to the best test bed, and is connected to a hydraulic actuator on the wheel side. These three items lie on the bench with the same position they have on the vehicle.

Fatigue test can be performed on universal benches. Tests should be repeated on many specimen as to obtain a sufficient statistic confidence; test results are positive if all specimen have accomplish their mission without ruptures of any kind.

The steering box is usually installed on a dedicated bench, reproducing the position on the vehicle and mechanical interfaces; the maximum mission torque is applied for a defined number of cycles (i.e. 100,000).

The mission torque is defined on a complete steering from stop to stop, with the vehicle standing on dry good tarmac. During this fatigue test functional tests are repeated from time to time to verify that performance is not affected; particularly mechanical efficiency must not decrease. At the end of the test there should be no degradation or rupture.

The bench is including an electric motor with gear box and reverse gear able to apply alternate rotations in the field between ±500°, at a frequency of about 0.2 Hz; the torque is applied through the steering column, while the steering rods or the spheric heads directly are loaded by a couple of pneumatic actuator applying a constant reaction force; they simulate the spin friction of the tire on the road.

The steering column is tested on a fatigue bench suitable for pulsating torque. Universal joints should reproduce working angles of the actual vehicle.

The test procedure is the same as for the steering box and, if convenient, these two tests can be performed together.

The hydraulic pump is tested on a dummy hydraulic circuit reproducing a realistic driving cycle, including only urban and bendy road driving; this kind of mission could be applied for about 50,000 km. At the end of the test there shouldn't be any rupture or loss of performance.

Similar tests can be designed for the remaining components. We will describe now some screening test used to verify design quality and to obtain input data to be used on mathematical models.

The main synthetic parameters describing steering system performance are:

- transmission ratio, between steering wheel and king-pin axis;
- dry and viscous friction or hysteresis;
- torque on the steering wheel as function of the cornering force.

A test system suitable to perform this kind of functional evaluation may include the following parts:

- a rotary hydraulic actuator, with angular displacement of at least 90° and radial load capacity of at least 30 N, with its dedicated hydraulic power generator and control system;
- an universal joint without angular plays to apply forces to the steering wheel, compensating for actuator and steering wheel misalignments; high torsional stiffness and reduced inertia are also important features;
- an angular displacement transducer (measurement field ±180°) to acquire steering wheel angle;
- four linear displacement transducers (measurement field ±20 mm) to acquire displacements of right and left wheels during steering;

Fig. 4.23 Driver seat of a car equipped with a measurement system to acquire steering wheel torque

- two reaction springs on the wheels able to simulate a reaction torque of about 60 Nm/°;
- two load cells (measurement field 0–1,000 N) to acquire actual steering torque at the contact patch;
- a static torsiometer (measurement field 0–50 Nm) to acquire steering wheel torque;
- signal acquisition system and dedicated data processing software.

During this test, front wheels are resting on a floating plate (floating on an air bearing) with reduced friction. Screening tests are performed imposing steering angle and measuring remaining magnitudes.

Figure 4.23 shows the driver compartment of a car equipped with such measurement system: we can notice the hydraulic actuator 1, the universal joint 2, the angular displacement transducer 3 and the static torsiometer 4.

Figure 4.24 shows, instead the wheel side; we can notice the reaction springs 1, whose effective load is measured by load cells 2.

Steering wheel actuation is done by sinusoidal or triangular waves of assigned frequency and amplitude, in the following test modes:

- wheels free on floating plates;
- one wheel free and one loaded;
- one wheel free and one blocked;

The following magnitudes are measured:

- steering wheel angle;
- wheels steering angle;
- wheels steering torque;

Fig. 4.24 Part of the same measurement system of previous picture dedicated to wheel rotation and torque acquisition. Wheels are resting on floating plates

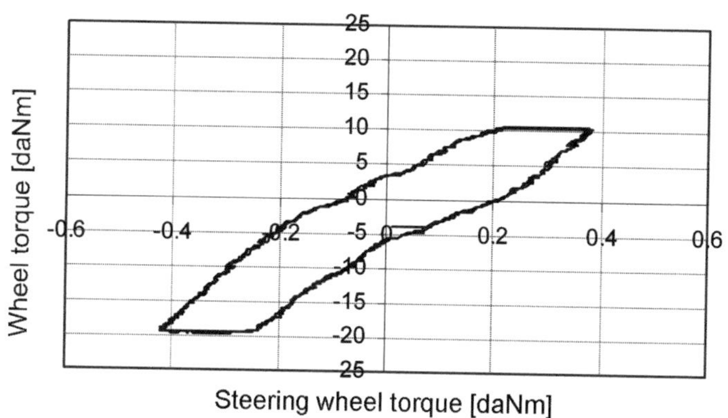

Fig. 4.25 Steering hysteresis cycle, referred to the steering wheel

- steering wheel torque.

The following relevant magnitudes can be calculated by data processing:

- steering system transmission ratio, as function of steering wheel angle;
- steering wheel torque, as function of angle;
- wheel steering torque as function of steering wheel torque.

This last kind of diagram is shown on Fig. 4.25. We notice an hysteresis cycle caused by mechanical friction and by viscous friction, mainly caused by the power assistance system.

Dry mechanical friction is due to relative motion and in localized on spheric heads, rack with sliding block and pinion and steering column bearings and joints. This part of friction is usually independent on steering angle, but is dependent on force.

Other major source of friction is the contact between tire and ground.

Viscous friction is due to oil choking and flow, localizing on hydraulically assisted steering boxes; also magnetic losses of electric power steering shows a similar effect.

The total amount of these frictions determines the hysteresis cycle; the area of this cycle shows part of the work performed by the driver, during steering and should be as small as possible.

Chapter 5
Braking System

5.1 Introduction

The braking system must accomplish three different jobs:

- to completely stop the vehicle; this function entails braking moments on the wheel as high as possible;
- to allow speed control, when the natural deceleration of the vehicle, due to mechanical friction and motion resistances is not enough; this function entails braking moments on the wheels not very high, but applied for a long time;
- to keep the vehicle stopped also on a slope.

Because of the nature of these jobs the braking system is part of the safety systems of the vehicle. As a consequence, the State Authority and, afterwards, the European Union have introduced regulations that describe design conditions of this system and its minimum operation requirements.

Vehicle manufacturers and their components suppliers are, therefore, responsible for adequacy of products to regulations, including correct fabrication and system reliability for a reasonable period of time; users too must play their role because many parts of this system are subject to wear and the safety functions cannot be assured without the necessary maintenance and part substitution. A periodic compulsory control is addressed to assess the correct operation of this system.

If regulation determine minimum performance for this system, each manufacturer considers this as a starting point, because more stringent requirements are requested by the market and are a remarkable selling points.

For this fact, braking systems have reached in the normal praxis levels of performance and reliability that are very high.

It should be remarked that the relationship between reliability and accident probability is not very evident; statistics on road accidents show, in fact, that less then 2% of road accident are caused by a not adequate operation of the braking system.

G. Genta and L. Morello, *The Automotive Chassis*, Mechanical Engineering Series,
https://doi.org/10.1007/978-3-030-35635-4_5

Within this total, 90% of accident is estimated to be due to a not sufficient maintenance and 10% to dynamic instability, consequent to a braking not compatible with transversal accelerations.

The wide application of anti lock systems (ABS) represent an improvement for braking safety, even if accident statistics not contain sufficient witness of this fact.

Studies about accidents in Germany, after introduction of this system in 1976, showed reduction on vehicles equipped with this system; most recent data show that the presence of this system leads user to overrate its contribution and, therefore, to be more exposed to dangers, in particular situations as icy roads or driving with reduced safety distance.

Again in Germany, a study on taxis doesn't show a relevant difference between cars with and without ABS.

Before starting the description on braking system components, we prefer to introduce some preliminary consideration on braking system design, that will be better explained in the second volume of this book, dedicated to system design.

We have seen in the chapter about tires, that the maximum longitudinal force exchanged with the ground depends on many factors, including, as first, the vertical load on the wheel. Secondly, this force is influenced by the ground nature, the vehicle speed and the coexistence of cornering forces.

The system maximum performance is, therefore, conditioned by many factors outside of the system itself; while the recognition of last factors is a drivers job, who is in charge of limiting vehicle speed and controlling the distance between close vehicle, vertical loads cannot be easily understood.

These are determined by different factors, as:

- payload and its distribution in the vehicle;
- road slope;
- longitudinal acceleration, in particular, the same braking acceleration.

To focus this fact, let us consider a vehicle of mass m, moving on a flat road, inclined of a slope angle α.

We assume that the vehicle is symmetric, with reference to its median plane and its center of gravity G is distant a from the front axle and b from the rear axle; the center G is high h_G on the ground (see Fig. 5.1); finally:

$$l = a + b \ ,$$

is vehicle wheelbase.

We now assume that all acting forces are negligible, but braking forces F_{x1} and F_{x2}, applied to the front and rear axle. Each of these forces represents the sum of the those acting on the wheels of the same axle, that in the assumptions we have made are equal.

If we indicate with F_{z1} and F_{z2} the total vertical force on each axle and with a_x the braking acceleration (it should be remarked than it is negative, by braking), we have:

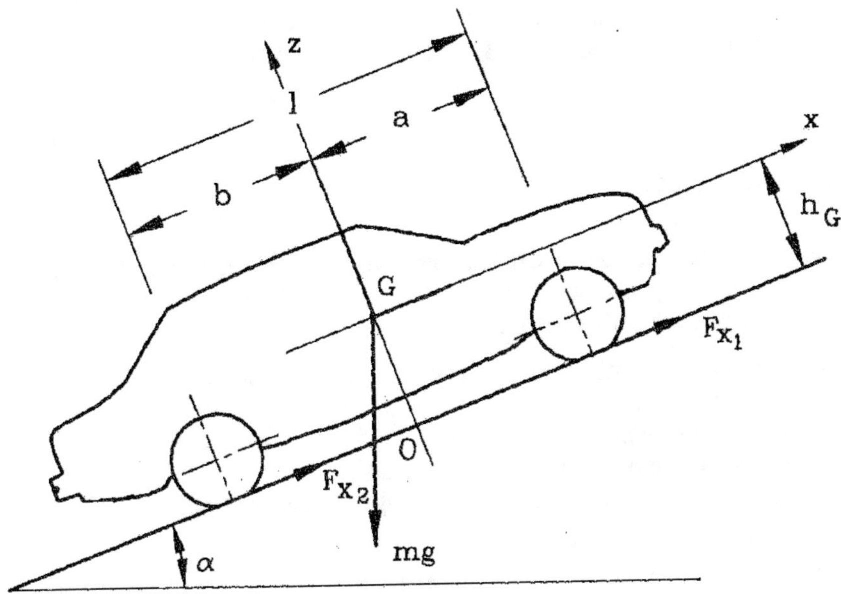

Fig. 5.1 Scheme of forces acting on a vehicle on a flat road with slope α

$$\begin{cases} F_{x1} + F_{x2} - mg \sin\alpha = ma_x \\ F_{z1} + F_{z2} - mg \cos\alpha = 0 \\ F_{z1}a - F_{z2}b + mgh_G \sin\alpha = mh_G a_x \end{cases} \quad . \tag{5.1}$$

From this system, we obtain:

$$\begin{cases} F_{z1} = \frac{m}{l}\left(gb\cos\alpha - gh_G \sin\alpha - h_G a_x\right) \\ F_{z2} = \frac{m}{l}\left(ga\cos\alpha + gh_G \sin\alpha + h_G a_x\right) \end{cases} \quad . \tag{5.2}$$

These equations indicate clearly that vertical reaction forces depend not only on the center of gravity position, but also on the vehicle braking deceleration.

In the chapter about the tire we have seen that the maximum obtainable longitudinal force is proportional, through the friction coefficient μ_{xp}, the peak value of the curve $\mu_x(\sigma)$, to the vertical load on the wheel.

As a consequence, the following conclusions can be draught:

- maximum braking forces are determined by load conditions through $m,\ b,\ a,\ h_G$;
- they are also determined by the braking deceleration a_x;
- braking forces distribution between the two axles is also determined by the obtained longitudinal acceleration.

Therefore, a braking system should be able to change its geometry to adapt braking forces to existing vertical loads. In particular the braking force on the rear axle should be progressively limited when the total braking force is increased.

This function is performed by the brake distributor valve, which will be described in the following paragraphs; the ABS system can also take care of this operation very carefully.

5.2 Car Brakes

According to the regulation, the functions mentioned at the beginning of the previous paragraph, are to by accomplished in a vehicle by three different systems, that cannot directly be matched with them; they are:

- *service braking system*, able to reduce the speed or stop the vehicle, in normal driving conditions;
- *emergency or secondary braking system*, suitable for the same above function, but to be used in case of failure of the service brake;
- *parking braking system*, suitable for parking only, also on slopes.

All these systems must exert an adjustable braking force on the vehicle.

Many components are common to the first two braking systems, but some of them are specific; redundancies guarantee reliability and availability of braking function.

5.2.1 *Service and Secondary Systems*

Figure 5.2 shows the scheme of a hydraulic braking system; it includes the service and secondary systems.

Front and rear brakes are actuated by two completely independent hydraulic circuits A and B; this separation is present also in the master pump C, actuated by the pedal and enlarged in the picture at the lower left.

This feature, de facto imposed by the law, allows to define the two circuits, when they operate together, in normal conditions, as service circuits and the same, when they operate separately, when one of the two has failed, as emergency circuits.

The kind of failure, considered in this approach, is the rupture of one of the flexible tubes connecting brake actuators, moving with the wheels, and the rest of the circuit, fixed to the vehicle chassis structure; if, for example, the tube connecting the left front wheel would fail, at the first braking, a defined quantity of oil may be squeezed out, but the rear circuit would be still able to work.

The homologation rules request that, on a perfectly paved road, for a car, this stopping distance must be obtained:

$$s \leq 0.1V + \frac{V^2}{150}, \tag{5.3}$$

where s is the stopping distance, measured in (m) and V is the car speed, measured in (km/h), at the beginning of braking.

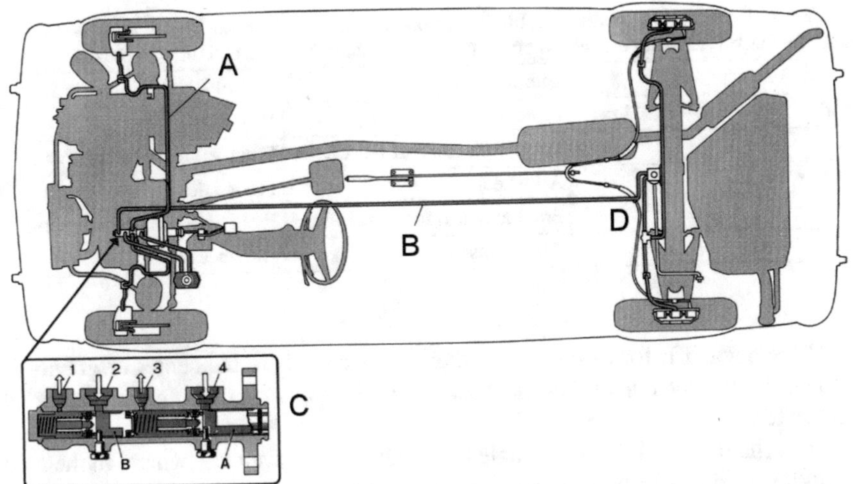

Fig. 5.2 Pictorial scheme of a car braking system; front and rear brakes are operated by two independent hydraulic circuits A and B. C shows an enlargement of the master pump and D the distributor

This condition must be applied for service system; an increased distance:

$$s \leq 0.1V + \frac{2V^2}{150} , \tag{5.4}$$

is allowed, when one of the circuit is broken, when the emergency system only is working.

Therefore, for a service circuit, an average deceleration:

$$a \geq 2.9 \text{ m/s}^2 ,$$

is accepted, while, with the emergency circuit it can be increased up to:

$$a \geq 5.8 \text{ m/s}^2 ;$$

these prescriptions are applicable for any load condition within those allowed by the homologation form.

The braking circuit can be organized in different ways; the schemes described on Table 5.1 are accepted.

Circuits 1 and 2 represent the possible configurations of the emergency circuit when one of the wheels has its brake hydraulic flexible pipe out of order.

The choice among these schemes must take into account the vehicle load breakdown on the axles.

Table 5.1 Braking circuits allowed by the regulations of the European Union; circuits 1 and 2 describe the still working brakes, when each one of the flexible tubes, at a time, is broken

Symbol	Circuit 1	Circuit 2
TT	Front axle	Rear axle
K	Front right and rear left wheels	Front left and rear right wheels
HT	All wheels	Front axle
LL	Front axle and rear right wheel	Front axle and rear left wheel
HH	All wheels	All wheels

The scheme TT, for example, is suitable to a car where the center of gravity is located about at the mid of the wheelbase, as can happen on a front engine, rear drive two seater.

The scheme K is, instead, suitable to a front wheel driven car, where, in the load condition of driver only, it is necessary that one of the front wheels, at least, can brake, to satisfy the requested performance.

The following schemes, as HH, have the advantage to grant the emergency braking system a complete, or almost complete, braking circuit at a slightly increased cost; this fact allows to reduce the torque applied by the braking forces unbalance to the steering wheel to a minimum.

Also the safety margin on *fading* conditions must be taken into account for this choice. We will discuss about fading later on; at this time we define fading as a loss of braking efficiency because a partial evaporation of the braking fluid, due to high temperatures.

When the fluid evaporates, it increases dramatically its elasticity and looses its capacity to transmit pressure by compression. This phenomenon occurs near a heat source and, therefore, near the wheel. For this facts circuits HT, LL, HH are more vulnerable when the fluid is evaporating near the front wheel.

Referring again to Fig. 5.2 we can see the valve D, including the pressure distributor, we will explain later on.

In addition to the prescription about stopping distance and braking acceleration, a further condition is requested about the control force applied to obtain such stopping distance; such force, applied to the pedal by the driver's foot must be:

$$F_p \leq 500 \text{ N} ;$$

this condition causes often the application of a power assistance device, we will explain later on.

5.2.2 Parking System

The parking braking system, operated by a hand lever or by a further pedal, is requested to keep the vehicle stopped on a slope when the driver is not on the vehicle.

The regulation requests for this system a mechanical, non hydraulic, connection between the drivers control and the brake; tie rods and cables are allowed.[1]

The rationale for this prescription is to build up a further emergency system, to increase system reliability, by adding a second chance in the very rare event of a contemporary failure of service and emergency braking systems.

The parking braking system must be designed to keep the vehicle stopped on a slope of:

$$i \geq 18\%,$$

or to reduce its speed, on a flat road, by an acceleration of:

$$a \geq 1.5\,\text{m/s}^2\ ,$$

after having applied on the control a force:

$$F_p \leq 500\,\text{N}\ ;\ \ F_m \leq 400\,\text{N};$$

the first value applies to pedals and the second to hand levers.

The braking system must be able to generate an adjustable force, that can be maintained also with to action on control; it works on the wheels of a single axle; brakes can be the same as for the service system.

Figure 5.3 shows a hand lever parking system: the control lever 1 moves the tie rod 17 and can be hold in any position by the spring loaded pawl 27 engaging in the ratchet sector 23; the disengagement button 26 is used to unlock the pawl 27.

The tie rod 15 pulls the cable 6, through the sliding equalizer 16 the divides in two equal parts the force on the two brakes. The cable is connected to the rear brakes, through the pin 8, connected to the two shoes.

The variant with pedal is similar; the pawl can be unlocked by a push-push pedal or by button on the dashboard.

The connection with drum and disc brakes is explained in the dedicated following paragraphs.

5.2.3 *Disc Brakes*

This brake includes a disc, rotating with the wheel, on whose two faces two braking linings can be pressed, made of high friction material.

On these linings, also called *pads*, work one or more hydraulic cylinders mounted on a suitable *calliper*.

---

[1]Quite recently, also the possibility of operating brakes through a bistable separate hydraulic or electric actuator has been allowed. In this case the braking force must be maintained indefinitely with no energy.

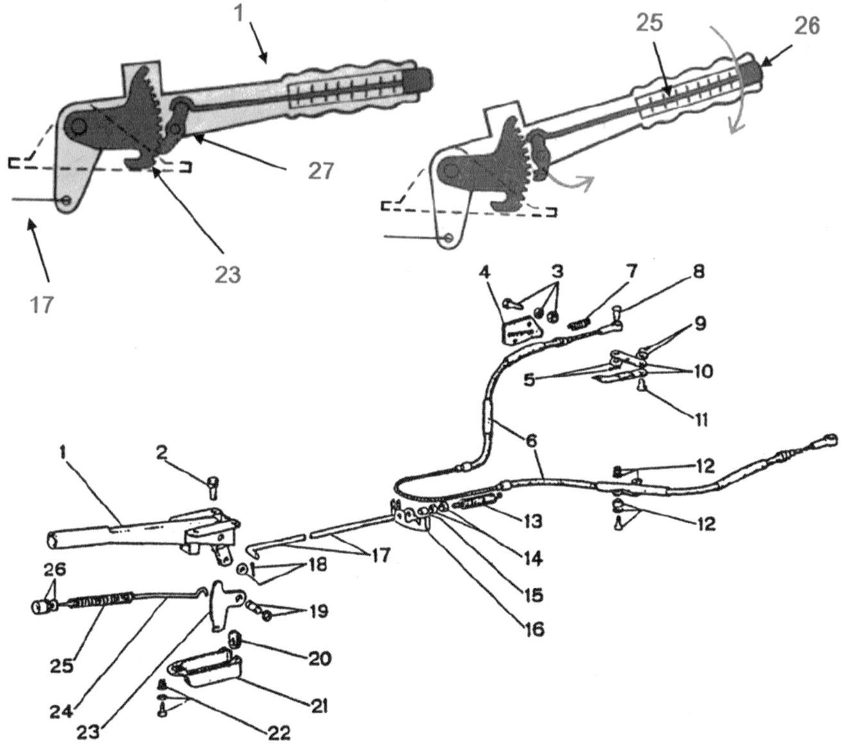

Fig. 5.3 Scheme of a parking system with hand lever control; on the upper part of the drawing, we can see the mechanism able to maintain the braking force also with no external force applied

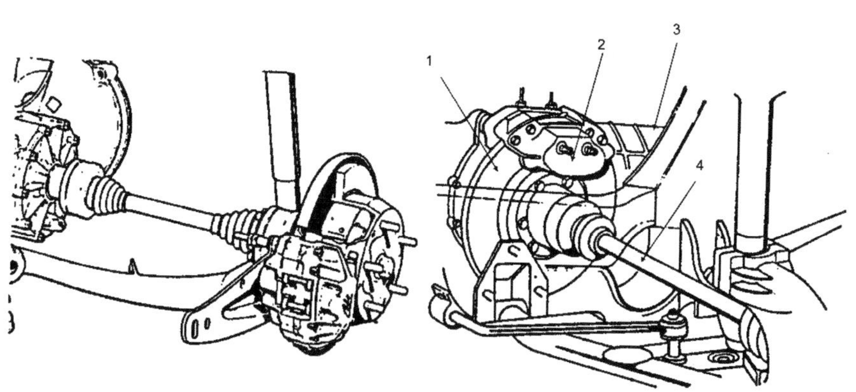

Fig. 5.4 Examples of application of a disc brake to a rear axle; the disc can be mounted on the wheel itself (left) or to the differential box and work on the half shaft (right), to reduce the unsprung mass

Fig. 5.5 Ventilated disc brake with fixed caliper; the two pads are pressed against the disc by four hydraulic cylinders, two for each pad

The disc brake, as the drum brake we will describe in the following paragraph, can be applied in two different ways, shown on Fig. 5.4, where the disc is fit together:

- on the wheel hub, directly;
- on an auxiliary hub on the half shaft, at the differential box.

The examples on this figure refer to a rear driving axle, but can be applied also to a front axle or to a rear idling axle, by adding in this case, a shaft dedicated to braking force.

The left figure shows the assembly of a disc brake on a guided trailing arms rear suspension. The brake caliper is fixed on the suspension strut, where the reaction to the braking force will be applied.

On the right figure, instead, the caliper 2 is fixed to the differential box, or to the transmission, if differential and gearbox are integrated in a front wheel driven vehicle; the reaction to the braking force is applied, in this case, to the car body directly or through a suspension. The disc 1 is fit to the shaft 4 that transmits both driving and braking torque.

This kind of expensive configuration allows a reduction of sprung mass, with benefits on suspension comfort.

There are two kinds of architectures for disc brakes:

- fixed caliper;
- sliding or floating calipers.

By fixed caliper architecture, shown on Fig. 5.5, pads are pressed against the disc by two couples of independent hydraulic cylinders, inside the caliper, connected in parallel to the same pressure source; the double cylinder arrangement can be exploited to separate service circuit from emergency circuit.

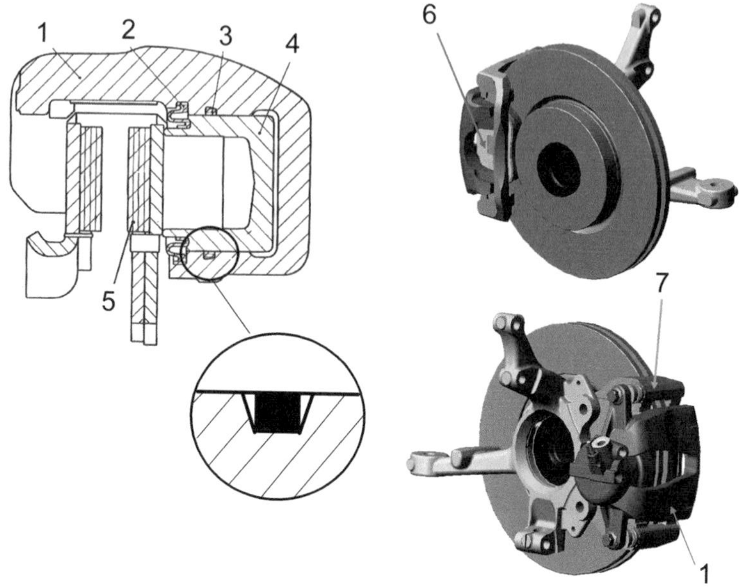

Fig. 5.6 Floating caliper disc; the caliper is assembled as to be able to move is a direction perpendicular to the disc surface; the hydraulic cylinder can be single. An enlarged cross section of the piston and cylinder is shown with the detail of the sealing

The disc shows inside semi radial channels of suitable shape; they behave as a radial ventilator which activates air for brake cooling. The transversal drilling on the disc improve the direct cooling of the disc too, on its working surface.

On lighter cars with less performance, therefore with smaller discs, a single couple of cylinders might be sufficient and ventilation channels and drilling can be avoided.

Discs are made by iron alloy; on high performance cars, the very expensive carbo-ceramic discs are appearing, for their property of surviving very high operation temperatures.

The floating caliper disc brake, on Fig. 5.6, shows, instead, a single cylinder 4, working on the inside pad 5, while the outside pad 6 is pressed on the disc by the caliper body 1, able to slide on its mounts 7 along a direction perpendicular to the disc surface. This solution has the advantage of reducing cost and dimension, allowing also to reduce king-pin offset to negative values.

Hydraulic cylinders are operated by the fluid pressure applied by the master cylinder; as soon as the pedal is released, the piston of the master cylinder returns in the initial position, because of a return spring, and the hydraulic pressure is set to zero.

The brake piston is pulled away of the disc because of the elastic force, determined by the lateral deformation of the sealing ring, invested on each piston with a certain radial load; the shape of the groove (see the enlarged detail of the sealing ring in rest position) allows piston motion without sliding on the sealing. A similar design feature is also applied to fixed calipers.

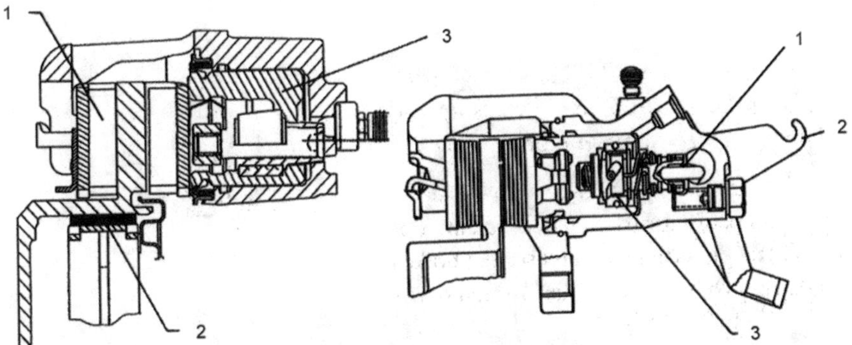

Fig. 5.7 Different controls for parking function on disc brakes; at left, the function is obtained by an additional coaxial drum brake; at right, a lever 2 and a cam 1 act directly on the braking cylinder

Floating calipers have the inconvenience that cross sliding can be impaired or blocked by mud sediments or by corrosion. For this reasons protection bellows and special coatings on slide pins are applied.

An impaired sliding of the caliper can cause different wear on internal and external pads; in some case the caliper can be blocked with asymmetric braking, affecting vehicle path.

By disc brakes, the parking function is obtained with an additional coaxial drum brake; this solution is necessary if the force to be applied to the disc pad results to be too high.

This solution is shown on Fig. 5.7, at left: inside the disc bell 1, two additional braking shoes are installed as by drum brakes; the control force application is done as it will be explained in the following paragraph.

The second solution, on the right of the same figure, combines service and parking functions in a single unit; brake pads are pushed by a crank and rod mechanism 1 (the crank is made by a cam fit to the control lever 2) operated by cables. Usually this control includes an automatic backlash adjuster.

By parking brakes a braking force difference on two wheels of the same axle is not very important for safety.

5.2.4 *Drum Brakes*

Figure 5.8 shows a cross section of a drum brake and some design details.

This brake is made by a rotating hollow cylinder 7 (at left) fit to the wheel hub; on its inside surface works two symmetric shoes 6, on which are riveted or bonded braking linings. Shoes are pushed against the drum by the double piston cylinder 1.

At one end of the shoe one of the pistons is pushing, while the other end is linked with a hinge or is resting on a suitable surface. The two shoes are kept away from the drum by two springs 5.

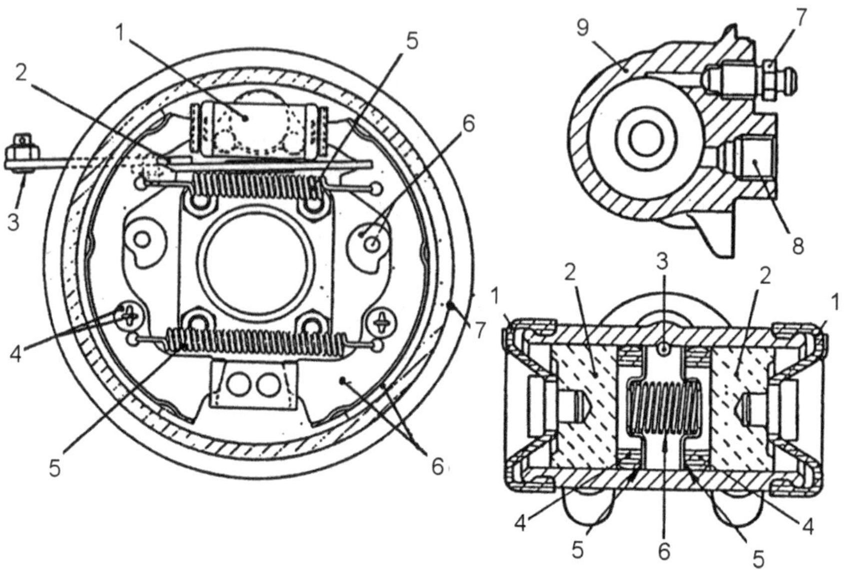

Fig. 5.8 Section of a drum brake (at left) and of its actuation cylinder (at right)

Because of the necessary machining tolerance of the drum surface and of the possible thermal deformations, the clearance between linings and drum must be well above of that between lining and disc of disc brakes, which, because of their simplicity, can be machined with higher accuracy and don't show thermal deformations in the pads motion direction.

On the same figure, at the right side, a cross section of the control cylinder is shown: two pistons 2, the oil feed drilling 3 and the return spring 6 can be seen; two pistons are assembled into the cylinder 9 and show their seals 4. The bellows 1 avoid water or dust contamination on the cylinder sliding surface.

The cylinder is connected to the oil circuit trough the nipple 8; the bleeding valve 7 is used to remove the air that could be entered into the circuit.

On modern cars drum brakes are applied to rear wheel only and have, therefore, the mechanical parking brake control. This is made by the push rod 3 and the crank 3, moved by cables.

The friction between linings and drum is causing their wear. This fact is obviously present on disc brake too, but the particular shape of the seal (see again the detail on Fig. 5.6) is eliminating the effect of the additional clearance caused by wear.

On drum brakes, where shoes are returned against a rest by a spring, this wear could cause an increase of the distance between linings and drums and a consequent increase of brake pedal stroke; this last could eventually become too large.

The rest position of the shoe must be therefore adapted to the actual wear, as to maintain a constant clearance between shoe and drum. This adjustment can be made by rotating manually the two cams 6 by a nut and a lock nut.

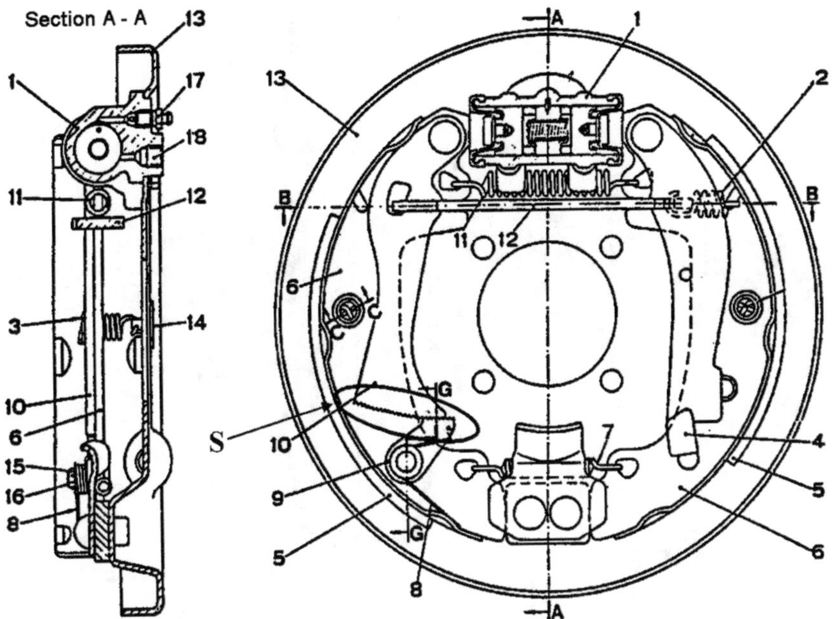

Fig. 5.9 Section and view of a drum brake with automatic clearance adjustment. This function is accomplished by two levers 9 and 10 gearing through a tooth sector, outlined by the oval line S

On most modern brakes this adjustment is made automatically. A possible solution is shown on Fig. 5.9.

According to this system (Bendix system), one of the two shoes is made by two pieces 9 and 10 with tooth sectors gearing together, according to the detail shown within the oval line S, on this figure. The rest of the left shoe is made by the push rod 12; the clearance between rod and shoes determines the clearance between lining and drum. Until this last clearance is higher as the shoe stroke, the two pieces 9 and 10 are working as a rigid system; as soon as the stroke is higher as the clearance, the upper part is retained by the push rod. This fact is causing two sectors to rotate to a different position, causing a reduction of the distance between linings and drum.

The same push rod 12 can be used for parking brake control.

5.2.5 *Control System Components*

Pump

The pedal works on the pump piston, through the push rod on the right of Fig. 5.10.

Between this push rod and the pump the power brake is set, we well explain on the next paragraph.

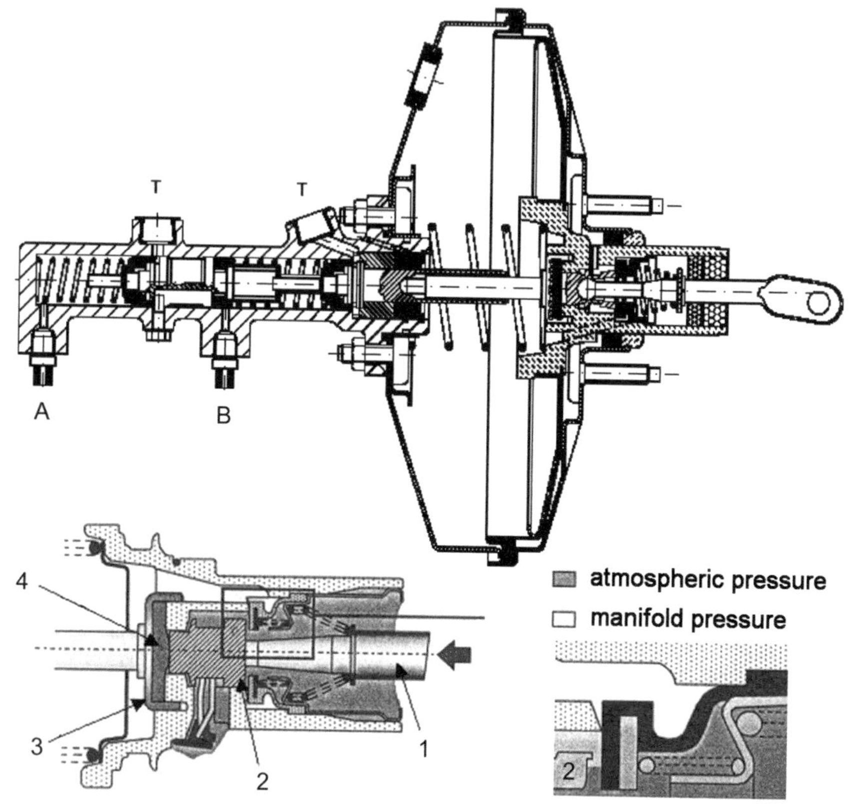

Fig. 5.10 Cross section of the master pump and of the vacuum power unit. Below is shown an enlarged detail of the valve between the vacuum chamber and the ambient, used to modulate the power assistance

The *master pump* or *tandem pump* is made by two pistons in series into the same cylinder; on this cylinder wall the openings T are connected to the oil tank and the openings A and B to the two braking circuits. This particular pump is feeding two completely separated circuits (service and emergency circuits) each of them can be connected to the brakes, according to the schemes on Table 5.1.

In rest condition, the two pistons are kept to the right by the coil springs shown on the figure; the two piston chambers are connected to the tank at the atmospheric pressure; in this way the additional fluid necessary to adapt to a clearance increase due to linings wear can be supplied.

As soon as the pedal is depressed the two holes T are closed and the pressure inside the circuits is increased, in proportion to the pedal force; this pressure will act on pistons working on pads or shoes.

This kind of arrangement for the pump guarantees the operation of one circuit when the other has failed (is spilling, as a consequence of a pipe rupture); in fact, if one of the circuit would spill the two independent pistons will contact each other and the pressure could increase in any case in the still working circuit. The increase in the pedal stroke is warning the driver about the failure.

Braking Fluid

As we have seen the transmission of the pedal force to the braking surfaces is made hydraulically.

The working fluid for this purpose must have particular features and satisfy to the following specifications:

- in normal working pressure conditions must be incompressible;
- its boiling point must be over a certain minimum value, in order to maintain its properties also after a long lasting braking;
- must have a low viscosity also at very low temperature, in the range of −40 °C;
- must have suitable lubrication properties for parts on relative motion (pistons, seals and cylinders);
- must be chemically stable and non aggressive to metal and elastomeric components.

These conditions are fulfilled by some organic oils.

These oils must be changed after a certain period of use, because their are hygroscopic; water is present in the air as humidity and molecules of water can contaminate the oil in the atmospheric tank, when the level is decreasing, as consequence of wear.

Water in solution decreases oil boiling temperature.

Because of brakes heating up, water in solution is changing to vapor; vapor bubbles decrease oil compressibility and increase pedal stroke at the same level of pressure. At critical conditions the pedal stroke can't be enough to allow the desired braking force. This lack of braking efficiency is called *fading* or *vapor-lock*.

Water absorption speed is largely dependent on climate; by humid and hot climate a percentage of water of 3% of the oil can be absorbed, with a consequent 80% reduction in boiling temperature.

The United States Department of Transportation certifies fluids called DOT3, DOT4 and DOT5 defining different boiling points as function of the water content.

Distributor

Because of vertical load transfer, due to the vehicle deceleration, the braking force applied to front wheels must increase as compared to the static value; for the same reason braking force applied to the rear wheels must decrease.

Also static load conditions are affecting braking force distribution between the axles, because of the different position of payloads on the vehicle, with reference to the axles positions.

The function of adapting the braking force share between axles is performed by the *brake distributor*.

The braking circuit is designed to grant rear wheels the maximum braking force, they need, usually at static full load; the distributor is designed as to reduce this pressure to the suitable value, corresponding to the actual static load and to the load transfer.

This target is achieved according to the following rules:

- when the circuit pressure is lower than a threshold value, the rear pressure is not reduced;
- when this threshold is overcome, the front pressure and the rear pressure increase, according to a preset value lower than one.

This function is achieved by the valve sketched on Fig. 5.11.

The nipple 7 is connected with the pump and the front circuit, while the nipple 6 is connected with the rear circuit. In this valve a moving spool 1 is sensible to the rear suspension stroke, through the tip 2.

The scheme at the right of this figure shows the installation on the vehicle; a suitable leverage pushes on the tip 2 when the suspension moves to compression (displacement in direction a) and vice versa in the extension direction (displacement in direction b). The vehicle suspensions acts as a dynamometer measuring the axle load through the suspension stroke.

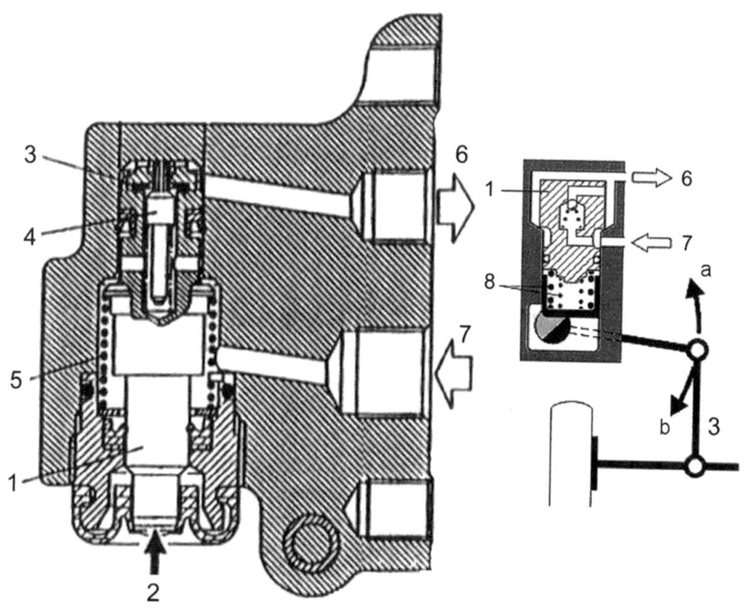

Fig. 5.11 Cross section of a brake distributor and of its installation on the rear axle

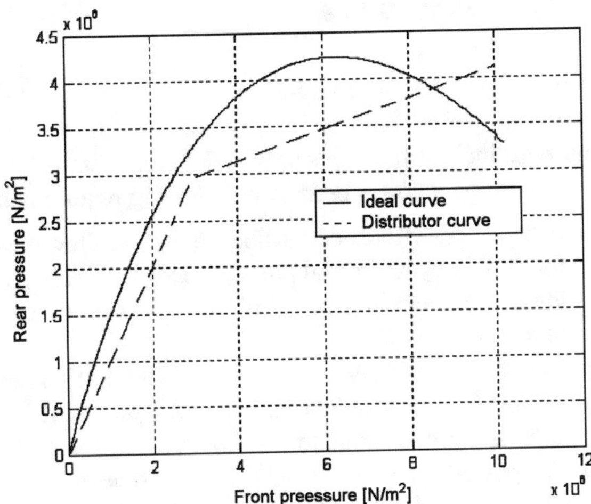

Fig. 5.12 Comparison between the ideal distribution curve of the braking pressure and the actual distributor curve

When the spool 1 is compressed to the upper direction, the valve 4 is lifted, opening the passage 3; when the spool is descending, at a given position the valve 4 closes, interrupting the connection between the pump circuit and the rear brake circuit. At this condition the pressure at the nipple 6 will be reduced with reference to the pressure at nipple 7, according to the ratio between the areas on the spool 1.

The stiffness of the spring 5 determines the suspension load at which the braking pressure is reduced, while the ratio between the surfaces determines the value of this reduction.

Figure 5.12 shows an example of comparison between the curve of the ideal braking pressure distribution and the actual distribution made by a valve of the kind of Fig. 5.11. By an ideal distribution, each wheel can brake at the maximum value of the friction coefficient. The parameter for this comparison is the pressure in the front and rear circuits.

The disadvantage of a valve of this kind is that the real distribution curve is at any condition lower as the ideal one and it doesn't allow the rear wheels to reach their maximum braking capacity; on the contrary, slip is completely avoided.

A further disadvantage is that is difficult to obtain a result of this kind for any of the possible load combinations of the front and the rear axle.

For reduced load variations (as on a two seater) simple pressure reducers can be used.

There is a third class of valves that are sensible to vehicle deceleration. In this case the change of slope of the distributor curve is determined by the braking deceleration. This kind of valve takes into account also the weight distribution change. A possible malfunction of this kind of valve is due to the effect of the internal friction of its mechanical accelerometer.

ABS systems perform also the function of limiting braking pressure on the rear axle, according to the actual vertical load; in this case the distributor valve is no long necessary.

5.2.6 *Power Brakes*

Power brakes allow to brake a vehicle with a reduced force on the pedal and a reduced stroke, that can be contained within acceptable limits. Their application purpose include braking safety and drivers comfort.

To design a power brake, the following requirements must be specified.

- A power brake must be sensible enough to allow the driver to modulate the braking force, also by low pedal pressure; internal friction must therefore be limited. The lower intervention point of the power system must be in the range of a pedal force of about 15–20 N.
- The effort exerted on the pedal is the feed-back of the braking system to the driver; is must be correlated to vehicle deceleration. The force on the pedal must be, in any case, proportional to braking deceleration.
- The power system response time should be lower as 0.1 s; the response time is the time necessary to reach the maximum assistance value, during a sudden braking, where the pedal is depressed at a speed of about 1 m/s.
- The passage from the assisted mode to the unassisted mode at the saturation point (see the definition of this condition later on) must be gradual to allow the driver to further force increments during emergency situations.
- Reliability of this system must be absolute; power system failures can panic the driver.
- Weight and volume must be limited to suit the installation condition into the engine compartment.

The *tip-in load* of a power brake is the minimum load on the pedal necessary to start the assistance of the system, during a braking.

The *saturation load* is, instead, the value of the load at which the diagram of force on pedal, versus braking pressure is changing its slope, because the assistance of the power brake has reached its maximum value. This load should be never reached in practical conditions; therefore it must be over the braking force necessary to stop the vehicle at maximum friction coefficient and maximum vehicle load.

Vacuum Power Brake

The power brakes is fit between the brake pump and the pedal and amplifies the force applied on the pedal by the driver, exploiting the difference of pressure between two chambers, one connected with the ambient and one with the intake manifold, for a throttled gasoline engine. When the vacuum manifold is not sufficient, as for diesel engines, a vacuum pump is driven by the engine.

With reference to Fig. 5.10, already used to explain braking pump, we can notice the dimension of the actuator, bound to the modest value of the difference in pressure between the manifold and the ambient.

The actuator includes a cylinder and a piston made of steel sheet, with a suitable membrane seal; there are a front (at left) and a rear chambers in the actuator.

The front chamber is always in communication with the intake manifold, downstream of the throttle valve or with the vacuum pump.

Three different situations may be identified or phases:

1. rest position, with pedal completely released;
2. pedal depressed;
3. pedal released.

Phase 1

When the pedal is released or the pedal stroke is zero (as in Fig. 5.10), the two chambers are set at the same pressure p_s. This pressure is equal to that of the vacuum source.

Because there is no pressure difference on the two faces of the power brake membrane, there will be no assistance.

On the same figure, on its lower side, is shown a detail of the shaft of the actuator's piston. On this shaft there is a valve which puts in communication the two chambers of the power brake; on this figure the valve is draft when the pedal is depressed at zero force, or any play is set to zero.

Phase 2

Let us assume now that a pressure is applied to the pedal. After a transient, the valve will cut the communication between the two chambers. On the detail on the left of the same figure, a plunger 1 is closing the communication between the two chambers by a rubber surface.

After a short while, the rubber element 4, compressed by the braking force, will assume a given deformation, opening a passage 2 between the rear chamber and the ambient pressure.

The pressure difference between the two chamber will determine the assistance force.

The opening of the passage is function of the deformation of the rubber element 4, sensible to the load applied by the pedal. The passage will close as soon as the driver reaches the desired load on the pedal; the pressure in the rear chamber will be therefore proportional to the rubber element deformation and to applied load. The rubber element 4 is measuring the desired load.

Phase 3

When, at the end of braking, the pedal will be released, the communication with the outside will be closed and that between the two chambers will be reopened. Both chambers will be set at the same pressure, with no force applied to the braking pump.

Table 5.2 Description of the states of a vacuum power brake, as function of the pedal force F and of its derivative over time $\frac{dF}{dt}$

F	$\frac{dF}{dt}$	Outside duct	Inside duct	Rear pressure	Assistance
=0	=0	Open	Closed	p_s	No
>0	>0	Closed	Open	$p_s < p \leqslant p_0$	Yes
>0	=0	Closed	Closed	$p_s < p \leqslant p_0$	Yes
>0	>0	Open	Closed	$p_s < p < p_0$	No

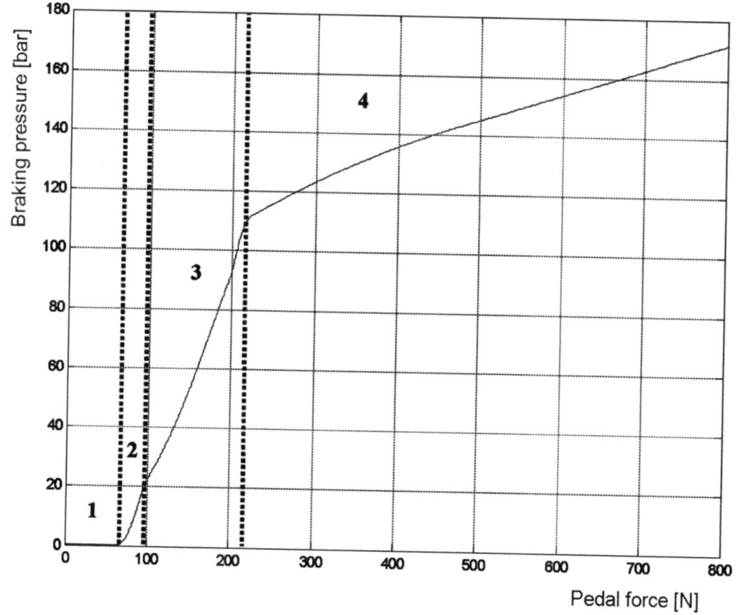

Fig. 5.13 Example of the characteristic curve of a power brake; the braking pressure is shown as function of the load applied on the pedal

In summary, the possible states of the power brake are shown on Table 5.2, as function of the load applied to the pedal and of its derivative over time.

Figure 5.13 shows an example of the characteristic of a power brake. This curve shows the braking pressure as function of the load on the pedal. The pressure can be calculated by the following formula:

$$p = \frac{F_s + F_p \tau}{A_p}, \tag{5.5}$$

where:

- F_s is the assistance supplied by the power brake;
- F_p is the force on the pedal;

- τ is the leverage ratio between pedal and plunger;
- p is the braking pressure;
- A_p is the useful area of the master pump.

This characteristic curve can be measured on a bench test, where the vacuum value is kept constant, while the braking force is set to different values.

On this diagram, four zones are outlined.

- Zone 1: the braking force is not sufficient to win the resistance of the springs that have the function of keeping the brake in rest position. The braking pressure is zero. The load value, where the power brake is starting to supply its contribution is called *tip-in load*.
- Zone 2: after the tip-in load is reached there is a sudden increase of the assistance force; this phase is called *jump-in* and *jump-in pressure* is the pressure value reached at the end of this phase.
- Zone 3: in this phase there is a constant amplification of the force applied by the pedal. The ratio G between the pressure and the applied load is called *power brake gain*.
- Zone 4: the power brake has reached the maximum value of the pressure difference, between the ambient pressure p_0 and the vacuum source p_s. The pressure increment in this area is only caused by an increase of the force applied by the driver to the pedal. This value is called *saturation pressure*.

Hydraulic Power Brake

By hydraulic power brakes the energy source is supplied by a pressurized fluid. In general, the pressure source is the same as for power steering system and the two circuits share the same fluid. The braking servo is completely similar to what we have seen, but the actuator.

The higher working pressure allows a reduction in system dimensions and makes this system available to heavy cars and medium size industrial vehicle, where the vacuum pressure is not sufficient.

The assistance system is made by a simple hydraulic cylinder, set in series with the master cylinder and is fed by the powersteering pump, through suitable valves.

A solenoid valve puts in communication the powersteering pump with the braking circuit. When brakes are in rest condition the pressure is available for the powersteering system. During braking, priority is given to this system. A second valve provides to modulate assistance pressure, according to the pedal force.

A pressure accumulator contains a quantity of pressure oil suitable for 2 or 3 braking, in case of failure of the pressure source or engine stall.

For heavier vehicles, where a larger storage of energy is necessary, an additional electrical pump is applied; this is used when the normal flow of oil from the steering pump is interrupted.

The power system fluid and the braking fluid are different and shouldn't be mixed; particular seals avoid contamination.

5.3 Industrial Vehicles Brakes

European homologation regulations impose to commercial or industrial vehicles a minimum stopping distance of:

$$s \leq 0.15V + \frac{V^2}{103.5}, \tag{5.6}$$

where s is the stopping distance in (m) and V the initial vehicle speed in (km/h). This formula must be applied for the service brake and, as we can see, the prescription is less severe than for a car.

With the emergency brake, the stopping distance can rise up to:

$$s \leq 0.15V + \frac{2V^2}{c}, \tag{5.7}$$

where c is 115 for industrial and commercial vehicles for goods transportation and is 130, for buses.

Therefore, with the emergency circuit the accepted mean deceleration is:

$$a \geq 2.2\text{–}2.5\,\text{m/s}^2,$$

while for the service circuit is:

$$a \geq 4.0\,\text{m/s}^2\,.$$

This formulae apply for every load condition within those allowed by the homologation form.

For industrial vehicles, the so called additional retarding device (*retarder*) is also considered by regulations; this system is applied to maintain a reduced speed by long descents.

The related test procedure requires that a full loaded vehicle is maintained for 6 km at a constant speed of 30 km/h on a downhill slope of 6%; after this test, the mean braking deceleration must be:

$$a \geq 3.3\text{–}3.75\,\text{m/s}^2;$$

the first value applies to vehicle for goods transportation, the second for buses.

Retarders are not part of the hydraulic or pneumatic braking systems, but are components integrated in the engine or in the transmission.

Those integrated in the engine are essentially devices that increase pumping losses during the intake and exhaust strokes, by suitable choking valves or by changing valves timing.

Those integrated in the transmission are electric or hydraulic machines that waste mechanical work, converted into heat, exchanged by dedicated radiators.

We will not comment on these devices that are usually discussed together with engines and transmissions; the second part of this volume contains the description of some hydraulic retarder.

Medium size, heavy trucks and buses have pneumatic brakes; this choice is justified by the weight of these vehicles that can't, at any rate, be braked by muscles force only.

By this kind of plants, the energy vector applied to actuation and assistance is compressed air, at pressures over 5 bar. A hydraulic actuated system with vacuum assistance would request, in fact, a too large dimension for the actuators. Nevertheless, the braking system features a heavy weight because of the modest value of the actuation pressure, as compared with the hydraulic one; as an advantage, this fluid is unlimitedly available in the atmosphere and allows a simple design if the working fluid is wasted at each braking.

In addition, compressed air is used also for other services on the vehicle, as automatic door opening, cabin opening for engine inspection, gearbox and clutch actuation, horns, etc.

This system is, for its nature, less quick as a hydraulic one, but with a suitable design is possible to cope with this drawback.

The driver's force on the pedal is used to modulate the air pressure, supplied to the brakes actuators. The force transmission from pressure chambers to brake shoes is made by levers or cams suitable to increase forces and to contain mechanical devices dimensions.

There are also hydro-pneumatic systems, where forces generated and controlled by compressed air are transmitted to braking pads or shoes by hydraulic pressure.

A pneumatic system is sketched on Fig. 5.14 in its simplest configuration, including a compressor 5, a control valve assembly 7, a reservoir 10, a distributor 13, a pressure gauge 16 and a number of braking actuators (4 and 8), one for each wheel.

Vehicles pulling trailers feature also a connection valve to feed the trailer braking system. In this case, regulations impose that the trailer is braked automatically with an assigned performance, in case of failure of the hook or of the air connection pipe.

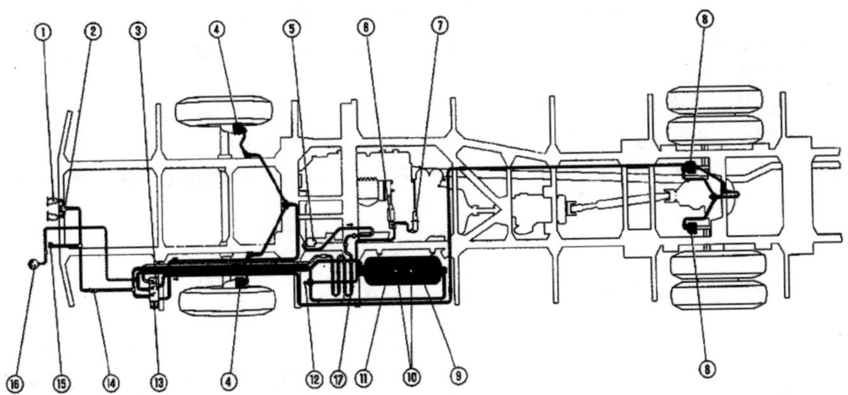

Fig. 5.14 Scheme of a hydraulic braking system applied to a bus

When the engine is running, the compressor feeds the control valve assembly and the reservoir. When a preset value for the reservoir pressure is reached, the air in excess is downloaded to the ambient and the compressor is disengaged.

The distributor is controlled by the brake pedal and is connected to the reservoir and the brake actuators; these operate on braking pads or shoes, through a mechanic or hydraulic transmission.

When the driver depresses the brake pedal, the distributor conveys the compressed air to the braking actuator in a measure and at a pressure that are dependent on the pedal stroke; when the pedal is released, the compressed air is downloaded and the brake actuator stop their function.

In the braking circuits there are two separate sections:

- automatic section, where the pressure is always the same as in the air reservoir;
- controlled section, where the pressure is present only by braking.

Suitable redundancies are provided to allow emergency braking also by failures.

On the same scheme, we can also see the reservoir, divided in two sections, for front and rear brakes 9 and 11 and the serpentine tube 17 for air cooling after the compressor outlet.

Other services relying on compressed air are present: horns 1, retarder actuator 6, an air tap to inflate tires and the pipe to feed windshield wipers 15.

The main components of this system are described in the following paragraphs.

5.3.1 Compressor

Piston volumetric compressors are used by this kind of systems; they feature automatic disc or blades inlet and exhaust valves.

Their main components are represented on Fig. 5.15, as the cylinder 6, the piston 5, the connecting rod 34 and the crankshaft 25; these compressors, quite similar to a small two stroke engine, apply roller bearing on all crankshaft pins.

The compressor is driven by the engine through a transmission belt. The opening of the intake valve 28 occurs when the cylinder pressure is less as the ambient pressure, while the exhaust valve 26 opens when the pressure of the cylinder is higher as the reservoir pressure.

As by thermal engines, compressor cooling can be made by water or air or can have an air cooled cylinder and a water cooled head, as in the drawing of the above figure; lubrication can be separated (splash type) or can be integrated into the engine system.

5.3.2 Control Valve Assembly

A control valve assembly is draft on Fig. 5.16; it includes a cleaning filter 7, a check valve 8, a pressure regulator valve 5 and a safety valve 19; it includes also a pressure gauge and a tap to inflate tires.

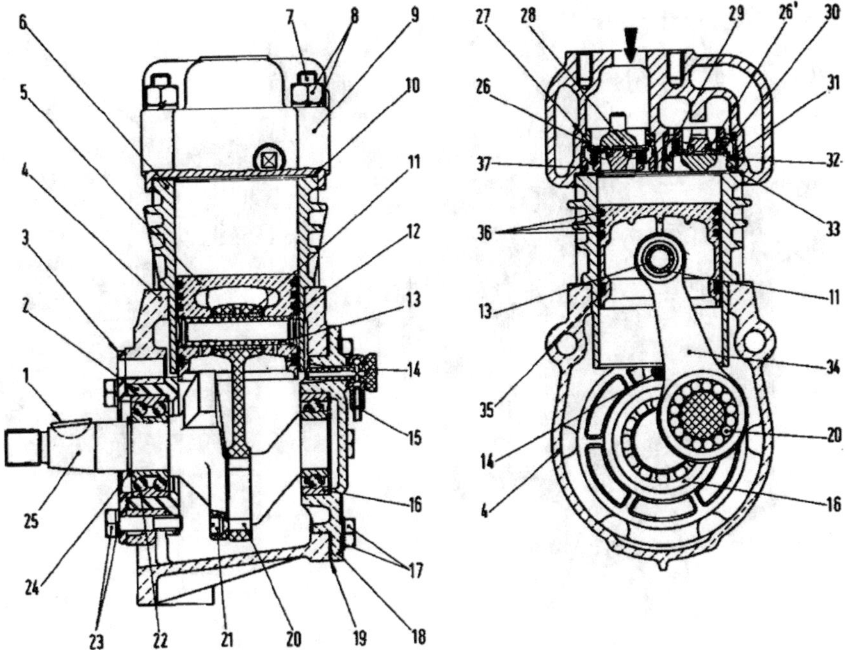

Fig. 5.15 Main cross sections of a piston compressor for a braking system, featuring automatic valves

The filter function is to retain contaminants present in the atmosphere, while the check valve avoids reservoir download to the compressor.

The pressure regulator valve, though the exhaust valve 13, has the job to maintain the reservoir pressure below a preset maximum value.

To avoid that pressure reaches dangerous values, a safety valve is provided: it opens as the reservoir pressure is higher of 1–2 bar as the regulator valve design pressure.

When the air pressure is, instead, lower as a safety threshold, a pressure gouge switches on a warning lamp on the dashboard.

It is also possible to use the compressed air to inflate tires; a connection pipe line, with suitable check valves can maintain tires to their design pressure automatically or the operation can be done manually, when the vehicle is stopped and the engine idling.

The way of working of these valves is the following; when the engine is running the compressor feeds reservoirs through the connection 1, the filter 7, the check valve 8, the chamber C and the duct 3; through the duct 9, the compressed air works on piston 5 of the pressure regulator.

If the air pressure is lower as a minimum safety value (usually set at 3.8 bar), the piston 5, pushed up by the spring 4 closes the switch 6, lightning the warning light on.

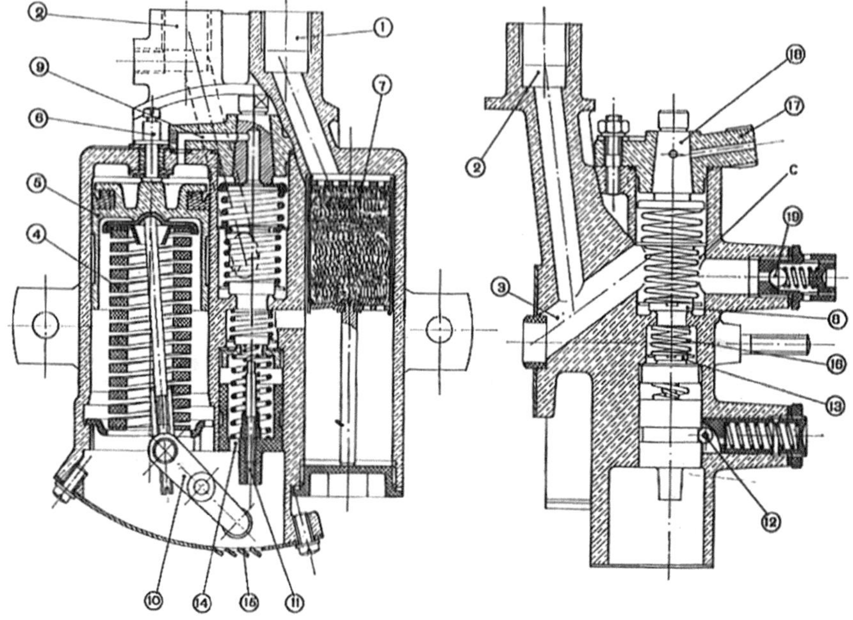

Fig. 5.16 Cross section of a control assembly valve; it maintains the system pressure below a preset maximum value and warns when a minimum safety value is reached

When the pressure increases, the force on the piston increases consequently: at a certain time the piston will be pushed down and the switch will be opened. When the pressure reaches its maximum value (usually set at 5.8–6 bar), the piston 5, through the yoke 10 and the push rod 11 will open the exhaust valve 13; from this time on the air coming from the compressor is downloaded to the atmosphere through the holes 14 and 15, while the connection to the reservoir is closed by the check valve 8.

When, as a consequence of braking, the reservoir pressure decreases, the piston 5 rises and the exhaust valve closes, connecting again the reservoir to the compressor.

The safety valve 19 avoids that the pressure goes, in any case, higher as 7–7.5 bar.

5.3.3 *Distributor*

Simple Distributor

Figure 5.17 shows the cross section of a simple distributor valve.

The distributor valve is made by a control piston 6, a regulation spring 7, a floating piston 8, an exhaust valve 3 and an intake valve 1.

The pressure on the pedal moves the control piston 6 and regulates the load on the regulation spring, which controls the motion of the floating piston, opening intake and exhaust valves.

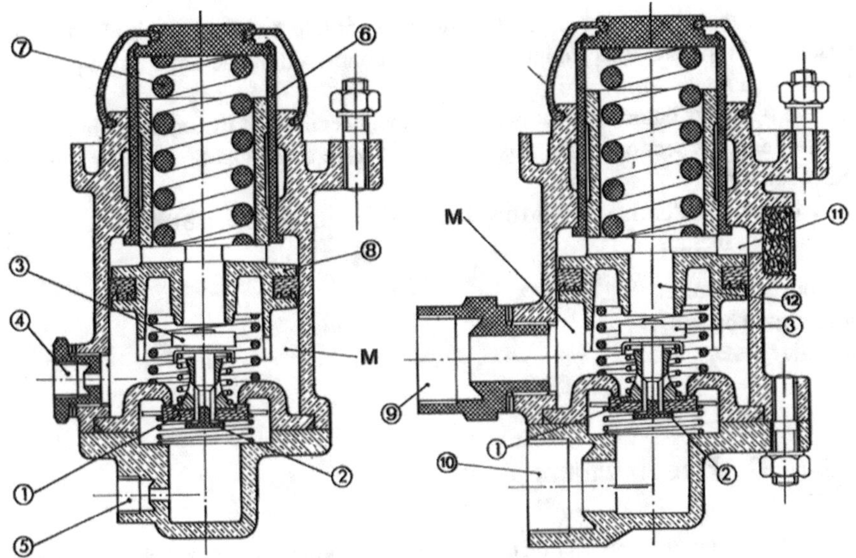

Fig. 5.17 Cross section of a simple distributor valve, to control the pressure on brake actuators

At each position of the brake pedal corresponds a pressure in the braking circuit; the driver has a feed-back on the actual braking force by the reaction on this piston of the pressure in the circuit.

Let us analyze the operation of the distributor valve; by depressing the brake pedal, the piston 6 is moved and, through the spring 7, also the piston 8 is moved.

The displacement of piston 8 closes the exhaust valve 3 and opens the pre-inlet and inlet valves 2 and 1, supplying the air to the braking actuators through the chamber M.

The opening of the intake valve is very quick; when the pressure in the chamber M (braking actuators pressure) is prevailing on the spring 7, the piston 8 is returning and the intake closes.

By reducing the force on the pedal, the force exerted by the air is higher as the force of the spring 7; as a consequence the piston goes back and opens the exhaust valve 3, downloading the air pressure in the chamber M, through the duct 12 and the hole 11.

The pressure on the braking actuator is decreasing until the spring 7 is again able to close the exhaust valve.

If pressure is bleeding in the pipes or in the brake actuators, the pressure decrease in chamber M can again close the exhaust valve and open the intake valve, until the regulation pressure is again reached.

Double and Triple Distributors

Some braking systems feature double or triple distributors, for sake of safety.

Double distributors are made by two equal simple distributors, similar to what described in the previous paragraph; they are controlled by the same brake pedal by a yoke or a rocker arm.

The two distributors control two independent sections of the braking circuit.

If one of the sections fails, it is possible to have a residual braking force by the other section.

Triple distributors have a third distributor added; in this case, the three independent circuits can include:

- front axle brakes,
- rear axle brakes,
- trailer brakes.

5.3.4 Braking Actuators

Figure 5.18 shows a cross section of a combined braking actuator suitable for service and parking functions; it is made by a cylinder and a piston 2, which is connected mechanically or hydraulically to the brake shoes or pads.

During braking, the pressure air controlled by the distributor works on the actuators of each wheel; pad or shoes pressure is proportional to the air pressure.

Return springs shift pistons in rest position when the pressure is set to zero.

Because by pneumatic plants, the energy source is not available for an unlimited period of time, the parking brake actuator is simply made by springs that keep the actuator normally in brake position.

Pneumatic actuators can have a second chamber with a compressed spring, as in Fig. 5.18; by this chamber the air pressure compresses the spring and relieves the brake.

The parking brake function is therefore not influence by the reservoir residual pressure.

Brakes not interested by parking function don't feature this second actuator.

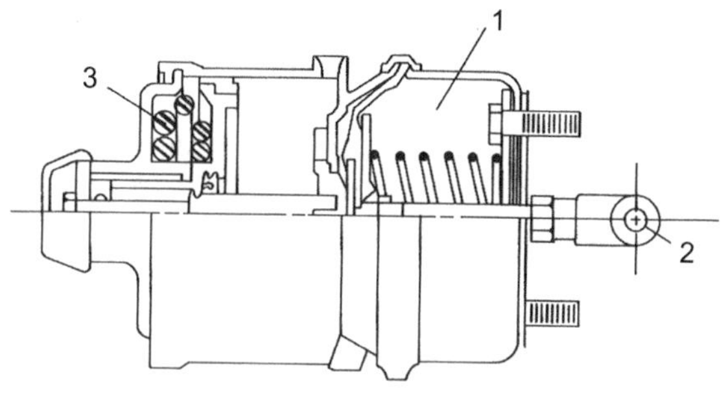

Fig. 5.18 Combined actuator for a pneumatic brake; the left actuator is for the parking function, the right one for the service function

5.4 Design and Testing

5.4.1 *Braking System Mechanics*

Deformations

It is extremely important that brake pedal stroke doesn't increase over a certain limit because of obvious space limitation and because excessive pedal displacements, during braking, are perceived by drivers as malfunction; finally, an excessive pedal displacement can obstruct the application of the maximum pressure by panic braking.

The stroke of the pedal, for example by a disc brake, is caused, during braking, by the elastic behavior of the caliper, of the disc, of pads and pipings: the oil absorption connected to those displacements contributes to increase the pedal stroke with reference to the theoretical value justified by pad clearance recovery.

By experimental tests results, it has been demonstrated that caliper deformations only are causing more than 50% of the total pedal stroke.

To model this mechanism a two degrees of freedom mechanical system can be considered, where the two independent variables are the absolute displacements of the caliper x_p and of the piston x_c. This model describing simply one half of the caliper, can be completed by a second half.

When braking pressure is set at the atmospheric value, the system is in rest position, where pads are set at a certain distance of the disc surface and:

$$\begin{cases} x_p = 0 \\ x_c = 0 \end{cases} . \tag{5.8}$$

When braking process starts and pressure rises, two different phases can be identified.

Phase 1

At the beginning, the only displacement of the piston is considered, until pads touch disc surfaces:

$$\begin{cases} x_p = 0 \\ m_c \frac{d^2 x_c}{dt^2} + r_c \frac{dx_c}{dt} + k_c x_c - pA_c = 0 \end{cases} , \tag{5.9}$$

where:

m_c is the mass of the piston,
r_c is the damping coefficient between piston and cylinder,
k_c is the compression stiffness of the piston,
p is the circuit pressure,
A_c is the piston cross area.

This equation system is applied until the following condition is true:

$$0 < x_c < c_a,$$

where c_a is the clearance between braking pads and disc surfaces.

Phase II

Over this limit, the piston is considered still, while the caliper is moving, according to the equation:

$$\begin{cases} x_c = c_a \\ m_p \frac{d^2 x_p}{dt^2} + r_c \frac{dx_p}{dt} + k_c(x_p - c) + k_p x_p - pA_c = 0 \end{cases}, \tag{5.10}$$

where k_p is the caliper stiffness.

After having calculated x_p and x_c, the product:

$$A_c(x_c - x_p)$$

is the volume of oil absorbed by the deformation of half caliper.

Therefore, the total volume of oil absorbed by calipers deformation V_p can be calculated by each half caliper contribution; we should remember that the rear brake pressure is, in general, different as the front brake one.

If we summarize in a single term the contributions of the three parts of the system (caliper c_p, pipes c_t and master pump c_{pd}), it is possible to evaluate the total pedal stroke c_{ped}:

$$c_{ped} = c_p + c_t + c_{pd} \, . \tag{5.11}$$

Calipers contribution c_p is calculated by dividing total oil absorption V_p by the area A of the master pump and by multiplying them by the transmission ratio of the pedal τ:

$$c_p = \frac{V_p \tau}{A} \, . \tag{5.12}$$

By determining the oil pressure in the circuit, it is easy to derive tubes radial and axial deformations; pump deformation may be evaluated by experimental tests or by finite elements analysis.

Figure 5.19 shows the result of one of these tests.

Dynamic Behavior

The dynamic response of a hydraulic system is characterized by a modest delay time between input and output variables, generally lower as 0.1–0.2 s.

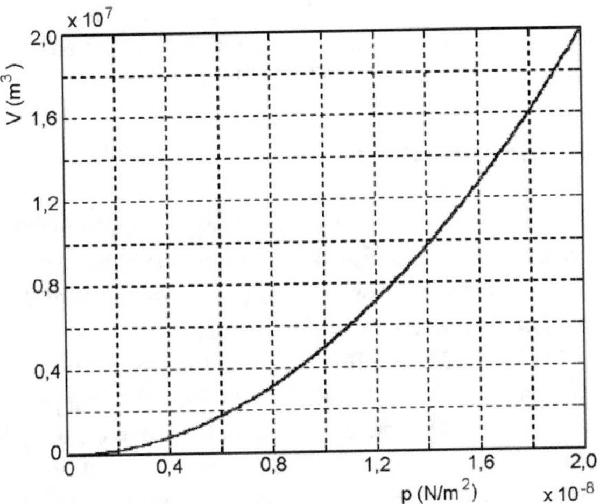

Fig. 5.19 Diagram of the absorbed volume V in a master pump of a hydraulic braking circuit, under the pressure p

The dynamic behavior can be described by a static and a transient components. This last is caused by quick changes of system variables, as circuit pressure, as consequence of a pulse on the brake pedal. The quasi static behavior is associated to slow variations of some characteristics as the friction coefficient between pads and disc, because of the wheel slowing down during braking.

Different contributions are considered in the following paragraphs.

Fluid

The flow conditions of the fluid in the hydraulic circuit, from master pump to actuator pistons, depend on viscosity, cross section and length of tubes. The parameters that determine flow speed are:

- fluid compressibility,
- tubes walls elasticity,
- flow resistance and
- fluid inertia.

Fluid viscosity contributes to increase the time lag between force application to the pedal and braking force rise. It will increase also, later on, the time needed to release the braking force.

By most vehicles, the tubes that feed brakes on the driving side are shorter as those on the other side, because of the position of the master pump. This fact leads to have the driving side brakes actuated more quickly as on the other side; this difference is difficult to be perceived for the low viscosity of the braking fluid, in normal conditions. But, if the ambient temperature is very low, the viscosity increase can cause a yaw torque to be applied to the vehicle. The level of asymmetry of braking is influenced by the speed of application of the force to the pedal.

A viscosity increase, in addition, can increase the time needed to displace the fluid, which can induce an increase of the braking force application time. These phenomena affect also ABS systems.

Vacuum Power Brake

The power brake brings a relevant contribution to the response delay of braking systems. The force amplification is supplied by the motion of many different components as pistons, valves, springs and push rods, each of them participates in determining the dynamic behavior of the system.

On Fig. 5.20, at left the transient response (quick braking force application) and the quasi steady response (slow braking force application) are shown, for a vacuum power brake; the input variable is the force on the pedal, the output variable is the pressure in the circuit. By a quick force application the pressure in the circuit is lower as that obtained by slow application and this difference is decreasing only after a certain time lag.

In general, this behavior is perceived by the driver as a pedal force increase as it usually happens when the engine is switched off.

Master Cylinder

The effect of master cylinder is generally negligible, if compared with the rest of the system, because masses are small and cylinder walls are very stiff.

Piping

It is possible to analyze the behavior of tubes of the braking system with equations describing the longitudinal vibrations of the fluid in the ducts.

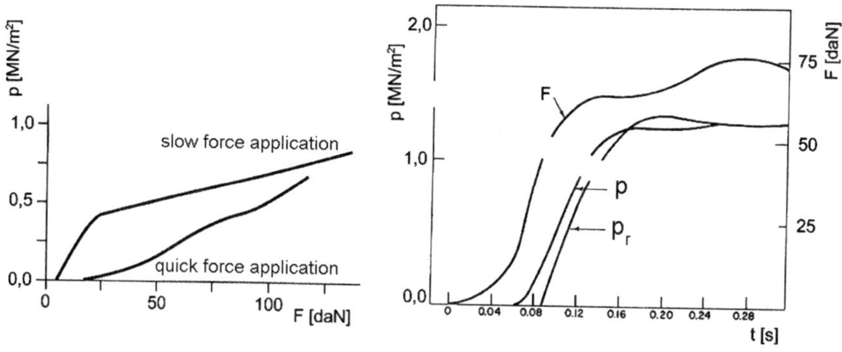

Fig. 5.20 Left side: quasi steady and transient response of a vacuum power brake. Right side: diagram of pedal force F, master cylinder pressure p and rear brake piston pressure p_r versus time

For small diameter tubes, the effect of viscosity is predominant.

A good correlation between calculated and experimental data can be obtained with a simple model, where a rod, representing the fluid, set in series with a spring, representing the system elasticity, receives, as input, a force pulse at one end.

Brakes

The dynamic behavior of a brake (either disc or drum brake) should be analyzed by different models, considering the following behaviors:

- thermal behavior, where the brake is considered as a device to convert kinetic energy into heat, to be wasted into the ambient;
- mechanic behavior, considering the relationship between the braking force, as function of the pressure and the friction between linings and brake, as a mechanical system including masses, elasticities and damping;
- friction coefficient variations.

Hydraulic Power Brake

Braking systems with hydraulic assistance, without pressure accumulator, allow, generally, shorter response time as vacuum assistance, because oil is less compressible as air.

The absence of the accumulator will cause, in any case, a time lag to build up the needed pressure, starting from a minimum.

By hydraulic power brakes with accumulator, the pressure is, instead, immediately accessible.

On Fig. 5.20, at right, the dynamic behavior of a hydraulic power brake is represented on a diagram; this diagram reports, as function of time, the force F applied to the pedal, the pressure p at the master cylinder and the pressure p_r at an actuation piston of a rear wheel.

The diagrams show that pressures follow quite well input force, with a small delay due to mechanical properties.

5.4.2 *Mechanical Design*

The mechanical design of brakes includes braking system sizing, mechanical resistance and deformation evaluation. As we have said, deformations and their consequent increase in pedal stroke should be as low as possible.

Thermal stresses are added to mechanical stresses: in fact the energy converted into heat is absorbed, as a significant part, by the disc or the drum and, than, wasted into the ambient.

In this field, finite element analysis can be fruitful applied.

On the following paragraphs we will describe shortly the design criteria for disc and drum brakes.

Disc Brakes

Modeling the system for function calculations of a disc brake is a particularly simple job.

Problem data are the following:

f friction coefficient between disc and linings,
R_1 pad outside diameter,
R_2 pad inside diameter,
Φ_c piston diameter,
α angular dimension of the pad.

The useful area of the piston is:

$$A_c = \frac{\pi}{4}\Phi_c^2 . \tag{5.13}$$

The useful area of the braking pad is, instead:

$$A_f = \pi(R_1^2 - R_2^2)\frac{\alpha}{360} , \tag{5.14}$$

where α is measured in degrees.

The braking pad exerts on the disc a pressure p_f that equals the total force F_c, acting on the pad, divided by the pad surface; if p is the pressure in the circuit, we obtain:

$$\begin{gathered} F_c = pA_c , \\ p_f = p\frac{\pi}{4}\frac{\Phi_c^2}{A_f} . \end{gathered} \tag{5.15}$$

By this equation p_f represents the average pressure between disc and pad and is useful for the evaluation of the stress on the pad.

If we assume that this pressure is constant in any point, the resultant force F_c will be applied at a radius:

$$R_m = \frac{2}{3}\frac{R_1^3 - R_2^3}{R_1^2 - R_2^2} . \tag{5.16}$$

The *brake efficiency*, usually indicated by ε, is the ratio between the braking moment and the force acting on pads; this parameter allows an immediate performance comparison between different brakes.

If we remember that the resulting braking force is the product of the normal force by the friction coefficient, if M_f is the braking moment, we have:

$$M_f = 2F_c f R_m \,,$$
$$\varepsilon = 2 f R_m \frac{\Phi_c^2}{A_f} \,. \qquad (5.17)$$

Drum Brakes

On a disc brake the pressure exerted by pads on the disc surface is constant as far as direction and intensity are concerned and formulae are consequently very simple.

By drum brakes, instead, the pressure distribution isn't constant along the contact arch neither in direction nor in intensity. Consequent modelling is more complicate.

Let us consider the drum brake on Fig. 5.9 where the wheel is rotating counter-clockwise; we refer to the scheme on Fig. 5.21.

The two shoes don't have the same behavior: by braking, tangent forces on the shoe are such as to push away the right shoe (which is called *trailing shoe*) and to attract the left one (which is called *leading shoe*), that, when approaching the brake, rotates in the same direction as the wheel.

In particular, the leading shoe receives a braking contribution by the friction force itself. In fact, if we consider the contribution of a lining element dS on the leading shoe, for the equilibrium we will have that:

$$F_c a = \int [p_f h dS - p_f f b dS]. \qquad (5.18)$$

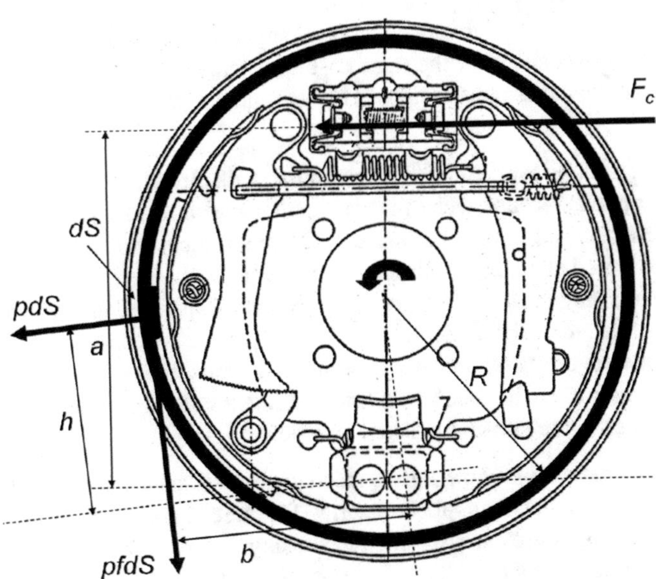

Fig. 5.21 Scheme for evaluating the self braking effect due to the leading and trailing shoe geometry

In this equation the force F_c is less as it would be for the only equilibrium of normal pressures acting on the shoe; the braking force itself is assisting the pedal force.

For the trailing shoe, the equation will have the positive sign and the conclusion will be opposite.

A good method to increase drum brakes efficiency is using two actuator cylinders, each of them working on a single leading shoe.

In the leading shoe the assistance is the higher as the higher is the friction coefficient; over certain limits the brake can be self locking when the second term in the above equations is equal or less than zero.

To avoid this inconvenience the actual friction coefficient of linings should be well below of the self locking value.

To these calculations, necessary to size brake shoes, other are added to verify thermal and mechanical stress.

By a drum brake in particular it is usual to verify that the shoe bending is not too high; a good practice is to admit an additional stroke of the pedal, for shoe bending, not higher as 20% of the total.

Materials

An important issue by brakes design is the friction coefficient between lining and metal. Recent studies have demonstrated that even small variation of the alloy content of some metals and titanium, in particular, can affect dramatically the friction coefficient, up to 20% with the same lining.

This fact has caused a common practice of applying on the same axle of a vehicle, discs and drums produced by the same casting batch.

Braking lining are usually composed by the following classes of component materials.

- Abrasive and solid lubricants; they impart the main physical properties; more than one are applied, because each of them is active in a narrow range of temperatures. This substances are diluted with fillers with mechanical and chemical resistance, but with limited abrasion or lubrication properties.
- Elastomers; they are added to modify physical properties, in order to increase elasticity and reduce brittleness.
- Metal powders or fibers; they improve thermal conductivity.
- Fibers; in addition to binders they allow to obtain a suitable mechanical resistance.
- Binders; they aggregate all the listed materials.

Discs and drums must show a good mechanical resistance and a notable capacity of wasting heat: the mostly applied material, for these characteristics, is grey iron.

Silicon content improves castability, increasing, as a drawback graphite granules size and brittleness, with a similar effect as carbon. Manganese, coming from metallurgical process, must be limited at reduced quantities; in combination with sulphur, produces manganese sulfite, that impairs machinability. The maximum allowed content should be less than 1%.

For high performance cars other materials can be used, as composites with carbon matrix; its cost discourages application to mass production. Also aluminum alloys, reinforced with silicon carbides, are considered for their reduced weight; at the state of the art, their limited thermal resistance reduces the potential advantages in weight and dimensions.

Because of heat generation, temperature rise on brakes is significant, especially when stopping a vehicle from a high speed. As a consequence of the temperature increase, the friction coefficient will decrease, affecting brake efficiency, that will reduce from the beginning to the end of the braking.

Figure 5.22 shows a diagram about this phenomenon. It is particularly relevant by descents, where temperature is continuously increasing because continuous braking doesn't allow the time necessary to cool down.

Efficiency can be reduced significantly and, as a consequence, pedal load can increase, with reference to the beginning of braking. This natural behavior should be partially compensated by a suitable selection of lining materials.

By heavy vehicles or high speed vehicles, self ventilated discs are applied, as that shown on Fig. 5.5.

Also drums cool down can be improved by the application of outside ventilation fins, along the circumference or parallel to rotation axis.

5.4.3 *Thermal Design*

Energy and Power

During braking kinetic and potential energy of the vehicle are converted into thermal energy, through linings and brake metal surface.

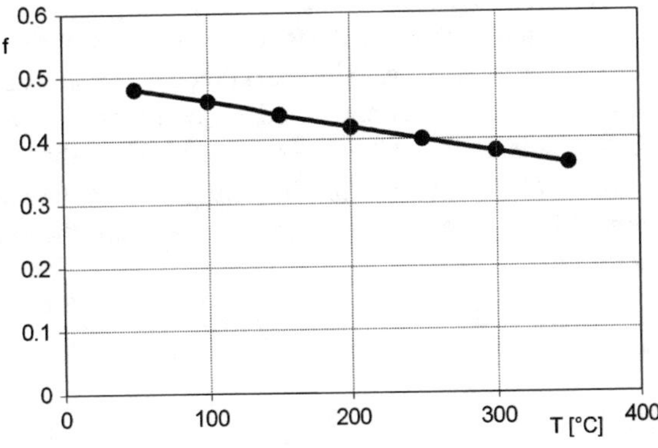

Fig. 5.22 Diagram of the friction coefficient between pad and disc, as function of the disc temperature

If we neglect, for simplicity, the braking force exerted by driving resistance and powertrain motoring, the thermal energy E to be wasted, during a braking from the initial speed V_1 to the final speed V_2, includes the following contributions:

$$E = \frac{1}{2}m\,(V_1^2 - V_2^2) + \frac{1}{2}J\,(\Omega_1^2 - \Omega_2^2) + mg\Delta h\,, \tag{5.19}$$

where:

m is the total vehicle mass,
J is the moment of inertia of rotating masses referred to wheels,
Ω_1 is the initial rotation speed of wheels,
Ω_2 is the final rotation speed of wheels,
Δh is the altitude difference between the start and the end of the maneuver.

As a first approximation, in case of normal conditions braking, tire longitudinal slip can be neglected and the rolling radius is equal to the loaded radius; therefore:

$$\Omega_1 = \frac{V_1}{R_{sc}},\ \Omega_2 = \frac{V_2}{R_{sc}}. \tag{5.20}$$

The pressure distribution between front and rear axles changes in function of the braking pressure and is set at a value suitable to maintain vehicle stability during braking, avoiding rear wheels skid.

If we call M_{f1} and M_{f2} the braking torques on the front and rear wheels, the equation:

$$E = 2L_{f1} + 2L_{f2} = 2\left(\int M_{f1}\Omega dt + \int M_{f2}\Omega dt\right) \tag{5.21}$$

defines the works L_{f1} and L_{f2} wasted by brakes; this equation can be equaled to the previous one.

Figure 5.23 shows, as an example, the diagram of wasted power versus time in a front brake of a car of 1500 kg of mass, during a stop braking from 100 km/h on a high friction coefficient ground.

The brake thermal analysis includes not only the determination of total energy wasted by brake, but also its breakdown between disc and pads.

This breakdown is correlated to the thermal resistance of discs (R_d) and pads (R_p), which are the elements on whose surface the heat is generated.

The heat transferred to the disc and to the liner flows through a network of different thermal resistances. In quasi steady conditions, the phenomenon can be represented by the ratio:

$$\gamma = \frac{Q_d}{Q_p} = \frac{\sum R_p}{\sum R_d}. \tag{5.22}$$

where:

Q_d is the flux absorbed by the disc and

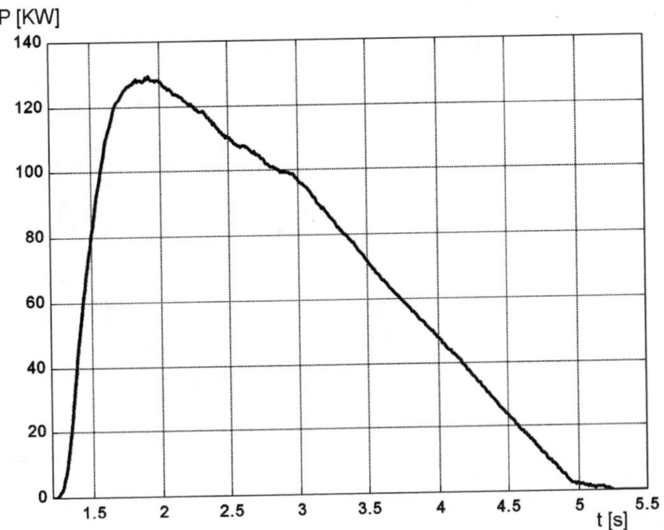

Fig. 5.23 Diagram of the thermal power P, as function of time t, wasted by a front brake of a car with a mass of 1500 kg, during a stop from 100 km/h

Q_p is the flux absorbed by the pad.

Previous equation implies that disc temperature and pad temperature are equal at their contact interface and that all generated heat during braking is absorbed by pad and disc only.

The ratio γ can also be written as:

$$\gamma = \frac{Q_d}{Q_p} = \sqrt{\frac{\rho_p c_p k_p}{\rho_d c_d k_d}} \tag{5.23}$$

where:

ρ is the density,
c is the specific heat,
k is the thermal conductivity.

Subscripts p and d refer, as usual, to the disc and the pad.

While we can assume that density and conductivity of disc and pad remain constant, this assumption isn't generally true for specific heats. As far as the pad is concerned, c_p is assumed to be constant, while c_d varies significantly in the temperatures field of interest.

The value of specific heat of disc material is shown on Fig. 5.24.

If the total generated heat, according to our assumption, is:

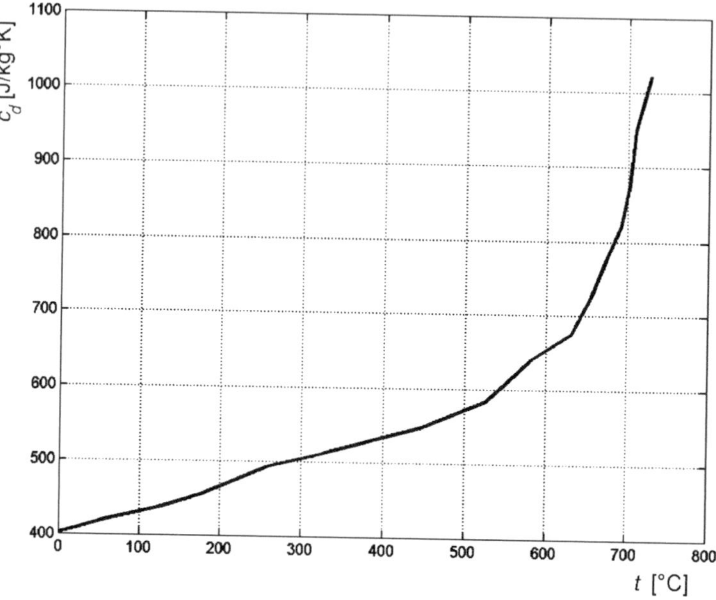

Fig. 5.24 Diagram of specific heat of disc material, as function of temperature

$$Q_d + Q_p \, ,$$

it derives that the part of heat absorbed by the disc is:

$$\frac{\gamma}{\gamma + 1} \, ,$$

while the part absorbed by the pad is:

$$\frac{1}{\gamma + 1} \, .$$

By introducing material characteristics of disc and pad, it derives that about 80% of the friction heat during braking is absorbed by discs.

This kind of breakdown is convenient, because a too high temperature in the pad would deteriorate the material and could increase fluid temperature as much as to induce local evaporation, with consequent braking torque decay.

The heat produced during braking should be wasted by the air in relative motion with reference to the vehicle.

Considering the heat breakdown it appears reasonable to modify the disc geometry only, in order to improve cooling. Suitable openings on fenders can be useful in most critical cases; in the mean time the wheel disc shape can improve air circulation.

While modeling this phenomenon to calculate temperatures, the following heat exchanges should be taken into account:

- convection, between disc and ambient air;
- convection between disc hub and ambient air;
- convection between caliper and ambient air;
- convection between pad and ambient air;
- conduction between disc hub and wheel hub;
- conduction between disc hub and disc;
- conduction between pad and caliper;
- conduction between caliper and disc hub;
- radiation from the disc.

In addition, part of the heat is absorbed by these parts as temperature increase.

Temperature Analysis

Simplified Approach

During short braking at high speed, with high deceleration, braking time is shorter as time needed to heat up rotating parts.

In these conditions convection doesn't contribute to cool down and all heat energy is absorbed by brake mechanisms and working fluid.

For drum brakes, the crossing time, necessary to heat to reach the outside surface is given by:

$$t = \frac{L^2}{5a}, \tag{5.24}$$

where L is the drum thickness and a is the thermal diffusion coefficient, given by the formula:

$$a = \frac{k}{\rho c}, \tag{5.25}$$

where c is the specific heat of drum material, k is the thermal conductivity coefficient, ρ is the density.

The same expression for the crossing time can be used to determine the time necessary for the flux to reach the middle of the disc of a brake of this kind. In this case L would be the half of the disc thickness.

For small drum brakes of for ventilated discs with small thickness, the crossing time will be lower, but in any case the heat wasted by convection will be lower as that stored in the rotor, during a short braking.

If we assume that braking power is linearly decreasing with time (constant braking deceleration), the surface temperature can be written as:

$$T(L,t) - T_i = \sqrt{\frac{5}{4}\frac{q_0}{kat}}\left(1 - \frac{2t}{3t_s}\right), \quad (5.26)$$

where T_i is the brake initial temperature, q_0 is the thermal flux to the drum or the disc, immediately after the start of braking and t_s is the vehicle stopping time.

We should notice that q_0 represents also the braking power for unit of surface, absorbed by the drum or the disc.

If we solve previous equation with reference to the time, the maximum surface temperature will be reached at:

$$t = t_s/2. \quad (5.27)$$

Therefore, the maximum temperature in this kind of short braking will be:

$$T_{\max,L} = \sqrt{\frac{5}{18}\frac{q_0 t_s}{\rho c k}}. \quad (5.28)$$

From the previous formula we can notice that, for a given heat flux, maximum temperature decreases as density, specific heat and thermal conductivity increases.

Table 5.3 reports the average properties of materials used on pads, shoes and rotors of disc and drum brakes.

Second Approximation Approach

A better approximation can be introduced by assuming the following hypotheses:

- the air temperature is constant and equal to T_∞;
- the wheel hub is considered as a sink at temperature T_m;
- convection coefficients are function of vehicle speed only;
- conduction coefficients are assumed to be constant;
- pad and disc specific heats are function of the temperature only;
- the surface between pad and disc is the only heat source;
- radiation invests ambient air only;
- the radiated heat in included into the convection term.

Table 5.3 Typical values of the thermal properties of material used for drum and disc brakes

	Pads	Shoes	Rotors
ρ (kg/m^3)	2,030	2,600	7,230
c (Nm/kg°K)	1,260	1,470	419
k (Nm/mh°K])	4,170	4,360	174
a (m^2/h)	0.00163	0.0011	0.0576

According to what we said, we should concentrate our attention on the disc, because the most relevant part of the generated heat is flowing through it.

To calculate convection the disc can be modeled with concentrated parameters. The disc is a single mass where temperature is everywhere constant; disc temperature is function of time only and not of space.

The convection coefficient is coming from tests on cooling discs with air. On these tests discs are mounted on complete cars to take into account the covering effect of car body.

Although the disc is exchanging heat by conduction with near parts, as wheel hub, most of the heat is wasted by convection.

The thermal balance is done by the following equation:

$$\rho c V_{eff} \frac{dT}{dt} = -S(H_a + H_i)(T - T_\infty) , \tag{5.29}$$

where:

S	is the disc surface;
V_{eff}	is the disc volume, relevant to cool dawn;
H_a	is the convection coefficient between air and disc;
H_i	is the radiation coefficient.

We recall that:

$$S = 2\pi(R_e^2 - R_i^2) + 2\pi R_e s, \tag{5.30}$$

where R_e and R_i are the outside and inside diameter of the disc and s is the disc thickness; therefore:

$$V_{eff} = \pi(R_e^2 - R_i^2)s. \tag{5.31}$$

Integrating the equation above, we can determine the diagram of the disc temperature as function of time and of the convection coefficient.

This last could be identified by minimizing the difference between the results of experimental tests and of calculations.

We can derive convection coefficient as function of vehicle speed.

We have chosen for this example a linear correlation between convection coefficient and car speed.

Results are shown on Fig. 5.25.

It can be noticed that front brakes convection coefficient is greater that rear one, because discs are ventilated and the cover effect of the body is less.

The assumptions made for rotors about convection coefficient cannot be applied to pads because most part of their absorbed heat cannot be waste to the air but is transferred to the calipers.

As a first approximation their convection coefficient has been set to the same value of discs.

To perform a mono-dimensional analysis of the disc by the finite differences method, the disc has been divided into five layers; those closer to pads are $\Delta x/2$

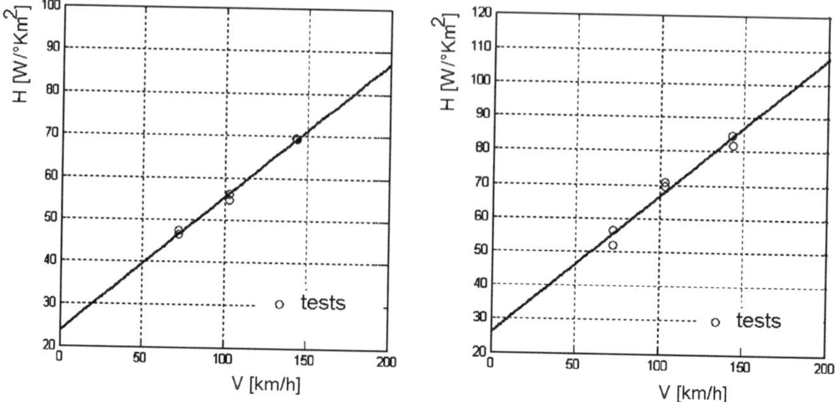

Fig. 5.25 Diagram of the total convection coefficient H for rear discs (left) and front discs (right), as function of vehicle speed

thick, while the three inner layers are Δx thick. The temperature will be uniquely function of the distance x from the contact surface and of time.

Outer layers have a half thickness to take into account the higher temperature gradient near the contact with the pad.

Applying the first law of thermodynamics to each layer, we obtain a system of differential equations, whose solution allows to identify temperatures for all layers at each time interval.

The thermal balance of the layers in contact with pads is:

$$\begin{aligned}\rho c \tfrac{\Delta x}{2} S \tfrac{\partial T_i}{\partial t} &= -H_a (S + \pi R_e \Delta x)(T_i - T_\infty) - \tfrac{4}{3} k S \tfrac{T_i - T_{i+1}}{\Delta x} + \\ &-k_c \pi R_i (\Delta x - s_g \tfrac{\Delta x}{s}) \tfrac{T_i - T_c}{\Delta x_c} + \tfrac{1}{2} \tfrac{\gamma}{\gamma + 1} Q.\end{aligned} \quad (5.32)$$

In this equation:

γ is the ratio of heat flux;
s_g is the thickness between disc and hub;
s is the disc thickness;
S is the disc surface;
H_a is the air convection coefficient;
T_c is the hub temperature;
T_i is the temperature of the layer i;
x_c is the thickness of the heat flux to the hub;
k_c is the thermal conductivity between disc and hub.

For elements of thickness Δx is, instead:

$$\begin{aligned}\rho c \Delta x S \tfrac{\partial T_i}{\partial t} &= H_a 2\pi R_e \Delta x (T_i - T_\infty) - k S \tfrac{T_i - T_{i+1}}{\Delta x} + \\ &-q_{i,i-1} - k_c 2(\Delta x - s_g \tfrac{\Delta x}{s}) \pi R_i \tfrac{T_i - T_c}{\Delta x},\end{aligned} \quad (5.33)$$

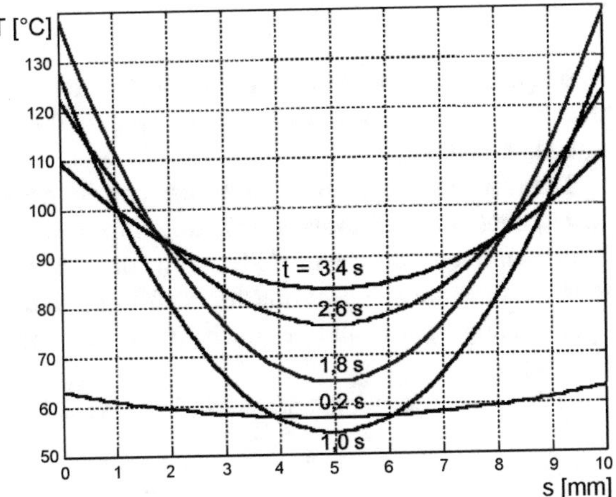

Fig. 5.26 Diagrams of temperature T inside brake disc, as function of time t and thickness s, in a stop brake from 100 km/h

where $q_{i,i-1}$ is the heat flux that the layer i receives by conduction from the near layer; it can be written in different way, if the element is closer to the outer or inner elements.

In the first case:

$$q_{i,i-1} = \frac{4}{3}kS\frac{T_i - T_{i-1}}{\Delta x}\ ; \tag{5.34}$$

in the second case:

$$q_{i,i-1} = kS\frac{T_i - T_{i-1}}{\Delta x}. \tag{5.35}$$

Because temperature distribution in a solid disc is symmetrical, the study can be limited, in this case, to half disc, considering the symmetry plane as an adiabatic surface.

The calculation method has been applied to a car in a stop brake from 100 km/h, lasting about 3 s.

The result of the calculation is shown on Fig. 5.26. The disc is 10 mm thick and $s = 5$ mm is the middle of the disc; it can be noticed that at the end of braking the temperature is almost uniform.

The temperature diagram as function of time can be read at $s = 0$ mm and $s = 10$ mm, where it is identical for the assumption that have been made.

A similar calculation can be made for the pad.

The hub and the caliper can be considered as concentrated masses; therefore their temperature will be function of the time only.

Fading

During ideal repeated braking, vehicle is slowed down from a higher to a lower speed; after it has slowed down it accelerates again to the initial speed and than braked for a number of times.

The reached temperature can be easily calculated in those conditions if braking power, cool dawn intervals and braking time are kept constant.

In addition, rotors are modelled with concentrated parameter elements, with uniform temperature and heat transfer coefficients and properties are kept constant.

If braking time would be much shorter the cool down time, this last could be neglected.

In this case rotor temperature will increase uniformly according to the equation:

$$\Delta T = \frac{Qt}{\rho c V}, \tag{5.36}$$

where Q is the braking power, t is the braking time, c is the specific heat, V is the rotor volume and ρ is the material density.

To describe the behavior after a braking, the following differential equation is used:

$$\rho c V \frac{dT}{dt} = hA(T - T_\infty), \tag{5.37}$$

where A is the rotor surface, h is the convection coefficient, T is the temperature at the time t and T_∞ is the ambient temperature.

Starting from an initial temperature T_0, the integration of the above equation leads to this solution:

$$\frac{T(t) - T_0}{T_0 - T_\infty} = e^{\frac{-hAt}{\rho c V}}. \tag{5.38}$$

Combining equations it is possible to calculate the temperatures at the second and following braking.

5.4.4 Test Methods

As other chassis components, brakes are tested also separated of the vehicle, to concentrate on particular aspects of their performance and to reduce development costs.

The test bench mostly applied is a dynamometer bench simulating also vehicle inertia, as shown on Fig. 5.27.

The bench includes an electric motor 1, having the task of launching the flywheels 2 at different speeds, simulating the initial vehicle speed at which the braking starts.

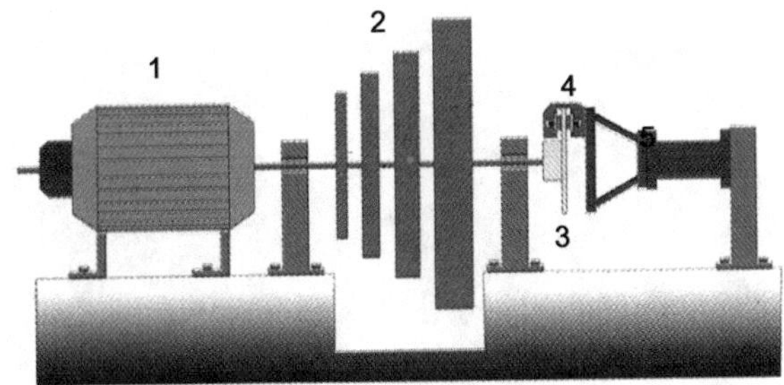

Fig. 5.27 Scheme for testing a brake separated by its vehicle. The test bench includes a dynamometer and a set of flywheels simulating vehicle inertia

Inertial masses of flywheels can be engaged to the bench, to simulate the vehicle apparent rotating mass[2] pertinent to the wheel, where the brake is applied.

After flywheels the brake rotor 3 is installed, while the fixed part of the brake (caliper or shoes) is bolted to the bench block.

A reaction torque meter can measure the braking torque.

The simulation of braking includes working phases and rests: it is requested that temperatures reach a defined value and, if braking is repeated a predefined rest time must be waited to simulate cool dawn; a ventilator may be necessary to obtain realistic results.

When the temperature is reached the rotation speed is adjusted to simulate car speed and braking are performed.

Braking performance may be controlled at constant braking torque or at constant circuit pressure.

During braking many parameters are recorded as speed, temperature, pressure, torque and the temperature of some element on which the attention is focussed.

Many test procedures are applied:

- homologation regulation procedures;
- EUROSPEC procedures, standardized between car and brake components manufacturers;
- procedure derived by the company know-how.

A particular aspect to be evaluated by bench testing are mechanical and thermal rotor deformations; this measurements can be made without interrupting tests, by using capacitive transducers, that work without contact.

After a cycle of different test simulating the brake life on the car, linings and rotors wear can be measured to check their performance on the vehicle.

[2] See the definition of apparent rotating mass on the second volume.

Chapter 6
Control Systems

This chapter is dedicated to control systems working on steering mechanism, brakes and elastic and damping elements of suspensions; sections are divided according to this classification.

Transmission control systems, working on gearbox and differentials will be examined in the second part.

This organization of subjects lays itself open to criticism, because the target that this systems should obtain is sometime similar, even if they are applied to different part of the chassis.

Most of this system, in fact, are addressed to improve the vehicle dynamic behavior through a suitable control of the forces exchanged between tires and ground; this control action can come from braking forces breakdown (through the brakes of each wheel), from vertical forces breakdown (through the suspension elastic or damping elements), from a different steering angle and from driving forces breakdown (through the axle differential or the central differential in four wheel driven vehicles).

The following sections are mainly addressed to explain how the cited chassis elements are modified to become actuators in a controlled system; these chapters will also outline the related control strategies, or the rules that the control system must follow to obtain a result of improving the vehicle dynamic behavior.

The study of the interaction of control systems with vehicle systems requires to model the entire vehicle, to foresee its dynamic behavior and modifications of vehicle functions. This study will be developed later, in the second volume.

6.1 Steering Control

A control system quicker and more precise as an average driver, working on steering angles directly can effectively stabilize and improve the vehicle dynamic behavior.

Steering angles calculated by the control system can be added or subtracted to those imposed by the driver; in the first case the function obtained will be an

G. Genta and L. Morello, *The Automotive Chassis*, Mechanical Engineering Series,
https://doi.org/10.1007/978-3-030-35635-4_6

improvement of maneuverability, in the second case, an improvement of stability or active safety. The control system can modify the characteristics of the existing steering mechanism or can work through a parallel mechanisms on rear wheels.

6.1.1 Rear Wheel Steering

RWS (*Rear Wheel Steering*) or 4WS (*Four Wheels Steering*) systems achieve an additional steering of the rear wheels, in function of different parameters; they can be front wheel steering angle, car speed, yaw velocity or lateral acceleration.

Schemes on Fig. 6.1 allow to understand the purpose of steering rear wheels, by using the kinematic steering plot we have introduced in the chapter on steering system for front wheel steering (scheme a).

In all cases we consider kinematic steering only, at speeds as low as sideslip angles can be neglected; plots outline the curvature radius R of vehicle path.

The addition of rear wheels steering angles allows a better handling at low speed, thanks to the lower radius that can be achieved applying an opposite steering angle, as that of front wheels. With reference to the scheme b it is possible to move the center of gravity of the car closer to the crossing point of perpendiculars to the wheels equator plane.

In addition, is possible to increase stability at high speeds, by applying a concordant steering angle to the rear wheels as front wheels; yaw velocity can be decreased, with vehicle sideslip angle decrease. Theoretically it is possible to deviate vehicle path with a pure translation motion with no yaw velocity, when perpendiculars to the wheels are parallel because all wheels have the same steering angle.

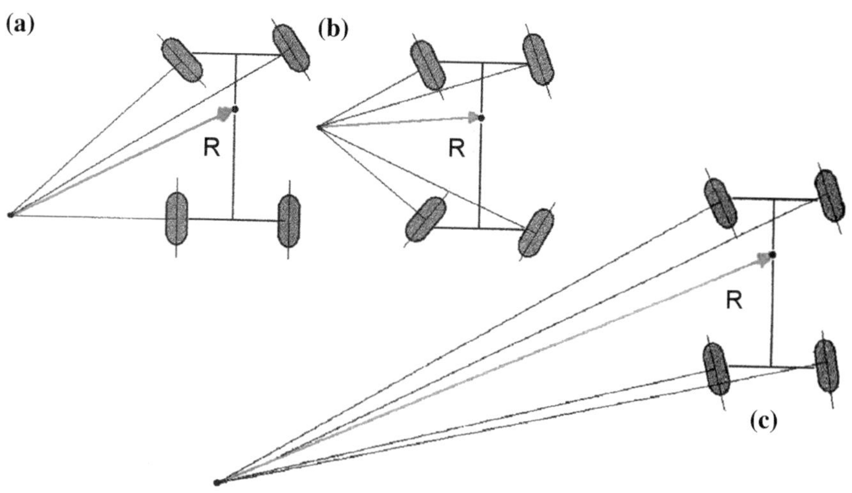

Fig. 6.1 Kinematic steering of a vehicle with the only front axle steering (scheme **a**) or with both axles steering in opposite directions (scheme **b**) or in same direction (scheme **c**)

There are different categories of RWS systems, each of them having particular effects on vehicle dynamics:

- *angle dependent*;
- *speed dependent*;
- *dynamic speed dependent*;
- *model following*.

This classification is more a trace of electronic control systems evolution on cars, than a response to really different needs.

Angle Dependent

In an angle dependent system, rear steering angle is only dependent on front wheel steering angle.

This condition can be obtained with a quite simple steering box, mechanically connected to front wheel steering box.

For small steering angles, as it happens at high vehicle speed, rear wheels steer in the same direction, reducing the vehicle sideslip and improving vehicle stability.

When the steering angle is over a certain threshold, typical of low speed maneuvers, rear wheel steering angle is opposite, to improve vehicle handling.

Typically the rear steering angle reversal occurs at about 200°–250° of steering wheel and the rear steering angle is less as 2° when in the same direction or 5° when in the opposite one. This open loop control strategy can be implemented with suitable mechanisms connected to the conventional steering mechanism.

An example of a purely mechanical four wheels steering mechanism appeared on the Honda Prelude in 1987. Figure 6.2 shows the effect of this steering mechanism on front and rear steering angles δ_1 and δ_2 as function of steering wheel angle δ_v.

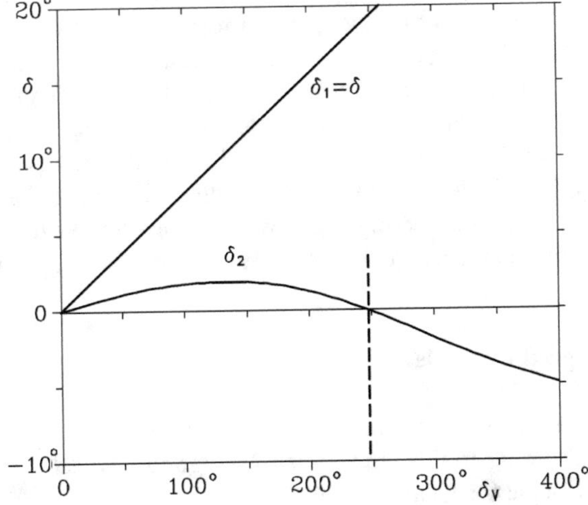

Fig. 6.2 Steer angle on the front axle δ_1 and on the rear axle δ_2 as function of the steering wheel angle δ_v

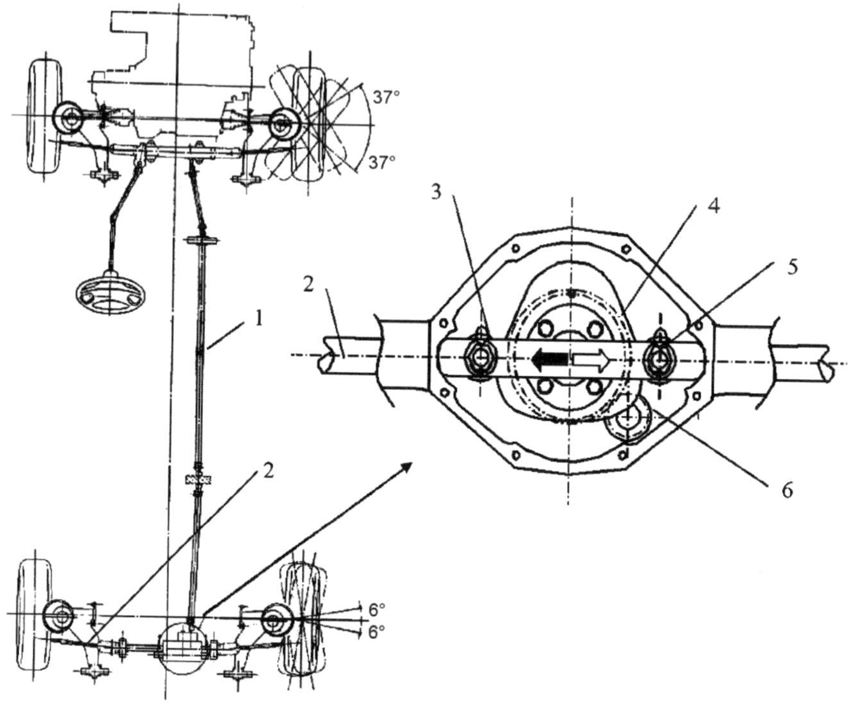

Fig. 6.3 Mechanical control of rear wheel steering angle, angle dependent type. We can identify: control shaft 1, rear steering push rods 2, roller contacts 3 and 5, steering cam 4 and steering gear 6 (CRF Patent)

Another example of angle dependent rear wheels steering mechanism has been patented in the same period by Fiat Research Center and is represented on Fig. 6.3.

Steering wheel rotation controls the rear axle steering box through a connection shaft and a gear 6. A suitable cam shifts two push rods 3 and 5 with contact roller and, as a consequence, the two rear steering rods 2, to which they are connected. These steering rods act on rear wheels.

The cam profile is such as to steer the rear axle in the same direction as the front one, for small angles; for a clockwise steering angle (seen from above) on front wheels, cam 4 rotates clockwise (in the enlargement at the upper right) shifting push rods to the left, with the lower profile, and then to the right with the higher profile.

There is a steering angle by which rear wheels don't steer; radial cam dimension is constant and equal to the distance between rollers on steering rods.

Speed Dependent

A slightly more complex control strategy is the so called speed dependent one; according to this strategy the average rear wheel steering angle δ_2 can be defined with the formula:

$$\delta_2 = K(V)\,\delta_1\,, \tag{6.1}$$

where K is function of the car speed only V and δ_1 is the average front wheel steering angle.

As obvious, a purely mechanical connection between steering axles is not viable without unacceptable complications (we can imagine, for instance, how car speed can be mechanically measured).

The most common target of a speed dependent rear steering system is to make vehicle sideslip angle as low as possible. In such case the control law can be derived by the mathematical models we will introduce in the second volume.

This approach is again an open loop one and makes reference to average vehicle driving conditions.

Many applications have been introduced applying mechanical or hydraulic actuators, that, today, are completely discontinued.

Honda Prelude, again, in 1992 introduced a system of this kind. In this case, rear wheel steering is achieved through an electric motor; sensors are installed on the car to measure steering angle and car speed. The control system maintains a quasi static vehicle sideslip angle at all speeds.

A similar system is produced by Delphi (*Quadrasteer*) and is the only one still present on the market.

Dynamic Speed Dependent

A more complete expression of a control law suitable to obtain $\beta = 0$ in all conditions must have the following form:

$$\delta_2 = K(V)G(t)\,\delta_1\,, \tag{6.2}$$

where $G(t)$ is a suitable function of the time elapsed from the starting point of a steering input.

A qualitative diagram of $K(V)$ is shown on Fig. 6.4. We see that at small speed (below 40 km/h) there is a negative gain, i.e. the rear steering angle is opposite to the front one, to improve handling; at higher speeds front wheels steer in the same direction to improve stability.

On the right part of the same figure a diagrams of steering angles versus time is also shown. The steady state angle δ_r is modulated by the dynamic function $G(t)$ as to obtain the response shown. In the first part of the diagram the opposite angle makes the beginning of turning quicker while the following values are addressed to regime stabilization.

To reach this target rear wheels are steered by a hydraulic actuator whose pressure is electronically controlled: the neutral position of steering mechanism is defined by high stiffness return springs.

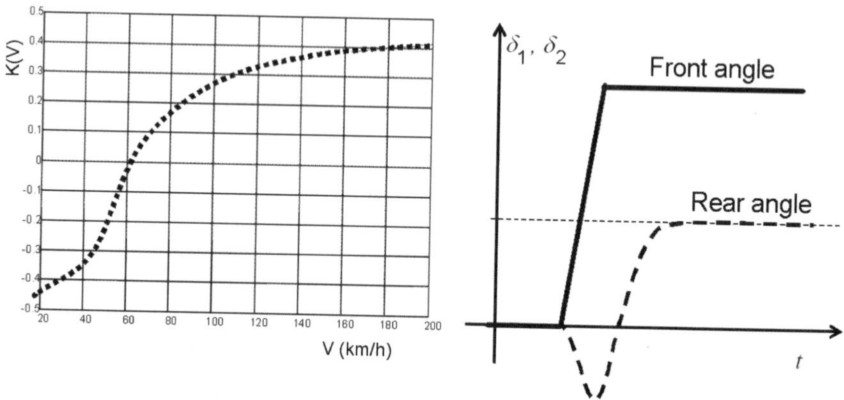

Fig. 6.4 Diagram of the front and rear steering angles versus time in a steering transient with a dynamic speed dependent system

Model Following

The control law calculating rear wheels steering angle includes two contributions.

An open loop control performs as the most sophisticated of the previously introduced strategies, while a closed loop control, is finalized to compensate for errors detected between the actual parameters and theoretical ones predicted by a real time mathematical model; these parameters can be yaw speed and vehicle sideslip angle.

In this way the control system can react to situations that cannot be foreseen by a mathematical model as:

- changes in tires characteristics (wear, inflation pressure);
- changes in vehicle mass and mass distribution;
- effect of external disturbs as wind gusts (i.e. like when exiting a tunnel) or changes of road friction coefficient.

Response is always calibrated to driving situation and handling and stability are consequently improved.

Figure 6.5 shows a cross section of the rear wheel steering actuator, applied to a prototype developed by Fiat Research Center for a model following RWS system.

This actuator includes two proportional electrovalves PV1 and PV2 to control the oil flow to the piston 5 to steer rear wheels and two on-off electrovalves (HSV1 and HSV2) to cut the oil flow to the cylinder chambers; a further electrovalve, not shown on the figure, opens the mechanical safety brake 4.

To move the actuator, all on-off valves must be open and the safety brake 4 released. This is made by a tapered bush around the steering rod 5. A set of bevel springs pushes the bush and keeps the brake naturally engaged; only the oil pressure, if available, can disengage the brake.

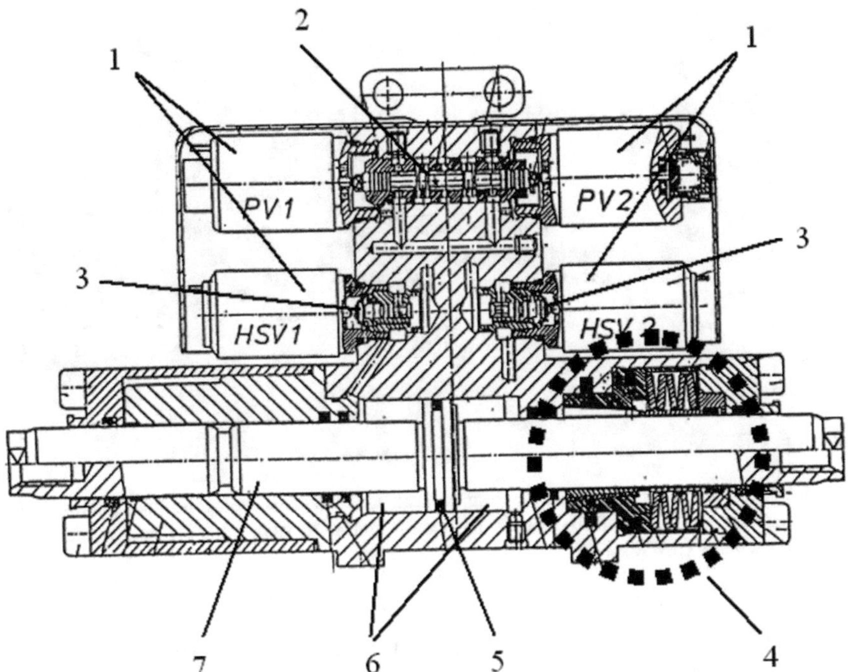

Fig. 6.5 Rear steering angle actuator for a dynamic speed dependent control. Proportional valves 2 and on-off valves 3 are shown on the drawing, together with the safety brake 4, to keep the system blocked in case of failure to the pressure circuit (FCA Patent)

It is important to outline the intrinsic safety feature of this system, required by regulations; the actuator guarantees steering angles to be fixed at the last actuation position; this feature alert the driver positively, in case of system failure.

Rear steering systems, after a period of relative interest during the ’990, at least on sport cars are now discontinued, because of the higher cost, as compared with VDC systems, paid by marginal advantages; handling advantages can induce to reconsider this system for vehicles with penalties on minimum turning radius, as front wheel driven cars with large engines.

6.1.2 Variable Ratio Steering Box

The variable ratio steering box represents a technology in evolution, particularly for high segment high performance cars.

BMW, in cooperation with ZF and Bosch has developed such system with a continuous variation capacity of the transmission ratio of the steering box.

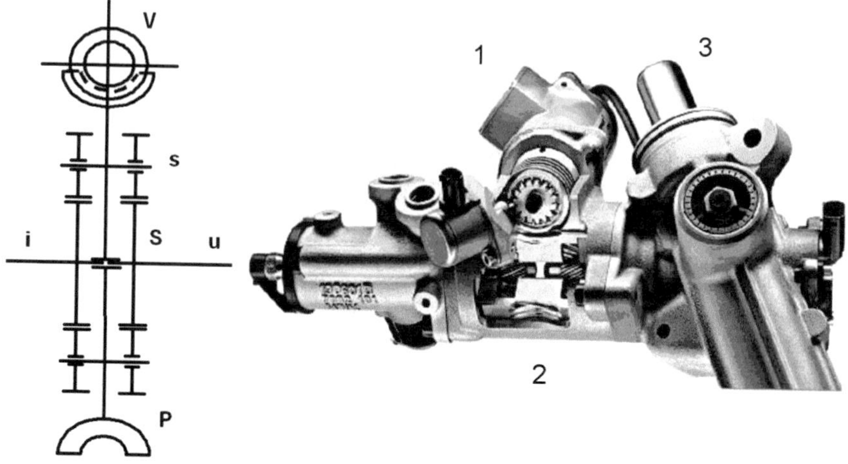

Fig. 6.6 Actuator for front wheel steering control, made with an epicyclic gear adding to the steering wheel angle the contribution of an electric motor. At right the steering box; at left the kinematic scheme (BMW)

Figure 6.6 represents, at right, the modified steering box; between steering column and pinion, a device made with an epicyclic gear 2 is applied, operated by an electric motor 1.

The scheme of the epicyclic gear is shown on the left of the same figure. The electric motor 1, controlled by an electronic circuit, acts through a worm gear V on the carrier P of an epicyclic gear 2, able to modify the transmission ratio between steering column i, connected to the steering wheel, and pinion u, geared with the rack 3.

The system preforms like a conventional steering box, in case of sudden failure, without affecting steering angle, as required by law. In fact, in case of stop of the electric motor, the worm gear gearbox, irreversible by its nature, keeps the carrier stopped in the last actuated position. The gear acts as a double reduction, with planets at fixed position, while their rotation axis s is blocked, with a transmission ratio of 1:1 between sun gears S, without motion reversal.

An electronic control box is able to correct vehicle path as function of yaw speed and lateral acceleration detected by sensors applied for the purpose.

At low speed the electric motor contributes to reduce the steering effort to improve handling.

A different variable ratio steering box has been developed by Honda.

The regulation of the transmission ratio is made by an electric actuator between the steering column and the pinion of the steering box.

For low ratio the actuator moves the pinion closer to the rack, to reduce its pitch radius; for a more direct ratio the rack is pulled away to increase pitch radius.

The control is able to change steering wheel rotation angle from stop to stop from a minimum of 1.4 turn to a maximum of 2.4 turns.

6.1.3 Steer by Wire

A *steer by wire* system, or a system where the mechanical connection between steering wheel and steering box is replaced by an electric or hydraulic transmission is able to obtain all functions described on previous paragraphs (auxiliary power, ratio variation, stability control) with a simpler mechanical system, at least from a conceptual standpoint.

The steering wheel drives a position sensor and is driven by a torque actuator able to replicate on controls the driving feed-back; wheels are steered by an angle actuator under the full authority of the control system.

The steering wheel can be substituted by different controls, as, for example, joy-sticks or control sticks. The actuation power can be hydraulic or electric, with a preference to the last for the easiness of interfacing electronic control circuits.

The expected advantages include also a better control layout with advantage in passive safety (reduce aggressiveness to the driver after crash) in roominess and drivers seat adaptation; controls could be easily moved from one side to the other of the vehicle.

These systems could be also easily adapted to the specific need of handicapped drivers.

This interesting conceptual simplification involves complication to guarantee the expected reliability; these systems are still not admitted by existing regulations that still request a default positive mechanical drive between control and steering box, in case of failure.

A system of this kind could easily integrate rear steering function.

With two different and independent control inputs (δ_1, δ_2), it is possible an independent control of two degrees of freedom as yaw speed and vehicle side slip angle.

Some additional potential advantages should be investigated by controlling each steering wheel separately on each axle.

The idea behind of such control should be to steer each wheel always in kinematic condition, improving vehicle handling especially on narrow turns.

6.2 Brakes Control

Longitudinal forces generated by braking system can affect not only longitudinal dynamics, but also lateral dynamics.

We suggest to look at Fig. 2.34, where the longitudinal friction coefficient μ_x is shown as function of the longitudinal slip σ. In addition, the application of a longitudinal braking force F_x changes, at a given side slip angle α and vertical force F_z, lateral force F_y and self aligning torque M_z, as shown on Figs. 2.63 and 2.64.

An empirical formula, commonly adopted to describe this phenomenon, has been proposed by Pacejika and commented on the chapter dedicate to the wheel.

Non symmetric braking torques applied to the wheels can produce yaw moments able to affect vehicle path.

Brakes are therefore actuators not only suitable to change vehicle speed, but also to control vehicle path and its traction capacity. The many control systems that are based upon this actuator are described in the following paragraphs, using for their identification Bosch trademarks and acronyms; they have entered in the common technical language.

6.2.1 ABS System

The objectives of ABS systems (*Antilock Braking System*) are, as already announced, avoiding too high longitudinal slips that can reduce braking capacity and lateral dynamics control.

This objective must be reached for every possible value of friction coefficient (in principle, different for each wheel) of road slope and vehicle load.

In the mean time, the control system must minimize any disturb coming from sudden variation of the friction coefficient (as driving on puddles, pits, manhole covers, etc.), suspension vibrations or tolerances of the fabrication or assembly process.

In addition, braking capacity should not be affected by transmission inertia (vehicle inertia at the driving wheels is affected by gear ratios and by the engagement or disengagement of the clutch); finally, yaw torque disturbs, coming from different friction coefficient on the wheels (the so called *μ-split conditions*), are to be reduced as vibrations on the brake pedal, caused by braking torque regulation.

Input parameters of this control system are wheel speeds, measured by sensors on wheel hubs, while the controlled parameter is the time derivative of the braking pressure at each brake. In fact, ABS controls are not able to set a given value of the braking pressure, but control increase, holding on or decrease of pressure.

Also if in the past simpler systems have been built, we will refer to the four channels system only, now universally applied; this system is characterized by having a speed sensor and an independent pressure control for each wheel.

To reach this function, an ABS system must include:

- a master pump with power system, same as by traditional systems;
- a brake actuator for each wheel (disc or drum) same as by traditional systems; we should notice that disc brake are preferred to drum brakes, because they are simpler to be regulated, being free of potential self-locking.
- a speed sensor for each wheel;
- a hydraulic pressure modulator, including an electronic control unit, regulation valves and a recirculation pump.

This components will be explained in a following paragraph; we will start now to explain the control strategy applied to fulfill the objectives.

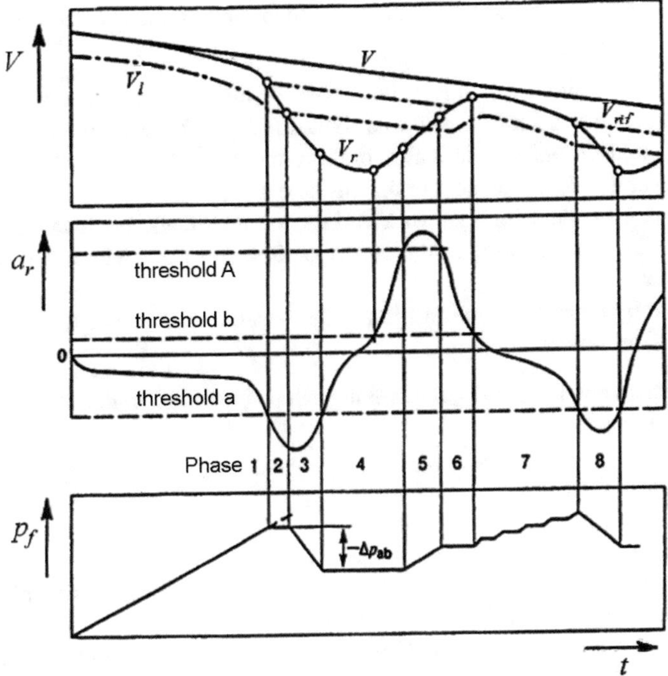

Fig. 6.7 Diagrams of speed, acceleration and braking pressure in braked wheel controlled by an ABS system

Figure 6.7 presents a typical diagrams for a wheel of some fundamental parameters, during a progressive braking on a friction surface, with the intervention of the ABS system.

The first graph above shows the diagram versus time of:

- vehicle speed V: it is not measured by the control system, but is shown on the graph for reference;
- reference vehicle speed V_{rif}: this parameter is estimated by the control system by algorithms that elaborate wheel speeds;
- wheel peripheral speed V_r: it is the product of the wheel speed Ω_r, measured by sensors, by the rolling radius R_0;
- limit speed V_l: it is the threshold of the wheel peripheral speed, below of which slip is too high as to guarantee control targets.

The second diagram (in the middle) shows the peripheral acceleration of the wheel a_r, while the third, at the bottom, is about the effective braking pressure on the actuator p_f as controlled by the electronic unit.

The electronic unit calculates the acceleration a_r, deriving the speed V_r measured by wheel sensors; particular techniques are applied to filter signals from magnetic pick ups, suppressing noise.

Some characteristic threshold values are identified on the acceleration diagram; they are:

- threshold a (value <0): this value certainly higher than the maximum vehicle acceptable deceleration shows a high difference between the actual braking moment and the maximum stable braking moment, due to a too high braking pressure or a sudden transient to the unstable area;
- threshold A (value >0): it shows a braking torque remarkably lower as the maximum stable braking torque; the corresponding slip has small value, below the maximum;
- threshold b (value >0): near to zero, underlines the return close to the maximum of the (F_x, σ) curve.

Seven characteristic phases of a braking can be identified, under the control of an ABS system.

Phase 1

The time $t = 0$ of diagrams corresponds to the start of braking. During all phase 1, there is no intervention of the control system.

In fact, in this phase the wheel speed V_r is greater than the threshold V_l and the peripheral acceleration a_r is negative, in the same order of value as the vehicle deceleration. Its value is lower as the threshold a, that indicates the transition to the unstable field.

As a consequence, in a progressive braking, the piston pressure is continuously increasing, according to drivers will and circuit delay. The acceleration will be maintained almost constant because the slip increase is such as to keep the friction torque increasing as the braking torque.

Before of reaching the threshold a, the acceleration will show a sudden change, underlining the boundary with the unstable area is near.

Phase 2

As soon as the threshold a is overtaken, phase 2 starts, where braking pressure is kept constant.

This occurs only after the threshold has been overtaken. In fact a sudden high speed braking, characterized by a high braking torque gradient, by tires with a low slope of the stable part of the characteristic, implies a slow increase of the friction torque. Therefore it is possible to reach the threshold a in stable condition and is not necessary to reduce the pressure in those conditions.

If the beginning of phase 2 represents the transition from stable to unstable area, the value of slip that has been reached is the optimum one, to be held constant during braking, to have a good performance.

In this phase the wheel deceleration keeps decreasing.

This explains why the braking torque is constant while the slip increases, with reduction of friction torque (the limit of the F_x-σ curve has been overtaken).

Phase 3

Below the value V_l the slip is very high, close to wheel block; therefore crossing the corresponding threshold for the wheel speed V_r implies the entrance in the phase 3, characterized by pressure decreasing.

The consequent decrease of the braking torque makes an inversion of the acceleration diagram possible, which even below the threshold a, starts increasing, while the wheel speed keeps decreasing, but with a reduced slope.

In phase 3 the wheel is still in the unstable area and the slip is still increasing. The friction torque decreases with a lower speed as the braking torque.

For this reason the acceleration start increasing.

Phase 4

As soon as the acceleration overtakes the threshold a, we enter the phase 4, characterized by keeping the pressure constant. While in phase 1 crossing the threshold a means the entrance to the unstable area, in this case crossing means the return to the stable area, because in the mean time the braking torque has been stabilized to a lower value, corresponding to Δp_{ab}, on the third diagram.

This fact allows the acceleration to continue increasing, till becoming positive. As a consequence wheel speed V_r will decrease in the first part of this phase till a minimum and than will start again to increase.

The acceleration growth is due to the fact that, while braking torque is kept constant, friction torque grows till it will become bigger as braking torque.

This growth occurs because of the slip decrease.

In the middle of this phase the deceleration overtakes again the threshold b. If this wouldn't happen, it should be necessary to proceed to a further decrease of the braking pressure.

This could happen because the pressure decrease in phase 3 has not been enough to allow the friction torque to become higher as the braking torque.

Phase 5

As soon as the acceleration overtakes the threshold A, the phase 5 begins; it is characterized by a pressure increase.

In the first part of this phase, the longitudinal slip decrease causes an increase of the friction torque and a consequent increase of the acceleration, until this increase is changing the trend of the acceleration.

During this phase, the wheel speed continues increasing, because the acceleration is, in any case, always positive.

Phase 6

Phase 6 starts as the acceleration overtakes, with negative slope, the threshold A, where the pressure is kept constant. In this phase the wheel speed continues increasing and the slip decreases, but because the slip is again in the stable region, also friction torque decreases, together with acceleration, until this overtakes, always with negative slope, the threshold b.

Phase 7

At this time the phase 7 begins, where the pressure in increased by steps, to avoid a steep increment of the braking torque.

This kind of increase allows a better exploitation of the tires and decreases the influence of wheel inertia. At the beginning of this phase, wheel speed continues increasing and slip further decreases with decrease of friction torque.

Acceleration continues decreasing till zero, returning then to negative values.

The value of zero corresponds to having both torques equal. As soon as the acceleration is negative, slip begins again increasing toward optimum values. They will be reached when the acceleration will decrease below the threshold a, starting a new cycle, as what we have described, starting from phase 3.

We observe that, by ABS controls, the actuator pressure cannot never overtake the pressure of the master pump. In other terms the braking torque cannot overtake the action of the driver on control.

In case of braking on low friction grounds some antilock systems, recognize this situation and adopt slightly different strategies.

Other strategies are addressed to improve vehicle dynamic behavior.

For instance, the strategy *Select Low*, designed by Bosch, provides that the pressure of rear actuators is determined by the wheel with lower friction, while front wheels are controlled with different pressures.

By this strategy the yaw torque (consequent to different braking forces, for instance on μ-split grounds) is decreased with a contained damage on braking efficiency.

Another strategy, again by Bosch, brakes front wheels with higher friction with a delayed pressure. In this case the yaw moment is only delayed, but to the driver is left a better chance to compensate for this disturb with a different steering angle. This delay should be calibrated to vehicle characteristics in order not to decrease too much braking performance.

6.2.2 *EBD System*

The acronym EBD derives of *Electronic Brake Distributor*.

As known, during braking, the inertia force applied to the center of gravity determines a vertical load transfer, increasing vertical load of front wheels and decreasing that of rear wheels.

If a braking moment simply proportional to the vertical static load would be applied, rear tires would loose their adhesion as first, compromising vehicle path stability.

To avoid this fact, on braking circuits without ABS control, a distributor valve has been introduced, we have described on a previous paragraph, addressed to limit the rear wheel braking pressure.

This function can be performed by an ABS system too and this added feature is called EBD.

For a given vehicle is possible to calculate ideal pressure distributions, as shown on Fig. 5.12; this curve represents the locus of points, on the p_{ant}-p_{post} plane that guarantees the maximum braking force on both axles.

This curve must be modified for each possible load combination on the vehicle. For instance the full load curve allows a higher rear braking as the empty condition, because, usually, load variations affect rear wheels load mainly.

EBD control tries to copy the ideal curve in a better way as mechanical brake distributors.

To reach this target, vehicle mass is estimated by comparison of the master pump pressure and vehicle deceleration; the value of the mass can be used to guess mass break down too.

Until the control action on the brake pedal is low the distributor follows the full load curve; additional slips of the rear axle are contained.

At higher master pump pressures, EBD control diversifies pressures using ABS valves.

It should be noticed that such strategy implies a precise calculation of the slip, which depends on vehicle speed. But this cannot be estimated precisely from wheel speeds. A control system keeping the difference between rear and front wheels slips is simpler and effective in the same way.

The components that are necessary to an EBD control system are the same as an ABS system; only software is affected, because specific algorithms are added for this function.

We should remember that EBD system, as a mechanical distributor, is subjected to regulations on minimum deceleration in case of failure; the system must include a recovery mode including a non dangerous brake down of braking forces in case of failure.

6.2.3 VDC System

VDC (*Vehicle Dynamics Control*) is also known through different commercial names, as ESP (*Electronic Stability Program*) or VSC (*Vehicle Stability Control*).

The objective of such system is to obtain a vehicle dynamic behavior always stable and predictable and to avoid that tires are working over their stable limit. The actuator of this system is again wheel braking torque applied as to obtain a suitable yaw moment on the vehicle.

Components necessary to a VDC control include:

- the master pump, with vacuum servo, as on a traditional system;
- brake actuators, as on a traditional system;
- wheel speed sensors, shared with ABS and EBD systems;
- hydraulic pressure modulator, including valves and load and recirculation pump, shared with ABS and EBD systems;
- a steering wheel angle sensor;
- yaw speed and lateral acceleration sensors;
- a connection line with the engine electronic control unit.

Steering wheel angle, yaw speed and lateral acceleration sensors are used to identify main parameters describing vehicle lateral dynamics.

The electric pump and valves are used for this system to generate pressure pulses for singles brakes, independently of driver's will.

When the vehicle is driven on a curve and the cornering force of one of the tires is less as what needed locally for the equilibrium with the centrifugal force, the vehicle deviates of the previous path.

If this deficiency is localized on the front axle, the phenomenon is similar to understeering[1] and the vehicle describes a path with less curvature as in normal conditions; if it is localized on the rear axle, the phenomenon is similar to oversteering and the vehicle describes a path with higher curvature as in normal conditions.

The VDC control system tries to recover the initial path, with a correction yaw torque, compatible with tires limits.

To do this, the electronic control unit calculates from steering angle and vehicle speed the expected yaw speed on the maneuver. If there are differences with the measured yaw speed, four braking torques are calculated that are useful to correct this situation or to reduce at least vehicle speed, if friction limits are overtaken.

To this speed reduction contributes also a reduction in engine torque; this is the purpose of the communication channel between engine control and VDC control systems.

Engine torque in reduced by spark advance reduction, by injected fuel reduction or by throttle valve reduction by engines driven by wire.

These interventions mustn't induce slips that bring the tire to side slip or longitudinal slip unstable region.

Two control loops are necessary. One calculates braking forces necessary to stabilization; an innermost loop controls wheel slip as to obtain the desired longitudinal and lateral forces.

A mathematical model must be suitable to predict vehicle motion consequent to load variation, at different vehicle load, road friction and tire wear.

---

[1] See definitions on understeering and oversteering in the second volume.

6.2.4 ASR System

This acronym derives of *Anti Spin Regulator.*

Critical driving situations rise not from braking or curves only. Also by accelerations, traction forces may cause vehicle instability.

ASR systems correct for these situations avoiding excessive slips on driving wheels.

This result is achieved either by temporarily reducing the torque applied by the engine to the driving wheels and by applying a braking torque by control system, that, also in this case, must be capable of generating torques autonomously.

ASR is again an additional function of ABS systems, implying a communication with engine electronic control unit.

ASR control unit is integrated with ABS control and shares with it a number of components, as wheel speed sensors and pressure control valves.

The components necessary to an ASR system are:

- brake actuators, as for a traditional circuit;
- wheel sped sensors, shared with ABS and EBD systems;
- hydraulic pressure modulator, including valves and load and recirculation pump, shared with ABS and EBD systems;
- a communication line with the engine control unit.

If throttle valve and brakes are available as actuators, the best performance can be obtained; since most engine controls are now driven by wire, there are no more difficulty for this option.

The control strategy can be outlined according to five phases.

Phase 1

The driving wheels longitudinal slip is calculated by comparison of their speeds with idle wheels speeds.

Phase 2

The beginning of this phase is characterized by the longitudinal slip of one of the driving wheels overtaking a limit threshold; a suitable braking force is calculated to reduce this slip and consequently the driving force.

Phase 3

The slip is again below the safety threshold and the braking pressure is decreased.

Phase 4

Phase 4 is similar to phase 1, but starts if an excessive slip is shown also by the other wheel.

Phase 5

During this phase, if all driving wheel were braked, also the engine torque would be reduced as to limit longitudinal slip.

ASR system can be implemented by the MSR system (*Motor Spin Regulator*), which provides to reduce the engine braking torque during gear box down shifts or at sudden accelerator pedal release.

A potential slip situation could occur on low friction grounds (snow, mud, etc.) by low gears. In this case MSR systems require an increment of the engine torque, as to reduce the engine braking effect.

6.2.5 BAS System

BAS systems (*Brake Assist System*) have the task of applying a constant full braking pressure, independently of driver's will, trying to decrease the stopping distance as much as possible, in emergency situations.

Such control system is useful by panic braking of non expert drivers; also expert drivers not suited to ABS systems may instinctively decrease the force on brake pedal, trying to avoid slips that will be, in any case, avoided by the control system. For impaired drivers, the pedal force could be inadequate to exploit the full braking capacity.

Two alternative strategies may be applied:

- when a panic situation is detected, the system generates the maximum braking force, independently of drive's will;
- when a panic situation is detected, a supplementary braking force is applied, modulated by driver's action on brake control.

A panic situation is usually detected when the braking pressure time gradient is over a threshold value.

By the second strategy the deceleration is modulated by drivers force and speed. An ampler field of intervention can be reached as for the first strategy.

The components necessary to a BAS system are again the same as for the ABS system, having modified the hydraulic circuit with a suitable storage capacity.

The pressure modulator maintains a certain quantity of high pressure oil in a reservoir. The BAS system can apply a larger pressure as that determined by the master pump.

6.2.6 *Brake Controls Hardware Components*

All brakes control systems apply the following sensors and actuators.

Sensors

Wheel speed sensors are toothed wheels with magnetic pick-ups.

The toothed wheel RF on Fig. 6.8 has a rectangular profile made by ferromagnetic material, facing a magnetic pick-up; wheel rotation changes the magnetic flux captured by pick-ups.

Pick-ups are made by a magnetic core 1 and a soft iron core 2. The magnetic field created by the magnetic core is periodically altered by the profile motion of the wheel and its value is measured by the coil 3, which will supply a voltage where period and amplitude are determined by rotational speed.

On the diagram at the top right the voltage is shown qualitatively as function of time, for a low revolution speed (dotted line) and for a high revolution speed (solid line). This signal is amplified and squared by an electronic circuit; peaks count is supplying the requested speed value.

On current wheel speed sensors, electronic amplifier and signal conditioner are integrated into the magnetic pick-up, that can be directly interfaced with the control microprocessor.

The yaw speed sensor (gyrometer) and the acceleration sensor (accelerometer) are usually integrated into a unit, including a solid state electronic circuit.

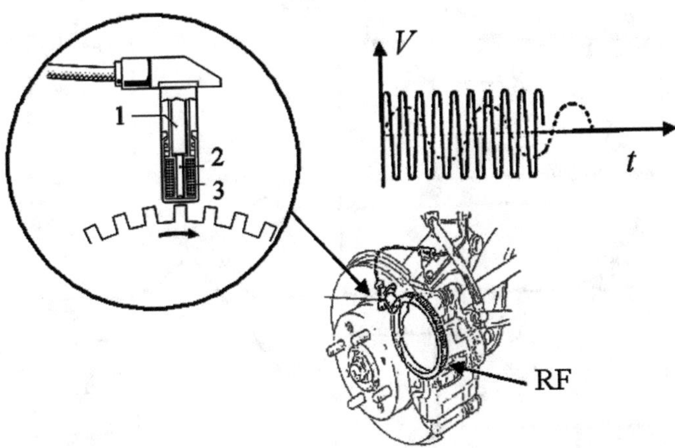

Fig. 6.8 Installation scheme for a wheel speed sensor; RF is the toothed wheel. The enlargement shows the detail of the magnetic pick-up; at right are shown two diagrams of the generated voltage at low speed (dotted line) and high speed (solid line)

The working principle of the accelerometer is simple and consists in measuring deformations of a small structure subject to the vehicle inertia forces. This measurement can be made by piezoelectricity, where an elastic deformation is converted to a voltage variation.

Also yaw speed is converted to an acceleration; it is measured by the Coriolis acceleration of the prongs of a diapason, brought to a forced oscillation by external means.

Steering wheel angle is measured by a potentiometer on the steering column or by an optical encoder. The requested precision is in the range of a tenth of a degree.

The encoder is still made by a toothed wheel, interrupting periodically the light beam received by a photo diode; the voltage pulse is measured and counted by an electronic circuit.

At the state of the art all those signals, converted to digital units, are made available to the CAN network of the vehicle that make them available to the brake control unit and to other services, as for instance, an electric power steering system.

Actuators

Figure 6.9 illustrates the functional scheme of a hydraulic modulator for an electronic brake control system.

Two independent hydraulic circuits can be identified, in this case, according to the X scheme, required to satisfy regulations on emergency circuit.

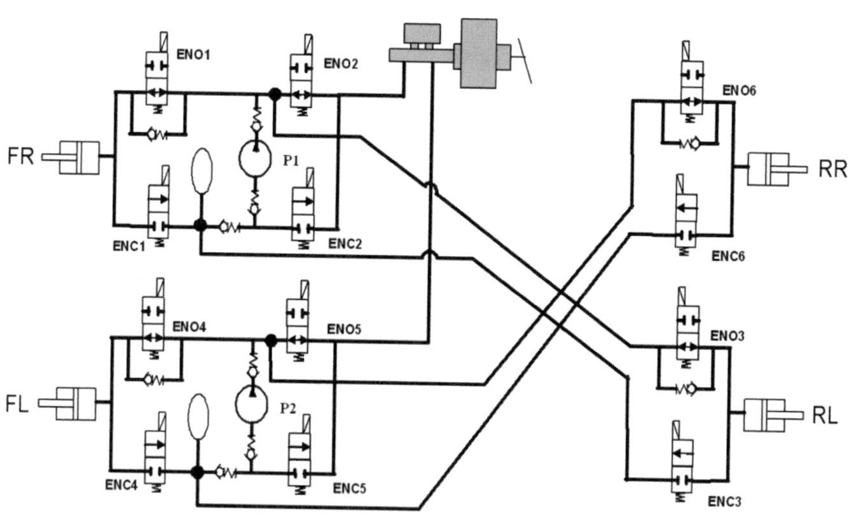

Fig. 6.9 Functional scheme of a hydraulic brake modulator for a four-wheeled vehicle. ENO valves are normally open; ENC are normally closed

Six electrovalves normally open (ENO1...ENO6; normally open means that the hydraulic circuit is open when the coil is switched-off) are present in the circuit and six electrovalves normally closed (ENC1...ENC6); the circuit includes also two pressure accumulators, two electric pumps used to speed up pressure decrease by ABS and VDC modulation and to create pressure during VDC and ASR modulation.

These pumps are made by two radial pistons moved by a cam and an electric motor.

We make reference to the front right wheel (FR) only; the default configuration, when no intervention is required to the control system and all electrovalves are switched-off is represented in the figure. There is a direct connection between the master pump and the brake actuator FR.

This condition allows a conventional operation of the hydraulic circuit also in case of control system failure.

When is necessary to maintain a constant pressure in the brake actuator, independently of the driver's will, the electrovalve ENO1 is switched-on and isolates the actuator from the rest of the circuit.

If it is also necessary to decrease the pressure, the electrovalve ENC1 will be also switched-on to allow to part of the oil in the actuator cylinder to return into the master cylinder atmospheric pressure reservoir, with the help of the electric pump, if necessary.

To increase the pressure independently of the pressure in the master pump (VDC and ASR regulation), the electrovalve ENC2 is switched-on putting in communication the pump P1 with the reservoir on the master pump; in the mean time also the electrovalve ENO2 is switched-on to avoid pressure oil return into the reservoir.

6.2.7 *Function Integration and Further Developments*

The number of functions assigned to the braking system has dramatically increased during the time; often, many of these new functions have been conceived as product options, originating a huge family of new add-on devices.

We think it may be useful to report an exhaustive list of the functions that are presently assigned to the components of the braking system.

Brake Functions

These functions modify the braking force exerted by the wheel to increase braking efficiency; they are performed by the ABS system and are known as:

- EBD or electronic brake distribution, modifying the brake pressure acting on rear wheels, in order to avoid their lock-up before front wheels:
- ABS or anti lock braking system, modulating the brake pressure acting on wheels, in order to avoid their lock-up with low longitudinal force coefficient.

- CDF or controlled deceleration function, slowing down the vehicle on request of other control system as, for example, adaptive cruise control.
- SMC or enhanced μ-split control, modulating the brake pressure acting on the wheels of one side of the vehicle, in order to avoid their lock-up when the longitudinal force coefficient has different value on the two sides of the vehicle.
- T-ABS or turning ABS, modulating the brake pressure acting on the wheels also in narrow turns.

Brake Booster Functions

These functions modify the boost efficiency, to enhance the braking force independently of the driver's intention; they are performed by the booster and known as:

- HBA or hydraulic brake assist, increasing the boost efficiency in panic stops.
- EPB or electronic brakes prefill, preparing an automatic sudden stop in case of imminent collision.

Stability Functions

These functions apply braking forces on some of the wheels automatically, to enhance the vehicle stability; they are performed by the ABS system and known as:

- VDC or vehicle dynamics control, avoiding yaw speed to become too different from its cinematic value.
- ERM or electronic roll mitigation, avoiding the risk of rollover in case of sudden lane changes or long turns driven at too high speed.

Traction Functions

These functions apply braking forces on some of the wheels automatically to avoid slippage; they are performed by the ABS system or by dedicated electric valves and known as:

- ASR or anti-spin regulation, avoiding the slippage of a driving wheel with low longitudinal force coefficient.
- HH or hill holder, avoiding the vehicle to move backwards during uphill starts.
- HDC or hill descent control, avoiding the vehicle to take speed, in downhill driving on dirty roads with low longitudinal force coefficient.

The opportunity offered by having a hydraulic pressure source in ABS systems that are from some years compulsorily installed on every car has suggested the idea to attribute all this function to the hydraulic brake modulator; in this case all braking actions are operated by the modulator and the brake pedal becomes simply a sensor able to interpreter the driver's intention.

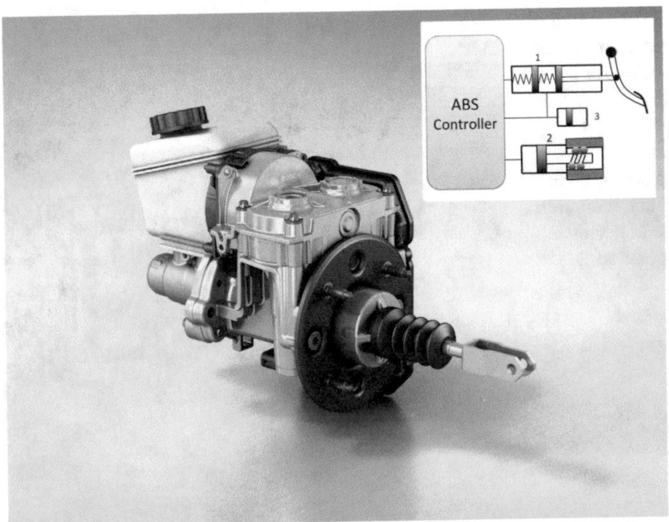

Fig. 6.10 A modern brake modulator including an internal electro-hydraulic one-piston pump to return the correct feedback to the driver. 1: master pump; 2: ball screw pump; 3: one piston pump

To the brake modulator is added an internal electro-hydraulic one-piston pump that pressurizes the brake fluid in the master pump according to the brake pedal travel/force, to return the correct feedback to the driver.

The circuit is design in such a way that, in case of failure of the electronic control or the electric actuators, there is hydraulic continuity between the master pump and the braking circuit, to allow at least unassisted operation.

Very compact modulators are now installed instead of the vacuum power brake that integrate also the master and the feedback simulator pump, as shown in Fig. 6.10.

The advantages of this system are:

- The brake assistance is vacuum independent, therefore is not affected by the engine operating conditions.
- Diesel engines don't need a vacuum pump, with consequent advantages in cost, reliability and fuel consumption.
- The pedal feel can be adapted to the driver's preference or to driving conditions.
- Regenerative braking in hybrid and electric vehicles can be obtained without intervention of mechanical brakes, with a seamless pedal feel between mechanical and electric braking.
- The cost is reduced and the braking plant layout simplified.

6.2.8 Hybrid and Electrohydraulic Circuits

Hydraulic braking systems, we have examined until now, have reached a high level of performance but have also become quite complex and expensive.

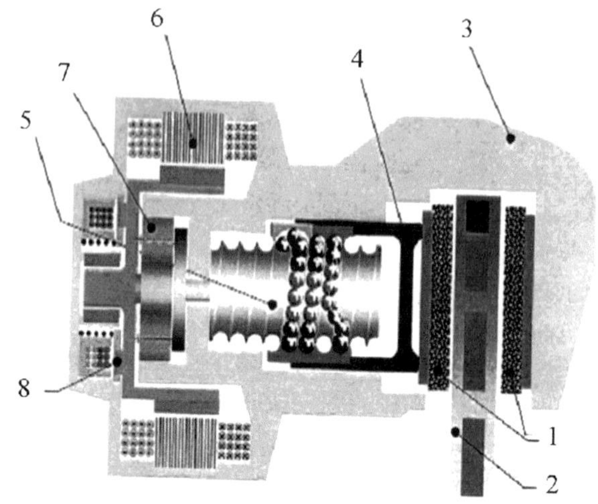

Fig. 6.11 Cross section of an electric actuator for rear brakes, also used for parking function. We can see: 1 disc, 2 linings, 3 caliper, 4 actuator cylinder, 5 ball recirculating screw, 6 electric motor, epicyclic reduction gear, 8 brake to keep the actuator in position

The transmission of the force from pedal to actuators is made using the hydraulic pressure; the need to contain this force within limits that are ergonomically compatible with the average driver has justified the addition of a vacuum actuator.

The vacuum pressure, sometimes not sufficient or unavailable as by diesel engines has required introducing electric vacuum pumps.

Control force generation only requires three different energy vectors.

Control systems require again oil in pressure and electric pumps; with two additional energy vectors. Electric energy looks to be the primary source for control force generation and regulation.

This irrational situation is justified by historic and by economic reasons, because any of the above devices have been born as optional accessory, to be implemented into a conventional system, and the high volumes of each of this components has discouraged the development of different, even more rational, solutions.

In some cases also regulations have required redundancies for reliability and safety, contributing to maintain this situation.

The diffusion of power brakes and ABS, now at 100% on new cars, encourages the development of new system architectures, addressed to avoid vacuum energy with hydraulic power source only or to avoid any other kind of energy, but the electric one, used for actuators and regulation.

Even if not largely present in series production, some vehicles feature so called hybrid systems that have a state of the art circuit on front wheels and electric brakes on rear wheels.

Hydraulic elements are eliminated for rear wheels only, on actuators and regulating circuits. The cost advantage is limited, probably interesting for electric parking brake application only; but it could be the starting point of the development of a new interesting technology.

Figure 6.11 shows the rear brake actuator for a hybrid system.

This system is completed by an electronic control for electric motors and a pressure sensor to interpret driver's intentions from master pump pressure; this sensor is already present on VDC control systems.

To have an adequate clamping force on rear pads with a small size electric motor a reduction gear with very high transmission ratio is adopted; in our case a double stage gearbox is used, including an epicyclic gear and a worm gear with balls recirculation: this feature is necessary to reduce friction and to guarantee a quick release of the braking force.

A position sensor, not shown on this figure is used to release the braking force maintaining the clearance between pads and disc at a preset value, independent of wear.

An electromagnetic brake 8 is actuated with inverted logics; when the electric motor is off, the electric brake in on, stopping the motor and avoiding pads motion. This condition occur at stopped vehicle, with the parking brake, or by failures. The maintenance of the state of the braking system before the failure is judged to be the best policy. When the electric motor is on, the electric brake is off, leaving pads free to move.

This kind of brake allows a further brake control system to be implemented, called EPB (*Electric Parking Brake*).

The purpose of this control includes a power assisted control of the parking brake (no energy is requested from the driver) and the HH (*Hill Holding*) function, where parking brakes are actuated automatically when the vehicle is stopped on a slope. The stopping condition is identified by having the brake pedal depressed once, the accelerator pedal released and the car speed zero. At next start up, the parking brake is released only when driving wheel are able to move the car, without undesired car move back.

Hybrid systems and electrohydraulic brakes, we will introduce later on, are the precursor of remote braking systems (*brake by wire*), that probably will be developed in the next years.

Electrohydraulic brakes (EHB, *Electronic Hydraulic Brakes*) include in a conventional electric control system also the function of power amplification.

Brake actuators are the same, but the master pump is without vacuum actuator, now no more necessary; the brake pressure is generated by the electric pump only, while the master pump guarantees the residual braking capacity of the emergency brake.

The pedal feedback is given by regulating the pressure from a pressure accumulator.

The signal from the pedal to the electronic control unit flows through two parallel redundant ways: a pedal position sensor and a pressure sensor at the outlet of the master pump.

For pressure regulation only proportional valves are used, that regulate the pressure coming from an accumulator according to the pressure on the pedal.

Two on-off valves cut the circuit of the master pump; these valves can put the circuit on pressure also when the pedal is not depressed for VDC and ASR function.

The ultimate goal will be a fully electric system with no hydraulic energy contribution, the so called EMB (*Electro Mechanical Brakes*).

The actuators could be similar to what we have presented on Fig. 6.11 and there will be no positive connections between the pedal and the actuators.

This kind of system will require the development of redundancies able to obtain a level of safety and reliability at least equal to that of present systems; this objective will not only affect sensors and actuators design, but also the communication network.

6.3 Suspension Control

Suspension systems must satisfy conflicting objectives in term of comfort and active safety. In fact, as it will be explained in the second volume, suspension able to guarantee a high comfort should be very soft and poorly damped, while to guarantee a constant contact of wheels with the ground should be rigid and damped.

Suspension systems made by springs and dampers are also called *passive*, because they react to forces coming from the road with energy contributions that are negative only; they can, in fact, only waste energy.

Passive suspensions are designed to offer a reasonable compromise between comfort and passive safety, considering the character of the vehicle in question.

The characteristics of vehicle suspensions depend in fact on precise design options. Sport cars have stiff suspensions, not suitable to absorb an important road unevenness, but finalized to a superior stability, also by high cornering forces; normal cars, instead feature softer suspensions, that allow a higher comfort with some limitation on dynamic performance.

The limit of these passive suspensions can be easily explained by the impossibility of managing two independent parameters, body vertical accelerations (related to comfort) and vertical force variations (related to active safety) with a single parameter, the suspension damping coefficient. The two objectives are independent and their optimum values are obtained with different damping coefficients.

At least conceptually, the best compromise can be obtained by adapting the suspension damping coefficient to the most important priority, changing from time to time.

Tangible improvements have been obtained by adopting suspension components able to adapt their mechanic characteristic to the changing needs of different driving situations.

Applying micro electronics and improved mechanical components, passive suspensions have evolved to *adaptive* or *controlled* suspensions; they need positive energy contributions from the outside; if these contribution are important they are called also *active* suspensions.

For controlled suspensions a classification is made considering the intensity of the energy contribution:

- *semi-active* suspensions have a limited energy consumption as by static vehicle trim control or by damping control; the energy contribution is not sufficient to modify the vehicle trim in a time consistent with suspensions oscillation period;
- *active* suspensions have a significant energy contribution as by dynamic vehicle trim control; the energy contribution is sufficient to modify vehicle trim in a time consistent with suspensions oscillation period; the control system is able to keep vehicle trim unchanged almost at any time, on any road.

Both suspension kinds should include the following components.

- An actuator suitable to apply a force depending on what is calculated by the control system. Sometimes electric valves are used that supply a defined fluid flow, corresponding to a speed of the actuator; therefore the actuator speed is controlled, more than actuator force.
- Sensors measuring the significant parameters of vehicle kinematics.
- An electronic control system.
- A power plant, feeding actuators through valves; this plant is always present with different size, according to system active or passive features.

Considering the objective of the suspension control system, the following alternative may be considered:

- *trim control*: a quasi static control provides for a constant vertical static displacement of the rear axle or of both axles, at any vehicle load;
- *damping control*: shock absorbers damping coefficient is adapted to different situations; static displacements are not affected;
- *roll control*: vehicle roll and roll speed are limited dynamically, as compared with conventional systems;
- *stiffness control: spring rate is changed during vehicle operation;*
- *full active control*: all above objectives are pursued in any dynamic situation.

The energy requested by the control system is significant in the third system and maximum in the fourth one.

6.3.1 Trim Control

Rear axle trim controls correct the height of the rear axle by decreasing the natural compression stroke when payloads increase. This effect can be obtained using hydraulic and pneumatic actuators.

Figure 6.12 shows the layout and functional scheme of this system on a car. This system includes an oil reservoir 1, at atmospheric pressure, a pump 2 driven by the engine, a pressure or trim regulator 3, connecting the two hydraulic actuators 5 with the pressure oil accumulator 4.

If the vehicle load has been changed, the rear suspension stroke variation is perceived, when the vehicle starts its new trip, and some oil is sent to the actuators in

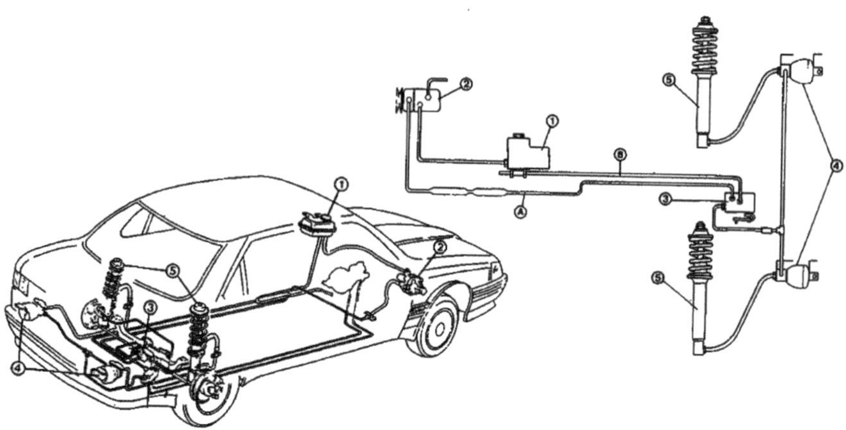

Fig. 6.12 Hydraulic rear axle trim control; we can identify: 1 reservoir, 2 oil pump, 3 regulator, 4 accumulators, 5 hydraulic actuators (FCA)

order to obtain the desired trim. Hydraulic actuators work in series with suspension springs and compensate for the spring compression.

A pneumatic system can be easily made by using elastic elements, similar to those introduced in Fig. 3.62; in this case, springs and actuators are integrated in a single component, the air spring. The control system can change the quantity of air in the elastic element, changing the pressure and obtaining the same position of equilibrium with an increased load.

It should be noticed that pneumatic systems change suspension flexibility too, as explained in the chapter dedicated to suspensions.

Another solution, quite popular on cars, is the Nivomat system, developed by ZF. It is made by single integrated units, that substitutes rear shock absorbers; they integrate pump, actuator and regulator.

The energy necessary to rise the suspension after a load increase is derived by the suspension strokes on an uneven road. Figure 6.13 shows a section of this unit and its functional scheme.

In a single unit, two chamber are built: the low pressure chamber 1, at the bottom, and the high pressure chamber 2, at the top; this chamber is divided in two sections by the diaphragm 3. The outside section is filled with compressed gas, while the inside section is filled with oil. At empty vehicle the two chambers are at the same pressure, about 25 bar; at full load, the low pressure chamber is at about 8 bar, while the high pressure one is at about 80 bar.

Inside the unit there is a conventional shock absorber 4, whose rod 5 is connected with the piston 6, bearing the choke valves. A drilled rod 7, at the bottom of the unit, moves inside a hollow cylinder 8 and works as an oil pump. On the outer surface of the rod 7 a hole 9 acts a trim (vertical suspension displacement) sensor.

Because of the vertical suspension stroke, produced by ground uneven, the oil in the low pressure chamber is aspirated by the pump, through the drilled rod 7. The

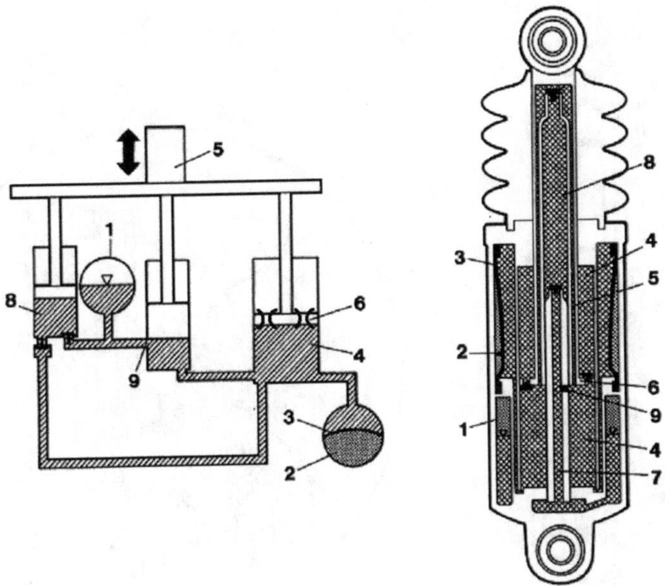

Fig. 6.13 Cross section of a Nivomat unit (at right) and its functional scheme (at left). On this scheme, we can see the low pressure chamber 1, the high pressure chamber 2, the shock absorber 4 (ZF)

pump sends the oil through the lower chamber of the shock absorber 4, into the high pressure chamber 4, compressing the gas separated by the diaphragm.

The higher pressure built, works on chamber 4, extending the rod 5 and rising the vehicle. When the hole 9 exit the cylinder, the desired trim is reached and the oil flows to the low pressure chamber, through the driller rod 7.

Some systems make front and rear axle trim control possible.

They are substantially two hydraulic or pneumatic systems working on each axle.

A characteristic system of this kind has been developed and adopted by Citroën. It uses a hydro pneumatic suspension, which matches air springs with high pressure gas (nitrogen) with original shock absorbers, integrated into a single hydraulic circuit.

6.3.2 Damping Control

The idea behind damping control systems is to apply shock absorbers with a damping coefficient that can be adjusted at different values; the adjustment can be continuous or at discrete levels. The damping coefficient is modified as function of the vehicle speed and vertical acceleration, with strategies that can be very simple or more sophisticated.

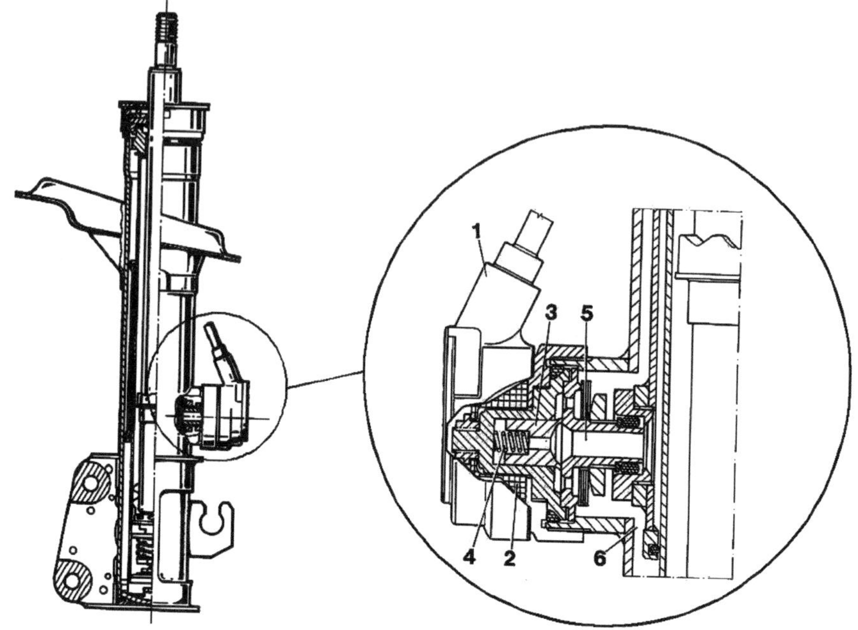

Fig. 6.14 Adjustable shock absorber; a solenoid valve (magnified at right) can change the choke orifice dimension and, consequently, the damping coefficient

The scheme on Fig. 6.14 shows a shock absorber with a solenoid valve used to change damping coefficient.

By controlling the current into the solenoid 2, is possible to change the position of the anchor 3, under the reaction of the spring 4; the position of the anchor determines the passage orifice and, therefore, the damping coefficient of the shock absorber.

Controlled dampers can be classified according to the kind of control strategies, as:

- adaptive;
- semi-active.

The operation principle of an adaptive control includes typically the following sensors:

- vertical accelerometers applied to the vehicle body, at least 1, with a maximum of 3;
- a lateral accelerometer on the front of the vehicle or, in alternative, a steering angle sensor;
- an pressure on-off switch on the braking circuit;
- a car speed sensor (may be the same as for the instrument cluster or the ABS system).

At very low speeds the damping coefficient is set to high levels as to limit car bounce at parking maneuvers.

To improve comfort, the damping coefficient is reduced very rapidly at higher speeds and kept at small values till about 120 km/h; at higher speeds is again increased to improve vehicle stability; high speed roads are, usually, even.

The longitudinal acceleration sensor is used to identify a medium speed sport driving, characterized by a frequent use of brakes and accelerator. Simple logic algorithms elaborating these two parameters are used to increase the damping coefficient, in order to reduce pitch angles. The detection of lateral acceleration over certain thresholds can increase further the damping coefficient to reduce roll angles and to better stabilize the vehicle in turns.

Semi active systems applied to vehicles are more complex but enable a better control of body motion, with a better performance; they are, usually, based upon the *sky-hook* damping theory.

According to this theory, the ideal situation could be reached when the vehicle body could be connected to the inertial reference system (the sky) by a shock absorber.

Since this architecture is not feasible, the control system provides to calculate a damping force proportional to the absolute vertical speed of the suspension, taking into account body and wheel speeds, with the formula:

$$F_{za,i} = C_s \dot{Z}_G + C(\dot{z}_i - \dot{Z}_G) \, , \tag{6.3}$$

where:

$F_{za,i}$	is the force exerted by the shock absorber of the wheel i,
$\dot{Z}_G$	is the absolute vertical speed of the vehicle center of gravity,
$\dot{z}_i$	is the relative vertical speed of the suspension of the wheel i,
C_s	is the sky-hook damping coefficient,
C	is the damping coefficient of the suspension of the wheel i.

The body speed is calculated by integrating the signals of three accelerometers on the car body, measuring the three components of the vehicle reference system. Relative suspension speeds are calculated by measuring the relative displacements of shock absorbers.

When directions of absolute and relative speeds are opposite and their difference is big, the ideal damping force $F_{za,i}$, to be applied, could have an opposite direction as the compression or extension speed of the actuator. In this case a full application of the sky-hook theory would mean applying double effect actuators with energy contribution from the outside. This fact would complicate very much the system complexity.

To avoid this problem, a simplified theory has been developed using conventional adjustable shock absorbers (semi active system).

An adjustable shock absorber allows to satisfy all conditions, where the requested force has the same direction as the relative suspension speed. By other situations, where force and relative speed are opposite, the force is set as low as possible, in order to minimize errors between the desired value and the actual one.

The results obtained with these techniques are quite interesting and significantly superior to those that are obtained with adaptive techniques.

New components are appearing on the market, based upon a new technology; they are the magneto-rheologic shock absorbers.

Magneto-rheologic fluids modify their viscosity when a magnetic field is applied. They are made by a suspension or ferromagnetic particles in a mineral or synthetic oil; additives are provided to reduce wear in restricted orifices and maintain suspension.

Particles in suspension represent a fraction between 20 and 60% of the total volume. With no magnetic field, the fluid behaves like a Newtonian one; we have:

$$\tau = \nu\gamma \,, \tag{6.4}$$

where:

τ is the shear stress,
ν is the kinematic viscosity,
γ is the deformation speed.

When a magnetic field is applied, particles tend to aggregate in chains along field lines and the fluid can be modelled according to Bingham; the shear is given by the formula:

$$\tau = \tau_0 + \nu\gamma \,, \tag{6.5}$$

where τ_0 is the limit shear induced by the magnetic field.

Under the action of the magnetic field, the particle aggregation makes the fluid similar to a solid body, until the shear stress is below a limit value τ_0; if this value is overcome, chains begin to break and the behavior is again modelled by Newton's formula.

This limit value is controlled by the magnetic field intensity. The magneto-rheologic shock absorbers have reduced response time, in the order of 5–8 ms, that allow a more punctual control, but need faster elaboration times, to take full advantage of this feature.

Magneto-rheologic components are conceptually simpler as electro hydraulic ones; on the contrary, they must have a stiffer structure to avoid damages because of the fluid, when at the solid state.

6.3.3 Roll Control

A first classification of roll control systems can be made considering the kind of actuators.

These actuators can be set on the anti roll bar, at its middle or at one end; the choice is decided by installation issues.

A second alternative is to intervene on elastic element flexibility, only by roll motion; this is possible by hydro pneumatic suspensions.

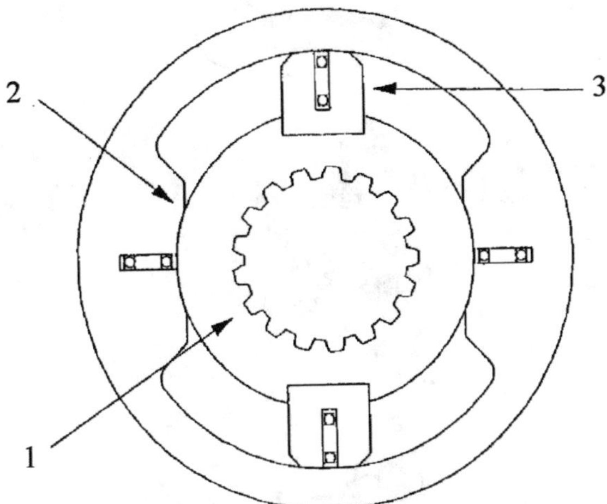

Fig. 6.15 Hydraulic rotary actuator to control an anti roll bar; this actuator is set in the middle of the central span

Figure 6.15 shows an example of actuator set in the middle of the central span of an anti roll bar. This actuator performs its function by imposing a preload to the two sections of the bar.

The body 2 and the rotor 1 are fixed to the two halves of the bar; the blades 3 divide the internal volume of the actuator in four different chambers.

Through a four ways solenoid valve is possible to send the oil in pressure (≅150 bar) to two opposed chambers and to exhaust it from the others or vice versa; two rotation directions are available being the angle proportional to the volume of displaced oil.

The actuator can be also made by a simple hydraulic cylinder working on one of the ends of the anti roll bar, between the bar and the suspension.

Figure 6.16 shows the scheme of a recent hydro pneumatic suspension system by Citroën.

This system, here represented for one of the two axles, includes the trim control function and the roll angle function.

The trim control function is obtained by charging or exhausting oil in the cylinders 6, communicating with the pneumatic suspension elements.

In addition to the four hydraulic units, one for each wheel, two supplementary elements 1 are provided, which increase roll flexibility and decrease damping, when put in parallel to the elements of the same axle.

The insertion of these supplementary elements is controlled by the solenoid valve 7, on the same figure. The solenoid valve moves a spool 3, communicating with the two main elastic elements; this spool is represented at left in open position, at right in closed position. The body of the supplementary unit includes the roll control valve (4 on the same figure).

This is a simple check valve that, in a curve, interrupts communication to the wheel inside the curve, limiting in this way the roll angle. The spool is locked in neutral position when the trim control is active.

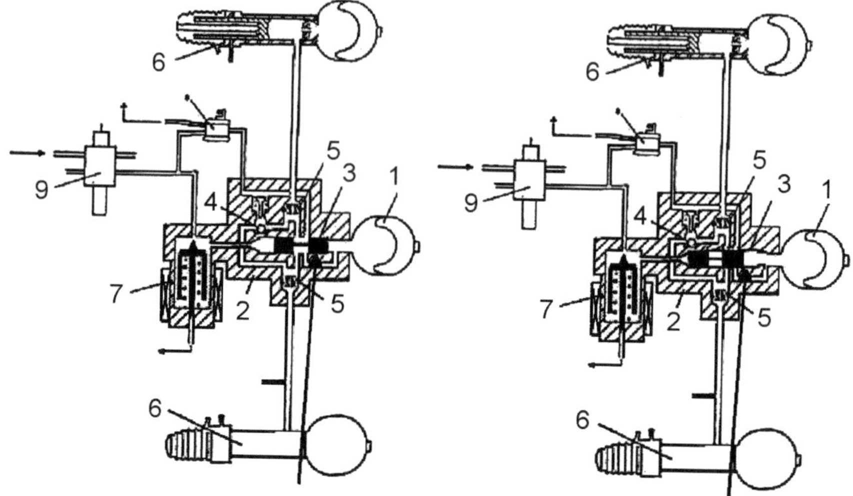

Fig. 6.16 Scheme of hydraulic pneumatic suspensions, able to control vehicle trim and roll stiffness; on A position (at left) the axle has a low roll stiffness, while on B position (at right), the stiffness is higher (Citroën)

Two different roll stiffness values can be obtained, in A and in B positions; at B position the flexibility can be further decreased by the valve 4.

A second classification criterion considers the type of control adopted; this can be active or passive.

By a passive control the system can have two configurations:

- low roll stiffness (free anti roll bar) for driving on a straight road;
- high roll stiffness (pre-loaded anti roll bar) for driving in a curve.

Actuator control is done by a simple on-off valve; the energy supplied to the system is very low.

By an active control the system features two proportional valves, one for each axle; this system can stabilize the vehicle dynamic response, with a higher energy consumption.

6.3.4 *Stiffness Control*

Despite the fact that designing the suspension stiffness is an inevitable trade-off between ride comfort and handling, where soft suspensions allow a top performance in terms of road unevenness absorption and hard suspensions the fastest and most stable response to steering, a variable or adaptable stiffness suspension could in principle satisfy the two opposite conditions or, at least, allow a vehicle to be comfortable or responsive at a driver's choice.

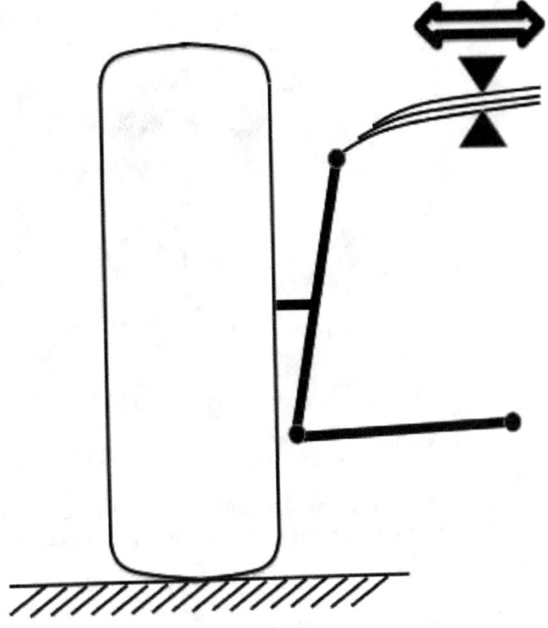

Fig. 6.17 A double wishbone suspension, with one of the arms made with a leaf spring with moveable bearing, could feature variable stiffness, as any spring where some device is provided to change its free length

A mechanical variable stiffness suspension, that should not be confused with a progressive stiffness suspension, could be very simple in principle; for example a double wishbone suspension, with one of the arms made with a leaf spring with moveable bearing, could have this property as any spring where some device is provided to change its free length, as in Fig. 6.17. Nevertheless a mechanism of this kind would be very difficult to be integrated in a car suspension and to be easily adjusted when the car is in motion.

A solution to the problem is offered by pneumatic spring suspensions if the air volume is changed. Porsche[2] presented an interesting solution of this kind, shown in Fig. 6.18, with a spring assembly featuring a main volume of reduced capacity, for responsive behavior, and two additional chambers adding capacity for improving comfort. These additional volumes can be put in communication with the main volume by fast response electro valves.

This enables a considerably large spread of stiffness. The chassis can be set to a lower basic spring rate for increased comfort, as the spring rate can be changed electronically in a fraction of a second where necessary—for example, during acceleration and braking or to reduce rolling motion. In addition, as any air suspension, it offers the benefits of self-leveling with the ability to choose different degrees of ground clearance. The stiffness can be decreased from 80 N/mm to about 34 N/mm.

[2]Gantikow M., Boyraz E., Kallert N., Legierski R., The new multi-chamber air spring by Porsche—future innovation in chassis mechatronics and integration, Proceedings of the 7th International Munich Chassis Symposium 2016.

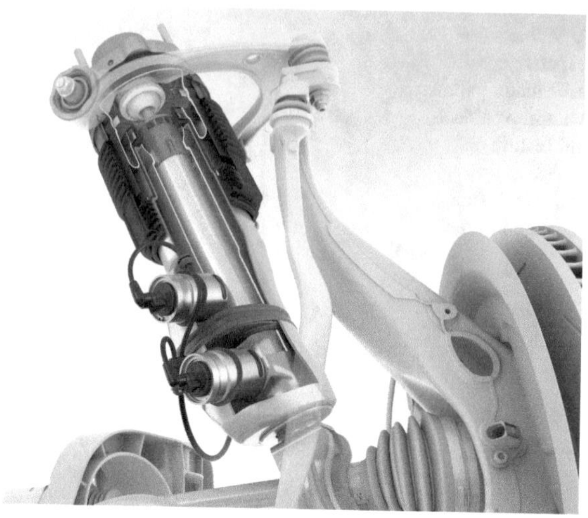

Fig. 6.18 Porsche air spring with a main volume of reduced capacity, for responsive behavior, and two additional chambers adding capacity for improving comfort

6.3.5 Active Suspensions

Active suspensions could avoid, in principle, the application of elastic elements and is inspired by the model of suspensions capable of retracting wheels, while climbing an obstacle, or extending them when descending.

This function is quite different of that of traditional shock absorbers; by them, in fact, the applied force oppose the suspension motion every time.

This kind of suspension is usually classified in two categories:

- slow active, or narrow band systems;
- full active, or broad band systems.

Slow active systems control 4 pressure regulating valves, one for each suspension.

These valves are able to control body motion (band in the frequency field between 0 and 5 Hz), while higher frequency motions are managed passively by pneumatic springs, in parallel with the suspension hydraulic actuator.

The active function is able to control pitch and roll motion of the body; the scheme is shown on Fig. 6.19.

The control strategy can follow the sky-hook damping theory for slow frequencies only; for higher frequencies, as the tire frequencies, excited by small obstacles they behave as passive suspensions with optimized damping coefficients.

In addition to sky-hook damping they can improve vehicle maneuverability and stability and perform anti dive and anti squat functions.

Full active systems are able to manage dynamic phenomena in the range between 0 and 25 Hz, having flow regulating solenoid valve; in this way also pulse maneuvers, like driving on an obstacle, can be managed.

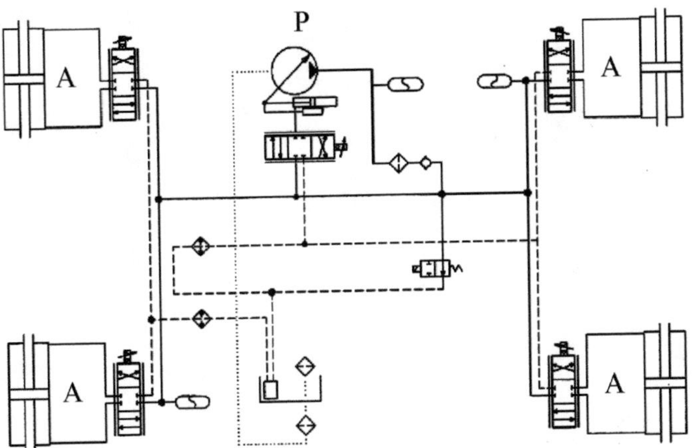

Fig. 6.19 Hydraulic scheme of an active suspension; the suspension motion is controlled by four double effect actuators in communication with a pressure circuit, through solenoid valves; peak pressures are faced with pressure accumulators

A peculiar characteristic of these systems is the high demand of hydraulic power from the engine (8–10 kW), at high pressure (150–180 bar); large pressure accumulators are necessary to face peak demands.

Actuators speed must reach, at least, 2–2.5 m/s.

Chapter 7
Chassis Structures

Chassis structures are stressed by internal and external loads.

External loads are coming from wheel-ground interface, through suspension mechanism and its elastic elements and from the aerodynamic field around the car body.

Internal loads are caused by vehicle and payloads masses (as passengers and baggage). Significant internal loads are caused by the reaction forces of the powertrain suspension.

Chassis structures can be *separated* from the body, as by industrial vehicles and by some commercial and off-road vehicles or can be *integrated*, as by unitized bodies. In this case, sometimes *auxiliary* structures are applied, to better distribute local loads to the body, as for carrying suspension mechanisms, engine or powertrain, transmission and final drive.

Chassis structures in these three cases are also respectively called *frames*, *underbodies* and *subframes*.

A typical example of front subframe, provided to carry-on lower suspensions joints, some of powertrain mounts and the steering box, is offered by Fig. 3.6.

7.1 Underbody

The name unitized body derives by the fact that the body structure is designed to bear all vehicle loads directly; the other case of separated body doesn't mean that the body is only one of the external loads stressing the chassis structure, the frame; separated body contribute to carry-on external loads too and the adjectives separated and unitized should be better referred to the assembly technology.

By unitized bodies, chassis structures cannot be disassembled after the production process, while by separated bodies they can be conceptually disassembled; separated

G. Genta and L. Morello, *The Automotive Chassis*, Mechanical Engineering Series,
https://doi.org/10.1007/978-3-030-35635-4_7

Fig. 7.1 Unitized welded body shell for a two volumes, two doors sedan; the body shell is what remains of a unitized body, after removing interior trimming, doors, hood and other element not relevant to the structural function (FCA)

body technology allows the vehicle manufacturer to supply a complete running chassis to a body manufacturer.

In this paragraph we will give a short description of the body of a two door mass production two volumes sedan; we will make reference to the most diffused technology of a spot welded stamped steel sheet solution. The purpose of this description is to identify chassis structures in a unitized body.

Other solutions are present on the market, usually dedicated to niche cars, adopting different materials or joining techniques. Considering the objective of these book we will neglect them.

Figure 7.1 shows the body shell; this name is used for calling what remains of a car body, after having separated all parts without structural function, as removable parts, interiors and glasses.

Removable parts are side doors, rear or trunk door and hood. In many cars, also front fenders and the front end can be easily removed, to make their substitution easier, in case of collision damages. These components have a limited structural function, as far as body stiffness and resistance are concerned and are not considered to be part of the body shell.

The same can be said for glasses, with the exception of the windshield, when bonded to the body; in this case its contribution to the body stiffness is not negligible.

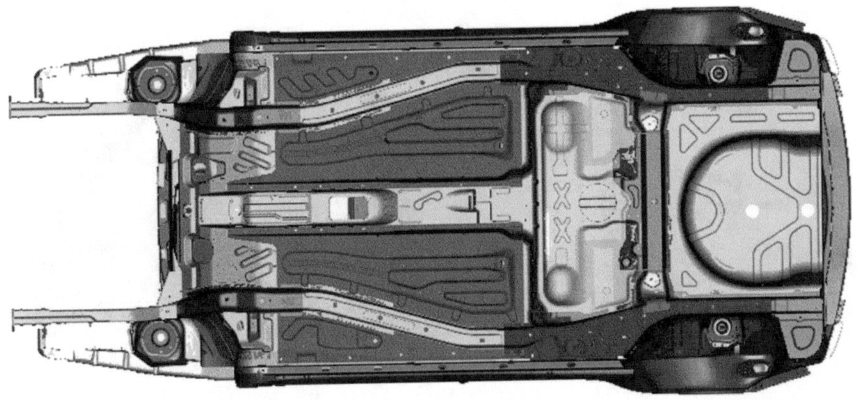

Fig. 7.2 View of the underbody of the body shell of the previous figure giving evidence of the integrated beams performing the function of the chassis structure

The body shell can be imagined as a space frame (all beams making up the frame are not lying on a single plane, as by conventional frames) with many thin panels, as fire wall, roof, sides and floor bringing a structural contribute with their shear stiffness.

A view from the bottom side of the same body shell is shown on Fig. 7.2 and puts in evidence the integrated beams performing the function of the chassis structure; longitudinal elements are still called side beams, and cross beams are the elements across. The layout of these beams is imposed by the requirement of offering stiff mounts to suspensions, powertrain, bumpers and other components of the chassis.

Their shape plays also a major role in determining the behavior of the body (deformation, energy absorption and occupants protection) in case of crash.

The underbody has a functional analogy with the chassis of an old car or an industrial vehicle; this subassembly is sometimes called *platform* and even if it is not able to perform any function without the rest of the body, it has some interesting technological and organizational feature.

In fact, this part of the body shell can be conveniently shared between vehicles that can be also very different in appearance, as two doors and four doors sedans, station wagons, minivans, coupés, roadsters, etc., as major components of the chassis with beneficial effects on investments and costs.

Having this objective in mind, modular underbodies are designed where interchangeable side beams and cross beams of different length can be assembled together to generate also platforms for different wheelbase and tracks, produced by common facilities.

A further technology feature is that this subassembly can have a production life longer as that of models based upon it.

The organizational issue, consequent to technological issue, is that the underbody can be designed and developed by a different team, together with other chassis components, as it is made now by many car manufacturers.

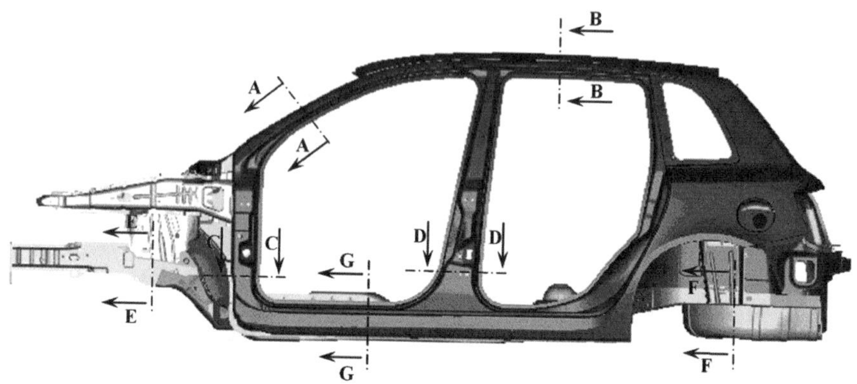

Fig. 7.3 Side view of a four door body shell showing references of the cross sections presented on the following figure

The underbody, represented on the pictures above, is made by two main longitudinal beams (which are the most important structural member of this assembly) long as the entire vehicle; floor panels are welded on them. The floor between the axles is trimmed by two additional side beams, part of the body sides.

Cross beams connecting these side beams feature a closed section that is obtained by welding an open stamped profile to the floor panel; by looking to the different section we can realize how is possible to obtain very stiff beams for the body space frame by joining thin stamped panels.

The body space frame is therefore composed by a ladder structure for the underbody, a top frame, below the roof, connected by vertical elements, the so called *pillars*; usually, are called A pillars, those ahead of front doors, B pillars, those behind front doors and C pillars those behind rear doors.

To understand how this space frame is made we can analyze the cross sections shown on Figs. 7.3 and 7.4; numbers shown on cross sections refer to same stamped elements.

Section A-A refers to the side beam of a frame surrounding the windshield integrated with the A pillar; this frame is composed by the side panel of the body 1 and by two reinforcements 2 and 3.

Section B-B cuts the roof side beam, part of the frame of the rear door; the beam is again made by the side of the body 1 and by an internal double frame 20 and 21; the roof panel 5 is also welded to this beam.

Section C-C refers to the lower part of the A pillar, where front door hinges are assembled: the beam is made again by the body side 1 and by an internal side panel 15 with a reinforcement 10; the hinge is bolted to the plate 12; to this element also the fire wall panel 16 is connected by welding.

Section D-D refers to the part of the B pillar where rear door hinges and front seat safety belts are connected; the beam is again made by the body side 1 and by the internal side 20, with a local reinforcement 21; the hinge is bolted to the plate 23.

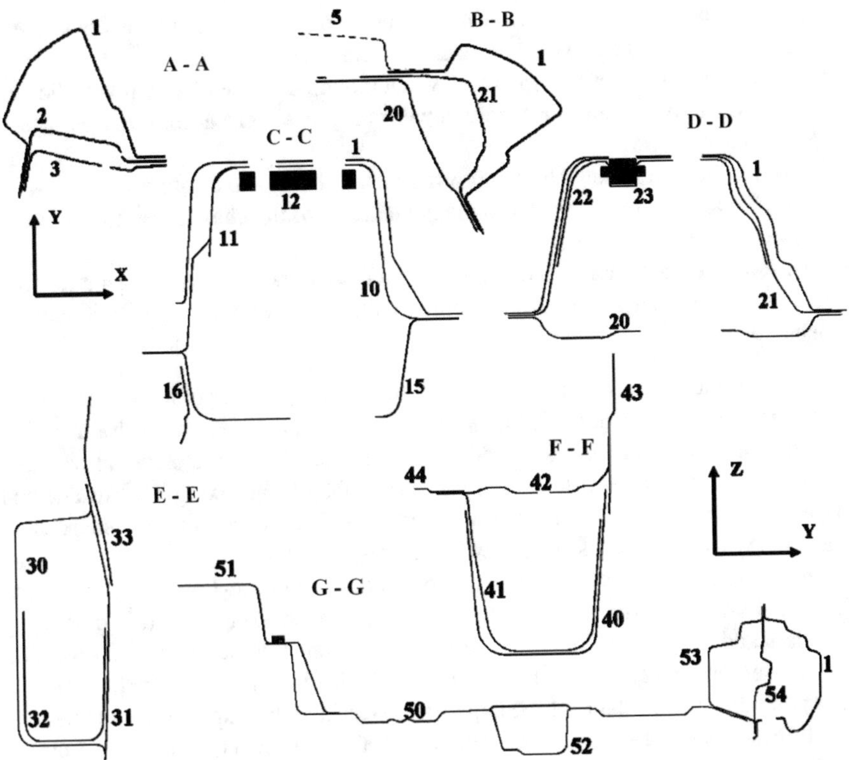

Fig. 7.4 Section A-A: vertical member of the windshield frame; section B-B: roof side beam; section C-C: A pillar; section D-D: B pillar; section E-E: front section of main side beam; section F-F: rear section of main side beam; section G-G: left half of the floor

Section E-E is cut on the front of the main side beam (see also Fig. 7.3): we can notice an inside panel 30, an outside panel 31, a reinforcement 32; to this beam is also welded the inside panel of the front fender 33, connecting to the upper pivot of the Mc Pherson suspension.

Section F-F represents the rear part of the main side beam (see also Fig. 7.3) connecting with the rear twist beam suspension; this beam is obtained by joining different panels as rear floor 44 and the wheel well 43.

Section G-G, finally, cuts the floor across and shows another section of the main side beam (element 52 and floor 50) and of the lateral side beam (elements 53, 54 and body side 1); a third side beam is shaped beside the tunnel, made by the floor 50 and a reinforcement 51.

7.2 Subframe

Subframes or auxiliary frames, on a unitized body structure, perform the following functions:

- they offer suspensions and powertrain mounts and they distribute the consequent loads to body area more suitable from a structural standpoint;
- they built up a secondary suspension, when mounted on the body with elastic elements, able to filter vibrations from powertrain and wheels, at frequencies critical to acoustic comfort;
- they contribute to manage body deformations following to crash;
- they offer an assembly support to many elements of the chassis, with benefits on work organization;
- their reduced dimensions, as compared with body dimensions, allow a better control of tolerance of suspensions mounts, with benefits on their elasto-kinematic behavior.

Subframes are often applied to the front of the car and offer mounting points for suspension lower arms, anti roll bar, steering box and part of the powertrain.

Figure 3.6, in the chapter dedicated to suspensions is a quite essential example of this kind of structure; it is made by a very stubby beam fixing the Mc Pherson suspension arms. This subframe allows to have a stiff connection for the suspension in an area where the body structure would be rather distant.

The same figure shows the anti roll bar mounts and the Fig. 3.9, in the same chapter, shows the installation of the rack and pinion steering box. It is quite easy to understand how is possible to guarantee for this small structure a high precision for all assembly holes, eventually with a multiple drilling machine. The precision of their positions is important for toe angles and steering wheel spokes alignment.

All these chassis elements are assembled on the subframe on a dedicated preparation line and this subassembly can be installed on the body in a shorter time, compatible with the main assembly line speed.

This subframe is connected to the body through two large mounts, visible in its rear area; this mounts are flexible rubber bushing, designed to filter powertrain and wheel vibrations. The remaining mounts, further of the passenger compartment are instead rigid (on the two vertical cantilevers).

The subframe is made by two stamped steel shells (see section A-A on Fig. 3.28), which are spot welded along a mounting flange. It is also possible to have a structure of this kind made with an single piece aluminum casting, with almost the same shape; the choice is bound to a trade-off between weight and cost.

The powertrain (including engine, transmission and differential) is mounted on the underbody main side beams through elastic suspensions; only a reaction rod (for powertrain reaction torque) loads the lower part of the subframe.

This kind of subframe is very diffused for front wheel driven small and medium size cars.

By larger cars a different kind of subframe is used, surrounding the perimeter of the lower part of the engine compartment, as shown on Fig. 7.5.

Figure 3.42 shown in the chapter dedicated to suspensions offers a partial view of the suspension matched to this subframe; the car is again front wheel driven with transversal powertrain. A similar solution could be also applied to rear wheel driven cars or to front wheel driven cars with longitudinal powertrain.

Fig. 7.5 Perimetrical subframe for large front wheel driven cars; the front of the car should be imagined on the left side of the picture (FCA)

With a perimetrical subframe of this kind is possible to avoid any direct connection to the body of lower suspension arms, anti roll bar and steering box even when the distance between these mounts is larger as in previous example; the powertrain too can have no direct mount on the body.

Figure 7.6 shows a view from the top of this kind of subframe (car front at right) with references of the most important cross sections.

The joining points with the body shell are set at the four vertices of the frame; sections 5-5 and 7-7 put in evidence the rubber bushings seats for an elastic suspension. The frame is a monolithic beam (see section 6-6), bent to a U shape, welded on a multiple elements cross beam, cut on section 2-2.

The monolithic beam is made by hydroforming. This process, whose implementation is relatively recent, provides that a steel tube, with no welding, is bent to the final U shape. This semi finished part is set an dye, reproducing inside the final external shape, and then pressed inside by a very high pressure fluid (about 1,000 bar); this pressure brings steel to its yield point and imparts the final shape to the beam.

The advantage of this process is rather evident; tooling is simpler as conventional stamps and the beam cross section is not weakened by welding or made heavier by flanges or overlaps. In addition, the cross section can change along the beam axis.

On the contrary the final shape has no aesthetic finishing (not important for this kind of application) and requires that the cross section perimeter is almost constant along the axis, to avoid excessive thickness decrease or rips.

The closure beam of the subframe, cut by section 2-2, is made by two shells, welded together and to the hydroformed beam through arc welding.

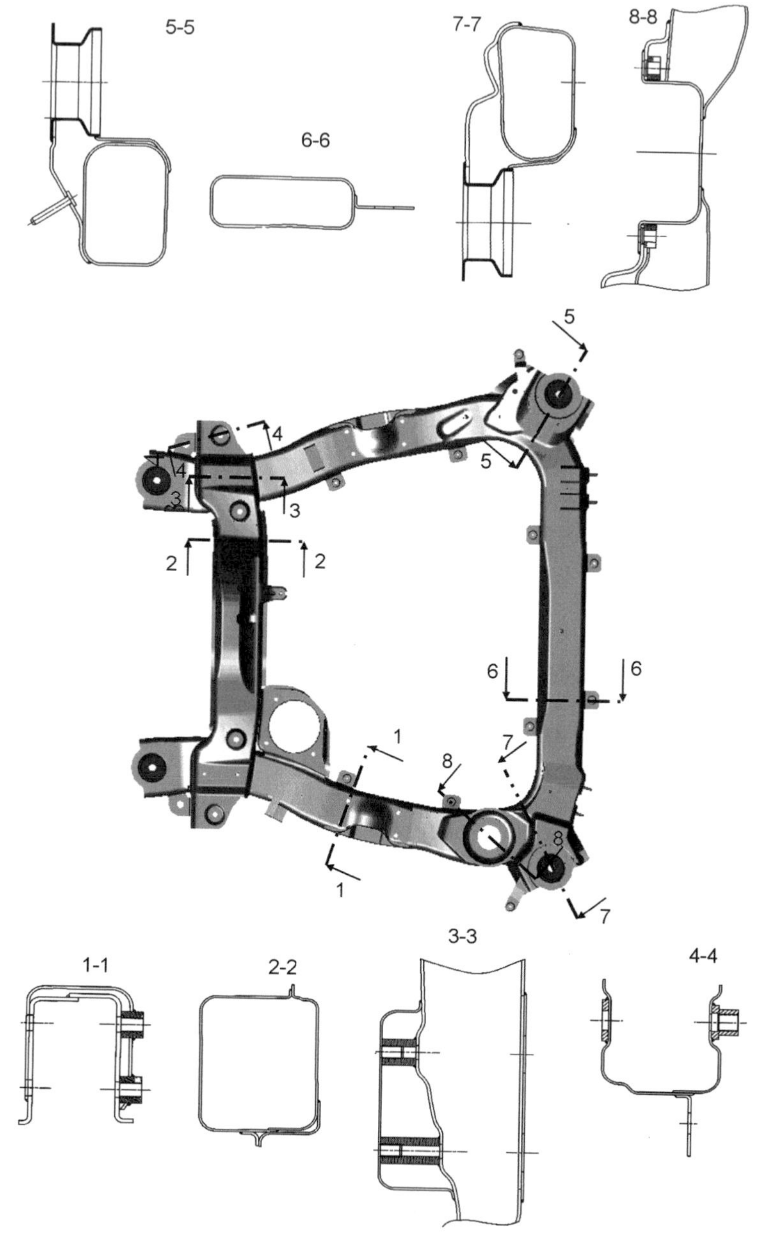

Fig. 7.6 Perimetrical subframe showing its most important cross sections

Fig. 7.7 Subframe for a multilink rear suspension for a large front wheel driven car; on the upper right image the complete assembled suspension is shown for reference (FCA)

The figure puts in evidence two powertrain mounts, for large rubber bushings; one of them is cut by section 8-8; powertrain mounts (4 in total) are all on the perimetrical subframe.

Where suspension arm rubber bushing are installed, the hydroformed beam is cut and reinforced; this feature is shown by cross sections 1-1 and 4-4. Openings allow to position arm rotation axis closer to the center of the car, with a more suitable elasto-kinematic performance.

Section 3-3 shows also one of the steering box mounts; all mounts are made with projection welded threaded bushings.

Many other holes allow a complete installation of powertrain accessories, harness and piping before of final assembly of subframe on car body.

Another example of subframe is shown on Fig. 7.7; this example is applied to a rear suspension for a large front wheel driven car.

Car front should be imagined at right.

The frame has in this case an H shape and features on the side beams of the H the mounts of suspension arms and spring and shock absorber seats.

The eyes at the tips of the longitudinal beams are used for rubber bushings installation for assembly with car body.

By small cars the rear subframe can be avoided and the suspension mounts can be directly applied to the body shell.

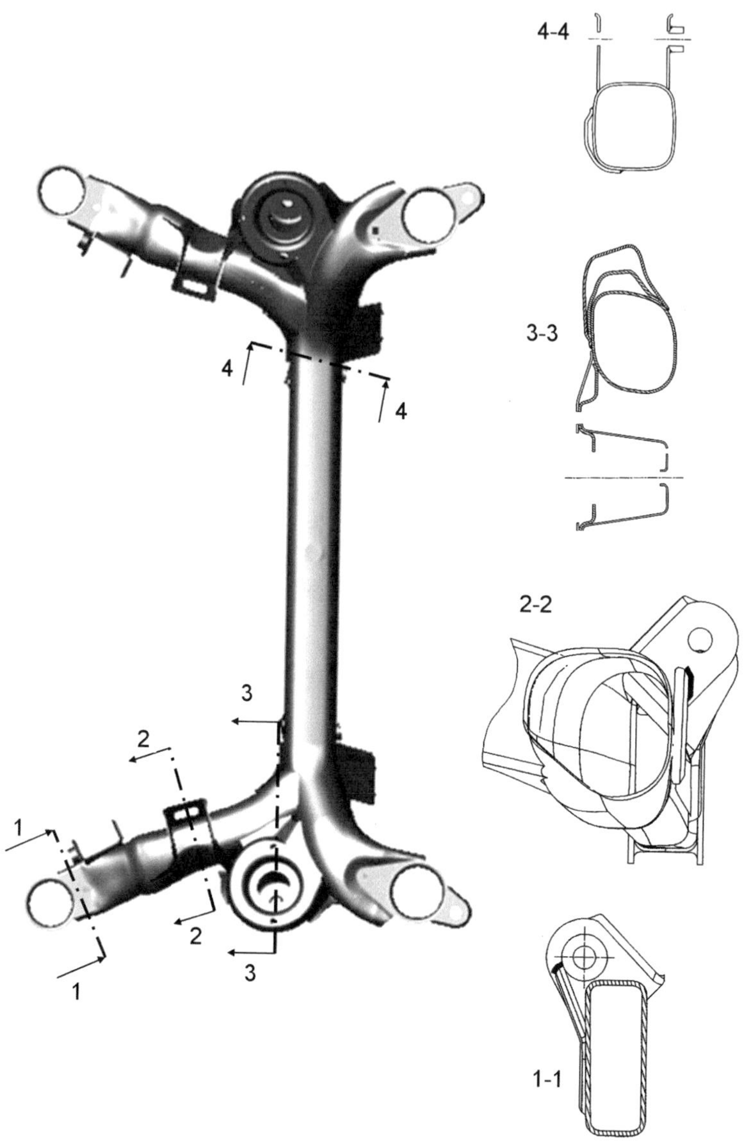

Fig. 7.8 Relevant cross sections for the rear subframe of the previous figure

By front wheel driven cars with suspended differential box, the subframe is modified to install differential and its dimension is increased longitudinally to offer a good reaction to the torque that is applied by the differential in the y direction.

Figure 7.8 shows some cross sections of the rear subframe of previous figure.

The simpler structure, as compared with the front subframe, can be obtained by bent tubes with local stamping, welded together; hydroforming is, in this case, not necessary.

A U bent tube (The U end are upwards in the figure) with two short beams compose the final shape. Section 3-3 shows an overlap area for welding the two elements; a reinforcement diaphragm is also added. The same section shows the coil spring seat.

Sections 2-2 and 4-4 shows the U shaped tube in two different points; section 1-1 shows the front mount of the lower suspension arm.

7.3 Industrial Vehicles Frames

Industrial vehicles frames have saved through the time the configuration developed for the first cars, the ladder or grillage structure, composed by side and cross members.

As a matter of fact all considerations reported in the chapter dedicated to design evolution still apply to industrial vehicles.

Bearing in mind the enormous diversification of final applications of these vehicles, body manufacturers are still dedicated specialists different than vehicle manufacturers; many bodies (as flat trucks, vans, dumpers, etc.) cannot be suitable for structural jobs, as they are described at the beginning of this chapter.

The cabin too, even if made by technologies similar to the car body have dimension too limited to carry on significant loads.

For all these reasons a real frame is provided to carry on all the loads and to connect the chassis components. Some exception is applied to buses and some tank trailer, whose study is outside of the scope of this book.

The frame is made by side members and cross members that should be fixed to the first ones as stiffly as possible; the rigidity of this fixation is responsible for the most important part of the torsional stiffness of the frame, that is the only element able to react to torques applied along the vehicle x axis.

Figure 7.9 shows an example of chassis frame for a heavy duty truck and some design details useful to understand how this structure works.

Side beams are shaped to different widths, in their top view; the narrower part is suitable to accommodate steering wheels and twin wheels, while the larger one is adapted to install the engine and the drivers cabin.

In addition, side beams are also tapered in their elevation view, to allow the suspension stroke of rear wheels, due to the large range of pay loads.

The side beam cross section is shaped like a C, even if an open section is not the most suitable one for a high torsional stiffness. Open cross sections, widely applied, can be easily bent, starting from high thickness, high resistance steel sheets. Where necessary, their cross section is reinforced with the application of additional larger C beam or by platbands.

Cross beams carry on the job of distributing local loads applied by suspensions and powertrain; an example of their layout can be seen on Fig. 3.69 and on others of the same paragraph, dedicated to industrial vehicles suspensions. In addition, cross beams contribute to the torsional stiffness of the frame.

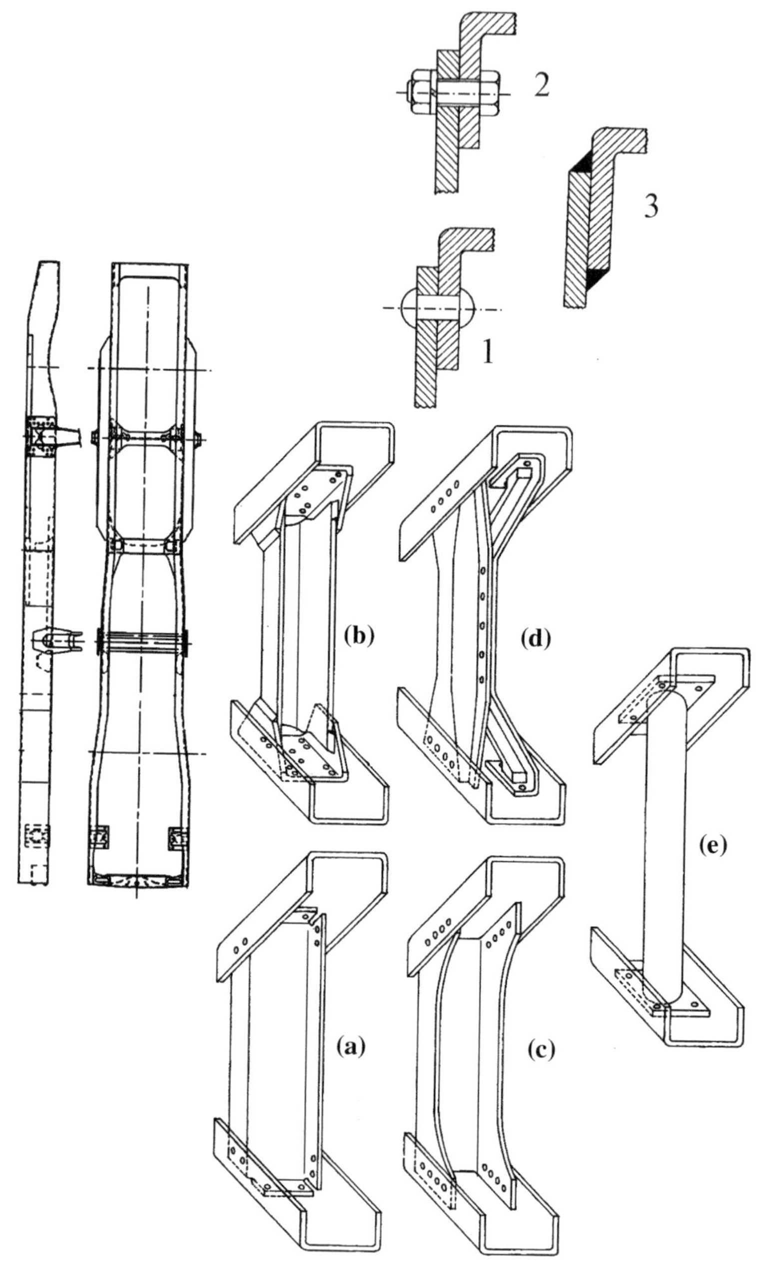

Fig. 7.9 Side and top view of a chassis frame for a heavy duty truck (at left). Details **a**, **b**, **c**, **d** and **e** show different types of joints used to connect cross with side beams. Details 1, 2 and 3 show different joining technologies, addressed to exchange high intensity shears in the side beams plane

As a matter of fact, if the two side beams were completely disconnected, the moment of inertia of the frame would be the double of that of a single section. Because this section is built up with an open profile, characterized by rectangular elements of thickness s, much smaller as their length h, we have approximately:

$$I_x = \frac{\eta}{3}\Sigma h_i s_i^3 \, , \tag{7.1}$$

where h_i and s_i are the dimensions of the three rectangles building up the side beam cross section; the parameter, in this case, is $\eta \approx 1, 12$.

Cross beams ideally constrain side beams to remain parallel and the torsional deformation of side beams is limited by the bending stiffness of cross beams, increasing the torsional stiffness of the assembly.

Cross beams can be made by a simple C profile, with their ends bent and deprived of platbands, inserted in the inside of side beams (as in figure a) or they can be reinforced locally with C profile cuts, welded at their ends, to improve the connection with side beams (as in figure b).

Further local reinforcements can be obtained by shaping the ends of the cross beams (as in figure c) or applying two Ω profiles shaped for the purpose and welded together (as in figure d).

When is necessary to reduce dimensions of cross beams, also tubular cross section can be applied, welded to flanges at their ends (as in figure e).

It should be noticed that all joints with side and cross beams are made with rivets or bolts if their are to be disassembled for repairs (details 1 and 2); this choice is caused by the large dimension of the frame. Arc welding, with its high heat introduction, would induce too large deformations and consequent residual stress, that are difficult to be eliminated by heat treatment.

Arc welding (detail 3) is limited to small components or to cross beams, during their preparation process.

7.4 Structure Tasks

Tasks of the chassis structure are substantially to bear forces and payloads and to contain consequent deformations.

Speaking of deformations, we shouldn't forget the this issue must be approached not only from a static point of view, but also from a dynamic one, considering vibration amplitudes and their consequent noise in the passenger compartment.

Some considerations on this topic are made in the second volume.

7.4.1 External Loads

External loads applied to chassis structure, during its life, can be classified under two different categories:

- instantaneous overloads;
- fatigue loads.

An example of the first category can be supplied by imagining what happens by driving of pits of large dimension, by hitting curbs of side walks, by sudden braking on high friction grounds or by starting up in low gears.

It is common practice to consider those loads as static loads, by introducing multiplication coefficients to their steady state values. For loads bound to friction between wheel and ground, prudential friction coefficient $\mu_{y,p} = \mu_{x,p} = 1.2{-}1.3$ are taken into account; as far as shocks are concerned, accelerations 5–7 times higher as the gravity accelerations are considered.

Emblematic fatigue loads can be imagined by driving on bumpy roads, with almost full loads; typical roads are those paved with cobble stones or with asymmetric waves.

The expected life of a car structure is at least 200,000 km of average use; each manufacturer has designed and built test courses that for their increased severity allow to simulate the vehicle average life in only 50–100,000 km; the completion of a mission like this requires at least few months of driving, with drivers organized on multiple shifts.

Loads recorded on these courses, that are empirically correlated to the car life, can be further synthesized, by eliminating loads under the structure fatigue limit; load histories are elaborated that allow to reproduce average life in shorter time.

Static Loads and Mass Properties

Vehicle weight is always present. The reaction to this load is coming from tires contact points with ground, through the elastic and structural elements of suspensions. Suspensions, according to their architecture, distribute reaction loads through their connections with the chassis structure, in all direction, not necessarily vertical.

We have seen, for example, in the Mc Pherson suspension, that the vertical reaction applied to the wheel, the lower arm and the shock absorber assembly generate a bending torque that is equilibrated by a lateral force applied to the chassis structure.

We should remember that unsprung masses (wheels, tires, struts, brakes and part of the mass of suspension arms, springs and shock absorbers) rest on ground directly, without interesting vehicle structure.

Vehicle weight and reaction forces distribution are function of passengers on board and of transported baggage; this set of parameters is called weight condition.

Table 7.1 summarizes axle loads and longitudinal center of gravity positions for different load condition of a medium size front wheel driven car, assuming a mass of 70 kg for each passenger, including driver. On this kind of car, passengers seating in front have their center of gravity almost at mid wheel base and their weight is shared at 50% by the two axles. Rear passengers are seated at about 80% of the wheelbase, while the baggage center of gravity is usually set at about 105% of the wheel base.

Each added mass shifts the position of the center of gravity, according to this formula:

Table 7.1 Axle loads F_{z1}, F_{z2} and center of gravity position a, measured from the front axle, divided by the wheel base l as function of load conditions, on a medium front wheel driven car. M is the sprung mass

Load condition	F_{z1} (N)	F_{z2} (N)	F_z (N)	a/l %	M (kg)
Curb weight	6,800	4,770	11,570	41.2	1,030
1 pass. (driver)	7,150	5,120	12,270	41.7	1,100
2 pass.	7,500	5,470	12,970	42.2	1,170
3 pass.	7,630	6,040	13,670	44.2	1,240
3 pass. + 30 kg	7,610	6,360	13,970	45.5	1,270
5 pass. + 50 kg	7,860	7,710	15,570	49.5	1,430

$$x_G = \frac{m_1 x_{G1} + m_2 x_{G2}}{m_1 + m_2} \,, \tag{7.2}$$

where x_{G1} is the initial position of the center of gravity of a vehicle of mass m_1 on the x axis and x_{G2} is the position of the added mass of value m_2.

The table shows also an estimate of the sprung mass, made by car body and interiors, powertrain, passengers and their baggage.

Load conditions affect also the moment of inertia of the sprung mass, useful to predict the vehicle dynamic behavior.

Considering the standard reference system x, y, z with its origin in the sprung mass center of gravity, the vehicle system is characterized by an inertia ellipsoid, whose main axis are generally not coincident with reference axis.

The vehicle inertia tensor is therefore characterized by three inertia moments J_{xx}, J_{yy}, J_{zz} and by three centrifugal moments J_{xy}, J_{xz}, J_{yz} often neglected, at least at curb weight condition.

Mass contribution of passengers and baggage modifies the moments of inertia according to the following formula:

$$J_{ii} = J_{ii,1} + J_{ii,2} + m_1 d_1^2 + m_2 d_2^2 \,, \tag{7.3}$$

where subscripts 1 and 2 refer to the original reference mass and to the added mass, J_{ii} is the moment of inertia of the added body, according to the i direction, through its center of gravity, m is the added mass and d is the distance between the center of gravity of the added mass and that of the reference condition.

With a similar formula also centrifugal moments can be calculated.

The experimental determination of the center of gravity position can be made by weighing. The ISO 10392 standard provides that, having measured sprung masses m_1, m_2, m_3, m_4 at the wheels positions (1 is front left, 2 is front right, 3 is rear left and 4 is rear right), the center of gravity position can be measured by the following formula:

$$x_G = \frac{m_3 + m_4}{m_1 + m_2 + m_3 + m_4} l \,, \tag{7.4}$$

where x_G is the distance of the center from the front axle and l is the wheelbase.

The transversal position of the center of gravity, with reference to the vehicle mid plane (positive to the right) is function of the front and rear tracks t_1 and t_2:

$$y_G = \frac{t_1(m_1 - m_2) + t_2(m_3 - m_4)}{2(m_1 + m_2 + m_3 + m_4)} \,. \tag{7.5}$$

Center of gravity elevation can be obtained by rising an axle and measuring the masses on the axles remaining on the ground m'. If the rear axle is raised, we can use the following formula:

$$z_G = \frac{l(m_1' + m_2' - m_1 - m_2)}{(m_1 + m_2 + m_4 + m_4) \tan \vartheta} + R_{l,i} \,, \tag{7.6}$$

where ϑ is the angle of rising, $R_{l,i}$ is the rolling radius of the axle remaining on the ground; an analogous formula applies when rising the front axle.

To measure moments of inertia, see the procedure described on the second Volume.

Driving Loads

Because of drivers maneuvers, the chassis structure is subject to loads, that change in time and are to be added to the static one. Control forces on steering wheel, accelerator and brake pedals cause additional loads on contact points between wheels and ground, because of side and longitudinal slip implied.

These loads again are transferred to the chassis structure by suspensions.

The final result are longitudinal and transversal accelerations that affect vehicle motion.

By modern production cars a transversal acceleration of 1 g is reached in bending at limit conditions, while about 0.5 is reached by longitudinal accelerations.

Inertia forces cause, in addition, pitch and roll angles of the sprung mass and transfer loads from one wheel to the other.

Figure 7.10 shows sprung mass transversal acceleration and some of front suspension loads, calculated for a car during an ISO overtaking maneuver. The simulation has been made by a multi body mathematical model of the vehicle, including a complete description of the quadrilateral suspensions.

Often loads are calculated from a simplified model (where suspensions are functionally described by their elasto-kinematic curves) applying then these loads to multibody or FEM models of the vehicle suspension.

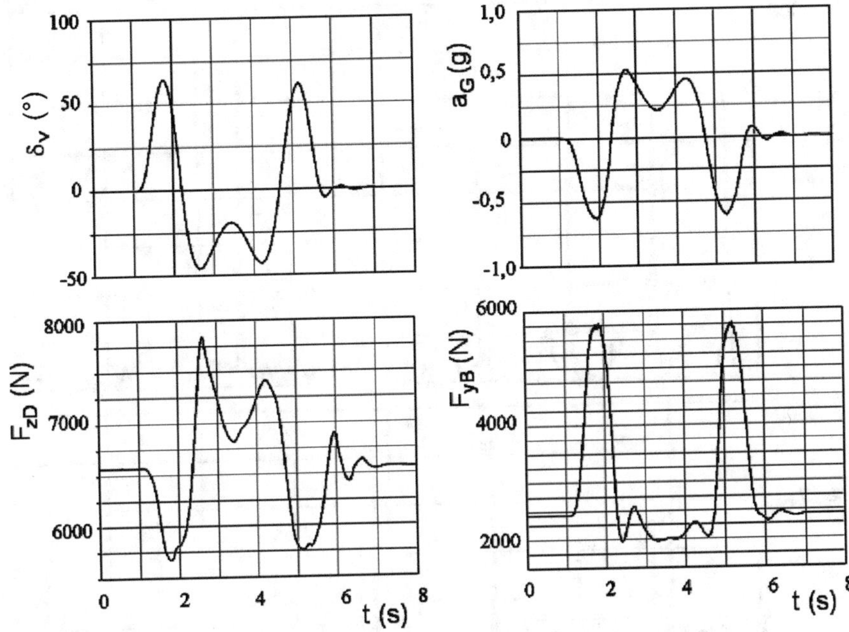

Fig. 7.10 Records of steering wheel angle (above left), transversal acceleration at the center of gravity (above right), vertical force on suspension upper pivot (below left) and transversal force on the lower suspension arm (below right) by an ISO overtaking test

Loads Due to Ground Uneven

Vehicle structure is subject to dynamic loads coming from ground uneven too. Under these forces, vehicle subsystems react dynamically and load their connections with chassis structure.

Figure 7.11 shows some results of a simulation of a car driving on an uneven road, used for fatigue life prediction.

The calculation is made by using a multibody model of the entire vehicle, including all mechanical details of suspensions and engine mounts. High level of forces are evident, either through suspension joints or engine mounts. The firsts are influenced by suspension stops, that for large displacements make suspension characteristic stiffer. Similar phenomena occur for the seconds.

7.4.2 *Internal Loads*

Vertical accelerations generate, as we have seen, inertial forces by powertrain mounts to the chassis. Also torque applied to driving wheels reacts on suspension, gearbox an differential. By front wheel driven cars all reaction torques are applied to powertrain suspensions.

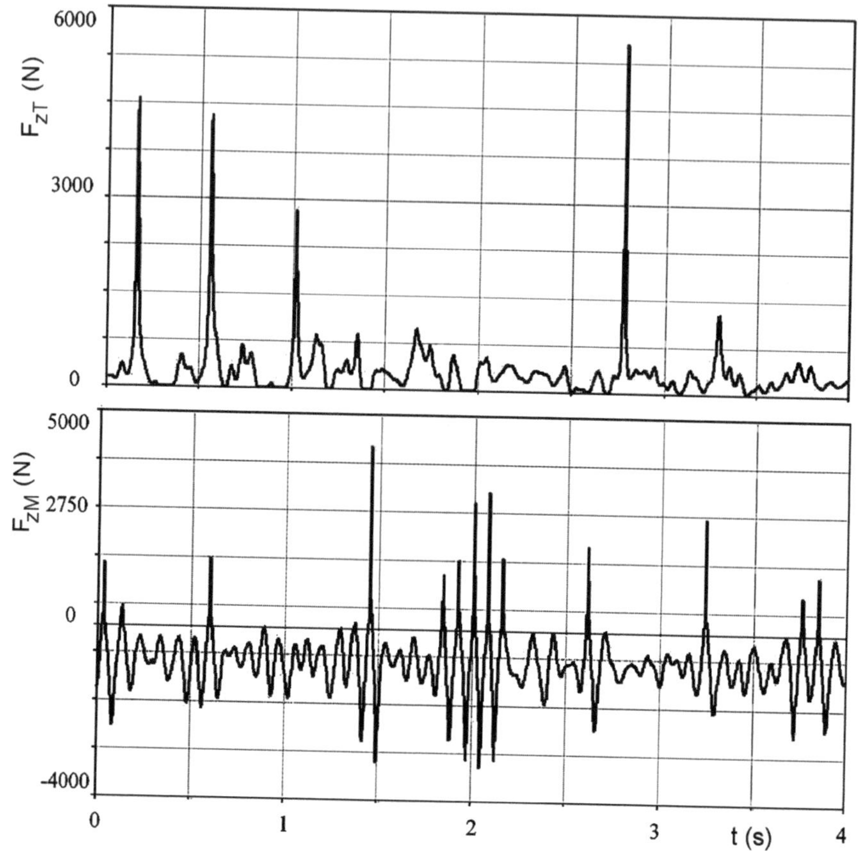

Fig. 7.11 Simulation of loads on the suspension stop of a rear axle and on an engine mount, during driving on an uneven ground

Other internal loads are due to reciprocating masses in the internal combustion engines.

The engine mechanism is made by a crank rod system, that applies a reciprocating motion to the piston; to represent forces applied to engine crankcase, we are used to decompose the reciprocating mass of the rod in two parts concentrated into the crank and piston pins (see second Volume). The masses of the two parts, together with part of the crank and the piston, define the so called reciprocating mass m_a and rotating mass m_r.

Having defined as:

Ω the rotary engine speed,
r the crank radius,

the rotation of the crankshaft will induce on each crank an unbalanced centrifugal force:

$$F_c = m_r \Omega^2 r \ . \tag{7.7}$$

By neglecting engine block flexibility, centrifugal forces will be composed with those of other cylinders, generating often a totally balanced system.

If we call with:

ϑ the rotation angle of the crank,
l the distance between centers of crank pins,
α the aspect ratio of the mechanism r/l,

each reciprocating mass will apply to the engine block a force:

$$F_a = -\, m_a \Omega^2 r (\cos\vartheta + \alpha \cos 2\vartheta) \ . \tag{7.8}$$

This formula approximates the piston acceleration with a Fourier series of two members only; also in this case forces applied to different cylinders must be composed, taking into account the different orientation of different cranks.

These are not the only moving masses; also other suspended masses, as the steering wheel or the exhaust pipe can generate internal non negligible forces.

7.4.3 Stiffness

Structure stiffness plays a fundamental role in vehicle driving and vibratory behavior.

In addition it is also important to limit deformations, because important loads may affect vehicle operation, for instance by preventing doors opening or closing or altering the kinematic behavior of suspensions.

Flexural stiffness K_f is defined by the ratio between a load applied to the mid of the wheelbase and the deflection of the same point; reaching acceptable values is usually not difficult, if other structural requirements are satisfied, but by very long vehicles.

Torsional stiffness K_t is instead the ratio between a roll torque applied to wheels hubs of front axle and the consequent rotation, when rear axle hubs are fixed to the reference system. In this ideal experience, elastic primary and secondary elements of each suspension are substituted with rigid elements of equal geometry.

To justify the importance of torsional stiffness on car dynamic behavior, let us consider the very simple model on Fig. 7.12.

The vehicle is modelled with four element with torsional deformations, respectively:

the body shell, with torsional stiffness K_{tc},
the frame, with torsional stiffness K_{tt},
the front axle suspension, with torsional stiffness k_{ta},
the rear axle suspension, with torsional stiffness k_{tp}.

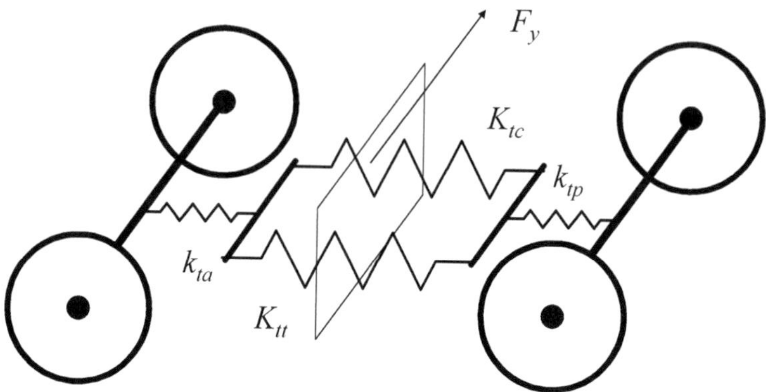

Fig. 7.12 Elementary scheme for modelling the torsional deformation of a vehicle: k_{ta} and k_{tp} represent the torsional stiffness of the two axle suspensions, while K_{tc} and K_{tt} represent torsional stiffness of body and chassis frame in an hypothetical vehicle with separated frame

In this simplified scheme, we suppose that axles are fixed to the frame on non deformable sections, as well as body shell to frame; in other words there is no other displacement but that induced by the considered elements. We assume in addition that the sprung mass is concentrated in its center of gravity and that its position is not affected by deformations.

If the vehicle is driven in a curve, if F_y is the centrifugal force, a roll moment will be applied to the vehicle as:

$$M_x = F_y h_G \,, \tag{7.9}$$

where h_G is the elevation of the center of gravity on the ground.

The roll torque must be equal to the elastic reaction of axles and of two structural elements of the vehicle, given by:

$$M_x = \varphi \left[\left(\frac{2}{K_{tc} + K_{tt}} + \frac{1}{k_{ta}} \right) + \left(\frac{2}{K_{tc} + K_{tt}} + \frac{1}{k_{tp}} \right) \right] , \tag{7.10}$$

where φ is the roll angle.

This formula considers the configuration in series or in parallel of the deformable elements: those in parallel as body shell and frame add their rigidity, while those in series, like suspensions and vehicle structures add their flexibility.

We have seen the importance of load transfer in determining cornering stiffness of an axle and we can argue, even if not explain yet, the influence of roll stiffness of the two axles in determining the understeering behavior of the vehicle.

As a matter of fact, the axle that is going to absorb the most important part of the load transfer will be subject to larger side slip angles at the same value of cornering forces; in our example the ratio between load transfers is given by:

$$R_t = \frac{\left(\frac{2}{K_{tc}+K_{tt}} + \frac{1}{k_{ta}}\right)}{\left(\frac{2}{K_{tc}+K_{tt}} + \frac{1}{k_{tp}}\right)} . \tag{7.11}$$

This ratio is controlled by suspension roll stiffness of both axles (ratios k_{ta} and k_{tp}) only if frame and body shell rigidities are much larger, as to make negligible first members of the numerator and of the denominator defining R_t.

In other word if the vehicle structure is not sufficiently stiff the understeering behavior will be determined by this last, instead of suspensions and anti-roll bars.

If we suppose now to cross with our car an asymmetric obstacle, as to impose, for instance to the front axle a roll angle φ, because of this fact a torque M_x will be applied to the vehicle as to satisfy the following condition:

$$M_x \left(\frac{1}{k_{ta}} + \frac{1}{k_{tp}} + \frac{1}{K_{tc} + K_{tt}} \right) = \varphi . \tag{7.12}$$

The torsion angle applied to the structure is given by the term in brackets; since frame and body shell must rotate of the same angle for their congruency, the ratio R_c between the torques insisting on the two elements is given by:

$$R_c = \frac{K_{tc}}{K_{tt}} . \tag{7.13}$$

The stiffer element will carry on the larger portion of the torque; stiffness and resistance must therefore be proportioned.

A very stiff body shell should be also well resistant; this explains why wooden bodies on steel chassis frame have been made with flexible joints in their skeleton or we can understand why, by a unitized body, a roof can crack because of a too flexible floor.

We can assume that a reasonable bending stiffness target should be set within the interval of 700–1,000 daN/mm, while a reasonable torsional stiffness target between 70,000–150,000 daNm/rad.

7.5 Structure Design

It is not our intention to supply in this paragraph a complete knowledge about vehicle structure design; as we have seen for modern cars, underbody and body shell are completely integrated and technical disciplines necessary to afford the design process are part of the body development teams.

On the contrary, we want to supply some highlights on methods to be easily applied already by concept development useful to predict the consequence on the vehicle structure of the application of a suspension or of a subframe.

The objective of this approach is to understand in advance if an assigned chassis component is compatible with the rest of the vehicle.

To be able to predict structure performance of a vehicle is important to verify the feasibility of the project targets.

The numeric analysis methods today available, together with the always increasing calculation capacity of computers are perfectly adequate to this purpose, but their need for detailed mathematical models of the structure makes their application difficult during the preliminary design phase, because many information they need will be only available later on. Vehicle structure is perfectly adherent to body visible shape, that is frequently modified, because body style is under development or many competing shapes are under development in parallel. In addition, aerodynamic performance optimization, to be also performed in parallel to this phase, introduces frequent shape changes.

During preliminary design of a vehicle, synthesis capacity and decision speed are essential.

This implies to be able to identify a set of critical performances to be predicted and taken under control within the project targets, starting from informations superficial, incomplete and frequently modified.

For this limited purpose we will introduce the *structural surface* method and the *beam model* method, partly interchangeable in this kind of application; the second one is particularly useful for separated frames or for industrial vehicles frames.

7.5.1 Structural Surface Method

The body shell structure can be idealized, to study its performance during preliminary design, with a system composed by beams shaping up a spatial grid and by closure panels; the contribution of these panels to the global structure behavior is quite important.

The structural surface concept introduces many assumption to simplify the model to the essentials.

A structural surface is an elementary flat panel, that because of its limited thickness can accept loads contained in its mean plane only; loads directed to other directions cannot be withstood because of its high flexibility.

The simplest structural surface is given by the rectangular panel shown on Fig. 7.13, defined by two sides of dimensions a and b and by its thickness s, which is assumed to be negligible with respect to other dimensions. For such panel the inertia moments of the cross sections are:

$$J_x = a\frac{s^3}{12}, \quad J_y = s\frac{b^3}{12}, \quad J_z = b\frac{s^3}{12}. \tag{7.14}$$

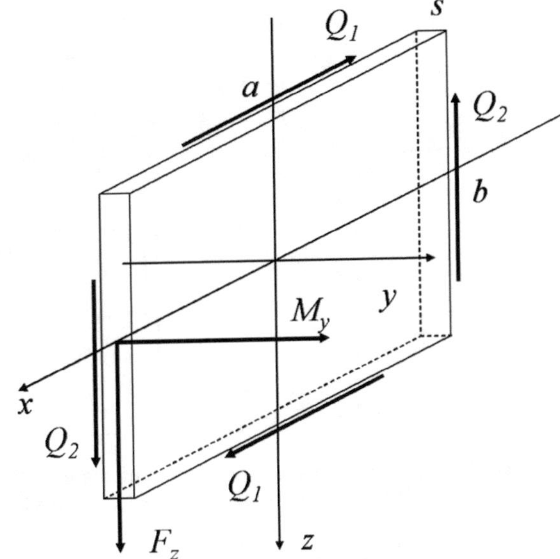

Fig. 7.13 Very simple structural surface composed by a flat rectangular panel

Because of our assumption, J_x and J_z are much smaller than J_y and the structural surface so defined will be able to withstand only bending moments along the y axis and forces parallel to the sides.

For example, the application of a force F_z to one side of this panel can be balanced by two shear forces Q_1 and Q_2 applied to the two sides, as to satisfy the equation:

$$\begin{aligned} Q_1 b - Q_2 a &= 0\,, \\ F_z - Q_2 &= 0\,. \end{aligned} \tag{7.15}$$

But in consideration of the limited thickness, these shear forces can be really applied if there are other constraints, able to limit shape instability of this thin surface on the compressed side (the lower one in our example). This constraint can be another panel, laying on a perpendicular plane, welded to the first or by a beam, probably shaped up by a closed section with the near panel.

The entire body shell can be modelled with a set of plane structural surfaces, which approximate its curved surface. The contribution of the beams can be also neglected, by retaining only their role for avoiding wall instability.

Panels can be made with different shapes, but in any case they are characterized by their capability of reacting with forces contained in their plane only.

Figure 7.14 shows some useful example for the following applications. The next Fig. 7.15 shows instead an example of a schematic model of a three volumes sedan.

A car body structure can be modelled for a simplified structural analysis as a combination of structural surfaces; now we can look at the consequences of these assumptions.

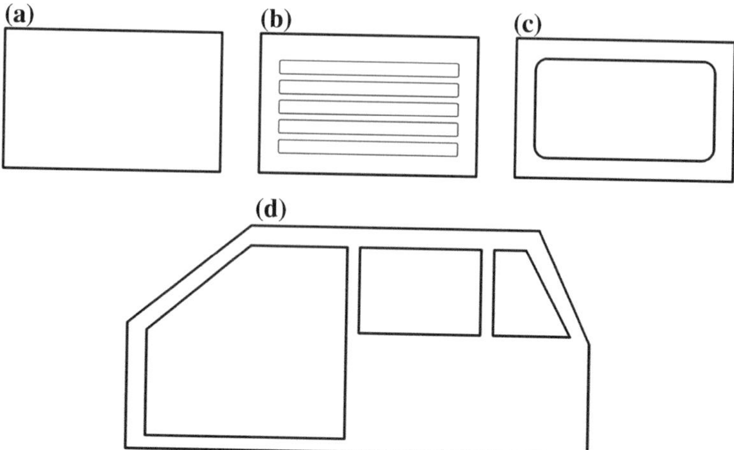

Fig. 7.14 Examples of panels for a structural surfaces model: **a** rectangular panel, **b** panel with reinforcement grooves, **c** panel with lightening hole, **d** door frame panel

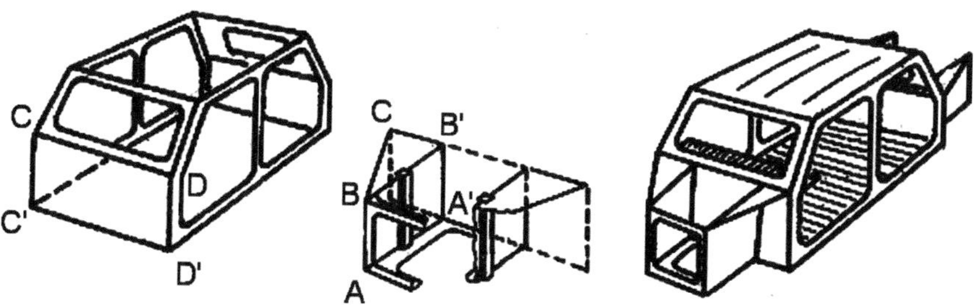

Fig. 7.15 Many kind of panel, necessary to build up a model of a two volumes four doors sedan

Let us consider a roll moment applied to the body shell of Fig. 7.15, at the front axle position; this moment can be balanced by two shear forces acting on panels simulating the interior walls of the front wheels wells.

For example the panel AA′BB′ receives a force contained in its mean plane and directed upwards; as for the panel in Fig. 7.13, it is possible to calculate the shear forces that must be applied to its sides by adjoining panels.

The shear reaction on the side of this panel will load the near CC′D′D panel and, from this, its near panels. It should be noticed that the neighboring triangular panel BB′C cannot bring any contribution, because is perpendicular to the force direction, as the roof panel or the floor panel.

A similar logical process can be applied to all forces that front suspensions apply to the body structure, through the front panel AA′B′B.

If we consider a lateral force applied to the suspension, because of a bend, of because of their arms inclination, we must conclude that the panel AA′B′B isn't able to react properly; in this case, the triangular panel BB′C will do the job.

Similar considerations can be made if we consider a concentrated load applied to the car floor, as a seat mount or a rear suspension arm. The structure will be able to react only if in the load application point a panel with its mean plane directed parallel to the force is applied.

This explains why floors must show also vertical panels (as side walls of side beams and lateral side beams) where loads are applied.

Even if resulting of rough approximations, this rule must be followed meticulously in order to design a good structure.

The structural surface method has the following advantages:

- allowing the behavior evaluation with simple calculations in two following steps;
- enabling design modifications of a single surface without necessarily affecting the neighboring ones.

The first step consists in defining loads on the sides of the structural surfaces. These loads can be kept unchanged, as a first approximation, even if their surface is going to be modified; this can speed up evaluations when design is changed frequently.

The second step consists in determining shear stress on each surface, where the previously defined loads act.

The many subassemblies as vehicle sides, with door frames, roof and floor can be studies separately as plane problems.

7.5.2 *Beam Model Method*

Beams are mathematical objects that allow a simplified description, in terms of stress and strains, of the vehicle structure under the action of external loads and constraints.

By structural analysis these mathematical models are defined according to the finite elements method (FEM). If we assume limited displacements, we can accept the usual relationship between forces and displacements:

$$\mathbf{K}\mathbf{u} = \mathbf{f}\,, \tag{7.16}$$

where:

K is the structure stiffness matrix;
u is the displacement vector;
f is the vector of forces applied to the structure.

By preliminary design the mostly applied numeric models are beam FEM models. In fact the beam is the simplest element: a segment joining two nodes with six degrees of freedom, to which the characteristic features (section area and moment of inertia) of a body shell closed section can be attributed. Examples of beams of this kind of approach can be designed from cross sections on Fig. 7.4.

The beam element formulation and the De Saint Venant theory are coincident.

By using elements of this type the structure can be described by a space frame made of beams only to them the characteristics of cross sections joining the nodes are assigned.

This approach is similar, even if more complex, to what is used to describe plane frames as those used by industrial vehicles or, as subframes, by cars.

A beam model of a car structure is composed by:

- beam elements describing the structure: they join the nodes in the space as elements with 12 degrees of freedom (6 for each node) and are characterized by appropriate sections and bending and torsional moments of inertia; panel contributions are neglected or included in the beam;
- nodes joining the beams that can be simulated by rigid or flexible joints.

During preliminary design of a new car, the informations available for a structural analyst are very few.

Usually only wheel base, tracks, suspensions and body shell components shared with already existing cars (carry-over); in some case the underbody is entirely carried-over.

In addition, a style development center has developed shapes for the external volumes, for openings cuts (doors, hood and hatch) and glasses; this informations are usually included in a mathematical model implemented in a computer aided styling system (CAS).

A beam model can be synthesized starting from:

- a CAS model on whose surface and opening cuts a structural space frame is adapted;
- existing models of carried-over elements;
- structural cross section; if missing they can be scaled from already existing cars of the same segment or from a data base prepared for the purpose; these informations can be again retrieved from drawings as those of Fig. 7.4.

The steps of this activity are the following.

- Repositioning existing carried-over structural models on the new style model.
- Adapting carried-over parts to the new dimensions.
- Identifying nodes positions of the new structure.
- Deriving from style surfaces beam center lines.
- Describing areas and moments of inertia of beam cross sections.
- Identifying relevant loads and constraints.

The result of this process is described on Fig. 7.16 for a small car body. The versatility of this approach is easily understood.

- A change in style can be introduced into the model by laying down beams below the new surface.
- Cross sections can be easily modified by their description tables.
- A node stiffness can also easily modified by their description tables.

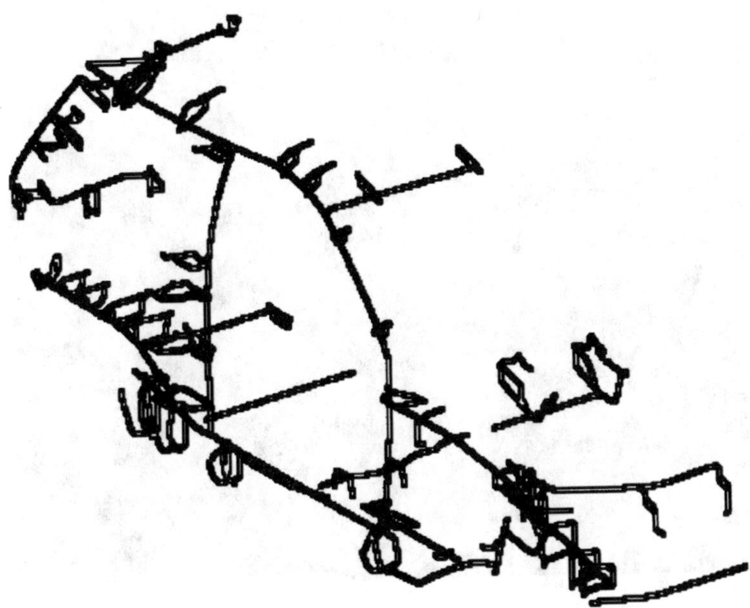

Fig. 7.16 Final scheme of a beam model; the shape of the beams is determined by lying down them below the styled exterior shape and by assigning to them compatible cross sections, scaled from existing data bases

A synthesis beam model allows, in addition, to draw first conclusions on how different parts of the body cooperate to the final result and to perform a first weight optimization. By scaling up dimensions of each section at a time it is, in fact, possible to affect system performance and to find out iteratively the most effective dimensions.

The beam model method assumes an opposite approach as the structural surface method, by neglecting the role of panels.

The two methods should not be considered as in competition to solve the same problem.

The first, for unitized bodies only, can be applied for concepts development, regarding the general architecture of the body. It is very simple and allows to consider many different alternatives in short time.

The second one consists in a better approximation and can be applied after the first for optimizing an already defined structure. As a difference with the first, the second method is sensitive to exterior shape modifications too. Finally, the second method is better suited to frames and space frames.

7.6 Structure Testing

Test benches used for structure fatigue evaluation are similar to those introduced while speaking of suspensions; for economy and considering that real suspensions are the simplest tool to apply loads to structures, in an accurate way, usually body shell, frames, suspensions and powertrain mounts are tested together.

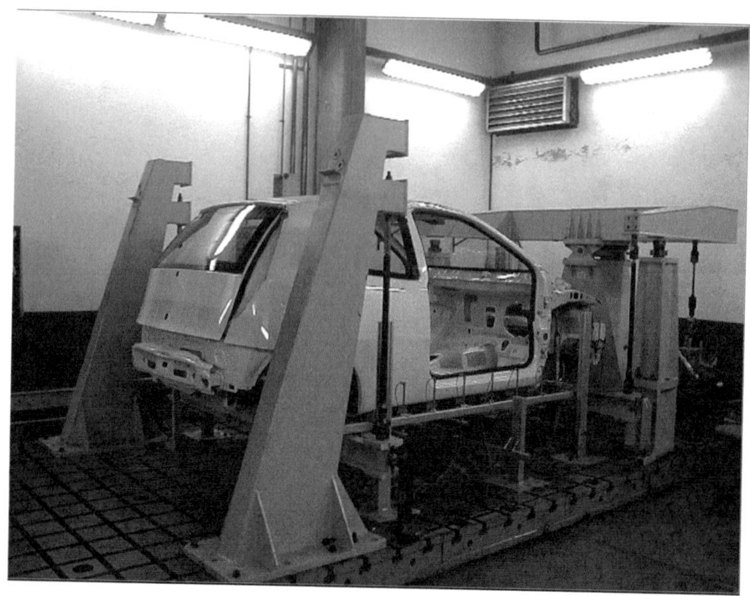

Fig. 7.17 Universal bench prepared for torsion test; the articulated frame in the front is used to apply a roll torque

The most important test, specific to chassis structure and to body shell is the overall evaluation of torsional stiffness.

In addition to what we already commented about inconveniences of an insufficient structure stiffness, we would like to point out that the analysis of local deformations by bending and torsion tests is useful to find out discontinuities on the structure function of the vehicle.

In fact, areas where deformation discontinuities appear evident represent potential rupture points during the on the road operation, because of the cyclic application of deformations by road obstacles.

We now concentrate on torsional and bending stiffness testing.

The torsion test simulates what happens when a vehicle steps on a side walk, while the bending test is oriented to what happens because of payloads application.

Both tests are performed on the same bench that is reconverted according to the purpose of the test.

The test bench (shown on Fig. 7.17 for torsion and on Fig. 7.18 for bending test) is made by a basement block characterized by a stiffness well above the values to be measured.

On this block additional structures are installed to fix the body or the frames properly; other mobile structures are use to apply loads.

In case of a unitized body, it is deprived by interiors (dashboard, seats and side panels), complete powertrain (engine, transmissions, exhaust pipe, etc.), bumpers and wheels. Suspensions are maintained, but springs and shock absorbers are changed with rigid spacers, in order to freeze the geometry under investigation.

Fig. 7.18 Universal bench prepared for bending test; the beam in the center is used to apply a local vertical load

It is common practice to verify the contribution to the total stiffness of removable parts (side doors, hatch door, trunk lid, hood, removable fenders) and of structural glasses (bonded windshield, bonded rear window glass) by testing the structure with and without their contribution; this allows also to understand if joining techniques are suitable for the purpose.

Constraints are similar for the two tests: wheels hubs are welded or fixed by brakes and are used to connect the structure to cross beams used for interfacing block and load application frames.

By torsion test, the rear axle is linked to the block while the front axle to the torque application structure; both links avoid longitudinal and transversal motion.

By bending test both axles are fixed to the block, but one is free for longitudinal motion.

Loads are applied by these two tests in a different way.

By torsion test, a pure torque is applied to the front axle through an articulated frame.

By bending test a vertical load is applied to the center of the structure, by an auxiliary beam.

In both cases precautions are taken to avoid the influence of local deformations where loads are applied.

Displacements are measured by linear transducers with vertical axis.

By torsion test a couple of transducers is placed for each suspension upper pivot.

By bending test one transducer is placed in the mid of the car and others are used to check local deformations.

Torsional stiffness is conventionally measured by the following formula:

$$K_t = \frac{M_x}{(\varphi_a - \varphi_p)} = \frac{M_x}{\arctan \frac{z_{da}+z_{sa}}{t_a} + \arctan \frac{z_{dp}+z_{sp}}{t_p}}, \tag{7.17}$$

where:

K_t is the torsional stiffness,
M_x is the applied torque,
φ is the roll angle at the two axles,
z is the vertical position variation of each wheel,
t is the track.

Subscripts refer to:

a front axle,
p rear axle,
d right side,
s left side.

To compare different cars, often torsional and bending stiffness are divided by wheelbase.

During this kind of test, also local deformations are recorded.

Many linear sensors are applied along main side beams and lateral side beams, in order to measure vertical displacements. In case of torsion, they are used to calculate local rotation angle of structure.

These displacements or these rotations are used to draw diagrams in function of the coordinate x of the body, to supply an indication of the stiffness distribution. Local variations of the slope of this curve, that can be detected through the first derivative diagram, put in evidence local critical areas (as local instability of panels where sensors are applied) or structural discontinuities, due to a not accurate welding or to design mistake.

During torsion test it is important to measure also deformations of frames, where doors and glasses are applied: usually the difference between the diagonal lines of the opening is taken as an indicator to be compared with project specifications.

Total stiffness is changing when parts are added to the body shell; usual references include body shell, body shell with fixed glasses, body shell with doors.

Table 7.2 Torsional stiffness K_t (daNm/rad), bending stiffness K_f (daN/mm) and masses M (kg) of body shells of two volume small cars of different years of commercial launch; the second subscript shows body conditions (n body shell, v with fixed glasses, c complete). The contribution of glasses to torsional stiffness can be noticed

M. Y.	$K_{t,n}$	$K_{t,v}$	$K_{t,c}$	$K_{f,n}$	$K_{f,v}$	$K_{f,c}$	M_n	M_c
1990	34,200	40,400	47,800	430	445	475	152	207
1995	42,800	57,900	68,300	580	600	620	193	271
2000	57,300	70,700	79,600	630	640	670	190	264

Table 7.2 shows some values of torsional and bending stiffness measured for bodies of different model years. They are small two volumes cars.

Fixed glasses are mounted on the body shell to examine their contribution; they are the windshield and the rear windows that cannot be opened; their contribution is due to the bonding joint, that limits the window frame deformation on the body. This contribution is negligible for bending stiffness, because of the limited deformation of glasses frames.

The body weight increase is due, in part, to increased dimensions, in part to more severe passive safety regulations.

Part II
Transmission Driveline

Introduction

Any kind of road motor vehicle needs a transmission, in order to convert the rotational speed and the torque of its engine.

Under the name of transmission are usually included start-up device (either clutch or torque converter), variable ratio gearbox, power takeoff used to operate equipments that are external to the vehicle, final drive, and differential that move the wheel; in other words, it includes the entire kinematic chain connecting wheels to engine.

The transmission function is to adapt the torque available of the engine, to the vehicle needs, which are imposed by the nature of the road, the will of the driver, and environmental requirements. The transmission is particularly relevant to determine some of vehicle functions, such as dynamic performance, fuel consumption, emission, drivability, and, last but not least, reliability.

Automotive transmissions (we include under this definition, cars, industrial vehicles, and buses) are characterized, in comparison with industrial transmission, by a higher level of stress. The specific mass (the ratio between the transmission mass and the processed power) of an automotive transmission is included in a range between 0.5 and 1 kg/kW; in an industrial transmission, same values are almost doubled. In addition, the number of available ratios of an automotive transmission is definitely higher than in an industrial one.

Transmission technology could be assumed as mature, but we don't mean that there will be no further evolution, particularly as automatic transmissions are concerned. As a matter of fact, a significant growth of this kind of transmission is expected in the European market. The particular needs of this market have justified the development of a new generation of robotized manual transmission that should not negatively affect the vehicle performance and should leave drivers the pleasure of driving manually at their will.

The fast and continuous development of electronics, both in terms of performance and cost, will have a significant impact on this kind of transmission and will cause a fast development.

We should also not forget that the possible growth of hybrid vehicle market will stimulate development of particularly sophisticated automatic transmission.

A last point relevant to transmissions is that their production and development capacity is leaving OEMs to become a specialists business.

This is true both for manual and automatic transmissions. It is particularly important both for those that will devote their career to vehicle design or to component design, to develop sound system knowledge of this kind of component.

This trend is confirmed by the growing importance in Europe of Getrag, ZF, GM, FCA Powertrain, and Graziano; the same could be said in the United States for Dana, Eaton, New Venture Gear, and Allison and in Japan for Aisin and JATCO.

A significant growth of automatic transmissions is expected in Europe.

The most diffused kind of this transmission is the power-shift type, with four or more transmission ratios; many manufacturers have decided to adopt on their car a continuously variable transmission, with steel belt and variable gouge pulley; beside some penalty on mechanical efficiency, they show a superior vehicle performance for the availability of an infinite number of transmission ratios.

Power-shift transmissions are leaving the pure hydraulic control, in favor of more sophisticated electronic controls that take advantage of the contemporary development of engine control system, including throttle automation.

On the other side, many applications with more than five transmission ratios were born and the lock-up function of the torque converter is extended to almost all gears.

Six-gear manual transmissions are starting to be diffused in the market, while five ratio configuration is standard on all applications; many of these transmissions have received an electro-hydraulic or electro-mechanic actuation system, with the aim to offer costumers a similar or superior performance, in comparison with the conventional automatic transmission, without affecting the traditional advantages of lower cost, high efficiency, and pleasure of driving.

The functions that can be obtained by this kind of transmission include the following:

- Clutch automation.
- Servo assistance in selecting and shifting gears, with positive effects on comfort and actuation speed.
- Availability of more ergonomic shifting commands of sequential type, also included on steering wheel.
- Total automation of the choice of the transmission ratio and of the shifting sequence.

Many other functions could be added, in consideration of the almost unlimited performance of microprocessors.

Transmission architectures are determined by the kind of traction that is adopted on the vehicle. On the other side, the choice of the kind of traction has a strong impact on the vehicle performance, in terms of handling, ride comfort, safety, and interior space organization.

The engine can be installed in the front or in the back of the vehicle; in the first case, the traction can be on the front axle, on the rear axle, or on both axles; in the second case, the traction can be on the rear axle or on both axles. Rear wheel driven cars, with longitudinal engine in the front, are usually called cars with conventional or traditional layout.

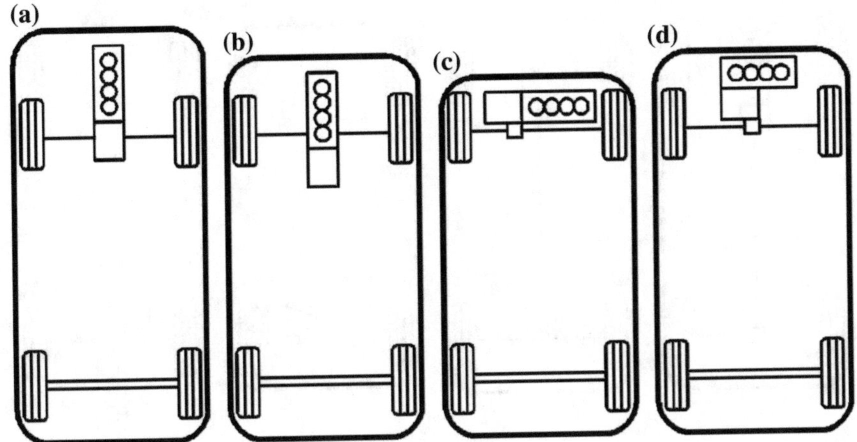

Fig. II.1 Different types of transmission architectures, available for front wheel driven cars; the impact on external dimensions is deliberately exaggerated

Figure II.1 shows all known configurations available for the front wheel drive.

Sketch a shows the longitudinal layout, with powertrain on driving axle; it also shows how some advantage can be obtained on the maximum steering angle of the wheel with some disadvantage on the front overhang. In this case, every transmission ratio can be obtained with a single stage of gears; the rotation axis of the gearbox output shaft must be turned of 90° with a pair of bevel gears.

Sketch b shows the same previous layout, where the front overhang is shortened by installing the engine over the front axle; in this case, an additional shaft and some modification on the oil sump are needed.

Sketch c shows the very highly diffused solution of the transversal powertrain with engine and transmission in line. Also, in this case, a single stage transmission is necessary; the output shaft is connected to the differential with a pair of spur gears. Vehicles can be more compact because the front overhang can be shorter; on the other side, the minimum turning radius is negatively affected. The solution is optimum if the engine is not too big.

Finally, the sketch d shows a not widely applied solution, where engine and gearbox are transversal and parallel; the gearbox can be in the front behind or beneath the engine. Advantages and disadvantages are in between, as compared with the previous case; an additional double stage transmission or a chain transmission is necessary to connect the engine shaft with the gearbox input shaft.

In all these cases, the subsystem including engine and transmission is called *powertrain* or *transaxle*.

Figure II.2 shows two possible architectures applied on conventional vehicles; the type a is the most diffused one, while the type b is applied primarily on sport cars, when a more uniform weight distribution is requested. In these cases, the gearboxes will be of the double stage type (also known as countershaft type), where input and

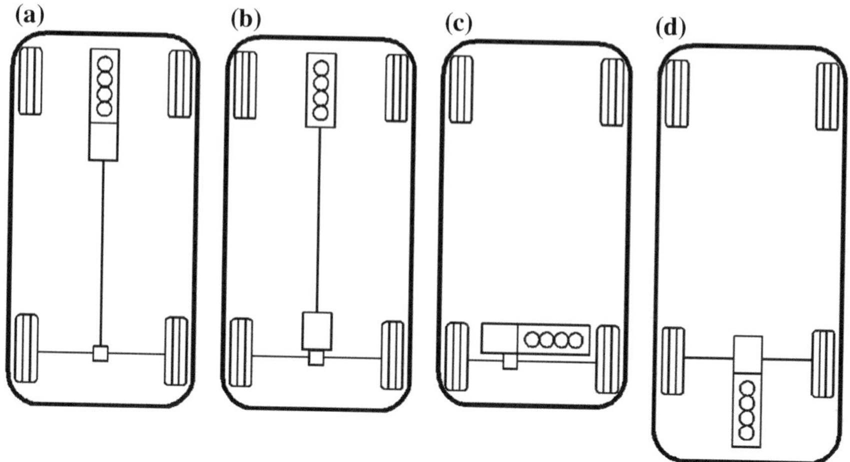

Fig. II.2 Different types of transmission, suitable for rear drives, with front or rear engine

output shafts are in line. A bevel gear final drive is also necessary. The front overhang can be reduced; the interior roominess is negatively affected by the transmission shaft under the floor.

Also, the rear engine layout, that is applied very seldom, shows two possible architectures, shown on the same figure, with transverse powertrain arrangement (c) and with longitudinal arrangement (d).

The kind of architecture is now only applied on sport cars, where weight distribution and yaw inertia are more important than interior roominess; the same considerations made on solution a and c of the previous figure apply on solutions c and d.

Architectures applied to commercial vehicles (people or goods transportation with GVW of less than 4t[1]) derived from car architectures d of the first figure and a in the second figure.

Figure II.3 shows the architectures that are known for four wheel drive vehicles; configurations a, b, and d derive from the corresponding layouts of the single axle vehicle. They are primarily applied on road vehicles or sport utility vehicles with permanent four wheels drive; solutions a and d are easy to be designed, because the transfer box can be integrated into the powertrain. Architecture c is, instead, applied on vehicles specialized for off-road mission; the transfer gearbox necessary to move the other axle can easily integrate a range-change unit and the differential lock.

Considering transmission configurations used on industrial vehicles (people or goods transportation with GVW of more than 4t), we must discriminate between trucks and buses, taking also into account that a low price market segment exists, where bus is strictly derived of a truck.

[1]This limit value is not coincident with the legislative one (3,5 t), but it is our guess about what is technically feasible with vehicles designed with car technology.

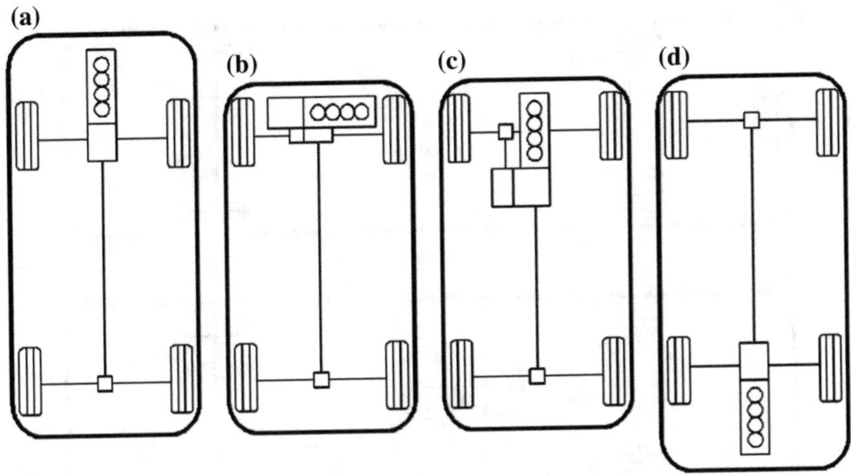

Fig. II.3 Transmission architectures applied to four wheel drive; the solution c is particularly suitable to off-road vehicles, because an additional range-change unit can be easily integrated into the transfer box

Architecture configurations for two-axle industrial vehicles are very similar to those described for conventional drive. The longitudinal powertrain is installed in the front in a position depending on the type of driver cabin that is adopted. In the same way, the four wheel drive configuration is similar to what explained is in Figure II.3c. In very heavy duty applications, a second rear driving axle is moved through the first.

Buses, initially derived from trucks, have recently received a dedicated architecture, finalized to assure an easy passenger access to the vehicle. This last depends on the height of door threshold; the desired value for this dimension has also differentiated the architecture of urban and suburban buses.

In the first case, the easy access must be obtained with no compromise for all doors and the corridor floor must not show steps or important ramps, because of the necessity of transporting standing passengers.

Figure II.4 shows two different examples; a represents a typical layout for urban buses, where b represents a typical layout for suburban buses. In the first, the engine and the transmission are installed in the back, under the last row of seats. The corridor floor is very low, and is only conditioned by the axle's shape, that is usually developed for this purpose; the slope of the corridor is limited and is also possible to have doors in the back of the vehicle.

Suburban buses or long range buses, represented in b, don't show a high need for a lower door threshold; on the contrary, they usually have the baggage compartment under the floor. Standing passengers are seldom; the most suitable layout has longitudinal central engine and rear wheel drive.

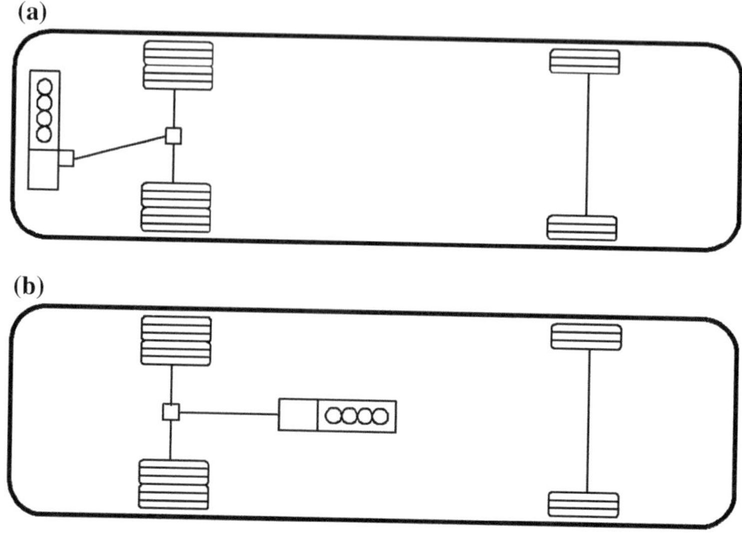

Fig. II.4 Transmission architectures designed for urban buses (high) and suburban buses (low); the engine shift-back lowers the vehicle floor behind the front door

Independent of transmission configuration, vehicle transmissions include the following components that are named starting from the engine:

Start-Up Device

Start-up devices are usually made with a friction clutch, with pedal operation, and in some cases assisted by an external source of energy (electric, pneumatic, or hydraulic); by most automatic gearboxes, start-up devices are made by a hydraulic torque converter, with lock-up clutch.

Gearbox

Gearbox is the transmission component devoted to transmission ratio change; it can be made with conventional or epicycloidal gears. Gear shifts can be made manually (with the only force of the driver), semiautomatically (with the help of auxiliary energy) or automatically. We can find in the gearbox the following mechanisms in addition to gears:

- *Gear shifting mechanisms*: They are classified into internal and external, with reference to the fact that they are contained in the gearbox or not. The function of gear shifting mechanisms is affected by powertrain suspension motion that could affect gearshift quality. These mechanisms are particularly important in manual transmissions, because they have a big influence on driver's feeling of precision and positivity.
- *Shifting sleeves* and *synchronizers*: They are addressed to engage and disengage the pair of wheels necessary to obtain the desired transmission ratio.

Range-Change Units

They can be considered as additional gearboxes with a limited number of speeds that are inserted on the transmission line after or before the main gearbox; they increase the total number of speeds by a number greater than the added gear pairs; under this category, we also include the so-called *splitters*.

Final Drive and Differential

Final drive and differential can be integrated into the gearbox; they further reduce the rotational speed of the wheel; the differential divides the torque for the two wheels theoretically in equal parts and can be free, self-locking, or manually locking.

Transfer Boxes

Transfer boxes are gearboxes that divide the torque for the axles of an all wheel drive or for vehicles with more than one driving axle.

Transmissions and Joints

Transmissions and joints connect the several rotating sections of the transmission; in conventional drives, the gearbox with the differential and in independent suspension, the wheel with the differential.

The second part of this volume, dedicated to automotive transmissions, begins with a chapter dedicated to the historical evolution of transmission design.

Main types of manual transmissions applied to cars and industrial vehicles will be described in the following chapter. Schemes and drawings will be explained for this purpose.

Particular attention will be addressed to internal and external shifting mechanisms and to their design and evaluation criteria.

Main start-up devices will be explained and the design and specification rules will be discussed. For the torque converter only, the methodology for calculating vehicle performance and consumption will be introduced.

As far as differentials are concerned, in addition to design criteria, a short study on the influence of controlled and not controlled units on the vehicle dynamic behavior will be afforded.

Some kinds of automatic transmissions in use on cars and industrial vehicles will be introduced; main control strategies that are implemented in their control systems will be also discussed.

The book is completed with some complementary notion on transmission components design and on methods for car and bench testing.

Chapter 8
Historical Evolution

Gearbox and clutch, or other start-up devices used as alternative, are essential to obtain from a reciprocating internal combustion engine a driving torque suitable to traction.

In fact, this category of engines is characterized by a useful torque almost constant with rotational speed and sometimes increasing with it; in an ideal case, on the contrary, the driving torque should decrease with vehicle speed.

We can say that having decided to install an engine of a given power, in view of the desired dynamic performance, this value of power should be made available at any vehicle speed; therefore traction force should be inversely proportional to vehicle speed.

This choice would also guarantee that any value of car speed would be stable, because a driving force decreasing with speed would balance a resistance force increasing with speed.

In addition, the internal combustion engine is unable to supply a positive torque if its rotational speed is less than a minimum value, whose amount is determined, as first approximation, to the torque period of the cycle and from the inertia of the crank mechanism; the vehicle, instead, must be able to start-up when stopped and should be able to exploit, at this condition, the maximum driving torque.

For these reasons it is necessary to have available a mechanical system able to adapt the transmission ratio between engine and wheels to the need of the vehicle and to the deficiencies of the engine. This transmission ratio should be very high, ideally infinite, at vehicle start-up.

Finally, the transmission has the purpose of transmitting the motion of the engine crankshaft to wheel hubs; as we know wheels and engine have a relative position not defined with precision, because of suspensions motion and of wheels steering.

By very first cars, the gearbox has been confused with a device for speed adjustment; from this misinterpretation has been born also the name of this device, almost equivalent in many European languages (for instance, *changement de vitesse*, in french), instead of torque converter or torque adaptor, more suitable to its real role.

G. Genta and L. Morello, *The Automotive Chassis*, Mechanical Engineering Series,
https://doi.org/10.1007/978-3-030-35635-4_8

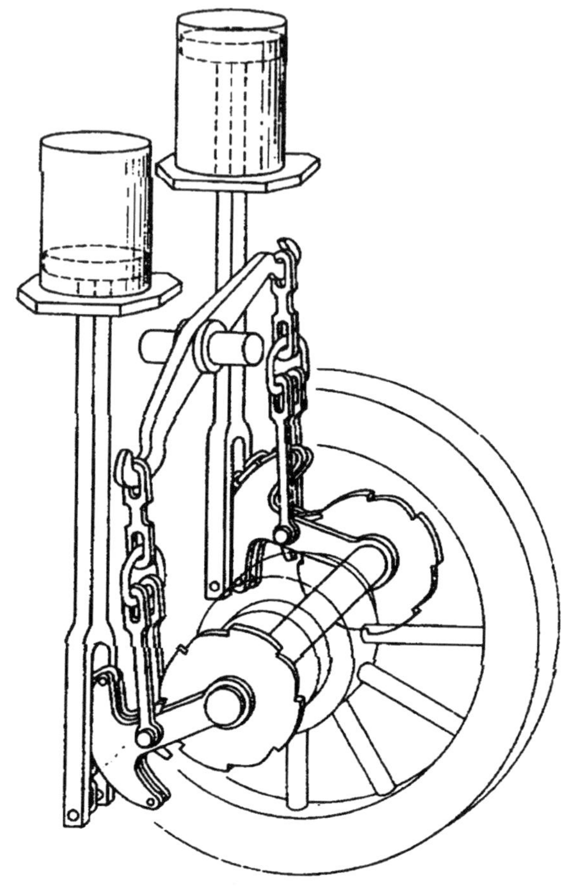

Fig. 8.1 Drawing of the transmission mechanism from the piston to the wheel by the steam engine cart designed by Cugnot in 1769

These considerations that are taken today for granted, were not so evident to the designers of the first cars, also because these were equipped with a steam engine, that among many disadvantage for the vehicular application, has the advantage of being able to supply a notable torque at crank stopped, very suitable for vehicle application.

This situation is testified from the Cugnot cart of 1769, unanimously recognized as the first self propelled vehicle.

Figure 8.1 shows a drawing of the transmission mechanism between pistons and driving wheel. The sole driving wheel is positioned in front of the vehicle and can steer together with the steam cylinders and the boiler (not represented); no suspension is applied.

Engine rods are directly connected to the wheel through ratchet gears that convert the piston reciprocating motion into wheel rotation.

This direct link, eventually made, more simply, with a crank mechanism was applied also to other steam coaches and widely on steam train locomotives.

The first marching internal combustion engine vehicle was probably introduced by De Rivaz in 1807. It adopts a similar transmission and exploits the ratchet gear mechanism to leave the vehicle marching under the action of inertia, during the non useful strokes of the piston.

8.1 Manual Gearbox

Patent documents of 1784 give evidence that Watt predicted the use of constant mesh gear box, with dog clutches, to improve traction performance of a steam engine; it is rather difficult to demonstrate how this idea might have influenced the design of following cars that didn't follow this scheme indiscriminately.

Unanimously the first commercialized internal combustion engine cars are those born of the efforts of Benz and Daimler in 1885 and 1886; the transmission problem received a complete solution using a scheme completely different of that proposed by Watt.

The complete Daimler's car transmission is shown on Fig. 8.2. There are two different transmission ratios made with leather belt transmissions with pulleys of different diameter; belts are always wound on their pulleys, but the motion is transmitted by only one of the couples, when one of the two tensioners 55 (for the first speed) and 56 (for the second speed) sets its belt at work.

Belt slip capacity is exploited, when the tensioner is not completely engaged, to start-up the vehicle when stopped.

Wheel suspensions are missing and the driveline is consequently simplified.

Many improvements of this scheme are applied to the following Benz car. The two speed transmission is inspired to contemporary workshops, where a single steam or water engine was moving a number of working machines. This kind of transmission was probably invented by Anderson in 1849.

The two driven pulleys (in the center of the lower drawing on Fig. 8.3) are coupled with as many idling pulleys (at the outside faces of the driven ones); these have a slightly smaller diameter than the driven ones and the active cylindrical surface of the driven pulleys is rounded to join the active surface of the idling pulleys.

The two driving pulleys (in the back of the car, aligned with the engine crankshaft) have an adequate with to bear the belt on both driven and idling pulleys. Two tensioners can shift the leather belts from matching driven pulleys to matching idling pulleys.

Belts are crossed in order to increase the winding angle on the pulleys; belt tension is adjusted periodically, by changing center lines distance.

The rounding on the surface of the driven pulleys makes belt shifting easier; the lower diameter of the idling pulleys decreases working tension of belts when they are not working, retarding the need for further tension adjustments.

The start-up function is performed again by exploiting the belt slip capacity; the almost vertical motion of driving wheels, already with suspension, is compensated by two chain transmissions, that connect driven pulleys with rear wheel hubs.

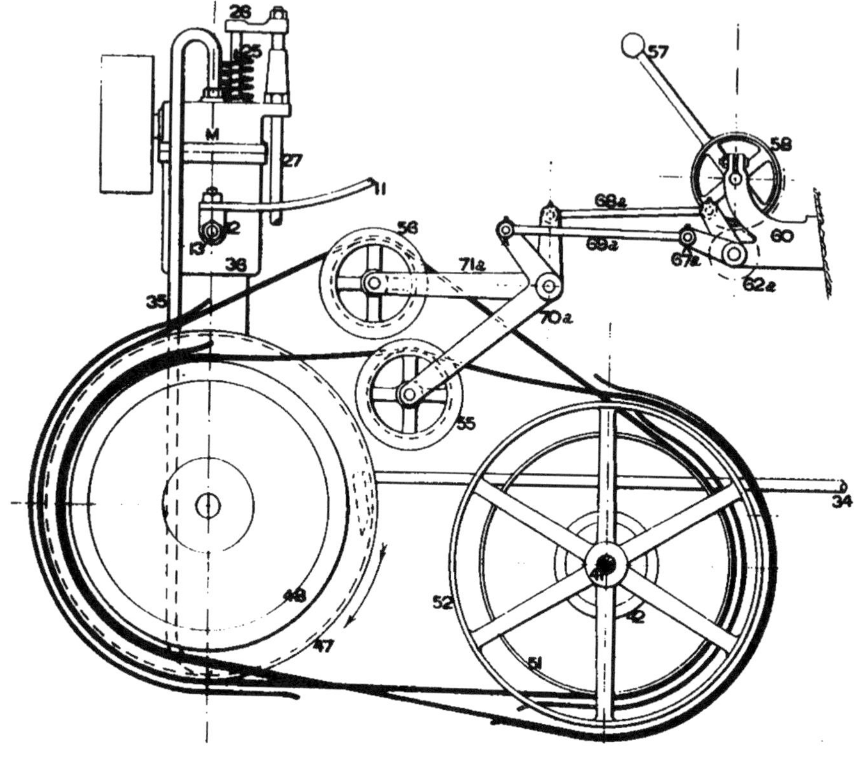

Fig. 8.2 Complete transmission of the Daimler's car of 1885. There are two transmission ratios available, made with a leather belt transmission. The slipping capacity of belts is exploited to start-up the car

It should be noticed that the literature on these and other cars of this time didn't suggest a sequential use of gears to accelerate the car, but starting-up was allowed with both gears, being their choice bound rather to the desired cruise speed, that to the necessary traction force.

The idea introduced by Watt will be applied on next generation design. This is usually commented by saying that the first cars demonstrated how the technical skills of their inventors were polarized on engines, neglecting some times other achievements that had been already obtained by the contemporary state of the art. This fact could be also explained by the objective difficulty of having exchanges of ideas throughout technical communities of different countries.

A typical example, more similar to contemporary ones, is represented on Fig. 8.4, where a drawing of the FIAT 8/16 HP of 1902 is reported.

The gearbox features three speeds and a reverse; on oldest car the reverse speed was missing.

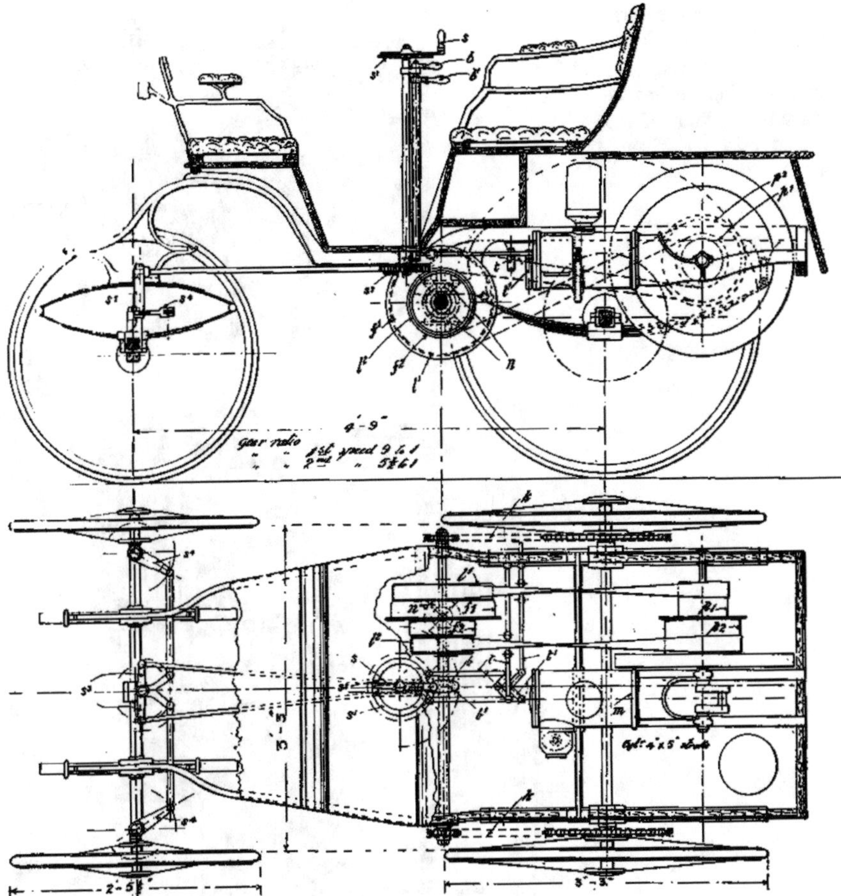

Fig. 8.3 Two speeds transmission of the Benz car in 1886, leather belt type. Two tensioners can shift belts from a position engaged with a driving pulley to that engaged with an idling pulley

The engine rotates the lower shaft, on the left view, that represents a side view of the gearbox. This gearbox has a single reduction stage, where all driving tooth wheels are aligned on the input shaft and all driven wheels on the output shaft; from the engine side, we see the reverse gear, as first, followed by first, second and third gears.

An idle tooth wheel can be noticed, to reverse the rotation speed of the output shaft on the cross section at the left of the side view.

Driving wheels are always meshing with driven ones and must, therefore, be idle with the output shaft; we will see later on how they are engaged. The output shaft rotates a bevel gear, which moves through a differential two shafts. These are coupled through drive chains to rear driving wheels, according to the scheme already commented on previous figure.

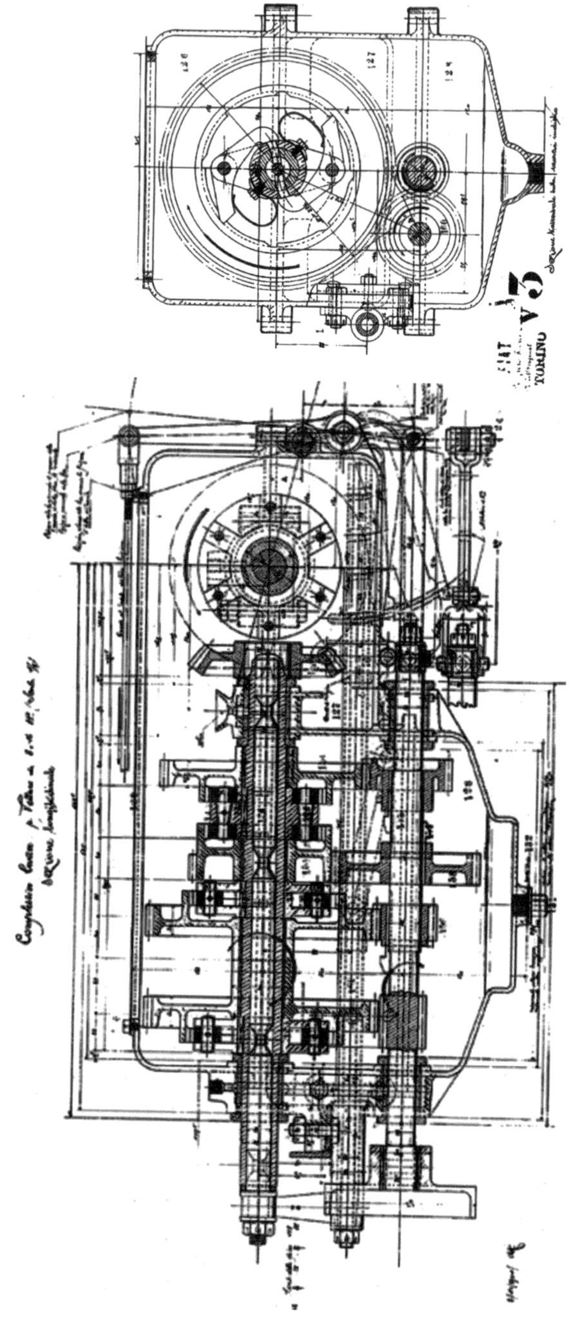

Fig. 8.4 Single stage, three gears gearbox of the FIAT 8/16 HP of 1902. The internal shifting mechanism is made by a cylindrical rod, mounted inside the driven shaft

A tapered friction clutch, not represented on this figure allows vehicle start-up and separates gearbox from the driveline, during gear shifts (see for reference Fig. 8.12).

We can notice three new mechanisms not yet introduced, as reverse gear, differential and friction clutch. Their origin is surely older as this car. The reverse gear was introduced by Selden in 1879, the differential by Pecquer in 1827 and the tapered friction clutch by Marcus in 1885.

During these years these three mechanisms are integrated in a transmission suitable for automotive application.

Let us look at the shifting mechanism now. It is rather sophisticated and is made by a shiftable cylindrical rod, mounted inside a cavity in the driven shaft. This rod shows some annular narrowing, that in certain positions allows two ratchets to engage with driven wheels; the detail of the reverse gears ratchets can be seen on the cross section on the left.

When one of the grooves is facing a couple of ratchets, two leaf springs provide for their engagement with the ratchet gear; when the rod is shifted, ratchets retract leaving the wheel again idle. Grooves position is congruent with a sequential shift stick, where the positions of reverse, first, second and third are following each other.

This kind of gearbox architecture (constant mesh gears) is applied to many cars of that time, but will be soon abandoned because of its complication and consequent fragility: the next scheme is that of *sliding gear trains* (*train balladeurs*, in french). Also this invention is older and born outside of the automotive world; its priority can be assigned to Griffith in 1821.

It is not the first time that a better performing architecture (constant mesh) is abandoned in favour of a less evolute one (sliding trains), because some component (dog clutch, synchronizer) has not been developed yet. Sliding trains will be soon abandoned again, in favor of constant mesh, as soon as the technological improvements will make it possible.

This innovation can be exemplified by the FIAT 60 HP car of 1904, whose gearbox is shown on Fig. 8.5.

The gearbox is still single stage type and presents four speeds and a reverse speed; tooth wheels, grouped in two trains, from the first to the fourth, can slide on the upper driving shaft. The engine (not shown) is on the left, while on the right we can see bevel gears moving the pinion of the chain transmission through the differential.

The two trains integrate respectively first and second speed wheels and third with fourth; these trains are mounted on a square section shaft that makes them always turning, being free to shift along the shaft.

Trains sliding is made by suitable sleeves that bring each wheel at a time to engage with their counterpart; gears are meshing only when their correspondent gear is selected.

Reverse speed is again made with train of idler gears only, not represented on this figure, meshing with the first speed gears, when their train is in idle position.

Sleeves are moved by forks fixed to sliding rods, partially visible under gearbox shafts; there are two rods for moving the first and second train and the third and fourth train, while an additional rod is dedicated to reverse speed.

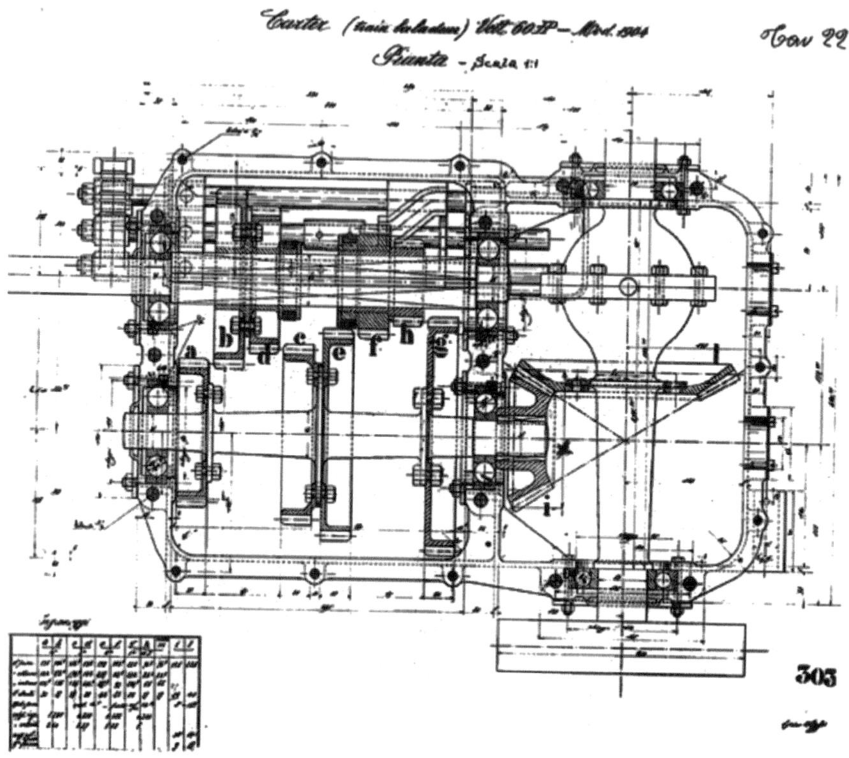

Fig. 8.5 Sliding trains gearbox with four speeds of the FIAT 60 HP of 1904. The two gear trains are made by the driving wheels of the first and second speed and by those of third and fourth speed

This kind of lay-out has caused the selection and shift control sticks, where gear shifting is no more sequential, but is characterized by two separated motions: one across the car, to select the train to be engaged, the second along the car, to engage the specific gear. This layout will consolidate in the praxis, arriving unchanged to our contemporary cars.

Sliding train gearboxes requested a particular skill to the driver, who had to try synchronizing wheels, by means of the engine during the idle time, before of engaging next gear (double clutching or *débrayage double*, in french); an imperfect maneuver was indicated by scratching of the gears in contact at an inappropriate speed.

Some manufacturers improved this architecture, coming back, in reality to the Watt's idea of constant mesh gears; an industrial vehicle gearbox of the end of the '910s is represented, to show this new lay-out, on Fig. 8.6.

In this gearbox three speeds are available and a reverse speed; it is connected to the engine on the right of the figure, while the output flange is on the left. The flywheel on the output shaft is not part of the engine, but is the drum of the transmission brake.

Input and output shafts are now coaxial; this lay-out is appropriate to the application of a universal joint transmission, starting to be applied in these years.

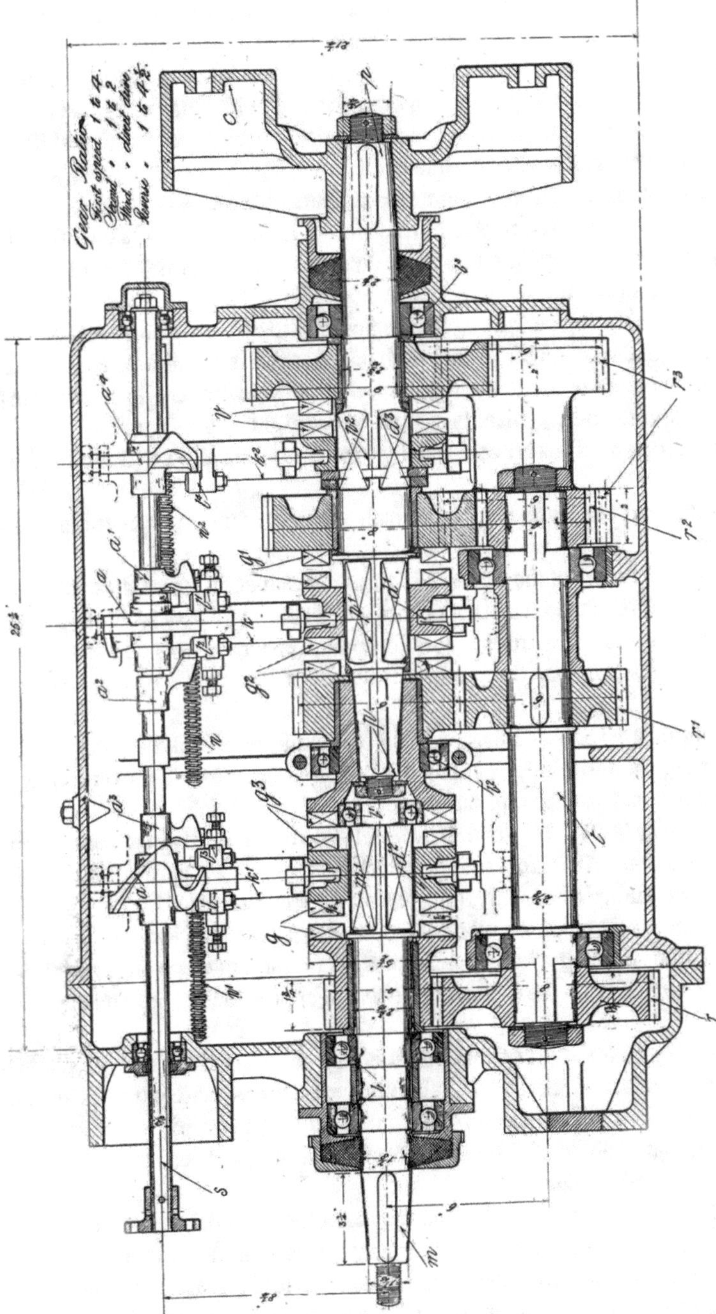

Fig. 8.6 Dog clutch truck gearbox of a vehicle made in the ’910s. Sliding trains are now reduced to small front tooth gears working as clutches

Tooth wheels are in constant mesh and have at their flanks a dog clutch; sliding trains are now reduced to the pairs of dog clutches.

The scratch problem isn't solved, but possible damages are localized on an auxiliary component, the clutch, that can be sacrificed, with smaller impact on the operation of the gearbox. Dog clutches teeth can also be rounded, making shifting easier, without drawbacks on wheels dimension.

We can observe that input and output shafts are made coaxial, being the upper shaft made by two parts, free to have different rotation angles. The left sliding train can alternatively engage the two parts directly or move a third shaft (*countershaft*), below on the same figure.

Only with this dog clutch engaged is possible to obtain first and the second speeds, with the second train on the figure; with the last train is possible to engage the reverse speed connecting the countershaft to the output shaft two a couple of idlers (in practice a second countershaft), partially visible on the right of the figure.

Sliding train motions are operated by front cams, that allow in this case a sequential control.

A similar architecture, but for control, is still present on front engine, rear wheel driven vehicles.

We shouldn't fall in the mistake of thinking that at this time these technical solutions were consolidated; as often occurs at the dawn of a new technology, deviations from what, for us, is the main evolution path were many; we can see that, after the engine, the gearbox was the preferred subject of innovation of the first automotive engineers.

A comprehensive classification of all tried solution is well outside of the scope of this chapter, even limiting to manual transmissions; we will consider only some of the more original solution, in our perception.

Frazer Nash built in England, between 1924 and 1938 different cars, all with sport performance, but with affordable price, in their category. Characteristic of these cars was the essentiality of the driveline, as can be seen on Fig. 8.7.

The gearbox is made by chain transmissions, three for forward speeds, one for reverse speed. Driving chain wheels (at right on the drawing) are moved by a bevel gear box, connected to the engine; this bevel gear box doesn't include any differential gear.

On the right half shaft of this intermediate shaft are fixed the sprockets for first speed and reverse; on the left half shaft there are the second and third speed.

We can notice the idler reverse system with chain and gears. Driven wheels are directly mounted on the rear one piece shaft; driven wheels can be moved along their shaft for chain alignment.

The lack of the differential made the car difficult to drive, but very manageable for an expert driver on unpaved roads, very diffused at those times.

Chain transmission was also used for compensating suspension motion.

Engagements were made by simple but sensitive dog clutches. Fans of this cars extolled the very easy road holding, the excellent gearbox maneuverability and easiness of substituting broken parts in the transmission.

Fig. 8.7 Frazer Nash gearbox of 1924, made by chain transmissions only; three speeds and a reverse are available. The driven shaft can be adjusted to obtain a good alignments and to compensate for chains wear

The entire mechanisms were lubricated by grease at the open air. We can also notice the rocker arm, on the bevel gear box, with particular slots, that match the engagement forks; this mechanism avoided the contemporary engagement of two speeds, also considering the effect of centrifugal forces and vibrations.

A car with different technical details, also original and uncommon, is the Sizair Naudin of 1907, also known for featuring one of the first independent suspensions.

Also in this case the design goal should have been to contain gearbox cost, or at least tooth wheels number, at that time very expensive.

This gearbox cut on Fig. 8.8 can be assimilated to a sliding train gearbox. Driven wheels are reduced to only one, with front teeth R. The only sliding train, made by wheels P1, P2 and P3 is installed on a bearing swinging around the lower shaft in the drawing; the bearing swinging motion is necessary to mesh driving wheel in use with the front teeth wheel, at different distances, depending on the driving wheel diameter.

The swinging motion of the driving shaft is compensated by a universal joints transmission between engine and gearbox. The cam C at the bottom of the drawing combines shift and swing motions and bears gearing forces.

Spur wheel should be instead bevel gears, for correct matching; teeth are, nevertheless, approximated with cylindrical teeth, accepting contact errors.

The reverse idler is also present on a dedicated shifting train.

A last example of amazing engineering ingenuity is given by the Turicum of 1904 (probably the only automotive trademark in Switzerland) shown on Fig. 8.9, with a

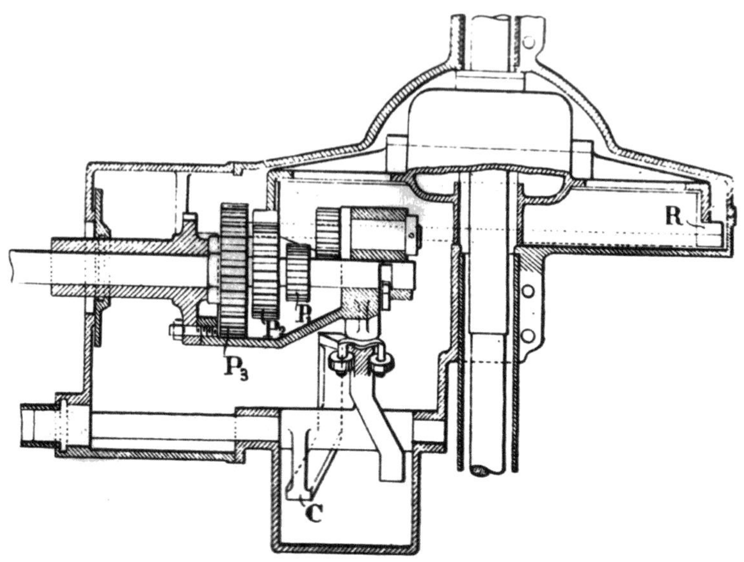

Fig. 8.8 Sizair Naudin gearbox of 1907. This gearbox can be assimilated to a sliding train one, with bevel gears. The single sliding train made by wheels P1, P2 and P3 is mounted on a swinging bearing to mesh driving shaft at different distances with the big output front teeth wheel

picture of the complete chassis; we can see one of the first continuously variable transmission in the history of the automobile.

Also in this case the car has no differential; the motion to the rear axle is transmitted with two friction wheels C and D, where the first is made of solid iron and the second has a rubber tread on its rim. The wheel D is fixed on the shaft E with a spline and groove, that allows the wheel to be shifted along the shaft; the shaft E rests on a swinging bearing G and maintains wheels D and C under pressure with the spring J.

By pulling the lever q is possible to change the contact point between the two pulleys, between the center of the wheel C (infinite transmission ratio: transmission idling) and its rim (transmission ratio about 1:1). In the idle position, the friction is eliminated by unloading the spring J; the possible slip between the two wheel is used to start the vehicle up.

These proposals were not imitated by other manufacturers and were probably abandoned by their own inventors. The evolution of manual gearboxes concentrated on perfecting countershaft or double stage architecture, that diffused unchallenged on all cars with front engine and rear drive.

An example of the evolution trend is offered by the gearbox of the FIAT Balilla of 1934 (four speed version, the first gearbox of this car in 1933 for a three speed one) shown on Fig. 8.10.

This gearbox shows two different sections: the rear one, for first, second and reverse speeds, features a sliding train with cylindrical straight teeth and a front one with helical gears (always meshing) and synchronizers.

Fig. 8.9 Turicum chassis of 1904; it shows a continuously variable transmission based upon two friction discs C and D, where the first is made of solid iron and the second has a rubber thread to improve contact friction

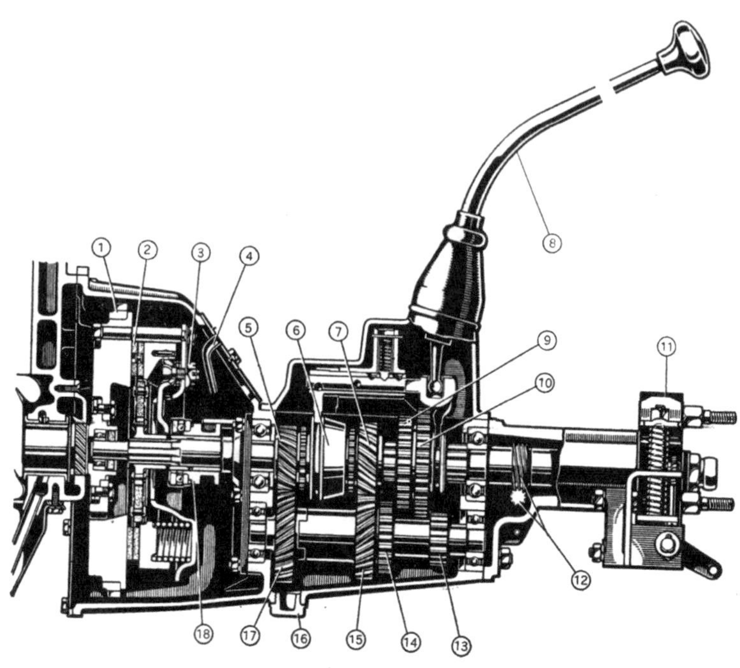

Fig. 8.10 Four speed gearbox of the FIAT Balilla of 1934. The drawing shows a longitudinal cross section; the rear part of the gearbox has a sliding train for first, second and reverse speeds; the front part features synchromesh gears for the third and fourth speeds

This compromise is justified by the high cost of synchronizers, to be considered high technology components at that time. Synchronizers are limited to the more frequently used speeds, that can benefit also of helical gears, with gearing noise reduction.

On this subject, we noticed that engineering manuals of this time suggested, as a good engineering practice, to design the top ratio (in this case the final differential ratio) with values slightly higher then what ideally necessary; this rule was addressed to limit the number of gearshifts, necessary to maintain the car at cruise speed and demonstrated the difficulty of drivers in changing speed with sliding train gearboxes.

We can therefore think that synchronizer brought benefits not only on shifting quality, but also on noise and fuel economy.

The gearbox of FIAT 1400 of 1950, shown on Fig. 8.11 adopts, as many cars of this time, synchronizers on all speeds, but the first, again for reasons of economy; the first is included in a sliding train mounted on the sleeve of third and fourth speeds. The reverse speed is made with an idler, not shown on the picture, meshing with the first wheels, when they are in neutral position.

During the following period of time, synchronizers were improved and made less expensive also thanks to higher volumes; since the '970s also economy cars received synchronizers on all forward speeds.

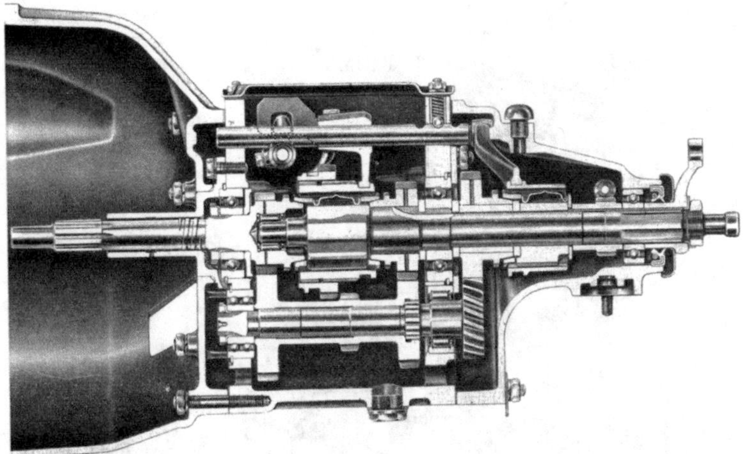

Fig. 8.11 Four speed gearbox of FIAT 1400 of 1950, with full synchronization, but the first gear. The reverse speed is made by a sliding idler, not shown on this figure

8.2 Friction Clutches

An important transmission component, the friction clutch, or, shortly, *clutch* posed many problems to designers about operation force and endurance.

On very first cars belt transmissions integrated clutch function into the gearbox. As we have already seen bevel clutches were already known at the beginning of the automobile era.

We see on Fig. 8.12 an example of bevel clutch of the first years of the past century.

The friction surface is covered by a leather lining, riveted on a bevel pulley in cast iron; although Frood had already invented the famous synthetic material called Ferodo in 1897 it became widely applied in the '920s only.

Leather has a friction coefficient not very different of modern friction materials, but has limited performance, as far as heat wasting capacity and endurance; these facts implied a large sizing of active surfaces. On the contrary, at those times, was cheap and easy to be repaired or changed.

The active surface of this application is single and shaped like a cone; this shape was chosen to limit the disengagement force on the pedal, depending on the cone diameter and the engine torque. Due the difficulty on integrating the leather lining into their support disc, active surface was usually single, and not double as by modern clutches.

If we refer again to the previous figure, we can notice that the engine flywheel features a short shaft bearing the reaction structure of the load springs working on friction surfaces.

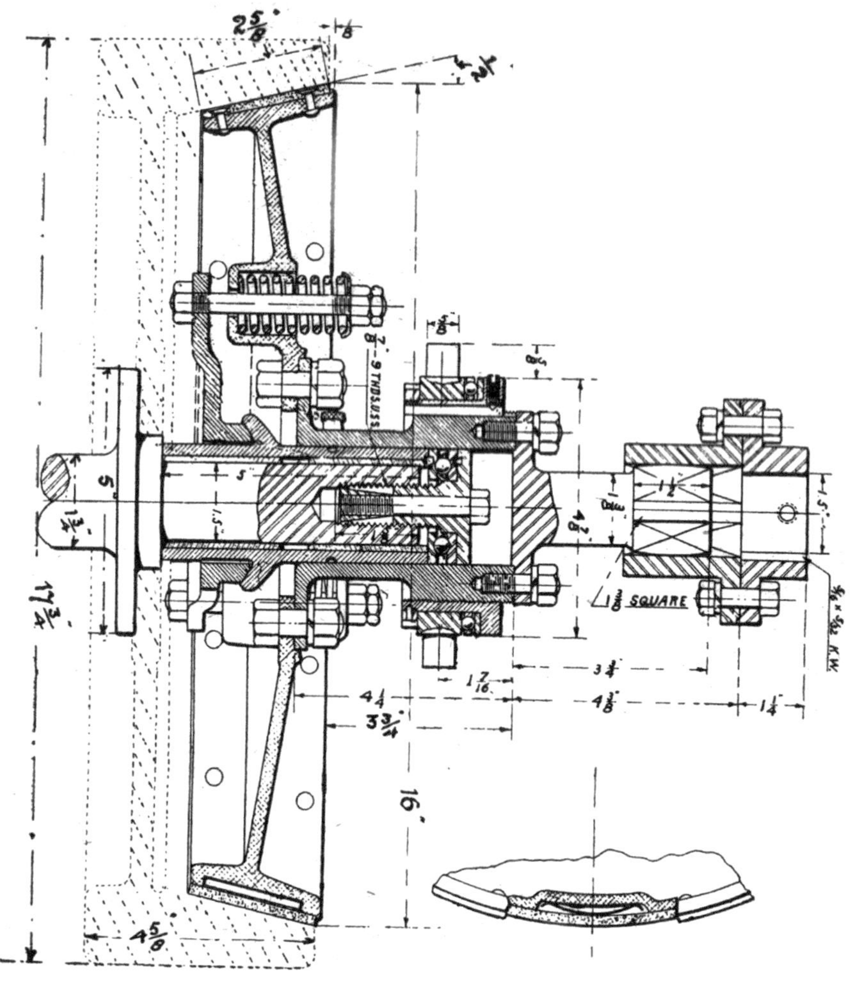

Fig. 8.12 Bevel clutch with leather linings. Clutch conicity (1:2; lower values could reduce pedal reaction) is limited by the available friction coefficient to avoid irreversible sticking of working surfaces, after engagement

Many coil springs (only one is shows on the cross section) push the tapered friction disc into the flywheel. The leather lining is riveted on this disc; very thin leaf springs are set between the lining and the disc to make the engagement more progressive.

With this kind of architecture, very different as the contemporary one, the gearbox input shaft must be able to slide on a square counterpart.

Friction conicity (1:2 on this drawing) is limited by the friction coefficient between leather and iron, in order to prevent irreversible sticking of the clutch, after engagement.

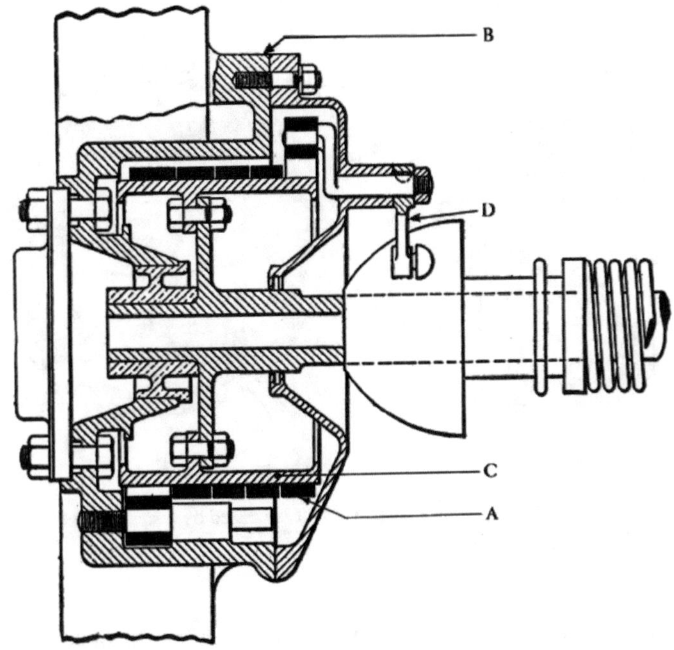

Fig. 8.13 Clutch with spiral friction spring. The clutch pedal moves the ogive that closes the spring onto the input shaft, shaping up a friction force; the same force is increasing the band tension. The resulting friction torque is function of the initial tension through an exponential function of the winding angle of the spring

The large engine displacement of many cars and the limited dimension of the flywheel made many clutches too heavy to be operated; for this reason other mechanisms were also developed.

The idea was to exploit the mechanical property of wound linings to reduce working forces, where the friction force is used itself to increase contact pressure as by leading shoes of drum brakes. This principle was applied by band clutches.

An application of this principle is shown by Fig. 8.13; a coil spring with rectangular section is installed in a cavity in the flywheel; coils are very close each other. One end of the spring is directly fixed to the flywheel, through the eye in the lower part of the figure; the other end is instead connected by a rocker arm D.

If an ogive is moved closer to the rocker arm, it is possible to twist the spring and reduce its internal diameter.

The gearbox input shaft C is surrounded by the spring with a little play; when the ogive is advanced through the clutch pedal the spring closes on the shaft with a resulting friction torque.

The friction tension along the spring coil, increases the tangential tension toward the eye, without increasing its reaction on the rocker arm; the resulting friction torque is an exponential function of the winding angle, that can be increased indefinitely.

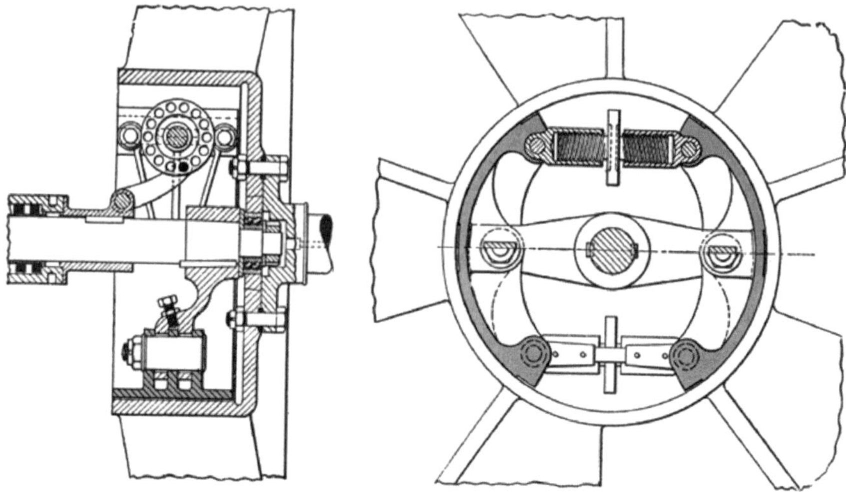

Fig. 8.14 Radial shoes clutch. The shoe displacement is caused by a screw mechanism, operated by the clutch pedal through a crank

With a modest friction coefficient between metals it is possible to transmit the desired torque with a reasonable force on the pedal; the counterpart is the brutality of the engagement maneuver, only in part softened by spring elasticity.

A very different configuration of the same principle is given by Fig. 8.14; the torque is transmitted by two shoes that expand in a drum, like in a drum brake.

The shoe motion is caused by a screw that is moved through a crank and rod mechanism; the disc shape of the crank is studied to allow a simple play adjustment, to compensate for lining wear.

Also in this case, facing a difficult problem to be solved, inventors investigated many different solutions before of consolidating and improving the best one; to solve these problems also electric and hydrostatic transmissions were investigated and applied.

The final solution was consolidated in the '930s with the single disc clutch with synthetic friction linings; one example of this time is shown on Fig. 8.15.

The friction surface is now flat and double; with the same force is possible to transmit a double torque. The friction disc is mounted within two surfaces (the flywheel and the pressure disc) that are compressed by a number of coil springs; a set of levers on pressure disc are used to release the clutch with the axial motion of a thrust bearing.

This kind of clutch received its last improvement by the application of cup springs; they were introduced at the end of the '970s and allowed many advantages as a further reduction of pedal force and a general simplification of this mechanism.

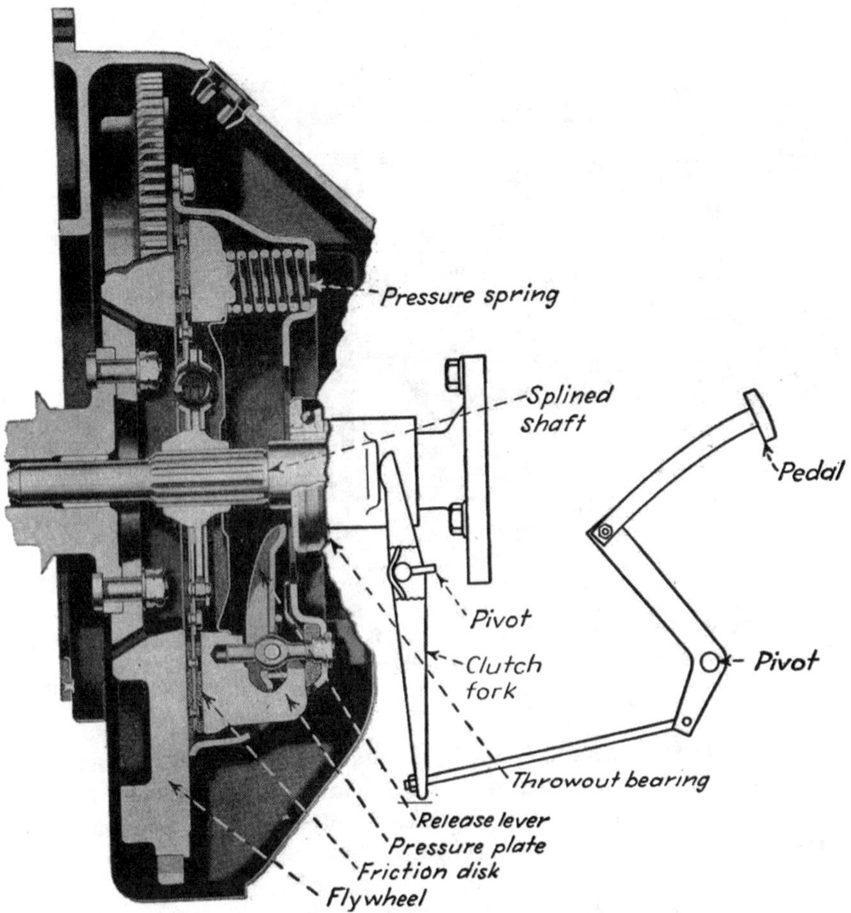

Fig. 8.15 Dry single disc clutch with coil pressure springs. A set of release articulated on the pressure plate is used to disengage the clutch, through the displacement of the throughout bearing

8.3 Automatic Gearboxes

Automatic gearboxes have had their own history that received a crucial contribution by the American industry.

We don't want to mean the Europe didn't contribute to this development; we will see, in fact, that many fundamental inventions were developed in this continent. Nevertheless the European market less reach and more fragmented didn't justify the mass production of this gearbox until very recently.

The problems to be solved, to develop an automatic gearbox, included a different mechanism to engage gears and to start the vehicle, easier to be interfaced with not sophisticated automatic controls, that could be mechanic (exploiting centrifugal forces) or hydraulic (exploiting the pressure variation of the oil in a rotary pump).

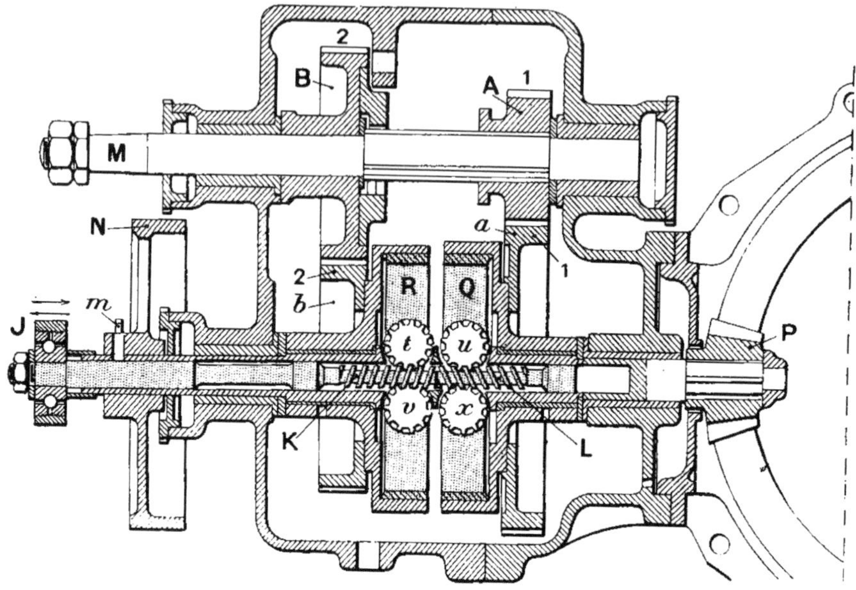

Fig. 8.16 The De Dion and Bouton gearbox can be considered as a precursor of power-shift gearboxes. By shifting the shaft with the bearing J, is possible to engage one of the two shoe clutches available on each gear; start-up clutch isn't necessary

Today this problem appears to have lost its sense, because electronic microprocessors allow an easy automatization of synchronizers and friction clutches, in use on manual gearboxes; many existing vehicles already testify this statement.

The first step was the development of gearboxes where speed shifts were possible without danger for tooth wheels and of parts to be synchronized.

From this point of view we con consider a precursor the manual gearbox of de Dion and Bouton, developed at the beginning of the past century and shown on Fig. 8.16.

This single stage gearbox has two speeds only; we can see on the left upper side the input shaft and on the lower right the output shaft which moves through a bevel gear the pinions of the chain drive.

The two gear are always meshing, with the driven wheels idling on the output shaft; the wheel engagement is made by shoe clutches, similar to that already commented on Fig. 8.14; they are controlled by the screws moving the gears *t*, *v*, *u* and *x*.

By shifting the shaft with the thrust bearing J is possible to engage one clutch and to disengage the other.

By this transmission start up clutch isn't used during gear shifts.

Even if it was developed for manual gearboxes, this kind of clutch is surely a relevant precursor of the powershift gearbox with band brakes and multi disc clutches.

A second gearbox of historical relevance is that of the Ford Model T of 1908, the first car that has been produced in millions.

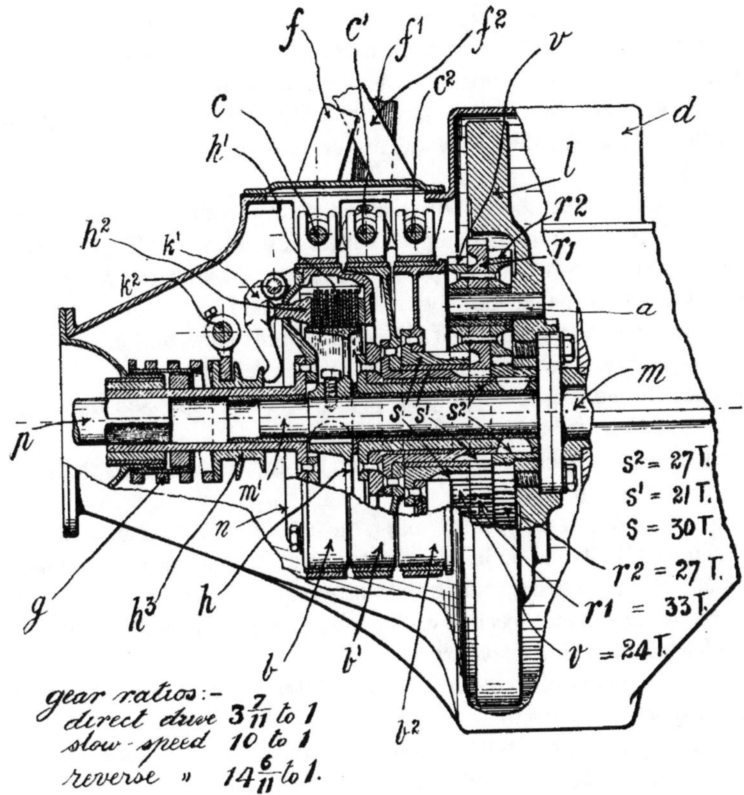

Fig. 8.17 Epicycloidal gears gearbox of Ford model T of 1908. One reverse, one reduced forward speed and a direct drive are available

Figure 8.17 shows a section of this gearbox; it is made with epicycloidal gears, instead of gears with fixed rotation axis.

These gears were not applied by Ford for the first time, because already known in other applications. The epicycloidal gearbox should have been invented by Bodmer in 1834, although there is evidence that these mechanisms were already known to the old Greeks in applications for astronomical computations.

On this figure we see the three satellites v, r^1 and r^2 (the unusual position for the subscript, not to be confused with an exponent, is brought from an original figure), rotating on a single carrier, fixed to the engine flywheel. They mesh with the corresponding sun gears s, s^1 and s^2.

If we imagine to keep the sun s stopped, by rotating the flywheel and the carrier we obtain a reduced output speed, in the opposite direction, at the sun gear s^2, fixed to the output shaft.

Vice versa, by keeping the sun s^1 stopped, we can obtain a reduced speed in the same direction, again at the sun s^2; on the figure a note reports the transmission ratios that were obtained, including the differential transmission ratio.

If the multi discs clutch h^1 is engaged, by shifting the sleeve h^3, it is possible to put the gearbox in direct drive, by fixing the hub h, rotating with the crankshaft, with the output shaft.

To obtain the different states of the gearbox, sun gears are rotating with the drums c, c^1 and c^2, that can be gradually stopped with their band brakes; band brakes control is made by front cams, moved by pedals on the car dash board; the lower part of these pedals is shown on the figure with letters f, f^1 and f^2.

Pedals have a spring system that make them stable either in released and in depressed position; each pedal raises if another one is depressed.

When engine and car are stopped, the pedal f must be depressed and the clutch h engaged; in this way the vehicle is in park condition.

By releasing the clutch h, through a lever, the engine is disengaged and can be cranked, the car is still stopped.

By depressing one of the pedals f^1 or f^2, the pedal f is raised, the car is left free to move and it will be started-up in low gear forwards or backwards; speed inversions can be made also by moving vehicle and start-up on slopes is made easier.

As soon as the suitable speed is reached, by engaging the clutch h, the pedal f^1 will be released, obtaining direct drive.

The gearbox is controlled by driver muscles, but clutches management is made automatically during gear shifts.

From this situation to a fully automatic gearbox the way was long, but these achievements made the final result closer.

The configuration of this gearbox allows to understand why epicycloidal gears were preferred, to the conventional ones, for the new automatic transmissions: for the easiness of integration of brakes and clutches.

A further step was made by Wilson, in England, in 1928, by proposing a gearbox made by two different epicycloidal gear trains in series, where the carrier of the first gear was connected to the ring of the next gear. With two gears is possible to obtain three speeds forward, one of them being a direct drive and a reverse speed.

The three speeds were obtained by braking drums with bands, as by Ford gearbox; a schematic example of the Wilson gear train is shown on Fig. 8.20, in the chapter dedicated to automatic transmissions.

These gearboxes, similar in use to the Model T ones, were semi automatic with manual preselection; according to this concept, a small lever nearby the steering wheel was used to select in advance the next gear to be used. At this time no gearshift was started, but the brakes mechanisms were arranged for the gearshift to be made; this occurred as soon as the driver depressed a pedal for this purpose, set in the position normally used for the clutch pedal.

The driver was supported, in this way, on executing a coordinate maneuver of the gear stick and of the clutch; the energy for this function was still produced by drivers muscles through a pedal.

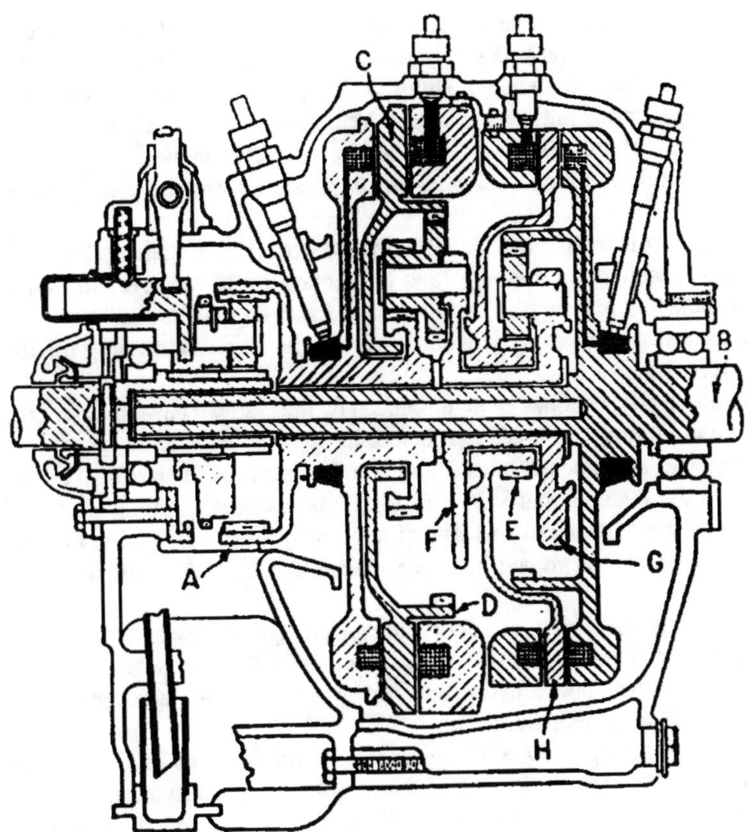

Fig. 8.18 Semiautomatic gearbox produced by Cotal from 1934. The different elements of epicycloidal gears are braked by electromagnets; the reverse is engaged manually

A semiautomatic gearbox particularly advanced has been produced by Cotal since 1934, in France; Fig. 8.18 shows a cross section of this gearbox.

This gearbox includes three epicycloidal gears; the engine is on the left, the output shaft on the right.

Some toroidal electromagnets can stop some elements of the gear train; particularly, the first puts the corresponding gear set in direct drive, by fixing together sun and ring gears; the second obtains a reduced speed. At right, the third electromagnet obtains a faster speed, while the last, again, a direct drive.

By energizing electromagnets in combination, two reduced speeds, a direct drive and an overdrive can be obtained.

A small switch with five positions, near steering wheel, allowed to obtain the four ratios automatically, without care of clutches, whose function was controlled by electromagnets timing and inertia of parts accelerated or slowed down, during shifts; the fifth position was for the idle gear, with all electromagnet circuits open.

The first epicycloidal gear on the right is operated, instead, manually, when the car is stopped and the transmission in idle position; a control lever moves the carrier back and for, which can engage with the ring gear, obtaining a forward speed or can be stopped, obtaining a reverse gear. Vehicle motion can be obtained, after this manual shift, with the first gear, controlled by its electromagnet.

The most relevant inconveniences of this gearbox were heavy weight and large size.

Semiautomatic Wilson and Cotal gearboxes were mainly used by European manufacturers specializing on luxury cars; the second world war crisis made many of these manufacturers disappear and these transmissions with them.

The final step towards modern automatic gearboxes was made by exploiting hydraulic torque converters.

The torque converter has been introduced by the German naval industry, after the invention of Föttinger in 1905, well before its application on cars.

He patented a torque transmission system by means of a centrifugal pump and a turbine, in the same hydraulic circuit. With this device, the torque transmission is obtained by the momentum variation of the flow through the rotating blades and is possible also when the pump (the engine) is rotating and the turbine (the vehicle) is stopped.

The idea was developed further designing an integrated device of reduced dimensions almost interchangeable with the conventional friction clutch.

In 1910 a patent about a hydraulic clutch was filed, which was simplified by the elimination of the reactor element.

Again in Germany, in 1928, the research consortium called Trilok developed the homonymous torque converter, able to obtain, with a single machine, the performance of the torque converter and of the hydraulic clutch, by mounting the reactor element with a freewheel.

The first automatic gearbox developed for a car is due to GM; called Hydramatic has been produced since 1939; a cross section of this gearbox is shown on Fig. 8.19.

We can see, on this figure, starting from the engine side on the left, the hydraulic clutch, followed by three epicycloidal gear trains, able to obtain three forward speeds and a reverse one; engagements and disengagements are obtained by two band braked (37 and 16) and two multi disc wet clutches (7 and 17).

Brakes and clutches are operated by oil pressure, generated by the gear pump 33, on the front side of the gearbox and modulated by servo valves and by a manual control on the steering wheel.

Gear shift automatization is based upon the comparison of oil pressure generated by this first pump (depending on engine speed) with a pressure generated by a second pump driven by the transmission output shaft (depending on vehicle speed). The difference between these two pressures is used to move gear shift servo valve; this valve is also sensitive to the accelerator pedal position, by a spring loaded mechanical link.

This system worked quite well on plain roads, upshifting speeds at higher vehicle speeds with higher accelerator compression; on slopes or on bending roads the automatic control had to be corrected by the manual selector.

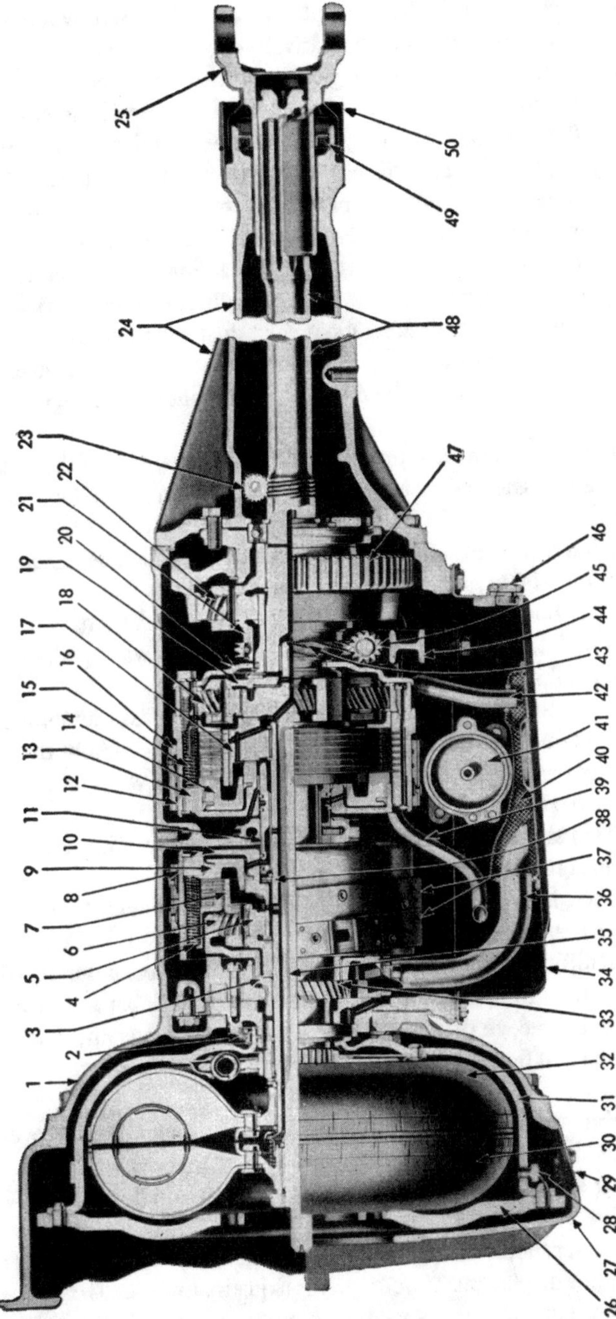

Fig. 8.19 The first automobile automatic gearbox, produced since 1939, is the Hydramatic by GM. We can see, starting from the engine side on the left, the hydraulic clutch, followed by three epicycloidal gear trains, able to obtain three forward and a reverse speeds

We can notice that the hydraulic clutch is always crossed by the useful power; it was used for start-ups and to dampen driveline torque vibrations.

The most important part of these gearboxes were absorbed by war production; in 1946 only it started its application on commercial cars receiving public appreciation.

The Dynaflow gearbox, again by GM, has been produced since 1948 and introduced many improvements with regard to the previous one (Fig. 8.20). The epicycloidal Wilson gear train, quite simpler, allowed also obtaining three forward and a reverse speeds, with two band brakes and multi disc clutch, used in combination.

The most relevant step forward is the introduction of a refined torque converter, featuring a two stage reactor on freewheels; with this device it was possible to start-up the car with a torque transmission ratio greater than two (instead of one, by definition on the hydraulic clutch), obtaining in the mean time the torque converter working as a clutch, with better efficiency, when input and output torques on the converter were equal.

This scheme is still present on automatic gearboxes, even if the need for a higher number of transmission ratios has justified the application of additional epicycloidal gear trains.

It is also interesting to remember the automatic gearbox designed in 1949 by the Dodge Division of Chrysler, with very original features.

Figure 8.21 shows the clutch of this gearbox; it includes a hydraulic clutch and a pedal friction clutch in series. Some gear shifts need always pedal clutch, but they are seldom, thanks to a particular automatization device.

The twin friction and hydraulic clutches allow transmission vibration dampening and a very smooth start-up, even if the pedal is released without particular skill; in addition the car can be kept stopped on a slope, simply working on the accelerator pedal and the next start-up is impressively easier.

Similar twin clutches were also applied in combination with conventional manual gearboxes on some European cars, as the Fiat 1900.

The gearbox is explained on the next Fig. 8.22; it shouldn't be confused with a simple counter shaft gearbox.

The main difference consists in the constant mesh wheel mounted with freewheel that allows torque to be transmitted to the countershaft, but not vice versa and on a disengagement sleeve of this freewheel on the countershaft; it is in disengaged position on the first figure on the left. The whole set of figures represents the different gearbox states.

The gearbox control features a manual lever working on the sleeve, on the right of the output shaft, through a suitable leverage; if this sleeve is set to the left, first and second speeds can be obtained; if it is set to the right, third and fourth speeds can be obtained.

This maneuver should be made by disengaging the friction clutch; but transmission ratios and engine displacement were such as to justify low speeds on slopes or in urban driving, while remaining gears were recommended on suburban driving, including related start-ups.

Upshifts from first or third gears and downshifts from fourth or second gears were made automatically, by a tachometer device shifting the freewheel sleeve. Notice that

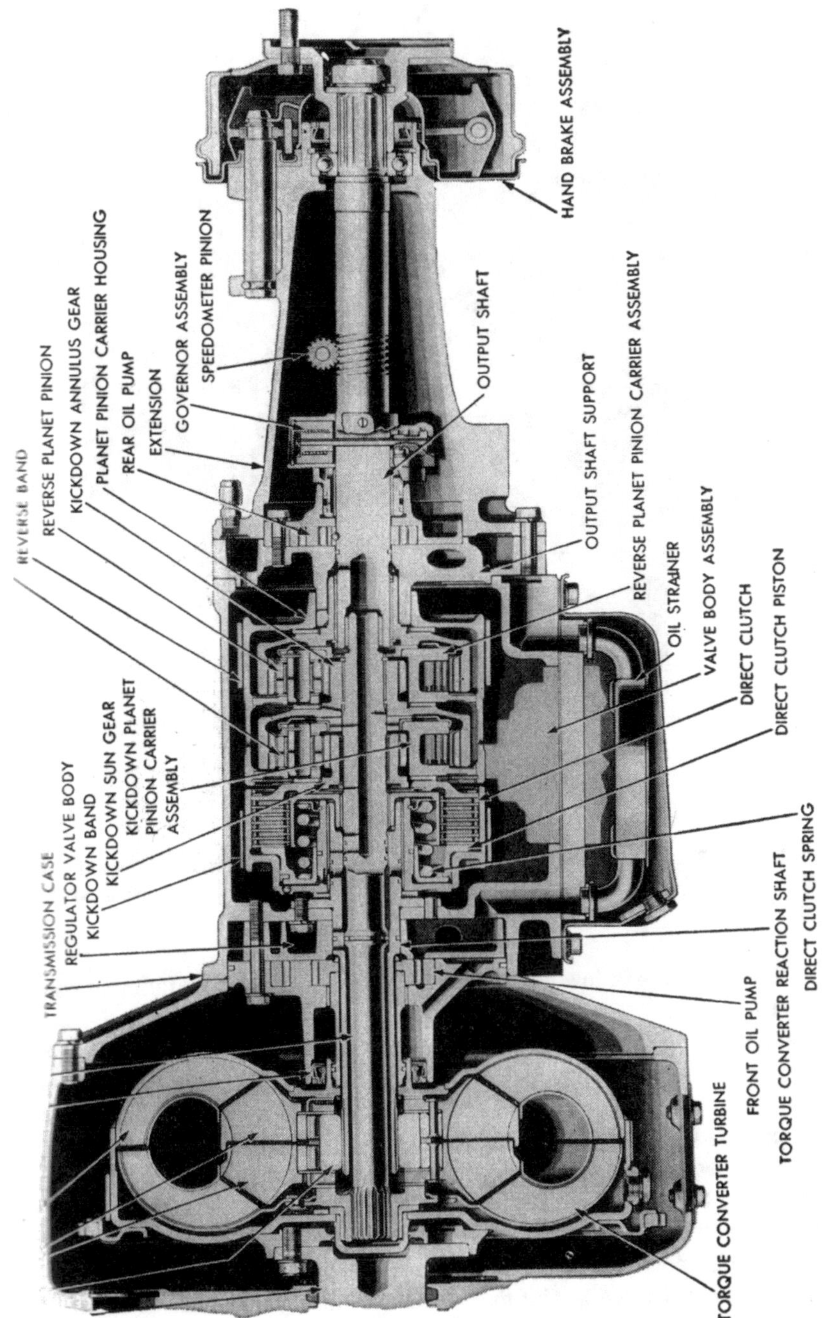

Fig. 8.20 The Dynaflow gearbox, again by GM, was produced since 1948 and can be considered as an improvement of the previous one. The Wilson epicycloidal gear train allows three forward and a reverse speeds

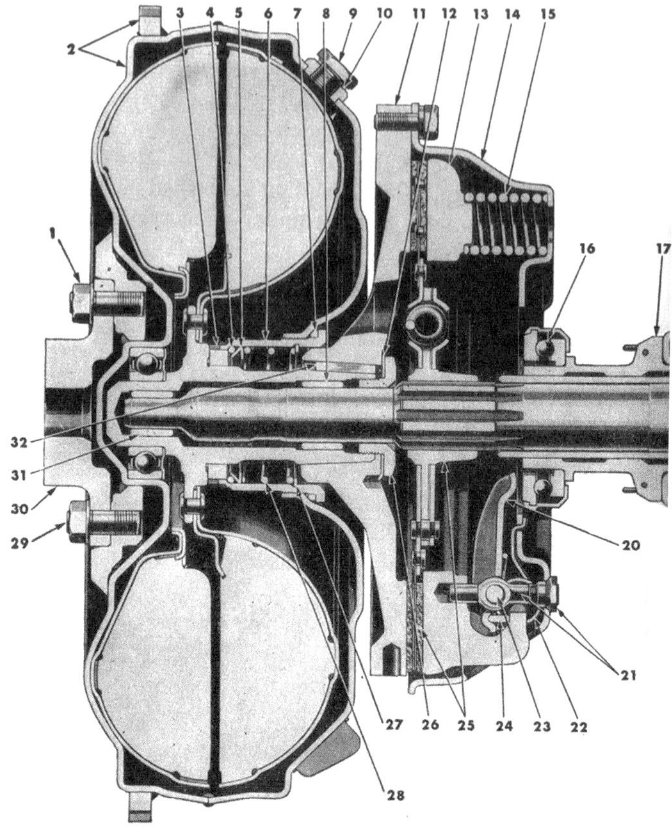

Fig. 8.21 Hydraulic clutch of the Gyromatic semiautomatic gearbox, produced since 1949 by the Dodge Division of Chrysler; we can see also a conventional friction clutch with pedal control

when the gearbox is in low gears, the third speed gears act as a different constant mesh gear.

These speed shifts didn't need clutch disengagement, because of the properties of the freewheel: the figures represent the positions in first and second speed, on the upper row and of third and fourth speeds on the lower raw; the first figure at left represent the idle position.

Dotted lines represent the power flow at the different speeds; the reverse speed was obtained by shifting an idler to engage the smallest wheel on the countershaft, with the largest on the output shaft.

A particular contribution of Europe to automatic gearboxes development was brought by DAF Daffodil, with the Variomatic transmission, introduced in 1950. This transmission was probably the first reliable application of the continuously variable transmission to a car.

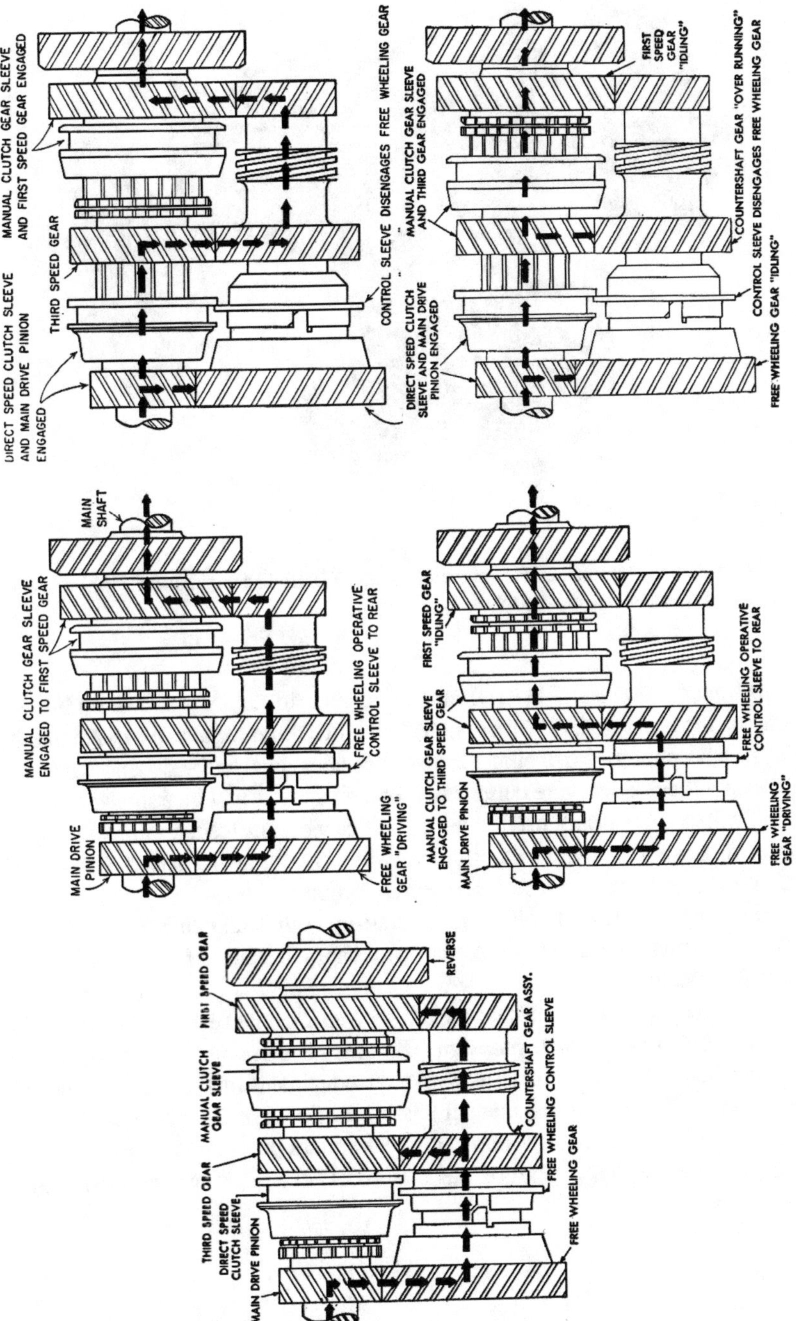

Fig. 8.22 Scheme of Chrysler's Gyromatic gearbox. The gearbox has a manual control to select a low (first and second speeds) and a high (third and fourth speeds) speed range. Speed shifts from first to second and from third to fourth and back is fully automatic

Fig. 8.23 Automatic Variomatic transmission of the DAF Daffodil of 1950. It is made by expandable pulleys and rubber belt, reinforced by cords

This transmission suitable for front engine, rear drive cars is shown on Fig. 8.23; the engine drove through a transmission shaft and a differential two expandable steel pulleys; these pulleys drove other similar ones connected to the driving wheels.

Driven pulleys sides were compressed by coil springs that guaranteed the correct friction with a rubber belt; driving pulleys sides were, instead, compressed by centrifugal masses and by engine manifold pressure. By this device speed ratio variation took into account engine speed and required torque.

A centrifugal friction clutch made car start-up completely automatic.

This transmission didn't receive other application because of its strong impact on vehicle architecture.

The concept was completely reworked by Van Doorne (the DAF holding Company), introducing a complete redesign. This study had as objective the development of a steel belt variable transmission of reduced dimensions, suitable to be interchangeable with conventional manual gearboxes. An experimental application was made by Fiat and by Ford and followed later by mass production.

This kind of automatic gearbox has now received a number applications on different car brands.

Chapter 9
Manual Gearboxes

9.1 Manual Gearboxes Classification

Gearboxes are normally classified according to the number of gear wheel couples involved in the motion transmission at a given speed; in case of manual vehicle transmissions, the number of gear wheel couples to be taken into account is that of the forward speeds only, without consideration of the final ratio, even if integrated into the gearbox.

Therefore there are:

- *Single stage* gearboxes.
- *Dual stage* or countershaft gearboxes.
- *Multi stage* gearboxes.

Figure 9.1 shows the three configurations for a four speed gearbox.

It is useful to comment on the rules of this scheme that are adopted generally. Each wheel is represented with a segment whose length is proportional to the pitch diameter of the gear; the segment is ended by horizontal strokes, representing the tooth width. If the segment is interrupted where crossing the shaft, the gear wheel is idle; vice versa, if the segment is crossing the line of the shaft without interruption, the wheel is rotating with the shaft. Hubs are represented according to the same rules, while sleeves are represented with a couple of horizontal strokes. Arrows show the input and the output shafts.

Single stage transmissions are mainly applied to front wheel driven vehicles, because in this case it is useful that the input and the output shaft are offset; on the contrary, in conventional vehicles, it is better that the input and the output shaft are aligned.

This is why rear wheel driven vehicles adopt usually a double stage transmission.

The multistage configuration is sometime adopted on front wheel driven vehicles with transversal engine, because the transversal length of the gearbox can be shortened; it is used when the number of speeds or the width of gears don't allow using a single stage transmission.

G. Genta and L. Morello, *The Automotive Chassis*, Mechanical Engineering Series,
https://doi.org/10.1007/978-3-030-35635-4_9

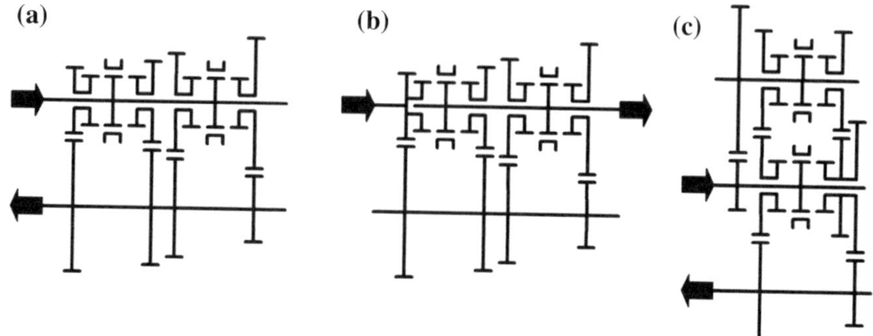

Fig. 9.1 Schemes for a four speed gearbox shown in three different configurations: **a** single stage, **b** double stage and **c** three stages

It should be noted that on a front wheel driven vehicle with transversal engine, having decided the value of the front track and the size of the tire, the length of the gearbox has a direct impact on the maximum steering angle of the wheel.

The positive result on transversal dimension of multi stage gearboxes is paid by higher mechanical losses, due to the increased number of engaged gear wheels.

It should be noted that on the three stage transmissions, shown on the picture, the axes of the three shafts don't lie on the same plane, as the scheme seems to show. In a lateral view, the traces of the three shafts should be represented as the vertices of a triangle; this lay-out reduces the transversal dimension of the gearbox. In this case and in other cases, we will show later on, the drawing is represented by turning the plane of the input shaft and of the second shaft on the plane of the input shaft and of the output shaft.

Gear trains used in the reverse speed are classified separately. The inversion of the speed is achieved by using an additional gear. As a matter of fact, in a train of three gears, the output speed has the same direction as the input speed, while the other trains of two gears only, have an output speed of opposite direction; the added gear is usually called *idler*.

The main configurations are reported on Fig. 9.2.

In the scheme a, an added countershaft shows a sliding idler, who can match two close gears, not in contact, as, for example, the input gear of the first speed and the output gear of the second speed. It should be noted that, in this scheme, the drawing doesn't maintain the real dimension of parts.

The scheme b shows instead two sliding idlers, rotating together; this arrangement gives additional freedom in obtaining a given transmission ratio. Also in this case the countershaft is offset from the drawing plane; arrows show the gear wheels that match, when the reverse speed is engaged.

The scheme c is similar to a, as the slider idler is concerned; it matches an added specific gear wheel on the output shaft and a gear wheel cut on the shifting sleeve of the first and second speed, when it is in idle position.

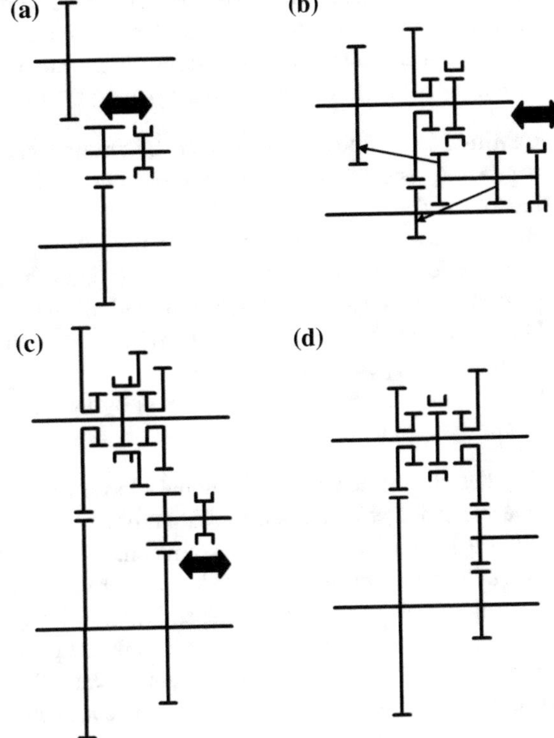

Fig. 9.2 Schemes used for reverse speed; such schemes fit every type of gearbox lay-out

The configuration d shows a dedicated couple of gears, with a fixed idler and a shifting sleeve.

The following are the advantages and disadvantages of the configurations shown on the figure.

- Schemes a, b and c are simpler, but don't allow the application of synchronizers (because the couples are not always engaged), nor allow the use of helical gears (because the wheels must be shifted by sliding).
- The scheme d is more complex but can receive a synchronizer and can adopt helical gears.
- Schemes a, b and c don't increase the gearbox length.

9.2 Mechanical Efficiency

The mechanical efficiency of an automotive gear wheel transmission shows values that are rather high, as compared with other mechanism performing the same function; nevertheless the value of this efficiency can't be neglected calculating dynamic performance and fuel consumption. The continuous effort of limiting fuel consumptions justifies the care of transmission designers in reducing the mechanical losses.

The total mechanical losses of a transmission are built up by terms that are dependent and independent of the processed power; the main terms are:

- Gearing losses; they are generated by friction between engaging teeth (power dependent) and by friction of wheels rotating in the air and in the oil (power independent).
- Bearing losses; they are generated by the extension of the contact area of rolling bodies and by their deformation (partly dependent and partly independent on power) and by their rotation in the air and in the oil (power independent).
- Sealing losses; they are generated by friction between seals and rotating shafts and are power independent.
- Lubrications losses; they are generated by the lubrication pump, if any, and are power independent.

All these losses depend on rotational speed of parts in contact and, therefore, on engine speed and selected transmission ratio.

Table 9.1 is reporting the values of mechanical efficiency to be adopted in calculations taking into consideration wide open throttle condition; they consider a couple of gearing wheels or a complete transmissions with splash lubrication; on the same table we can see also the efficiency for a complete powershift epicycloidal automatic transmission and for a steel belt continuously variable transmission. For the two last transmissions the torque converter has to be considered as locked-up.

It looks more correct to make reference to power loss measured as function of rotational input speed, instead of efficiency. Figure 9.3 shows the example of a double stage transmission, in fourth speed, at maximum power; the different contributions to the total are shown.

This kind of measurement is made eliminating step by step some of the transmission components, eliminating, this way, related loss.

In the first step all synchronizers rings are removed, leaving synchronizer hubs only; mechanical losses of non engaged synchronizers are, therefore, measurable. The loss is due to the relative speed of non engaged lubricated conical surfaces; the

Table 9.1 Mechanical efficiency of different transmission mechanisms

Mechanism type	Efficiency (%)
Complete manual gearbox with splash lubrication	92–97
Complete automatic transmission (ep. gears)	90–95
Complete automatic gearbox (steel belt; without press. contr.)	70–80
Complete automatic gearbox (steel belt; with press. contr.)	80–86
Couple of cyl. gears	99.0–99.5
Couple of bevel gears	90–93

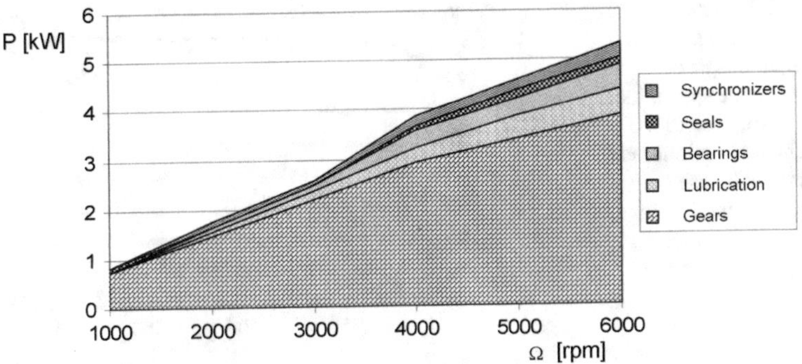

Fig. 9.3 Friction loss of a single stage gearbox designed for 300 Nm; on the horizontal axis is the input speed, on the vertical axis the power loss

value of this loss depends, obviously, on the speed and on the selected transmission ratio.

In the second step all rotating seals are removed.

In the third step the lubrication oil is removed, and therefore, the main part of the lubrication losses is eliminated; some oil must remain in order to leave the contact between teeth unaffected.

By removing the gear wheels not involved in power transmission, their mechanical losses are now measurable.

The rest of the loss is due to bearings; the previous removal of parts can affect this value.

A more exhaustive approach consists in measuring the complete efficiency map; the efficiency can be represented as the third coordinate of a surface, where the other two coordinates are input speed and engine torque. Efficiency calculations can be made by comparison of input and of output torque of a really working transmission.

Such a map can show how efficiency reaches an almost constant value at a not very high value of the input torque; it must not be forgot that standard fuel consumption evaluation cycles involve very modest values of torque and therefore imply values of the transmission efficiency that are always changing.

Figure 9.4 shows a qualitative cross section of the aforesaid map, cut at constant engine speed. It should be noted that efficiency is zero also at input torque values slightly greater than zero; as a matter of fact, friction implies a certain minimum value of torque, below of which motion is impossible.

A good approximation to represent mechanical efficiency is using the dotted broken line, as an interpolation of the real curve.

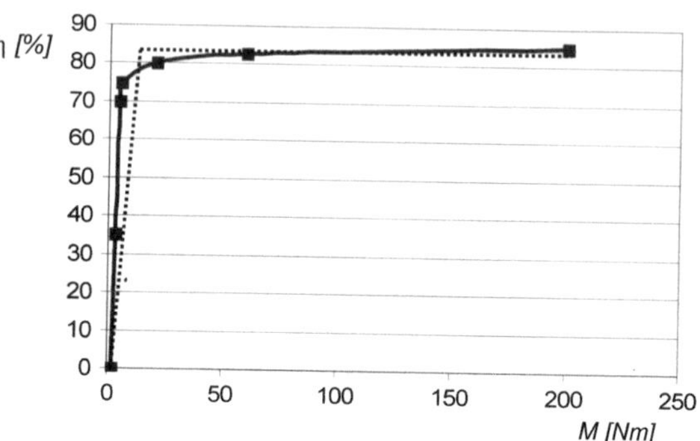

Fig. 9.4 Mechanical efficiency map, as function of the input torque at constant engine speed; the dotted line represents a reasonable approximation of this curve, to be used on mathematical models for performance and fuel consumption prediction

9.3 Manual Automobile Gearboxes

9.3.1 Adopted Schemes

In manual transmissions the process of changing speed and clutch engaging and disengaging are made with the contribution of the driver force only.

This kind of gearbox is made with helical gears and each speed has a synchronizer; some gearbox doesn't show the synchronizer for the reverse speed, particularly in economy minicars.

We reported, previously, a first classification; additional information is the number of speeds, usually between four and six.

Single step gearboxes are used on transaxles; this condition includes, with some exception, front wheel driven cars with front engine and rear driven cars with rear engine; this is true with longitudinal and transversal engine.

In all these situations the final drive is included in the gearbox that is therefore called also transmission.

Countershaft double stage gearboxes are used in conventionally driven cars, where the engine is mounted longitudinally in the front and the driving axle is the rear one. If the gearbox is mounted on the rear axle, in order to improve the weight distribution, the final drive could be included in the gearbox.

In multistage transmissions, some gear wheel could be used for different speeds. The number of gearing wheels could increase in some speed; this normally occurs in low speeds, because the less frequent use of these speeds reduces the penalty of worse mechanical efficiency on fuel consumption.

Cost and weight increase are justified by transmission length reduction, necessary sometimes on transversal engine with big displacement and more than four cylinders.

In all these gearboxes synchronizers are coupled for following neighbouring speeds (e.g.: first with second, third with fourth, etc.) in order to reduce overall length and to shift the two gears with the same selector rod.

We define as *selection plane* of a shift stick (it is almost parallel to the xz coordinate body reference system plane for shift sticks on transmission tunnel) the plane on

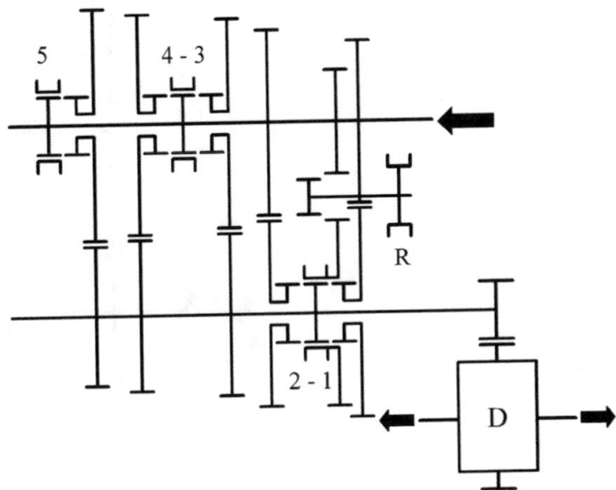

Fig. 9.5 Scheme for a five speed single stage transmission, suitable for front wheel drive with transversal engine

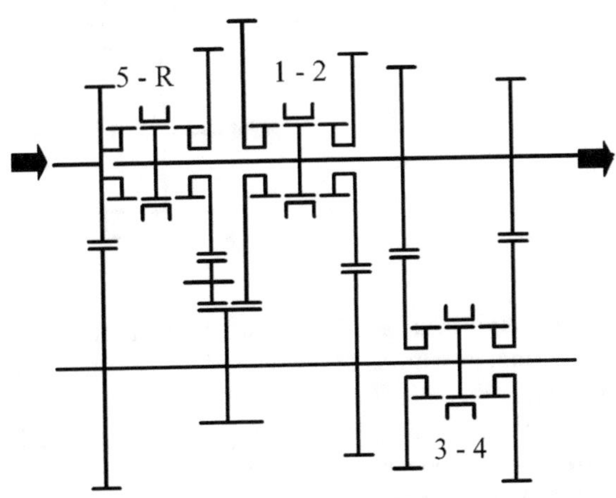

Fig. 9.6 Scheme of an on-line double stage gearbox for a conventional lay-out

which the lever knob must move in order to select two close speed couples. For instance, for a manual gearbox according to the FIAT scheme, first an second, third and fourth, fifth speed are organized on three different selection planes; the reverse speed can have a dedicate plane or share it with the fifth speed.

Figure 9.5 shows a typical example of a five speed single stage gearbox. First speed wheels are close to a bearing, in order to limit shaft deflection.

In this gearbox the total number of tooth wheels pairs is the same as for the double stage transmission shown on Fig. 9.6.

But, while in the first gearbox there are only two gearing wheels for each speed, in the second there are three gearing wheels for the first four speeds and none for the fifth. This property is given by the presence of the so called *constant gear* wheels (the first gear pair at the left) that moves the input wheels of the first four speeds; the fifth speed is a *direct drive* because the two parts of the upper shaft are joined together.

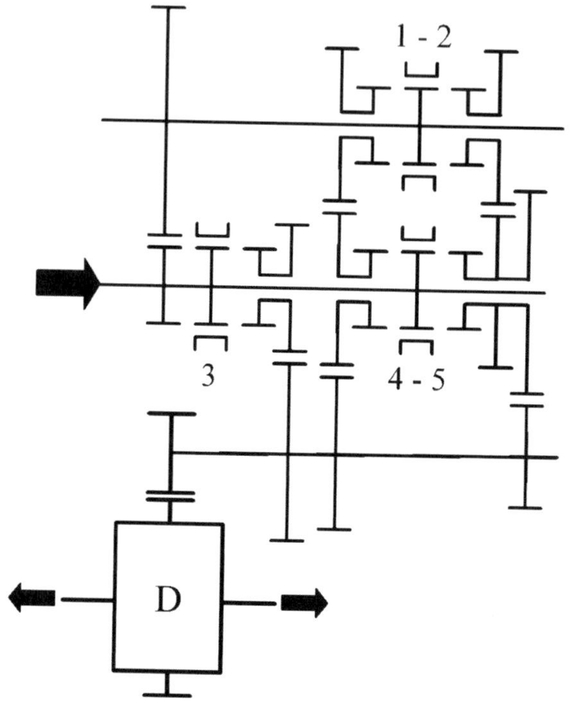

Fig. 9.7 Scheme of a triple stage five speeds gearbox, suitable for front wheel driven car with transversal engine

This single stage gearbox shows the fifth speed wheels couple positioned beyond the bearing, as witness of the upgrading, starting from an existent four speed transmission; in this case the fifth speed has a dedicated selection plane.

The double stage gearbox is organized in a completely different way but shows also the first speed pair of wheels close to the bearing. The direct drive is dedicated to the highest speed; also in this case the fifth speed shows a dedicated selection plane.

Six speeds double stage gearboxes don't show conceptual changes in comparison with the previous examples; synchronizers are organized in such a way as to leave first and second, third and fourth, fifth and sixth speeds on the same selection plane.

As we already noticed, the multistage configuration, shown on Fig. 9.7 allows a reasonable reduction of the lateral length of the gearbox. In this scheme, first and the second speeds only benefit of the second countershaft; power enters a countershaft through a constant gear pair of wheels and flows again to the input shaft at a reduced speed. Third, fourth and fifth speed have a single stage arrangement. Reverse speed is obtained with a conventional idling wheel.

9.3.2 Practical Examples

Four speed gearboxes represented the most diffused solution in Europe till the '70s, with some economy car with only three speeds.

With the installed power increase, with the improvement in aerodynamic performance of the bodies and with the increasing attention to the fuel consumption, it was necessary to increase the last speed, having the first speed always on same values; as a matter of fact car weight continued to increase and engine minimum speed didn't also change significantly.

To obtain satisfactory performance all manufacturers developed five speed gearboxes; now this solution is standard, but many examples of six speed gearboxes are available on the market, not only on sport cars.

Figure 9.8 presents an example of six speed double stage transmission with the fifth in direct drive; also in this case the first and the second pair of wheels are close to the bearing.

This rule is not generally accepted; on one side having the most stressed pairs of wheels close to the bearing allows a shaft weight containment; on the other side having the most frequently used pairs of wheels close to the bearing reduces the noise due to shaft deflection.

Synchronizers of fourth and third speed are mounted on the countershaft; this lay-out reduces the synchronization work, improving the shifting quality at the same dimension of the synchronizing rings. Synchronizers of first and second gear are, vice versa, on the output shaft, because of their diameter, larger than that of the corresponding gear; the penalty on the synchronization work is paid by the adoption of a double ring synchronizer.

Synchronizers on the countershaft cause a further advantage: in idle position the gears are stopped and don't produce the *rattling* noise; this subject will be better focused later on.

Figure 9.9 introduces an example of a single stage gearbox for a front longitudinal engine. The input upper shaft must jump over the differential that is set between the engine and the wheels. The increased length of the shafts has suggested adopting a hollow section. Because of this length the box is divided in two sections; on the joint between the two sections of the box additional bearings are provided to reduce the shaft deflection.

The input shaft features a ball bearing close to the engine and three other needle bearings that take only care of the radial loads. The output shaft has two tapered roller bearings on the differential side and a roller bearing on the opposite side. This choice is justified by the relevant axial thrust coming out of the bevel gears.

First and second speed synchronizers are on the output shaft and feature a double ring.

Reverse speed gears are immediately after the joint (the idler gear isn't visible) and has a synchronized shift. The remaining synchronizers are set in the second section of the box on the input shaft. The output shaft ends with the bevel pinion, part of the final ratio.

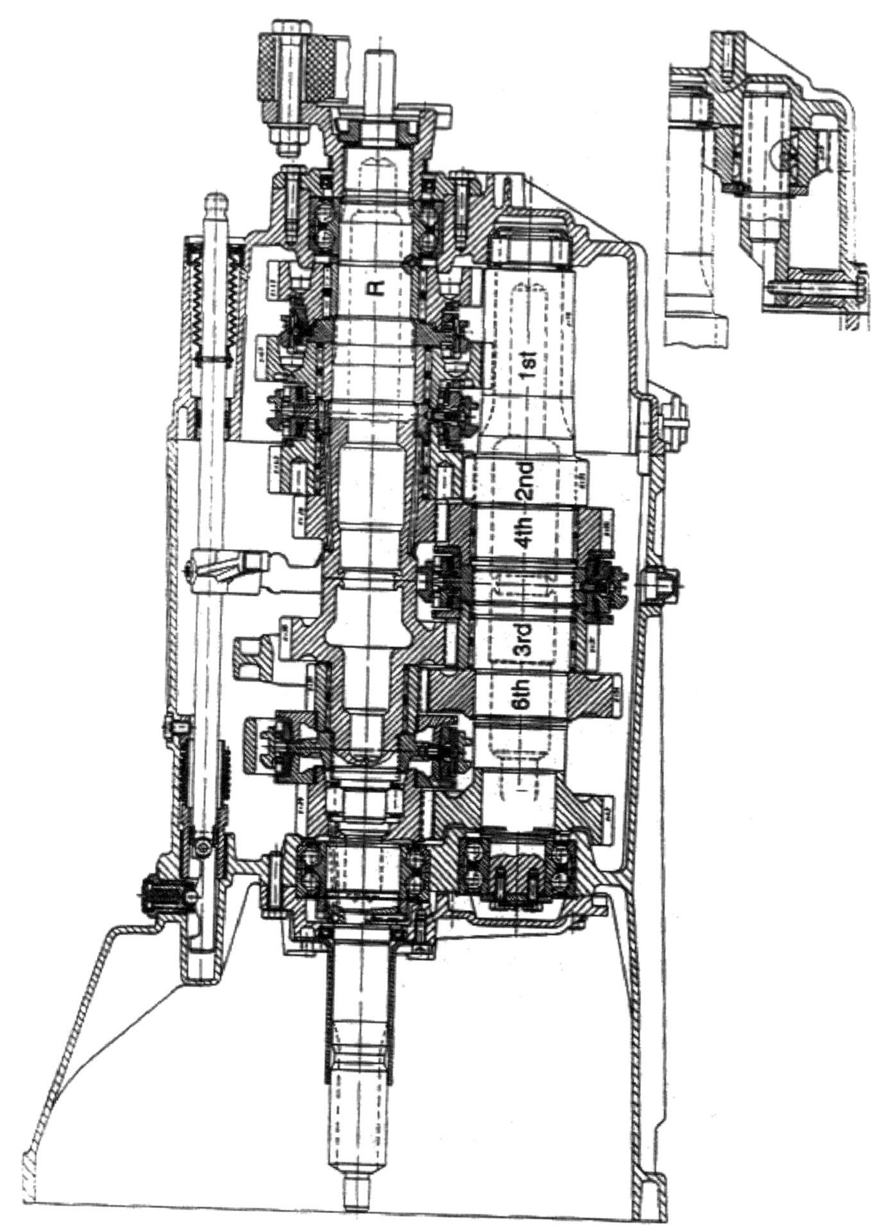

Fig. 9.8 Double stage six speed gearbox (GETRAG)

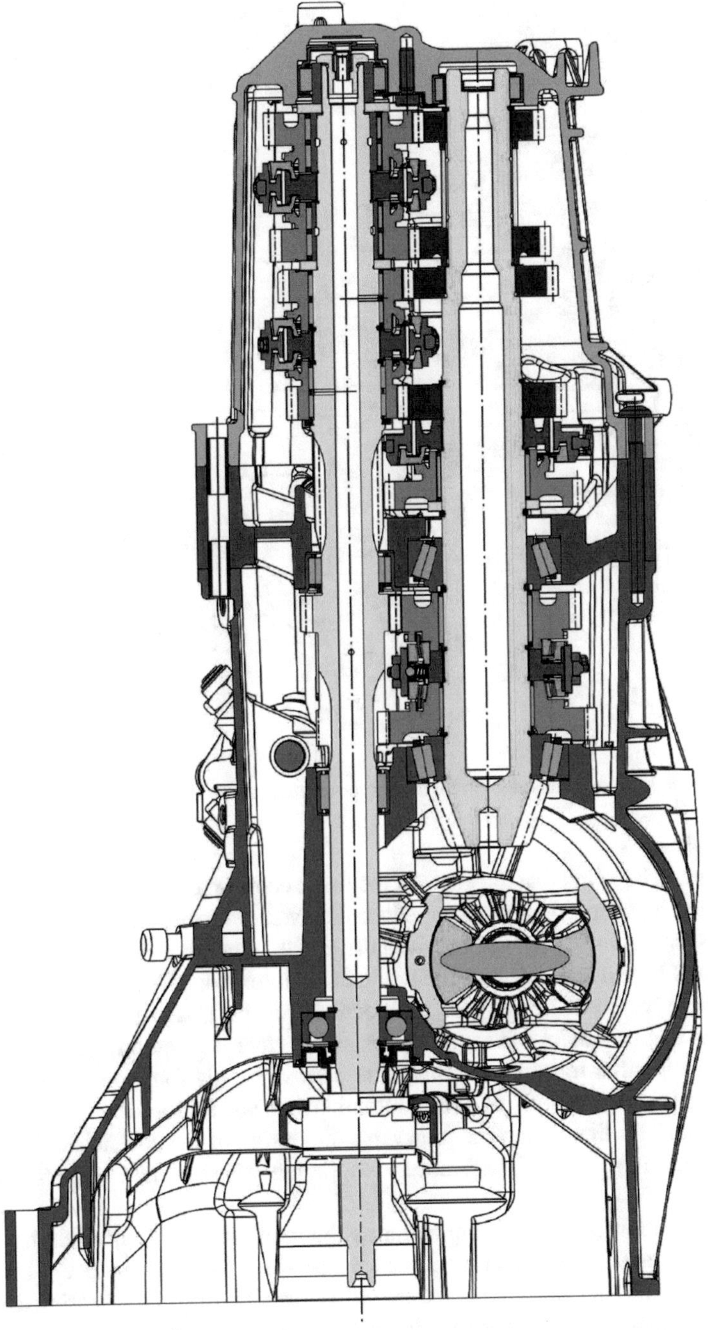

Fig. 9.9 Single stage six speed gearbox for longitudinal front engine (Audi)

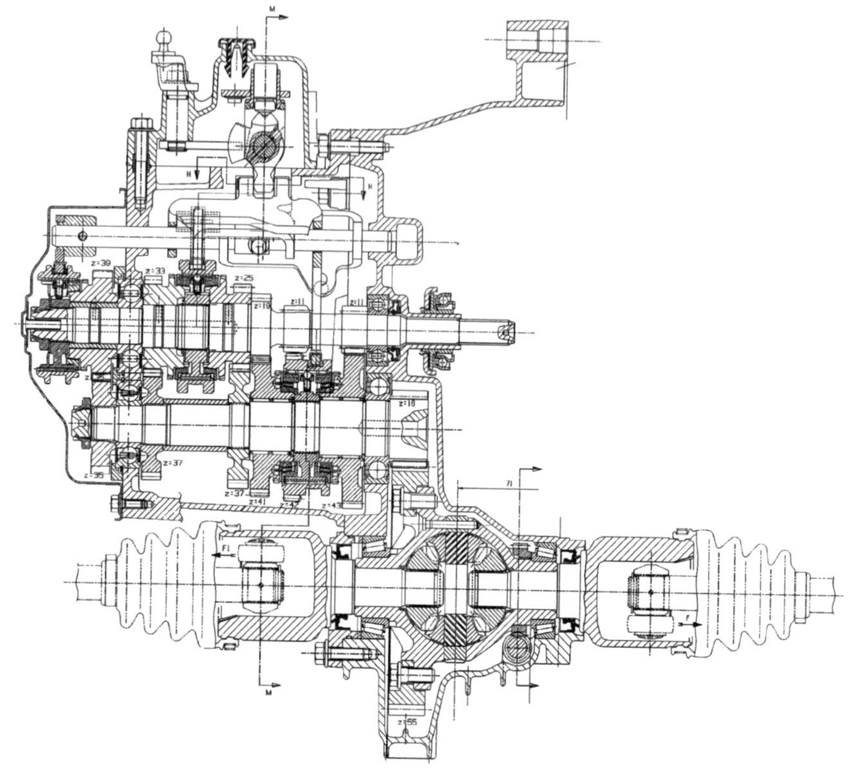

Fig. 9.10 Five speeds transmission for a transversal front engine (FCA)

It should be noted that the gears of the first, second and reverse speed are directly cut on the input speed, in order to reduce the overall dimension.

Most cars show today a front wheel drive with transversal engine; the number of gearboxes with integral helical final ratio is, therefore, dominant.

In these gearboxes geared pairs are mounted from the first to the last speed, starting from the engine side. An example of this architecture is given on Fig. 9.10.

As many other transmission that were born with only four speeds, it shows the fifth speed segregated outside of the aluminium box and enclosed by a thin steel sheet cover; this is to limit again the transverse dimension of the powertrain, in the area where there is potential interference with the left wheel in the completely steered position.

This solution is questionable as far as the total length is concerned but shows some advantage in the reduction of the span between the bearings. Every bearing is ball type; on the side opposite to the engine the external ring of the bearing can move axially, to compensate for thermal differential displacements.

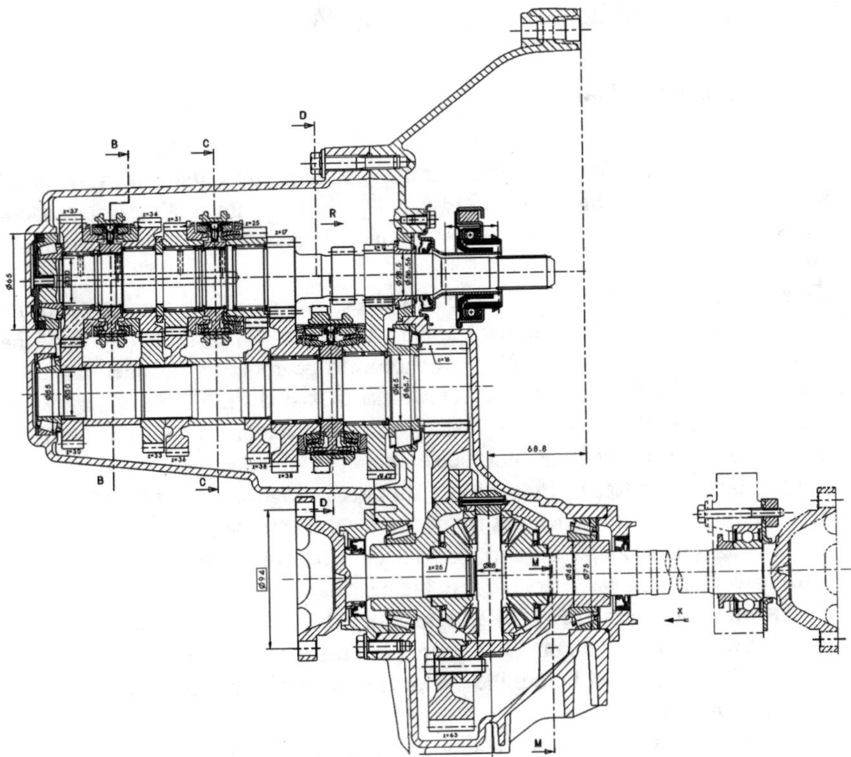

Fig. 9.11 Six speeds transmission for a transversal front engine (FCA)

One of the toothed wheels of the reverse speed is cut on the first and second shifting sleeve. The box is open on both sides; one of them is the rest of one of the bearings of the final ratio.

A big cover closes the boxes and, in the mean time, contains the second bearing of the final ratio and the space to install the clutch mechanism; it is also used to join the gearbox to the engine.

In this gearbox synchronizers are part on the input shaft and part on the output shaft.

Figure 9.11 shows the drawing of a more modern six speed gearbox, in which it was possible to install all the gears in a conventional single stage arrangement, thanks to the not very high value of the rated torque.

Starting from the engine, we see the gear pairs from the first to the sixth; as we have already said this arrangement is caused be the objective of minimizing the shaft deflection. The synchronizers only of first and second speed didn't find place on the input shaft; they are double ring type as for the first speed.

The reverse speed is synchronized and benefits of a countershaft not shown on this drawing.

9.4 Manual Gearboxes for Industrial Vehicles

9.4.1 Adopted Schemes

The gearboxes we are going to examine in this paragraph are suitable for vehicles of more than 4 t of GVW; lighter vehicles, usually called commercial vehicles, adopt gearboxes that are derived of automobile production, according to what was said in the previous paragraph.

Gearboxes used on industrial vehicles feature synchronizers too; they can be shifted directly, as in a conventional manual transmission, or indirectly with the assistance of servomechanisms. Non synchronized gearboxes are sometimes used on long haul trucks, because of their robustness. Assisted shifting mechanisms are diffused because of the easy power media availability. Automatic or semiautomatic transmissions are also used; the first type especially on busses.

For gearboxes with four up to six speeds, the double stage countershaft architecture represents a standard; the scheme is the same as we saw before.

The constant gear couple is used for all speeds but for the highest. Also in this case the lowest speed wheels are close to the bearings.

As shown on schemes of Fig. 9.12, highest speed can be either obtained in direct drive (scheme b) or with a couple of gears (scheme a); in this last case the direct drive is used for the speed before the last one: these architectures are called *direct drive* and *overdrive*.

In the figure, only the last and the first before the last speed are represented.

The choice between the two alternatives can be justified by the different vehicle mission; almost the same gearbox can be used on different vehicles with different speeds of more frequent use (for instance a truck and a bus).

Sometime the constant gear pair is set on output shaft, after the different speed gears; this configuration offers the following advantages:

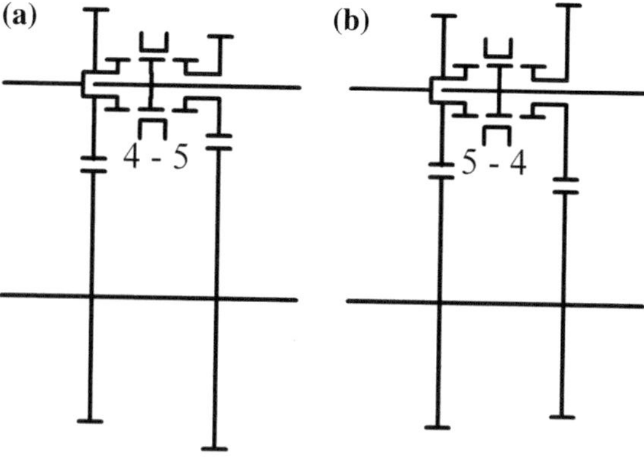

Fig. 9.12 Alternative constant gear schemes with last or first before the last speed in direct drive

- Reduction of the synchronization work, because of the smaller gear dimension, at the same torque and total transmission ratio.
- Less stress on the input shaft and on the countershaft.

On the contrary, there show the following disadvantages:

- Bearings rotate faster.
- Constant gear wheels are much more stressed.

This applies for the single range transmissions.

Multiple range transmissions feature, in addition to the main gearbox, other gearboxes that multiply the number of speeds of the main gearbox by their speeds number. With this architecture the total number of gear pairs might be reduced, at a given number of speeds, and, sometime the use of the gearshift lever can be simpler on a given mission.

This arrangement is used when more than six speeds are necessary. A multiple range transmission is therefore made out of a combination of different countershaft gearboxes, single range gearboxes or epicycloidal gearboxes.

Each added element is called *range changer* if it is conceived as to be able to use the main gearbox speed in sequence, in two completely non overlapping series of vehicle speeds; for example if the main gearbox has four speeds, the first speed in the high range is faster than the fourth speed in the low range.

It is called *splitter* if it is conceived in order to create speeds that are intermediate to the main gearbox speeds; in this case, for example the third speed in high range is faster than the third speed in low range, but slower than the fourth speed in low range.

We call main gearbox the gearbox with the highest number of speeds; the splitter and the range changer will be set in series before and after the main gearbox.

Figure 9.13 shows the scheme of a gearbox featuring a splitter and a range changer. The splitter is made out of a pair of wheels that work as two different constant gears for the main gearbox. The countershaft can therefore be moved at two different speeds, according to the position of the splitter unit. Because the main gearbox has four speeds, this splitter unit can make total of eight speeds, one of them being in direct drive.

At the output shaft of this assembly, there is a range changer unit made as a two speeds double stage gearbox with direct drive; this unit multiplies again by two the total number of obtainable speeds. The range changer is qualified by the significant difference of the two obtainable speeds.

The range changer can be made with a countershaft gearbox or with an epicycloidal gearbox with direct drive; the advantage in this case is the possibility of an easier automatic actuation, by braking some of the element of the epicycloidal gear.

It is also possible to have the range changer before the main gearbox and the splitter unit after the main gearbox.

A different way of defining functions of range change units, is to say that the splitter is a gearbox that *compresses* the gear sequence, because it reduces the gap

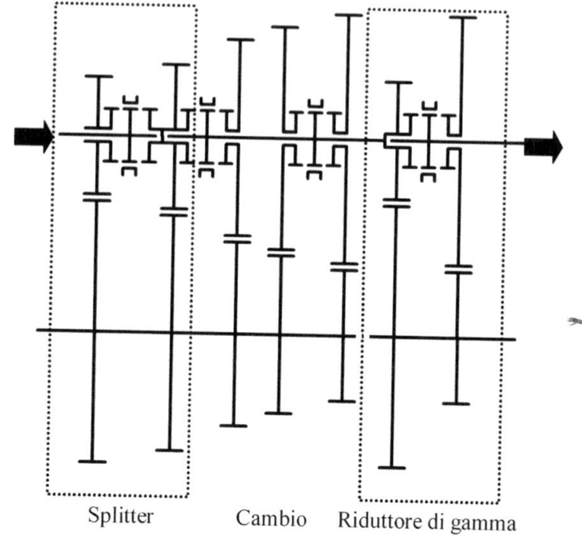

Fig. 9.13 Scheme of a sixteen speeds gearbox for industrial vehicles; it is made with a four gears main gearbox, a double speed splitter and a double speed range changer with direct drive

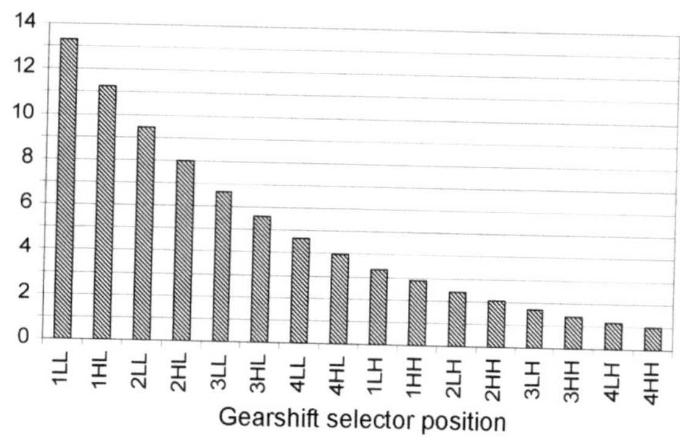

Fig. 9.14 Transmission ratios obtained with the scheme of transmission shown on Fig. 9.15; speed identification show the main gearbox speed with the number, the splitter position with the first letter, the range changer position with the second letter; L stands for low, H stands for high

between speeds, while the range changer is a gearbox that *expands* the gear sequence, because it increases the total range of the transmission.

Figure 9.14 explains the concept of compression; bars represent the ratios obtained in all shifting levers positions. Ratios obtained with the splitter unit in L position (first letter in the speed identification; L stands for lower ratio) are set in between the ratios obtained with the splitter unit in H position (H stands for higher ratio, in this case 1:1) and reduce the amplitude of gear steps of the main gearbox.

The same figure explains also the concept of expansion, showing on the same graph the ratio obtained with the range changer in H position (second identification letter) and in L position; the gear step between the first in low gear and the first in high gear is as big as the range of the main gearbox, and the total transmission range is widened.

The range changer is therefore used seldom, when the mission changes suddenly, as, for example, when leaving a normal road for a country road to be driven slowly, or when a strong slope with fully loaded vehicle is afforded. The splitter allows, instead, improving the dynamic performance of the vehicle, making available the optimum transmission ratio to obtain the desired power. The splitter is used therefore frequently: for example in a fully loaded vehicle all split ratios can be used in sequence, during full throttle acceleration from standing still.

The range changer and the splitter are usually made as modular units that can be mounted at both ends of the main gearbox, or changed with simple covers, in order to satisfy all application needs with limited total production investments.

The generalization of these concepts could suggest building transmission made of more range changing units arranged in series. They could be conceived as made only of splitter units with direct drive.

In such a case, with n pairs of tooth wheels it is possible to obtain a total of z transmission ratio, given by the formula:

$$z = 2^{n-1} . \tag{9.1}$$

The formula expresses the number of possible states that can be obtained of $n - 1$ pairs of gears; one unit is subtracted because one pair must be a constant gear to move the countershaft.

For example with four pairs of gears, four speeds can be obtained in a double stage gearbox; while in a cascade of splitters eight different speeds could be obtained. It must not be forgotten, while defining the best architecture, the goal of good shift manoeuvrability and the implications on mechanical losses.

Figure 9.15 shows the scheme of the sixteen speed transmission with splitter and range changer we already described. On this picture are represented the spans of shafts under torque; the dotted line shows where upper input and output shafts are loaded, while the solid line shows when the lower countershaft is loaded. The two lines are joined where a pair of wheels is gearing.

A totally different approach is shown by the double countershaft transmission on Fig. 9.16 (Fuller scheme); the power flow is divided between two countershafts by two constant gears and exits through a single output shaft. This configuration has been conceived with the objective to shorten the gearbox, because it is possible, in this way, to divide the torque on two gear wheels working in parallel. The teeth width can be reduced by about 40% at the same level of rated torque.

On the other side the transmission is much wider; this choice can represent a favorable compromise for certain vehicles such the road saddle tractors.

On this scheme, after the main gearbox, there is a three speeds splitter; one splitter speed id direct drive, one is over drive, the last under drive; the total number of speeds is therefore twelve.

It can be noticed that the reverse speed is obtained with the same wheels of the first speed and with two small idlers. With this arrangement is possible to use the splitter also in reverse speed.

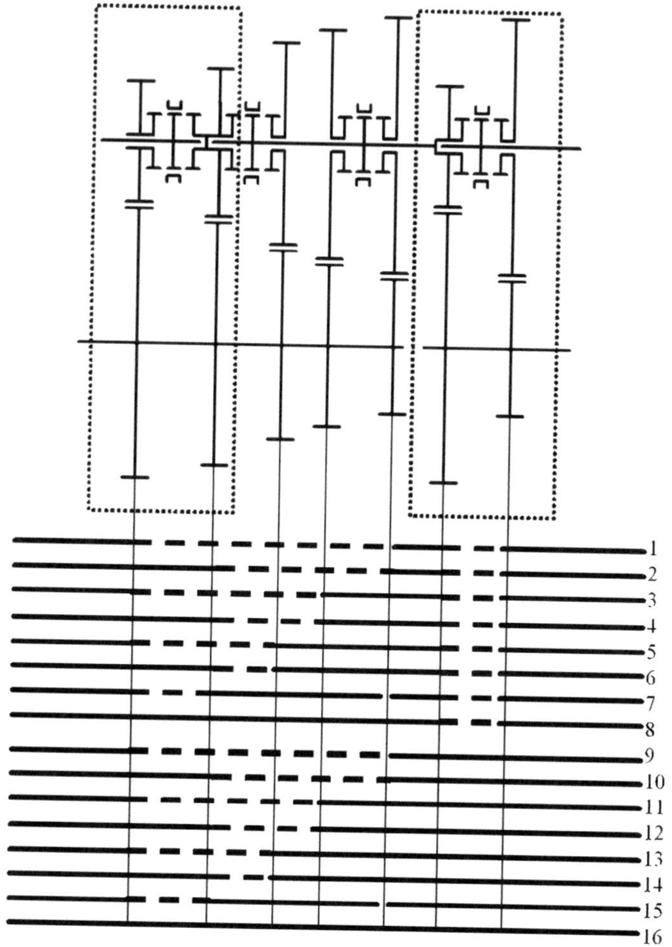

Fig. 9.15 Power flow scheme in the sixteen speeds of a gearbox; lines are dotted when the torque flows through the countershaft; they are solid when the torque flows through the upper shafts

9.4.2 *Practical Examples*

A practical example of a gearbox for a medium duty truck is shown on Fig. 9.17; in this example a double stage four speeds main gearbox is joined to a two speeds splitter, offering a total of eight speeds. The splitter unit with his direct drive can obtain the same transmission ratios of the main gearbox, while with the reduced speed can obtain transmission ratios that are in between the ratios of the main gearbox.

The section A-A on the lower right side shows a detail of the idler of the reverse speed; also the reverse speed is doubled by the splitter.

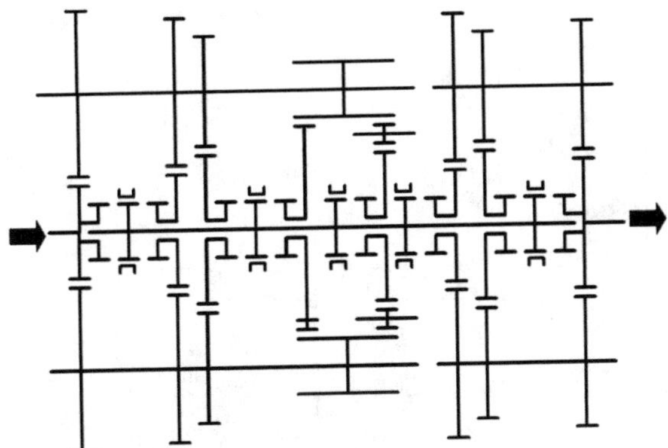

Fig. 9.16 Scheme of a Fuller gearbox featuring twelve speeds, made out by a main four speeds double countershaft gearbox and by a splitter gearbox with three speeds; one of these last is direct drive; the two countershafts in the main gearbox and in the splitter allow reducing the gear wheel width and therefore the total length of the gearbox

In the main gearbox the wheels of the first and reverse speeds are close to the rear bearing; the wheels of the following speeds are set in increasing order from the left to the right. The eighth speed is direct drive.

The synchronizers of the first and second speed show a double ring, while the reverse speed has no synchronization. This gearbox can receive a conventional manual shifting mechanism or a power assisted mechanism that can be fully automatic.

A three elements gearbox with a total of sixteen speeds is shown on Fig. 9.18. A four speed main gearbox (the same as in the previous example) is joined with a two speeds splitter and a two speeds range changer. The wheels in the main gearbox are ordered in increasing speed from the rear bearing.

The sixteenth gear is direct drive. The range changer is made with epicycloidal gears; when the rear shifting sleeve is moved to the left, the epicycloidal gear is blocked and acts like a locked joint.

When the rear shifting sleeve is moved to the right the annulus wheel is blocked with the casing and the carrier speed will be reduced, with the same direction of the input speed.

The reduced speeds stay in a range that is fully separated of the normal speeds range; they will be used when a very high torque or a very low speed are needed. The main gearbox shows ball bearings and tapered roller bearings, while the epicycloidal gear train, where the radial thrusts are self equilibrated, shows only needle bearings and a ball bearing.

The main gearbox countershaft is rotating also in idle speed; it shows a spline end that can be used to move auxiliary equipments such as a hydraulic pump useful to operate a tilting loading plane.

A practical example of the Fuller scheme is shown on Fig. 9.19; in this example the gearbox has a total of sixteen speeds and is made of a four speeds main gearbox, a two speeds splitter and a two speeds range epicycloidal gear changer.

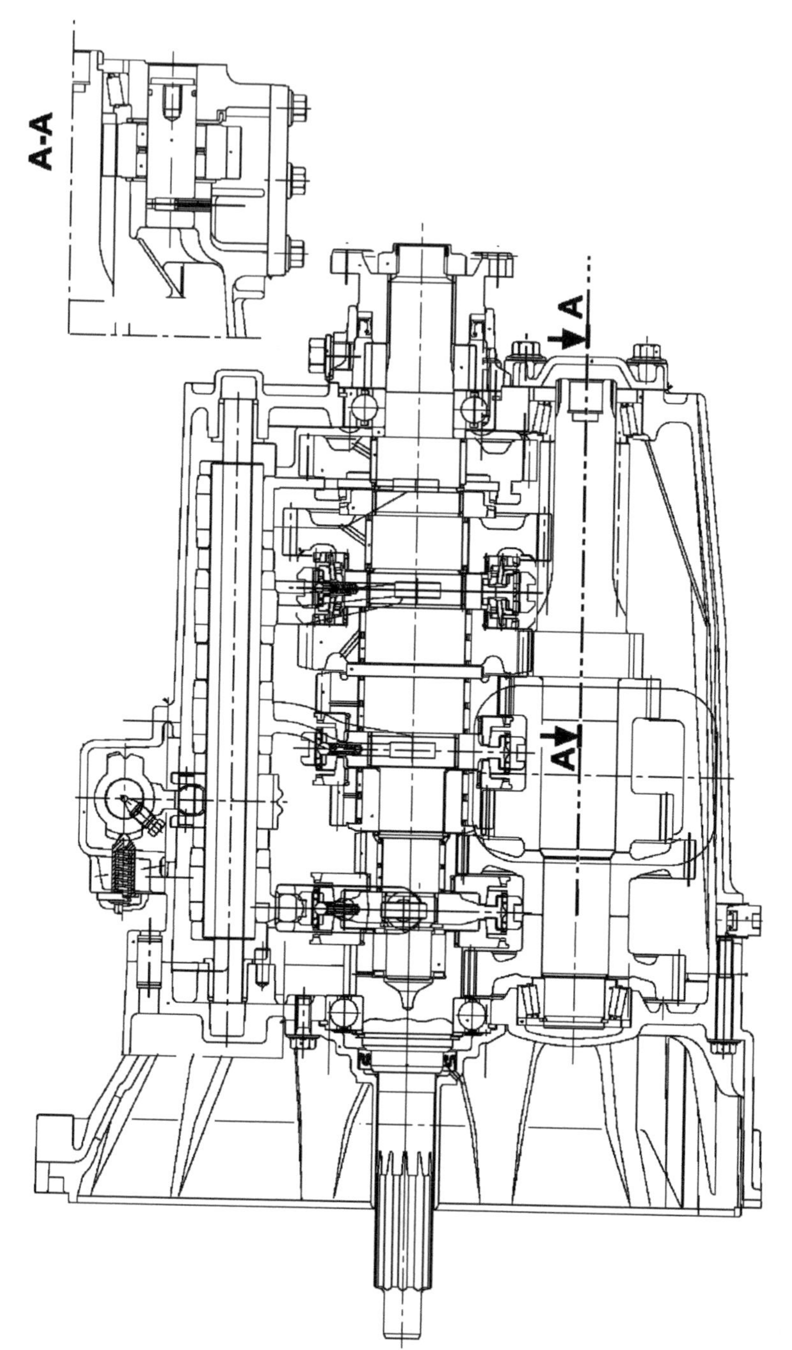

Fig. 9.17 Truck gearbox with eight speeds; on the lower side there is the reverse idler (IVECO)

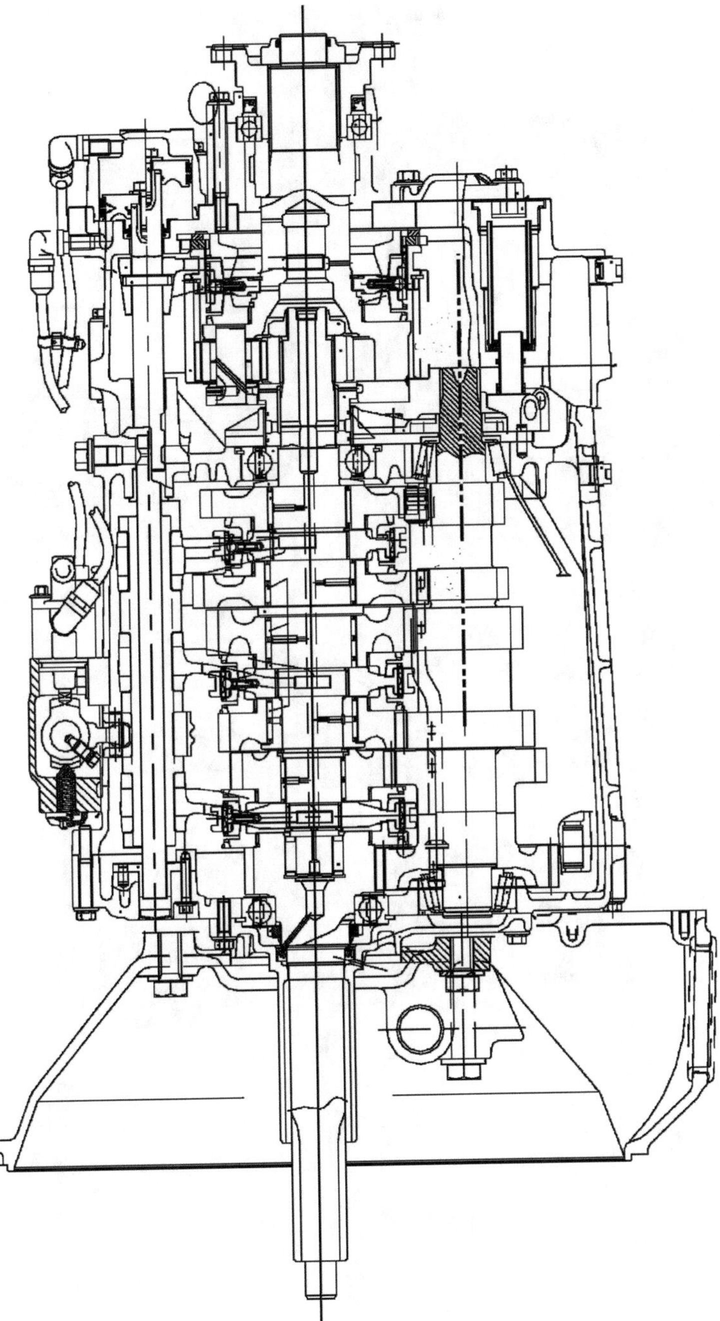

Fig. 9.18 Truck gearbox with sixteen speeds including splitter and range changer (IVECO)

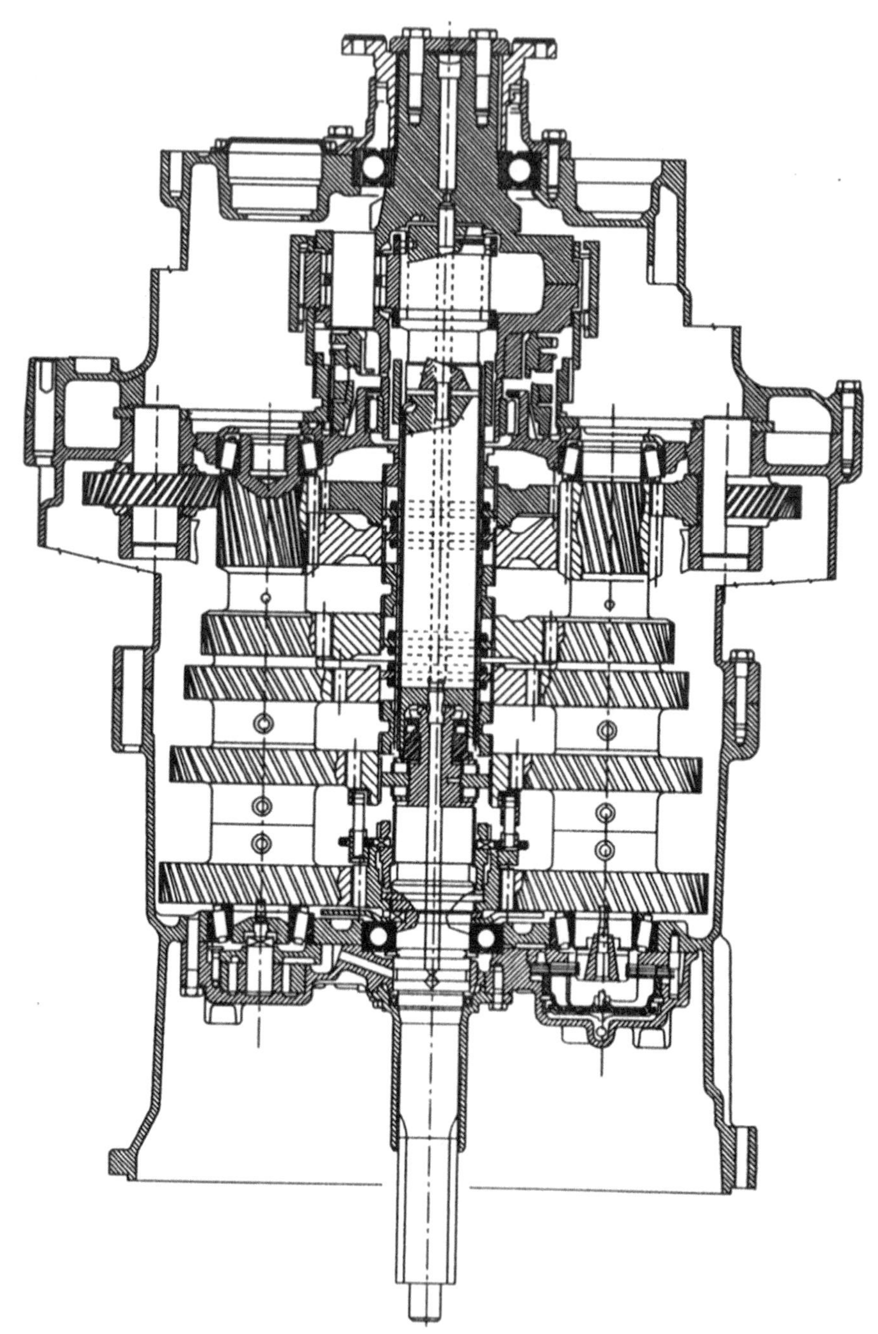

Fig. 9.19 Fuller gearbox (IVECO)

The two reverse speed idlers can be noticed. The splitter shift is synchronized, while the main gearbox features dog clutches.

The gear shifts are semiautomatic, manually preselected; in this case the gear shift lever doesn't move the shifting sleeves mechanically, but prepares an automatic sequence, where electric valves operate pneumatic actuators. The selection and shift motions don't occur when the lever is moved but when the power is cut off, because the accelerator pedal is released or the clutch pedal is depressed.

Chapter 10
Shifting Mechanisms

10.1 Internal Shifting Mechanisms

Mechanical devices that enables the driver shifting the pair of gearing wheels in such a way as to obtain the desired transmission ratio, are called *shifting mechanisms*. They are called *internal* if they are contained in the gearbox casings; they are called *external* when they mounted partly on the gearbox, partly on the vehicle body and connect the gearshift lever with the internal shifting mechanisms.

Shifting mechanisms perform a very import function, because they are responsible for the feeling, the driver receives from the shift stick, and therefore for the easy use of the gearbox itself. An ideal feeling should be positive and precise.

From an ergonomic standpoint, friction and clearance between moving parts of the shifting mechanisms can increase the shifting time and make the identification of the engagement positions of the shift stick more difficult. These positions, in fact, are not shown by any device but are learned by the driver through the feeling coming from the shift stick.

Internal mechanisms include selector bars and selector forks; these move the shifting sleeves and the synchronizers as it will be explained in a next chapter.

10.1.1 *Plunger Interlocking Device*

In order to understand how an internal mechanism works, we will refer to one of the gearboxes that we have seen in the previous chapter; we won't consider, at this time the synchronizer, since its presence doesn't change the nature of shifting manoeuvre.

As we have seen, shifting sleeves engage, through an internal spline, both with a hub on the gearbox shaft and with another hub on the idle gear wheel. Idle wheels are kept in the correct axial position by spacers and diameter transitions on shafts and are free to rotate; a gear wheel is locked on his shaft by shifting the sleeve on its hub.

G. Genta and L. Morello, *The Automotive Chassis*, Mechanical Engineering Series,
https://doi.org/10.1007/978-3-030-35635-4_10

Each sleeve has three remarkable positions; with reference to the sleeve on the lower right of Fig. 9.8 for the first and second speed:

- It is in neutral position, when engaged with no wheel hubs, as shown in the figure.
- If shifted to the left, it locks the second speed gear to the output shaft and applies the second gear ratio.
- If shifted to the right, it locks the first speed gear to the output shaft and applies the first gear ratio.

The gearbox has four different *sleeves*: one for the first and second speeds, one for the third and fourth speeds, one for the fifth speed and the last for the reverse speed. It should be noticed that in this gearbox the reverse speed is made by shifting an idler, not shown in the figure; when the reverse speed is engaged the idler is slit into mesh with the toothed sleeve of second and first (in neutral) and with a specific wheel on the input shaft.

A typical mechanism suitable for this task is shown on Fig. 10.1. It is made by three selector bars (4, 5, 6), each of them with a fork that engages with a groove on the shifting sleeve; by moving one bar at a time, the desired speed can be shifted. Each bar is moved by a gate 8, existing on each *fork*.

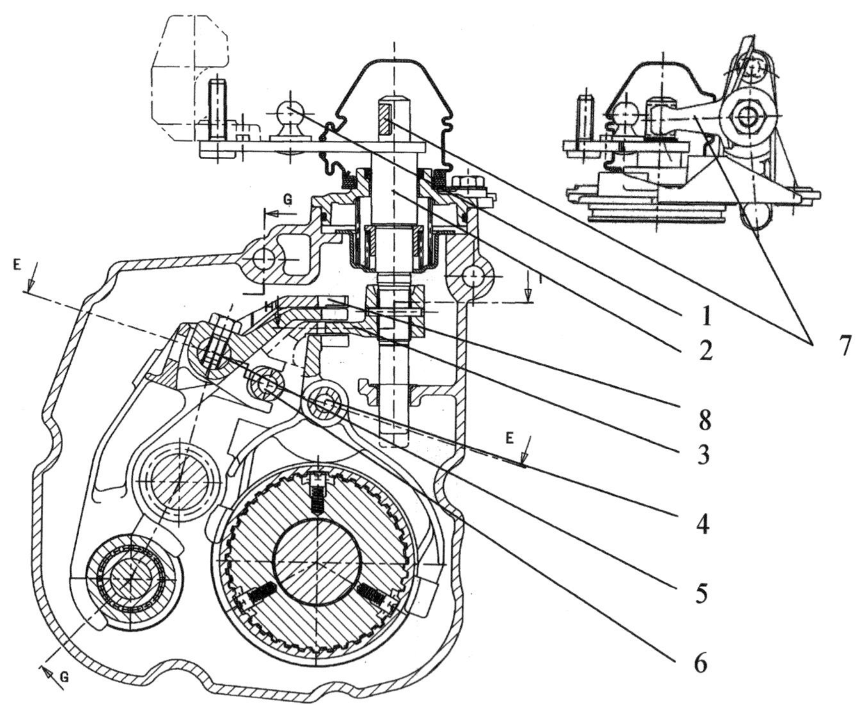

Fig. 10.1 Cross section of a transversal single stage gearbox on the internal shifting mechanism; 4, 5 and 6 are the shifting bars and forks moving the sleeves; a section of the sleeve on the output shaft, a section of the input shaft (upper left) and of the idling shaft of the reverse speed can be also seen (FCA)

A *finger* lever 3 on a shaft 2 can either rotate around the shaft center line or shift along this centre line; when shifted it selects the correct selector bar; when rotated, in one of the three gates, it shifts the desired speed. The finger matches with a suitable low clearance with the gates on the forks.

It should be noted that, in this gearbox, the reverse fork slides on the selector bar of the fifth speed but features its own gate. Acting on the lever 1 and 7, shifting and selection motions can be done.

When the gearbox is in neutral, all gates are aligned and the finger is set in the gate of the third and fourth fork.

In order to have a proper operation, there should be auxiliary devices suitable for:

- Maintaining the mechanisms in the selected position, even if the driver abandons the shift stick.
- Avoiding that more than one bar is moved in the same time, with fatal consequence for the gearbox integrity.

The first function is performed by plungers with spherical ends, matching spherical grooves on selector bars. Plungers are pushed into their grooves by springs that keep the bar in one of the three remarkable positions. Figure 10.2 shows the detail of such grooves (detail 7); the central groove if for the idle position, the other two for shifting the desired speed.

The second function is performed with *plungers* that are similar to those previously explained; they have their center line on the same plane of the bars center line (1 and 3 on Fig. 10.2); these plungers are set in two holes in the box, between the bars of first and second speed and of fifth and reverse speed. The bar of third and fourth

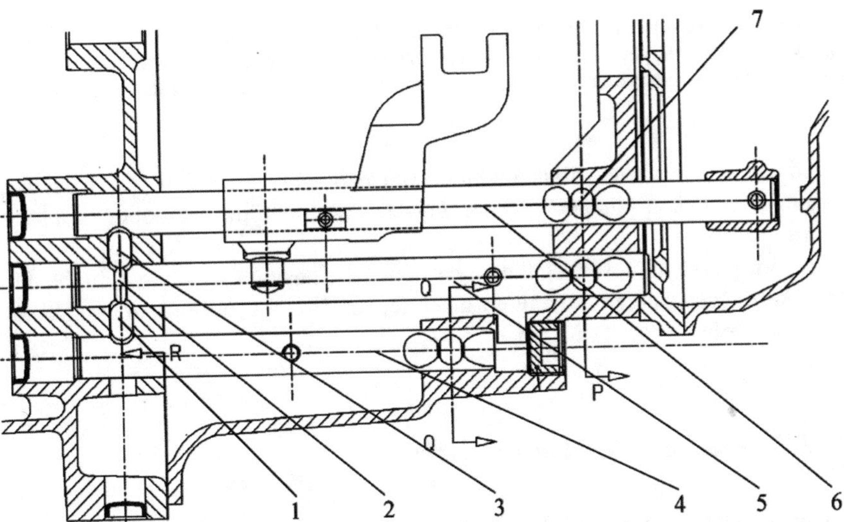

Fig. 10.2 Detail of the plunger and pin interlocking device, applied on the selector bars, in order to avoid the contemporary shift of two bars; if one bar is moved the plungers 1 and 3, through the pin 2 keep the remaining bars in idle position (FCA)

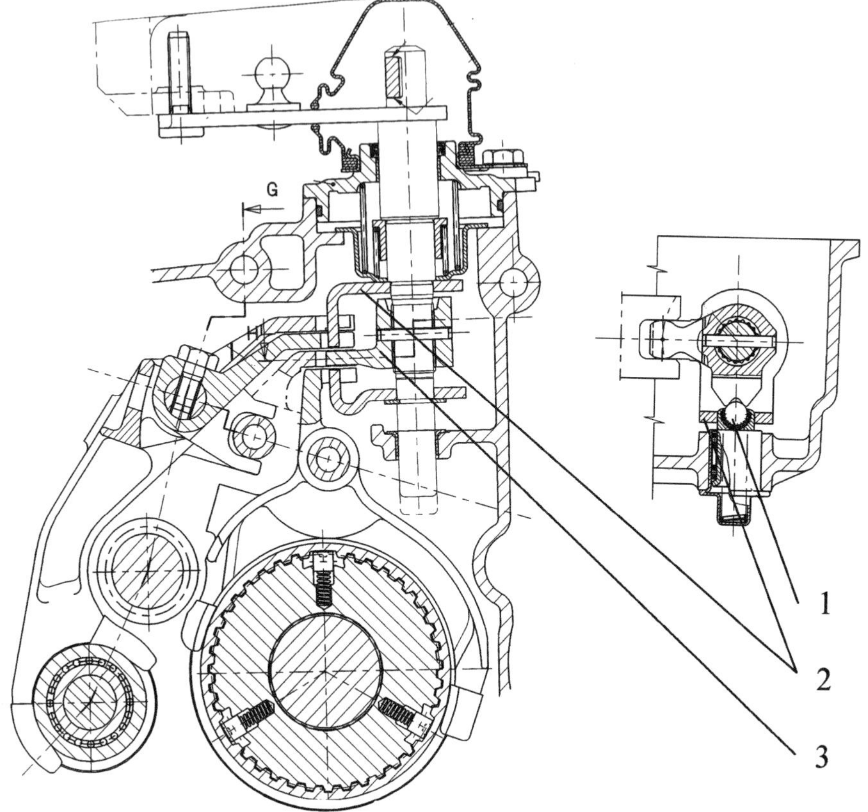

Fig. 10.3 Cross section of a transversal gearbox with calliper and finger interlocking device; the two prongs of the calliper 2 avoid that the finger moves contemporarily more than one fork; in fact if the finger could match more than one fork, the calliper prongs would block the selection (FCA)

speed contains in a hole, aligned with the others in neutral position, with a needle 2 inside; the length of plungers and needle is designed to allow the motion of a bar at a time.

A spring system must keep the finger in the idle position, as we said, in the third and fourth speed gate; stroke limiters must stop the finger at both ends of the selection stroke, in order to avoid sticking.

10.1.2 Calliper Interlocking Device

Clearance between bars and their bearings, shape errors between plunger and their spherical seat, wear between sliding parts can negatively affect the effectiveness of

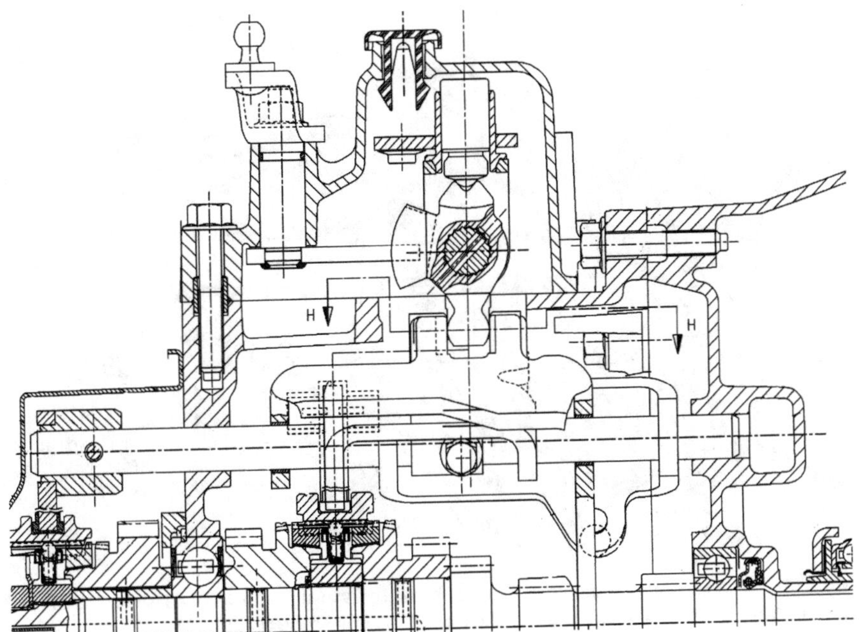

Fig. 10.4 Cross section of a transversal gearbox; the forks of first and second speeds and of third and fourth speeds can slide on the bar of the fork of the fifth speed (FCA)

this mechanism and, even if the contemporary motion of two bars does never occur, there can be sticking of parts that can make the selection difficult or unpleasant.

Figure 10.3 shows an improved solution. There are neither plungers nor grooves. The finger 3 that moves the bars shows on his view on the right of the drawing three spherical seats that can match a ball pressed by a spring (only one is shown). The three corresponding positions fit the neutral position of the three bars.

A U shaped bracket 2, called *calliper*, with two prongs, matches the cartridge of the ball; the calliper can slide up and down in the neutral position, but can't rotate. The two prongs match the end of the finger with no clearance.

The calliper allows the motion of one bar at a time, while locking the remaining bars in neutral position.

If the finger is moved in an intermediate position, the calliper would block in neutral position all the mechanism.

A possible simplification of this mechanism is shown on Fig. 10.4; in this case there is only one bar.

As a difference with the previous gearbox, where each fork has his bar, in this case each fork, made of stamped steel sheet, is sliding on the bar of the fifth gear. Each fork shows a gate that can match the finger of the selection mechanism.

These gates are aligned when the forks are in neutral position; devices, not shown on the drawing, avoid any rotation of forks around the bar, leaving only the sliding

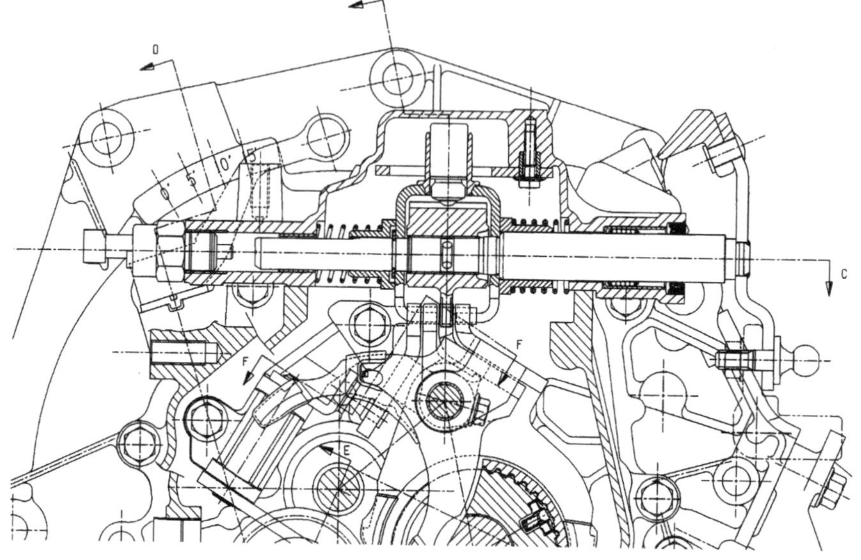

Fig. 10.5 Cross section of a transversal gearbox; the forks of first and second speeds and of third and fourth speeds can slide on the bar of the fork of the fifth speed (FCA)

motion possible. In this case the only bar can slide on its bearings and has the double function of moving one fork and guiding the remaining two.

As in the previous gearbox, the calliper shown on Fig. 10.5 moves with the finger only horizontally and avoids the contemporary selection of two forks.

In order to produce a positive feeling, the bar of the finger is sliding on ball bearings.

The same care is devoted to the positioning ball of the finger (1 on Fig. 10.4) that has balls either for its rotation and its translation.

10.2 External Shifting Mechanisms

As we already said, external shifting mechanisms are used to move the internal shifting mechanism as the driver moves the gear shift lever.

This problem finds an easy solution when the engine is in the front of the car in longitudinal position and the traction is applied to the rear axle, because the mechanism is very close to the natural position of the hand of the driver. Selection and engagement motions are coherent with those of the shifting bars and forks; the engagement positions of the gear shifting lever were historically defined by the simplest solution of the most common architecture.

The solution is quite more complicate when engine and gearbox are mounted transversely or when the lever is not mounted on the tunnel, but on the dashboard or on the steering wheel shaft.

- To convey the motion of the lever in a different direction; in fact, for example, in a front wheel driven car with transversal engine, the engagement motion in the gearbox is transversal, while the corresponding motion of the shift stick is longitudinal, when the lever is on the tunnel or on the dashboard, or almost vertical, when the lever is on the steering wheel shaft.
- To maintain the engagement position of the shift stick unchanged, with reference to the neutral, even if the powertrain is moving, because of the vehicle vertical acceleration or because the torque variation; the problem is particularly important, because on the front wheel driven cars the powertrain is reacting also to the wheel traction force.
- To guarantee a precise and positive feeling on the lever with a limited shifting effort.

We now try to define the attributes given on the previous sentence:

- *Precision* is the capacity to maintain unchanged the engagement positions of the shift stick, in any working condition or, at least, unchanged with reference to the neutral position.
- *Positivity* is the capacity to react to the drivers hand in a consistent way; drivers appreciate limited efforts for the selecting motion and for the first part of the engagement stroke; higher efforts are accepted and expected at the end of the engagement stroke, but they must quickly vanish, when the gear has been engaged.
- *Smoothness* is the capacity to limit the variation of the reaction force with reference to an ideal archetype; the reaction force must be not only small but must also show small variation in different manoeuvres; the opposite of smooth is a shift stick sticking and slipping.

Stick and slip occurs when two elastic parts are in contact and between them is present a relevant friction: an example is a rubber eraser on a drawing paper. The resulting motion between the two elements is characterized by a series of small displacements, with an initial part due to elastic deformation and a following one with real slip, that interrupts when adhesion is again reestablished.

10.2.1 Bar Mechanisms

A bar mechanism is a mechanical system made of bars articulated by means of spherical heads. Some bars in the longitudinal direction replicate the motion of the lever in a position close to the engine; other bars in the transversal direction connect the end of these bars with the internal shifting mechanism.

The widest motions of the powertrain (rotations caused by the reaction torque and vertical oscillations due to the road uneven surface) are contained in the xz plane of the vehicle reference system; these motions won't push the transversal bars.

Figure 10.6 shows an application example to be matched to a single bar calliper and finger interlocking device gearbox.

The shift stick is connected to the bar 1 for the selection motion, with a pivot with transversal axis; the selection motion will rotate the bar around its axis and will leave the lever free to perform the engagement motion. Bar 1 is mounted longitudinally.

A second bar 2 is linked to the shift stick in a different point, again with a pivot with transversal axis; the engagement motion of the lever shifts this bar longitudinally.

The two motions are decoupled; the end of the bar 1 shows a finger that moves transversely a second bar 3.

A rocker arm 4 converts the longitudinal motion into transversal motion of the bar 5. Bars 3 and 5 move the internal shifting mechanism in the gearbox.

All articulated heads of bars 1 and 2 and of the rocker arm 4 are mounted on the plates A and B, fixed on the car body.

If bars 3 and 5 have a sufficient length, all powertrain motions contained on the plane xz will move the lever in a negligible way. The powertrain motion on the yz plane is usually very small, because the engine suspension in this direction is very stiff, since there are no vibrations to be filtered.

The behavior of this mechanism can be adequate to his mission if clearance on pivots is limited and the lubrication of bearings is acceptable.

This target is sometime difficult to be reached because parts are exposed to dust and splashed water. In addition to that the sealing on the plate A, that insulate the passenger compartment from dust and noise, where bars are crossing the firewall, can introduce stick and slip on their motion. These mechanisms are progressively abandoned in favour of the cable mechanism.

10.2.2 Cable Mechanism

Figure 10.7 shows a cable mechanism. In this example the shift stick is mounted on a spherical joint, made on the plastic element 3; two different cable ends are present: 1 for the selection motion, 2 for the engagement motion.

The end 2 is fixed to the lever and copies the engagement longitudinal motion; the end 1 through a rocker arm articulated in 5 moves again longitudinally, but copies the transversal selection motion.

It should be noticed that the lever knob has a sliding collar 6 that in his natural position, imposed by a spring, avoids the selection of the reverse speed, because of the interference with the nose 7; reverse speed can be selected only if the collar is voluntarily pulled up.

Internal shifting mechanism is the same as in the previous paragraph. The two basic motions of the lever are transmitted to the internal shifting mechanism by two flexible cables (detail A).

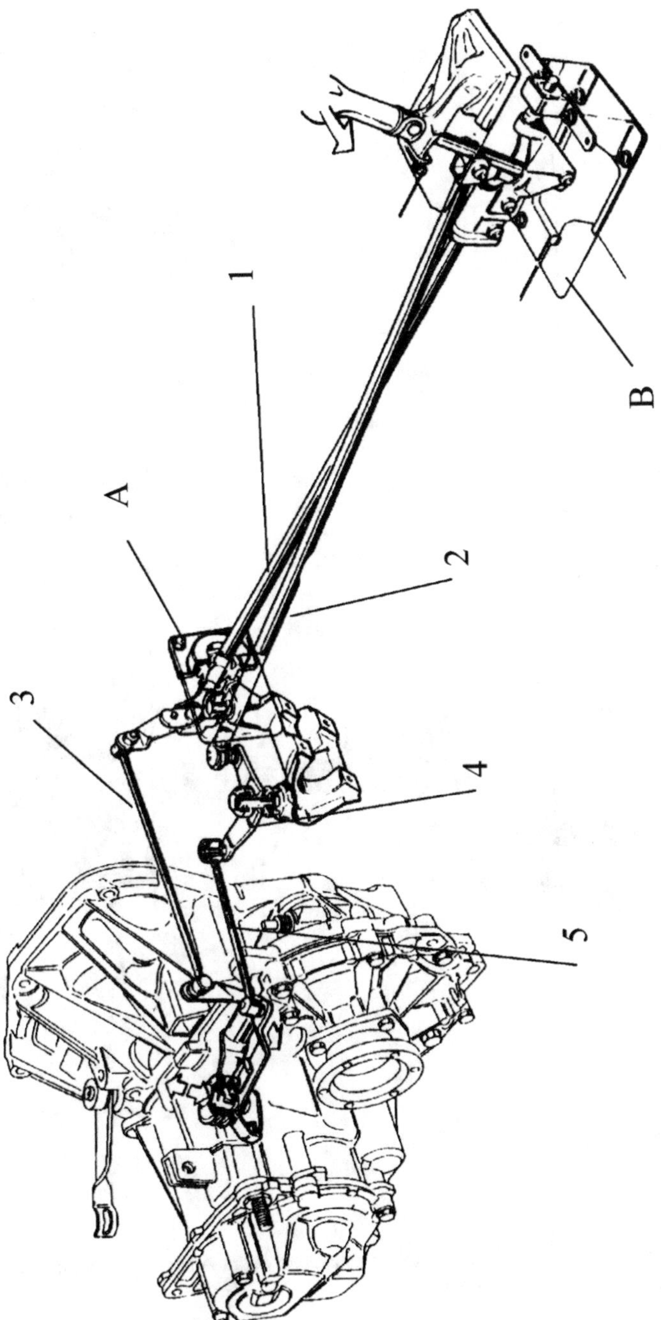

Fig. 10.6 View of a bar external shifting mechanism; bars 3 and 5 are not affected by the suspension motions of the powertrain; therefore the suspension motion don't affect bars 1 and 2 and the engagement position of the shift stick (FCA)

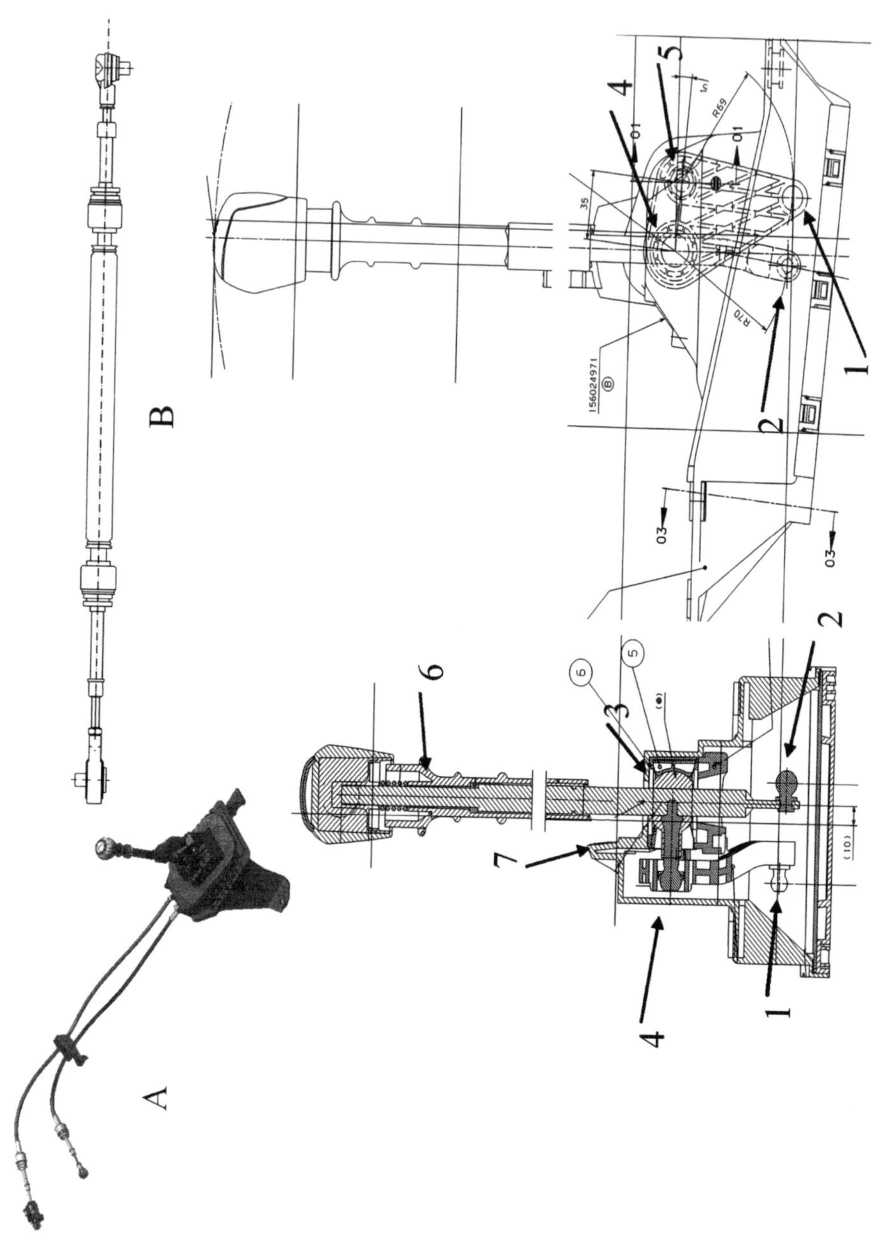

Fig. 10.7 Detail of a shift stick with cable mechanism; detail A shows the complete lever, while detail B shows the cable itself (FCA)

A flexible cable (detail B) is made with a multilayer steel wire sliding inside a sheath. The wires are wound on a cylindrical helix and, thanks to the good lubrication granted by the sheath protection are very resistant but very flexible too.

Cable shows a mechanical resistance higher than that of a wire of the same cross section, but his rigidity is only multiple as that of a single wire. The sheath is made with a spirally wound flexible sleeve, inserted in a plastic tube completely sealed from the outside.

The sheath is comparable to a structure which flexible in bending but very stiff in compression; it must be installed with bending radii not smaller than about hundred times the cross section diameter and with lengths a little bit larger than what necessary, in order to compensate for the powertrain motion.

If this last can happen without changing the sheath length the motion of the internal shifting mechanism will replicate the shift stick knob motion very accurately. Smoothness is conditioned by friction between cable and sheath.

Precision and positivity are also affected by mountings stiffness at the sheath ends; the stiffness of these mountings is conditioned by powertrain vibrations.

In fact, stiff mountings could be very efficient in transmitting noise and vibration to the drivers hand and to the passenger compartment; but it is very important that brackets sustaining the cable ends are very stiff, to not affect negatively the trade-off between manoeuvrability and comfort.

Mechanical efficiency is affected by contacts between cable and sheath; contacts are caused by bending diameter and length. Lubrication conditions and local contacts at the cable ends may also influence mechanical efficiency.

Chapter 11
Start Up Devices

11.1 Friction Clutch

11.1.1 Clutch Functions

In clutches friction is applied to transmit the torque between an input and an output shaft. It is built with three discs; two of them are connected to the engine crankshaft and one, between the other two, is connected to the gearbox input shaft. These discs are respectively named *driving plates* and *driven plate*.

Driving plates can change their distance and their load applied to the driven plate, through a suitable mechanism; through this mechanism the plate can be disengaged or engaged with a variable torque.

A spring system, as a matter of fact, provides the pressure force between driving plates and driven plate, in such a way as friction forces can be established between the discs; the intensity of the thrust determines the value of the transmitted or transmittable torque.

The distinction between *transmittable* and *transmitted torque* refers to two different conditions:

- Clutch completely engaged, without relative motion between the discs.
- Clutch partially engaged with relative angular speed between the driving and the driven discs.

In the first case the transmitted torque is equal to the engine torque and can be at maxim equal to the friction torque between the discs; in the second case the torque is determined by the spring load and is equal to the friction sliding torque between the discs.

The function of the clutch is the following:

- To transmit the torque from the engine crankshaft to the gearbox input shaft, when there is a difference of speed between the two parts, in particular when vehicle is stalling or moving back.

G. Genta and L. Morello, *The Automotive Chassis*, Mechanical Engineering Series,
https://doi.org/10.1007/978-3-030-35635-4_11

- To connect positively the two shafts, once they are synchronized, in such a way as to transmit all the engine torque.
- To separate engine and transmission, when transmission speed must be changed or the car stopped, without stopping the engine.

In addition to these functions, these functions have been added, in recent times:

- To absorb torque pulses caused by the engine polar inertia, in case of misuse of the clutch (main *torsion damper*).
- To control the driveline torsional stiffness, in such a way as to avoid vibration and noise when harmonic frequencies of the engine torque match transmission vibration modes (secondary *torsion damper*).

The springs of the clutch, formerly of helical type, are now *diaphragm* type, because of the advantages they offer:

- Reduction of powertrain length, at the same level of torque.
- Reduction of parts number.
- Easy reduction of weight unbalance, because of geometrical simplicity.
- Simplification of the internal engagement mechanism.
- Wear doesn't affect maximum transmittable torque.
- Reduction of disengagement load on the clutch pedal.
- Diaphragm springs are insensitive to centrifugal forces; on the contrary, helical springs with cwnter line parallel to rotation axis are loosing load because of the centrifugal acceleration deflection.

Figure 11.1 shows the cut-out view of a diaphragm spring clutch; the engine shaft 1 is flanged to the flywheel 2 which presents one of the active surfaces of the driving plate. Cover 3 is also flanged to the flywheel and puts in rotation the second driving plate, pushed by the diaphragm spring. On the driven plate 4 are mounted friction linings; the driven plate has a spline on the gearbox input shaft 5.

The thrust bearing 6 can be axially moved and reduce the reaction thrust of the spring, until the driven plate is disengaged. The bearing can slide on the tube 7 and is moved by the fork 8. The assembly of the moving driving plate, also called *pressure plate*, of the cover and the related elements utilized for the disengagement are called internal *disengagement mechanism*.

11.1.2 Disengagement Mechanism

Figure 11.2 reports a schematic cross section of a double mass flywheel clutch that will be explained later on. If we neglect for this time the increased complexity of the driving plate, we can consider the clutch as composed by two main parts: one part is the working surface machined on the engine flywheel and a second working surface facing the first one, rotating with the first, but free to move axially.

The picture represents on the left the engaged clutch, on the right the disengaged clutch. Some important details can be observed: the gearbox input shaft is centred

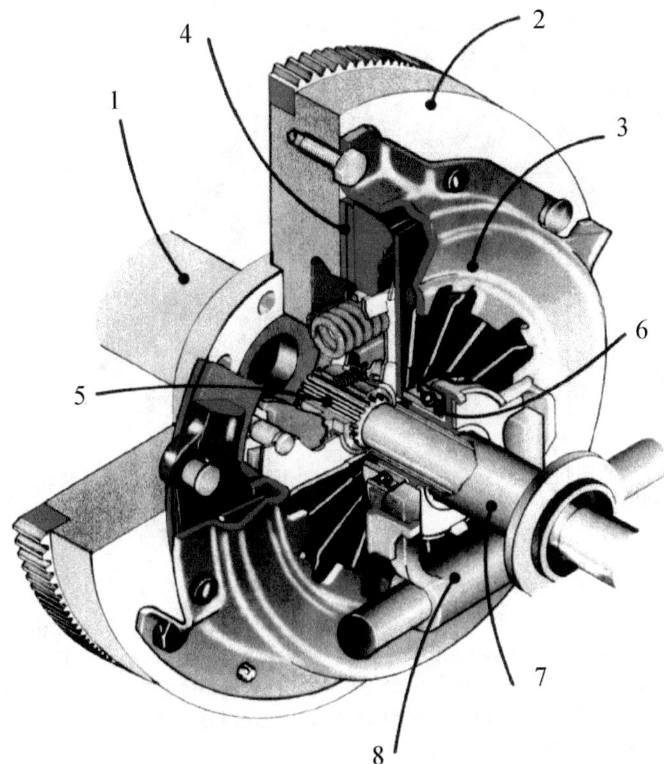

Fig. 11.1 Cut-out view of a diaphragm spring clutch (Valeo) 1: engine shaft. 2: engine flywheel and driving plate. 3: cover and internal engagement mechanism. 4: driven plate. 5: gearbox input shaft. 6: thrust bearing. 7: thrust bearing guide. 8: disengagement fork

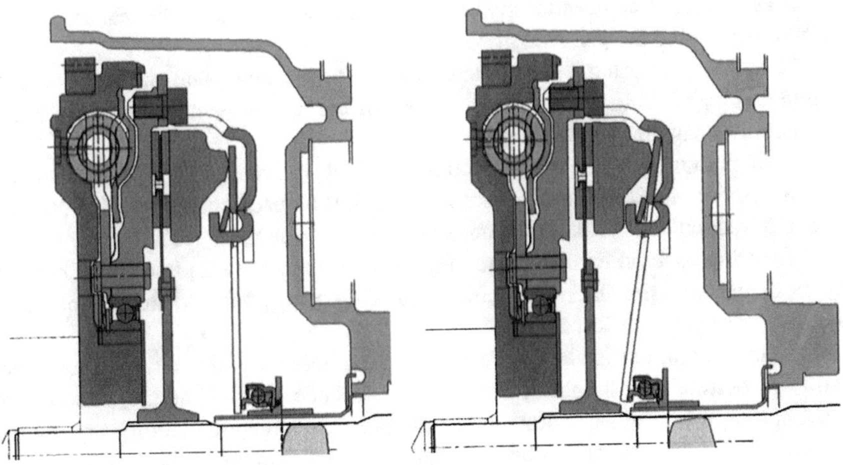

Fig. 11.2 Cross section of a double mass flywheel clutch, represented at left in engaged condition, at right in disengaged condition (Valeo)

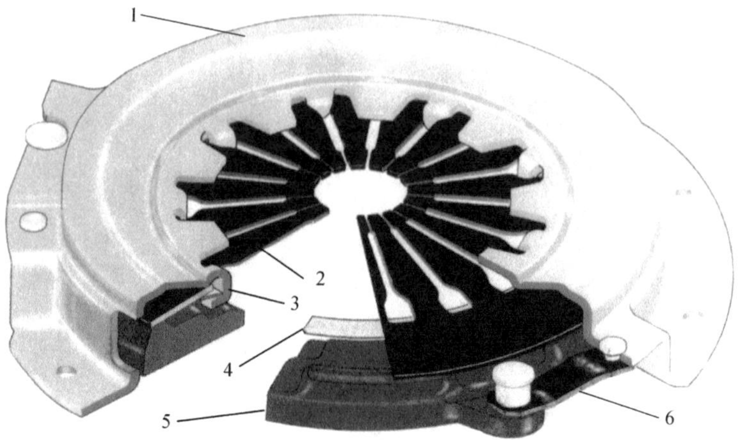

Fig. 11.3 Cut-out view of a clutch internal disengagement mechanism (Valeo). 1: cover. 2: diaphragm spring. 3: fulcrum of the spring. 4: ring retainer of the spring. 5: pressure plate. 6: pressure plate strap

and aligned with the engine crankshaft. The two driving plates should be connected by a device (not shown on this picture) that should be able to transmit the torque but that should leave the axial motion free; the driven plate is mounted with a spline shaft on the gearbox input shaft, because it should adapt its axial position. By moving axially the thrust bearing, the spring diaphragm can be deflected until the force on the pressure plate is brought to zero.

The second element of the driving plate, the pressure plate, shown on the Fig. 11.3 with the number 5, is made by a moving disc connected to the stamped steel cover by means that can transmit the necessary torque, but can leave the disc free to be moved axially.

This goal can be achieved using three flexible straps 6 (3 at 120°) that are riveted to the cover and to the pressure plate.

The diaphragm spring 2 is mounted between the pressure plate and the cover and is retained by the ring 4 in the fulcrum 3. All the above elements are part of the internal disengagement mechanism.

The diaphragm spring is represented on Fig. 11.4; it is made by a tapered steel disc with a number of radial cuts to increase its flexibility; each cut ends with a circular fillet that reduces the local stress and is also used as a centring reference.

The spring rests on its larger circumference on the pressure plate and is pressed by the cover through a fulcrum circular area. The spring conical deflection provides the necessary pressure force.

On the same figure is shown the elastic characteristic of the spring. The aspect ratio h/t between the height of the conical reference surface and the thickness of the diaphragm is used as a parameter of the pressure force-elongation diagram; the diagram isn't linear but shows an "S" shape with a maximum and a minimum. It is possible to use this feature to avoid the force reduction on the pressure plate, due to the driven plate wear, which is typical on the coil springs.

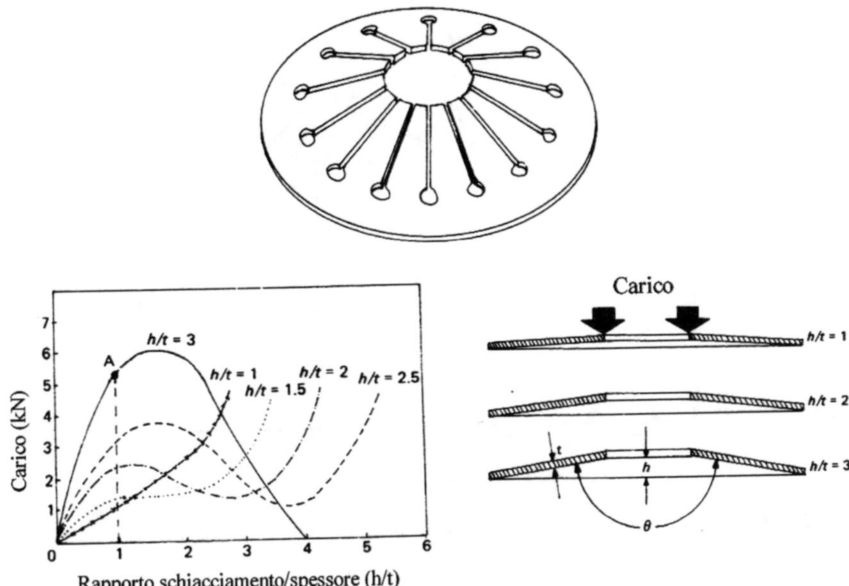

Fig. 11.4 Diaphragm spring. On the lower right graph is shown the elastic characteristic of springs of different aspect ratio, between the thickness t and the height h

In fact, let us compare the diagrams shown on Fig. 11.5 for a simple coil spring set and a diaphragm spring. The friction linings wear decrease the driven plate thickness, during its useful life.

In case of coil spring, if we assume that the thickness of the new plate is 8 mm and the maximum design wear is 2 mm, it is possible to see that, starting from the new clutch design point, the force on the pressure plate is reduced by spring deformation loss.

At a certain life time pressure can be as much reduced as to cause clutch slippage and driven plate wear-out. On an equally specified diaphragm spring, it is possible to maintain the same pressure also at the end of the lining useful life, with no slippage danger. There is a second advantage too; if we assume that the disengagement stroke is also equal to 2 mm, the force on the mechanism is increased by about 1 kN with the coil spring, while it will be reduced by almost the same value with the diaphragm spring.

It should be noticed that as the driven plate wear increases, the spring inner circumference will become closer to the thrust bearing; this last must therefore be retracted, in such a way as the bearing is never put in motion at released pedal.

The fulcrum between spring and pressure plate can be made according to different solutions, shown on Fig. 11.6 for a *pushed spring* (first two figures) and for a *pulled spring* (third figure) version; push an pull refer to the thrust bearing motion during the disengagement stroke.

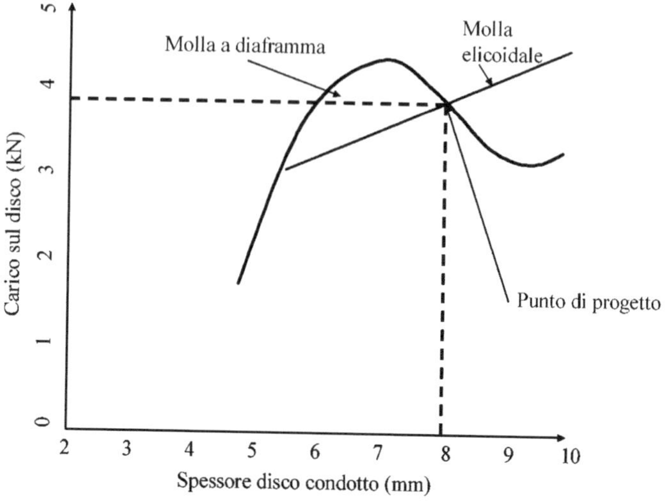

Fig. 11.5 Qualitative diagram of the pressure plate force, as a function of the driven plate thickness reduction, caused by wear

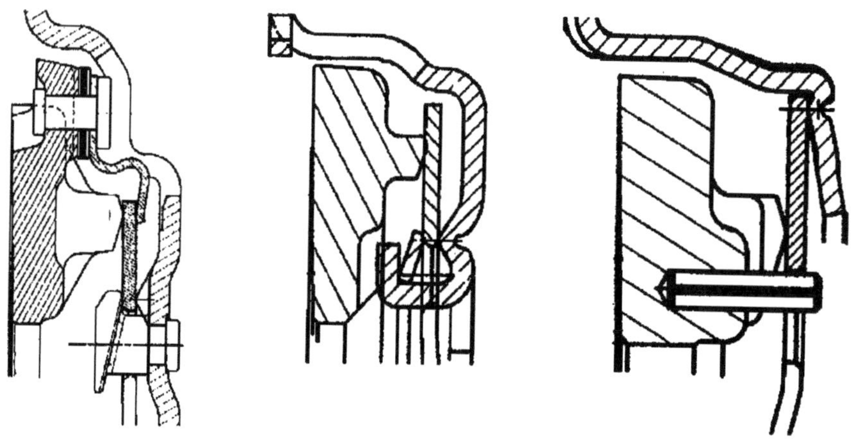

Fig. 11.6 Internal disengagement mechanisms with pulled disc (the first two at left) and pushed disc

In the first case the fulcrum circumference is internal to the sitting surface of the spring on the disc. The internal leaves of the spring must be pushed against the flywheel to disengage the clutch; the fulcrum can be made with brackets bent on the cover and a ring spacer (second figure) or with riveted pins inserted on the fillets of the spring (first figure).

In the second case the fulcrum circumference is external to the sitting surface and the leaves of the spring must be pulled away of the flywheel, in order to disengage the clutch. The advantage of this second arrangement consists of a better use of the available space inside the cover; in fact, at a given outside diameter of the flywheel, the steel cover is less stressed by bending, because the offset between the flange and the sitting circumference is reduced.

This advantage is paid buy a higher cost of the thrust bearing and makes this solution suitable with high torque engines of reduced displacement.

11.1.3 Driven Plate

Figure 11.7 shows a drawing of a driven plate. His design is quite complex, especially near the hub where there are the torsional dampers.

The plate is built with several steel discs. The first, marked with number 1, is the real hub and is mounted on a spline that fits the gearbox input shaft. Two other elements marked with 2 and 3 are made with two steel discs riveted with spacers; the element 4 that supports the friction linings is also riveted to the disc 3.

The disc 4 has a particular shape (look also at the detail on the lower right). This disc shows a number of separate sectors: the friction linings are riveted on them; the sectors are bent off the main disc plane and the rivets are set on parallel planes. With this arrangement friction linings are spaced and the two friction surfaces can displace elastically along the axial direction.

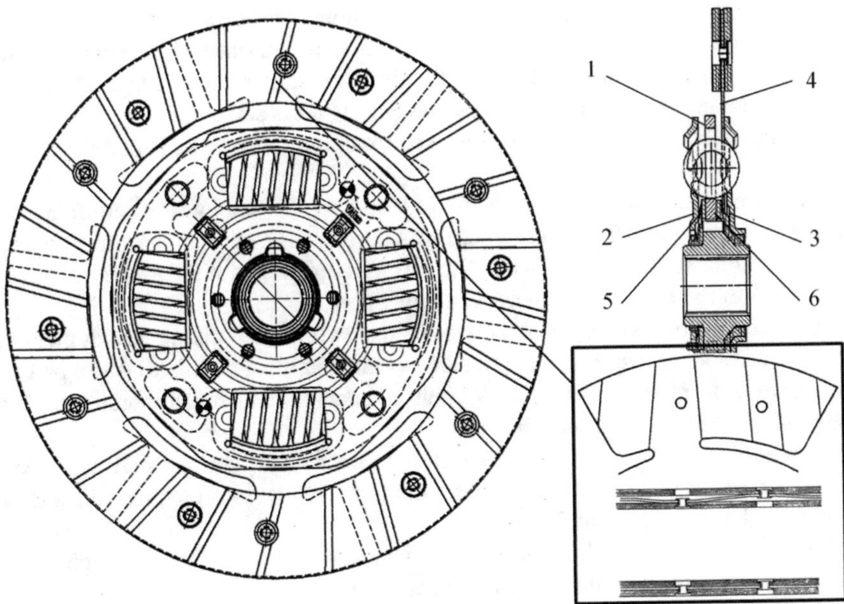

Fig. 11.7 Driven plate (Valeo). The cross section shows the main and the auxiliary torsion dampers. The detail on the right shows how friction liners are mounted on the disc, in order to make the engagement progressive

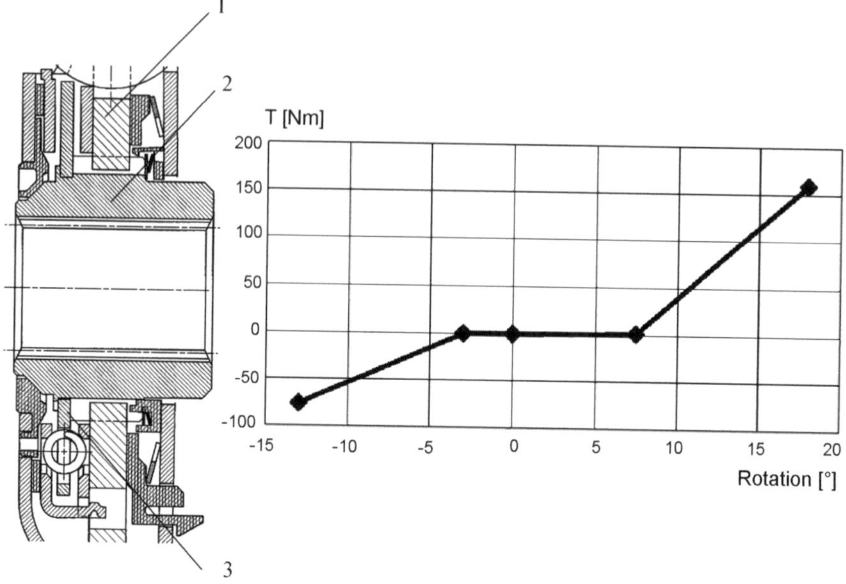

Fig. 11.8 Magnification of the driven plate hub (Valeo). The diagram on the right shows the angular displacement of the hub, as a function of the torque

The result of that is that the pressure force is limited by the disc elasticity and the clutch engagement is more progressive. In addition to that, since the thermal gradient on the pressure plate can cause a conical deformation, linings can better match this shape, maintaining a distributed contact on driving discs surface.

The assembly of discs 2, 3, 4 is free to rotate of a certain angle, with reference to the hub 1.

The three elements 1, 2 and 3 show four rectangular windows, where four coil spring are press fitted. If the torque acting on the disc is lesser than the total precompression moment of the springs, the disc acts as a rigid body; if the torque is higher, disc and hub can show a certain angular deflection.

Between the rotating elements, there are also two friction discs 5 and 6, pressed through a diaphragm spring; when the hub is moving it is possible to waste part of the elastic deformation work of the springs and damp the torsional vibrations. This assembly is the main *torsion damper*.

In the following Fig. 11.8 a magnification of the hub 1 is shown. We can see that it present the secondary torsion damper, similar to the previous one; the stiffness of this second damper is much lower.

Only the hub 3 is fixed on the input shaft, while the hub 1 can show significant angular displacements, limited by a number of teeth.

On the same picture is shown a possible elastic characteristic of the hub assembly; there is a central part of the diagram very flexible (secondary torsion damper), continuing in two stiffer ends (main torsion damper). The two ends of the diagram

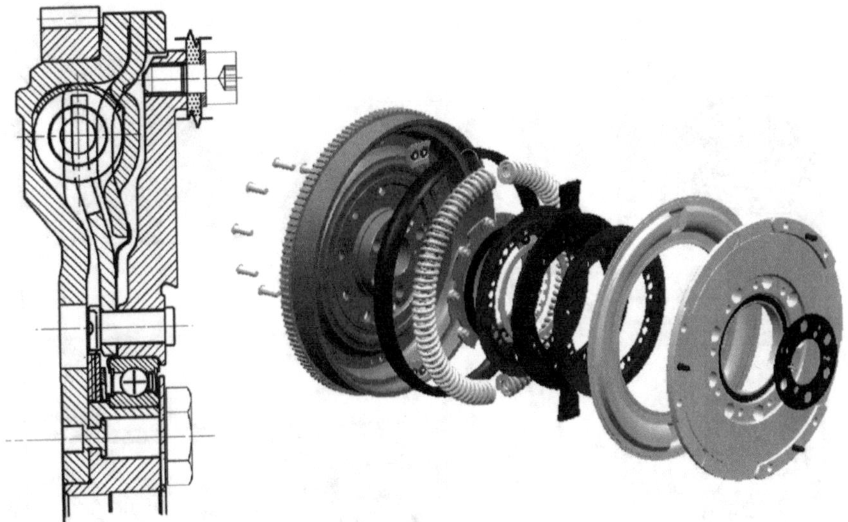

Fig. 11.9 Cross section and exploded view of a double mass clutch and flywheel. The torsional damper is now included in the driving plate; the driven plate is very much simplified (Valeo)

can be asymmetric if coil springs in the rectangular windows are double (one inside the other, of different length).

The stiffer damper is finalized to the idle rattle noise, while the softer one is finalized to the rattling noise at low speed operation.

In some clutch these functions are not present on the disc, that is very simple, but are fitted on the driving plate, on the engine side; an exploded view of this plate is shown on Fig. 11.9. The flywheel and the driving plate are made with two masses free to rotate independently on a ball bearing. Two long coil springs lowers the value of the natural frequency of one of the masses; a damping device is also present.

The result that can be obtained with a suitable tuning is that the lowest natural frequencies of the transmission, usually between 800 and 2000 rpm, can be shifted below the engine idle speed and over the maximum speed where the transmission noise can't be detected by human ears; the results are particularly interesting in high displacement engines.

The total inertia of the system is the same as that of the reference flywheel; therefore there are no effects on the engine torque fluctuation.

11.1.4 Thrust Bearing

The thrust bearing, as we have seen, slides on a tube flanged to the gearbox housing, coaxial to the input shaft; this tube integrates usually the gearbox oil seal cover; see part 7 on Fig. 11.1.

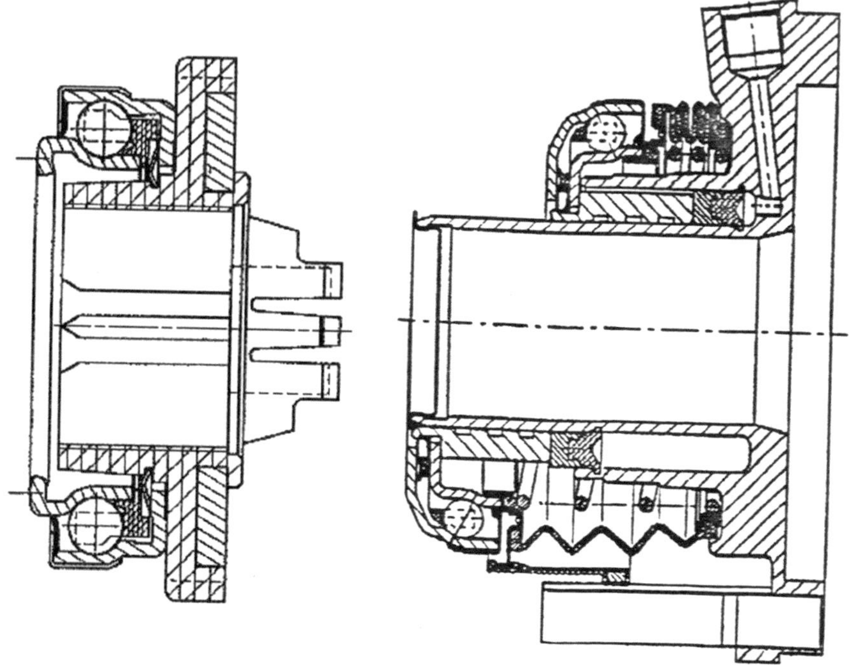

Fig. 11.10 Conventional thrust bearing (on the left) and bearing with integrated hydraulic actuator (Valeo)

The thrust bearing is moved by a finger lever on the gearbox housing and is moved by the clutch pedal through an external mechanism. The bearing on the left side of Fig. 11.10 has a very simple shape and acts on the leaves of the diaphragm springs directly.

The external disengagement mechanism can by made with a cable or a hydraulic actuator.

The cable mechanism is similar to what we explained for the external shifting mechanisms of the gearbox. As we have said, the cable assembly must be installed with no sharp bends that negatively affect mechanical efficiency; a poor mechanical efficiency can increase the disengagement pedal force and can introduce a hysteretic behavior that can make pedal return disagreeable or difficult.

The progressive friction linings wear on driven plate must be compensated with a corresponding adjustment of the thrust bearing position; this adjustment shifts the useful stroke of the pedal closer to the driver, with possible negative effects on disengagement ergonomics. To avoid this inconvenience and to have the stroke always correctly positioned, an adjustment nut must be provided on one of the rests of the cable sheath.

Hydraulic circuits are progressively substituting cable mechanisms. A hydraulic cylinder is linked to the pedal and operates through a pipe a hydraulic actuator, acting

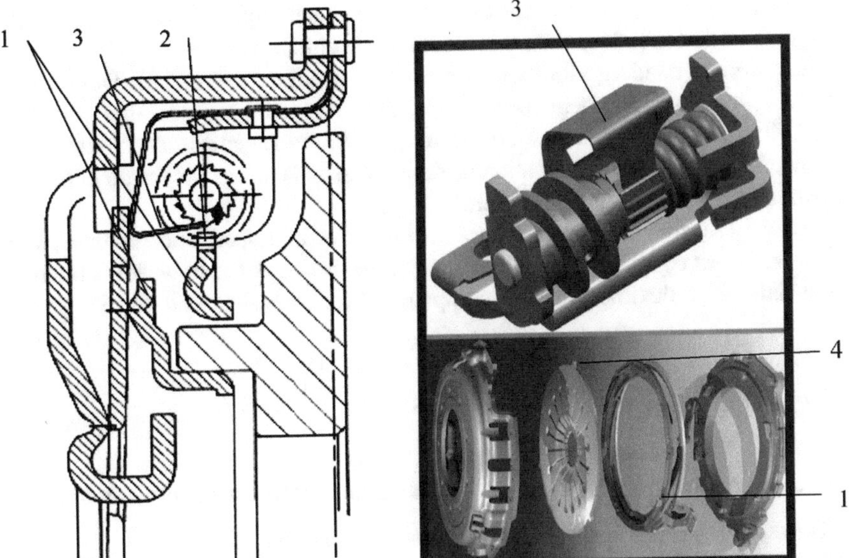

Fig. 11.11 Self adjusting mechanisms compensating for the effects of wear on the driven plate thickness (Valeo)

on the disengagement fork of the thrust bearing; sometimes is possible to integrate this actuator on the clutch housing in such a way as to directly push on the thrust bearing, with a simplification of the mechanism, as it is shown on Fig. 11.10.

The hydraulic mechanism has the advantage of a better mechanical efficiency almost independent on the piping lay-out; with a simple check valve between the oil reservoir and the operating piston on the pedal side, it is possible to obtain an automatic adjustment of the pedal stroke position easily.

Driven plate wear, also in case of pedal stroke adjustment, has the inconvenience of increasing pedal actuation force, because of the elastic characteristics of the diaphragm spring that was shown on Fig. 11.5; in addition to that, the inner space left inside the clutch cover must take into account the adjustment of bearing stroke caused by wear.

A definitive solution to all these problems is presented by the self adjusting internal mechanism shown on Fig. 11.11. In this mechanism the diaphragm spring pushes the pressure plate through a spacer 1, made of two steel sheet discs connected with flexible straps; the contact surface with the pressure plate is a helical ramp, so that by rotating the spacer the driven plate thickness decrease can be compensated.

The spacer is turned by means of teeth 2 that match with the worm gear that is magnified on the right side of the picture; if, during a clutch engagement the driven plate has decreased its thickness, the leaf spring 3 contacts the diaphragm spring and rotates the worm gear until the contact between the two springs is lost, because the worn thickness has been compensated.

With this device it is possible to keep mechanical clutch characteristics at the design value and to reduce the axial length of the clutch cover.

Driven plate bending vibrations are caused by engine crankshaft. During high engine speed disengagement manoeuvres, they can cause disagreeable noise and pedal vibrations; both are transmitted through the disengagement mechanism.

This fact justifies the application of rubber elastic bearings on sheath ends or of a hydraulic capacity on the oil circuit.

In case of severe vibration problems, in example on high speed sport engines, when the noise is propagating through other ways, the crown of the flywheel is mounted on its hub with a flexible plate that compensates for crankshaft end fluttering.

11.1.5 Design Criteria

While designing a clutch many points of attention must be taken into account:

- The disengagement force on the pedal shouldn't be too high.
- The friction coefficient must be as constant as possible in the different operating conditions.
- The wear of the working surfaces must be sufficiently slow.
- The heat produced during engagement manoeuvres must be taken away.
- The clutch must operate correctly also with reasonable wear.

The friction coefficient between discs and friction linings is ranging between 0.27 and 0.30 at acceptable temperatures.

Taking into account this value the diaphragm spring can be designed in such a way as maximum engine torque can be transmitted with a safety margin of about 15%, with new and completely worn linings.

It is possible to calculate pedal force with new and worn linings from spring characteristic; transmission mechanical efficiency must be also taken into account.

Typical values for a good cable mechanism range between 0.65 and 0.70; for a hydraulic mechanism with piston and disengagement fork and finger the value can reach 0.80. For an integrated hydraulic actuator the efficiency can be 0.90.

Friction is therefore not negligible in the calculation of the maximum disengagement thrust (must be kept below 300 N); this value can't be reduced by working on the transmission ratio between pedal and bearing only, because the pedal stroke must be kept under 150–200 mm; the elastic deformation of the mechanical chain between pedal and thrust bearing must be also taken into account.

A useful design parameter is the mechanical energy wasted at start up; it can be referred to the clutch useful friction surface and compared with similar values obtained in successful cases; the same can be said for the temperature at the end of a start up manoeuvres at maximum slope.

11.2 Start-up Devices for Automatic Gearboxes

Gearbox automation requires using devices that can start up the vehicle smoothly, without the help of the driver feeling; for this reason, devices different from the clutch were developed, utilizing intermediate non mechanic energy, such as electric or hydraulic one.

Today this issue is not so important, because it has also proven possible to automate a conventional friction clutch, designed for pedal actuation, by using an electronically controlled electro hydraulic actuator, with satisfactory results.

The hydraulic torque converter and the electromagnetic powder clutch, this second on very few applications, are used with automatic gearboxes, for their intrinsic comfort characteristics; they can only be matched with gearboxes where speeds can be changed without torque interruption; as a matter of fact, the very high rotating masses of these devices will cause a conventional synchronizer to work with a very long engagement time.

The powder electromagnetic clutch is made with two ferromagnetic coaxial rings that build the poles and the anchor of a magnetic circuit; the magnetic field can be set up by an electric coil on the outside ring, fed with sliding contacts.

The air gap is few millimeters thick; in the air gap a quantity of iron powder with suitable granular dimension is inserted. The magnetic field which builds up when the current is switched on is causing the iron grains to align along the radial field lines, between the two poles; grains exchange friction forces each other and with the polar surface. Friction brings the two poles to block while exchanging a torque depending on the current intensity.

The electromagnetic clutch is easy to be controlled; on the other side the long response time and the high inertia aren't suitable to a synchromesh gearbox. The use of this clutch is only justified by the reduced cost in comparison with hydrodynamic clutches and torque converters.

The torque converter is a particular hydraulic machine that allows connecting two shafts with a continuously variable transmission ratio. Opposite to gearboxes, the input and the output torque values of the torque converter are not bound by the energy conservation low, but by more complex relationships depending on the transmission ratio.

If we name with Ω the angular speed, with M the torque and with P the power and we associate subscripts 1 and 2 to these magnitudes on torque converter input and output shafts, we define:

- *speed transmission ratio*:

$$\nu = \frac{\Omega_2}{\Omega_1} .$$

Sometimes $1 - \nu$ or *slip* is used, in place of the speed transmission ratio;

- *torque transmission ratio*:

$$\mu = \frac{M_2}{M_1} .$$

- *transmission efficiency*:

$$\eta = \frac{P_2}{P_1} .$$

Because, by definition:

$$P_1 = M_1 \Omega_1 \text{ and } P_2 = M_2 \Omega_2 ,$$

as a consequence:

$$\eta = \nu \mu . \tag{11.1}$$

Therefore, also neglecting mechanical friction, mechanical efficiency can be less then one in a torque converter; transmission ratios are governed directly by the internal geometry of the machine and cannot be changed easily by means of external devices.

In order to understand the converter operation could be useful to think about a transmission made by a hydraulic pump and a hydraulic turbine connected one to the other with suitable piping.

The pump is rotating with the engine and moves a certain flow of oil from a suction pipe to an outlet pipe; pipes are also connected to a turbine which is mechanically linked to the gearbox input shaft.

A transmission like this has been used on ships to connect engine and propellers.

As a difference with a mechanical transmission, there is no positive link between the two shafts, but the transmission ratio is determined by the inertia of the oil flow in the piping and in the hydraulic machines.

It is therefore possible to have the engine running when the transmission shaft is for any reason stalling; as a matter of fact turbine blades cannot stop the oil and can, in the mean time, receive a torque from the flow.

Such a kind of transmission shows the following advantages:

- The transmission ratio can be continuously changed as a function of the torque ratio.
- There are no wearing parts.
- There is a high damping capacity of torsional vibrations.
- There is no danger of stalling the engine as a consequence of a too high torque insisting on the gearbox input shaft.

On the contrary we can think about the following disadvantages:

- Low transmission efficiency in all conditions.
- Increased transmission complexity.

As we have already commented, the converter cannot be easily cut off of the driveline and has a significant polar inertia; this makes the application of conventional synchronizers impossible: it is therefore necessary to use powershift gearboxes, where shifting speed can occur in longer times, because there is no torque interruption.

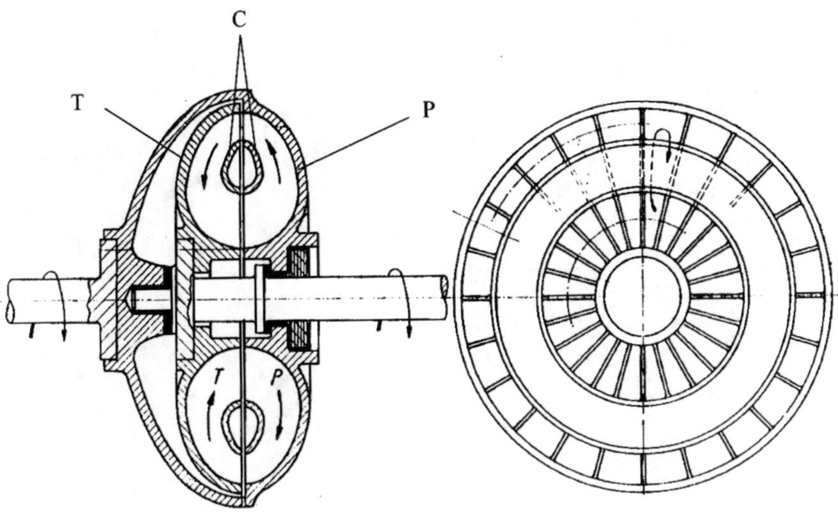

Fig. 11.12 Hydrodynamic clutch scheme; P is the centrifugal pump and T the turbine. The arrows show the direction of flow when the pump speed is greater then the turbine speed

The kind of lay-out we have imagined could be too bulky for vehicle application; H. Föttinger had the idea to integrate the pump and the turbine in a single compact machine, avoiding connecting pipes.

This result was obtained by applying turbine and pump wheels of similar radial dimension. In this case the wheels can face each other creating the hydraulic circuit without additional elements.

Figure 11.12 shows a scheme of this lay-out; the pump is made with a radial bladed wheel P, where the disc supporting blades shows a cavity, where the turbine T can be installed.

The pump is connected to the engine crankshaft. The volume where the oil flows is limited by the rotating discs, by blades and by the closure surface C, which can also be avoided.

Turbine and pump are almost identical; the two wheels must be positioned with the same center line and a rotating seal must be provided, to keep the oil inside. The proposed scheme doesn't show any stationary wheel suitable to receive a reaction torque. Input and output torque must in this case be equal; this particular machine is called *hydraulic clutch* and acts like a friction clutch.

This machine cannot transmit any torque if there is no speed difference between pump and turbine; in fact, the oil flow rate determines the transmitted torque and it becomes zero when speeds are equal. If wheels had a different geometry, the transmitted torque would be zero when the flow rate would stop.

Figure 11.13 shows instead the scheme of a *torque converter*; as a difference with the previous one that includes only the pump P and the turbine T, there is a third stator bladed wheel S, call also *reactor* wheel, connected to a standing element, like the gearbox housing.

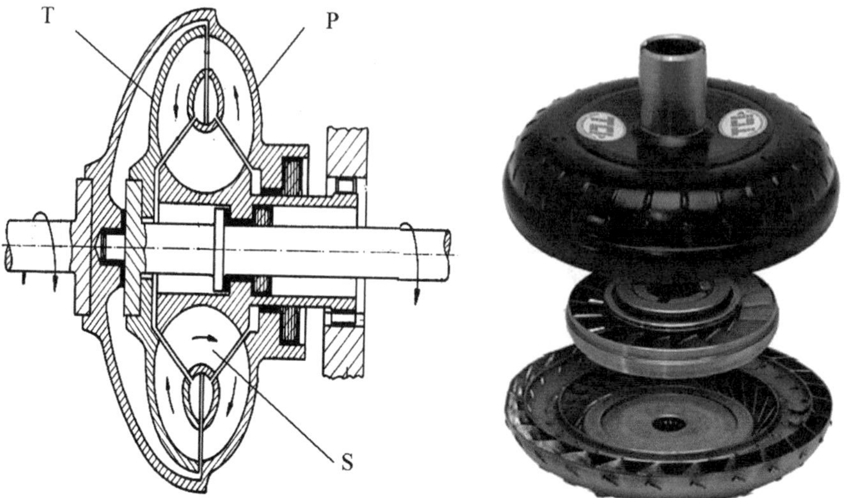

Fig. 11.13 A torque converter scheme is represented on the left. A picture of the elements of a torque converter is shown on the right

In this case the torque on the pump and on the turbine might be different, because the reacting element can equal the algebraic difference of torque.

If, as it occurs more frequently, the output torque is greater than the input torque, the reaction torque must have the same direction of the input torque. On the opposite, if the input torque should be less than the output torque, the reaction torque should have an opposite direction as the input torque.

If the stator is connected to the gearbox housing with a freewheel that can only withstand a torque with the same direction as the input torque, the machine could multiply the input torque or behave like a hydraulic clutch, when the output torque equals or is greater than the input torque.

The picture at the right of the scheme shows a disassembled torque converter completely made with stamped steel sheet parts.

11.2.1 Hydraulic Clutches and Torque Converters

In order to understand the operation of a rotating bladed wheel, we consider a wheel with a single, almost radial, channel whose walls show angles β_1 and β_2 at the outer intake radius and at the inner exhaust radius, as shown on Fig. 11.14.

We call r_1 and r_2 the outer and the inner radii of this wheel.

The wheel is struck by a mass flow rate Q, coming out of a nozzle that is inclined of the angle α_1 with respect to the tangent to the outside diameter of the wheel. We call c_1 the output velocity of the oil from the nozzle; for sake of simplicity we suppose the speed to be constant at any point of the cross section of the flow.

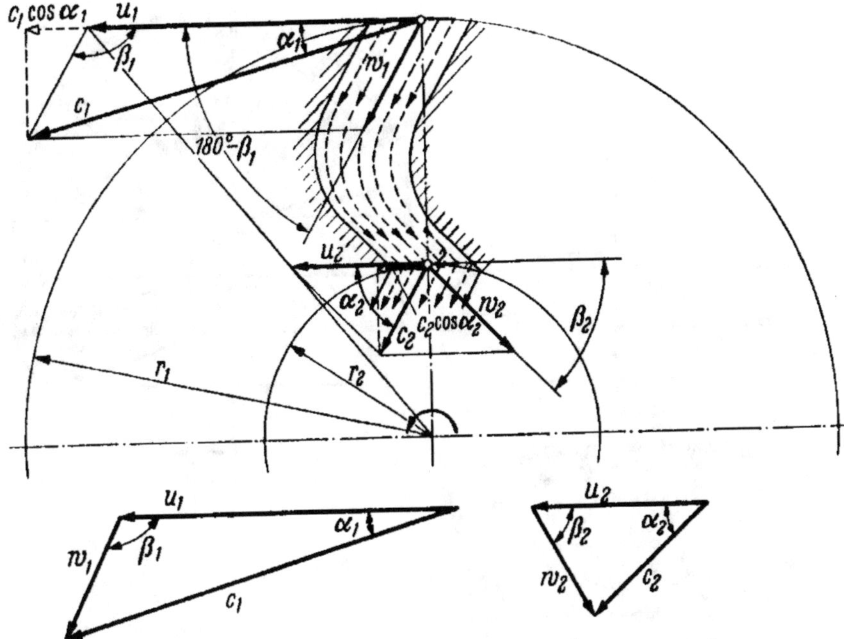

Fig. 11.14 Speed triangles on a bladed wheel invested by a nozzle with an output speed c_1 inclined of an angle α_1

The torque insisting on the bladed wheel will be given by the difference between the mass flow momentum at the intake and the exhaust of the channel on the wheel. The two speed triangles at the input and output cross section of the flow allow us to calculate the absolute speed of the flow when leaving the wheel.

Particularly the input speed c_1 can be considered as the vectorial sum of the oil speed relative to the channel and of the speed of the oil considered as part of the channel; the first speed is w_1, whose direction is parallel to the walls of the channel at the intake cross section of the channel, the second speed is the peripheral speed of the wheel at the radius r_1 that we call u_1.

The absolute speed at the exhaust cross section c_2 can be calculated starting from the speeds u_2 and w_2 at the exit of the channel; it can be noticed that the first is defined by the radius r_2 and the second from the continuity of the mass flow in the channel: its direction is again tangent to the walls of the channel and its module will be equal to the module of w_1 multiplied by the ratio of the areas of the two cross sections.

The momentum at the intake of the wheel is:

$$Qc_1 \cos \alpha_1 r_1,$$

while the momentum at the exit of the wheel is:

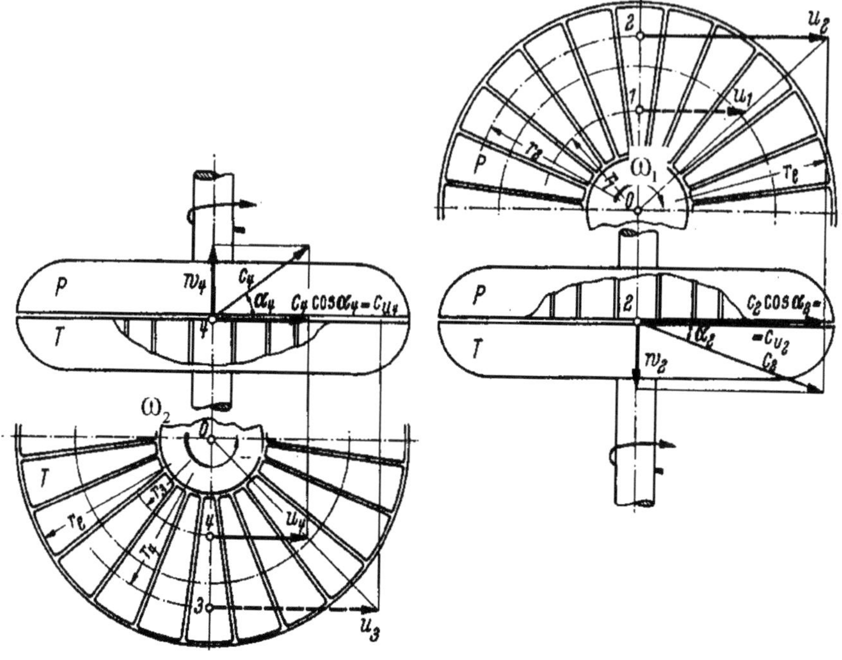

Fig. 11.15 Speed triangles in a hydraulic clutch

$$Qc_2 \cos \alpha_2 r_2.$$

The torque on the wheel will be therefore:

$$M = Q(c_1 \cos \alpha_1 r_1 - c_2 \cos \alpha_2 r_2). \tag{11.2}$$

It is sufficient to know the intake and exhaust angles of the blades and the related radii to determine the torque on the wheel with the approximations that we have introduced. The case of the hydrodynamic clutch and of the torque converter is complicated by the fact that the intake absolute speed is unknown, because it is bound to speed of the preceding wheel.

Let us consider for instance the case of the hydrodynamic clutch that is introduced on Fig. 11.15. P is again the pump, while T is the turbine; the blades are perfectly radial.

We assume that the oil is incompressible and the current line are perpendicular to the separation plane between the two wheels and, therefore, parallel to the wheels axis. The fluid is moving according to organized vortices with their axis perpendicular to the axis of the wheels.

If we consider the separation plane between the two wheels, we could draw on it a circumference (dots and lines on the figure) separating two constant flows of opposite direction.

The cross section between the blades that are inside and outside of this circumference will have the same area, because the oil is incompressible.

If the speed is constant on the cross section we can calculate the speed on the centres 1 and 2 for the pump and 3 and 4 for the turbine; the calculation can be repeated with same results for all the other blades.

In a generic operation point the speeds Ω_2 and Ω_1 of the turbine and of the pump won't be equal; more likely, because the pump is the moving element his speed will be greater than that of the turbine. On the facing points 1 and 4 at the intake of the pump, the speed c_1 will be determined by the two components:

$$w_1 \text{ e } u_1 = \Omega_1 r_1 .$$

Pay attention not to confuse the subscripts of the angular speeds bound to turbine and pump, with those of oil speeds that are bound to the passage sections.

In the same way the speed triangle at the facing turbine section can be built; it is shown on the figure and takes in to account that the two components w_1 and w_4 are equal because the cross sections are equal.

The vectorial difference:

$$\mathbf{c}_1 - \mathbf{c}_4 = \mathbf{u}_1 - \mathbf{u}_4 \tag{11.3}$$

represents a part of the linear momentum of the pump that can't be utilized by the turbine, because of the difference of rotational speed of the two wheels.

In the same way we can proceed on sections 2 and 3, if we remember again that the components w are equal to those of the sections 1 and 4.

At the intake of the pump, on section 1, there is an energetic loss due the fact that the fluid is given by the turbine at a peripheral speed lower than that of the pump and this last must provide for a theoretically sudden acceleration; this type of loss is called *impact loss*. In the same way at the intake of the turbine, on section 3, there is a sudden deceleration of the flow, due to the same cause; if we take in mind the impact losses the speed triangles of the two wheels will be the same on points 1, 4 and 3, 2.

The torque on the wheels will be:

$$M_1 = M_2 = Q\left(u_2 r_2 - u_4 r_1\right) = Q u_2 r_2 \left(1 - \frac{u_4 r_1}{u_2 r_2}\right) = Q u_2 r_2 \left(1 - \frac{\omega_2 r_1 r_1}{\omega_1 r_2 r_2}\right),$$

or:

$$M_1 = M_2 = Q u_2 r_2 \left[1 - \nu \left(\frac{r_1}{r_2}\right)^2\right]. \tag{11.4}$$

If we calculate the power from this last equation, we obtain again the Eq. 11.1, where $\mu = 1$:

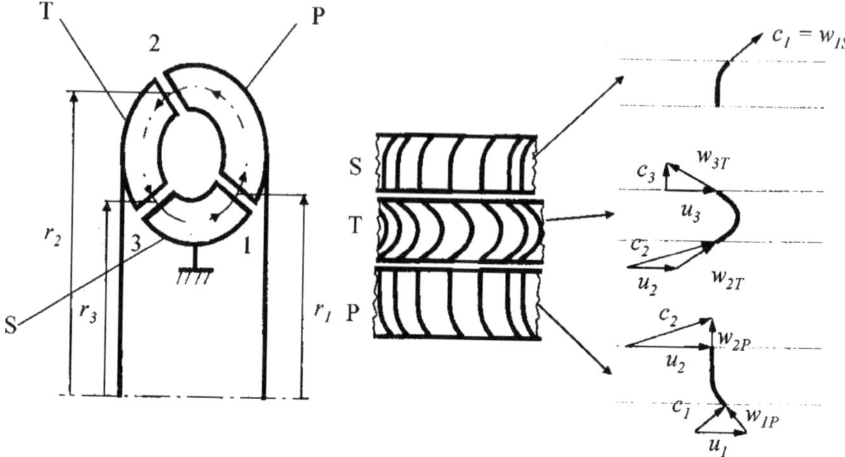

Fig. 11.16 Speed triangles in a torque converter

$$\eta = \frac{P_2}{P_1} = \nu.$$

The above formulas don't take into account friction losses in the oil and windage losses on wheels. These last are negligible until the transmitted torque is high. Friction losses in the oil, said *hydraulic losses* can be imagined as a function of the square of the flow speed w, according to the formula:

$$P_a = kQw^2, \tag{11.5}$$

where k is a constant depending upon the nature of the surface of the blades.

In the case where wheels have the same rotation speed, pressures values they generate on the fluid at the outside boundary between the wheels are equal an opposite. In this case there won't be any oil circulation and the flow rate must go to zero; as a consequence the processed torque and power are also zero.

In the same way the case of a torque converter could be treated.

On Fig. 11.16 the scheme of a torque converter is drawn; on the schematic cross section on the right of the figure the dotted lined represents the center line of the channel between to next blades, joining the centres of the different cross sections.

These centres are determined by discs and blades of the pump P, the stator S and the turbine T and by the internal walls that are now present; the position of these walls is such as to obtain a circulation speed w almost constant.

We imagine drawing on the rotation surface, generated by this line around the converter axis, traces of blades and developing this surface on a plane. We can come to the scheme on the centre of the figure relative to the blades.

On the right side of the same figure there are the speed triangles that can be built in the same way they were built for the hydrodynamic clutch.

The sections 1, 2, 3 are relative to the transition points between stator and pump, pump and turbine, turbine and stator. To the speed w is given a second subscript, according to the converter element they are referring to.

On the transition point between an element and the following one, speeds c are conserved, speeds u are calculated from the radius r of the centre of the section they refer to, speeds w are tangent to the blade walls.

Because the module of w is determined by the continuity equation of the mass flow, only in one condition that is shown on the figure there will be no impact losses. This condition is called *design point*. At these conditions the efficiency will be maximum and determined only by hydraulic losses. At conditions different as the design point, the geometry of blades will cause also impact losses.

It is usual for torque converters and hydrodynamic clutches to use the principles of similarity.

The reference geometric dimension is a diameter, usually the outer diameter of the largest wheel D; if ρ is the mass density of the fluid we can define a performance coefficient λ defined by the formula:

$$M_1 = \lambda \rho \omega_1^2 D^5 . \tag{11.6}$$

Two machines are *similar* if their linear dimensions are in scale and blade angles are equal; if two machines are similar, the torque absorbed by the pump will be defined by the same *performance coefficient* λ.

The non dimensional entities such as η and μ in similar machines will be also equal at the same value of ν.

11.2.2 Characteristic Curves

Figure 11.17 represents in a qualitative way the *characteristic curves* of a hydrodynamic clutch; these characteristics include the performance coefficient λ and the efficiency η. It should be observed that the performance coefficient is multiplied by 10,000.

λ is decreasing with the speed transmission ratio, because the processed torque is depending upon the flow, determined by the speed difference of the wheels; when the wheel are synchronous, $\nu = 1$ and the torque is zero. The efficiency should be always equal to the speed transmission ratio but at $\nu = 1$. At this point it goes also to zero.

This fact is justified if we take into account windage losses and rotating seal friction; since the processed power at synchronous speed is zero, even small losses will cause the efficiency to go to zero too.

In a hydraulic clutch, the torque transmission ratio is equal to 1, by definition.

Figure 11.18 represents instead the characteristic curves of a torque converter; characteristics include in this case λ, η and μ, the torque transmission ratio. It should be observed that, also in this case, λ is multiplied by 10,000.

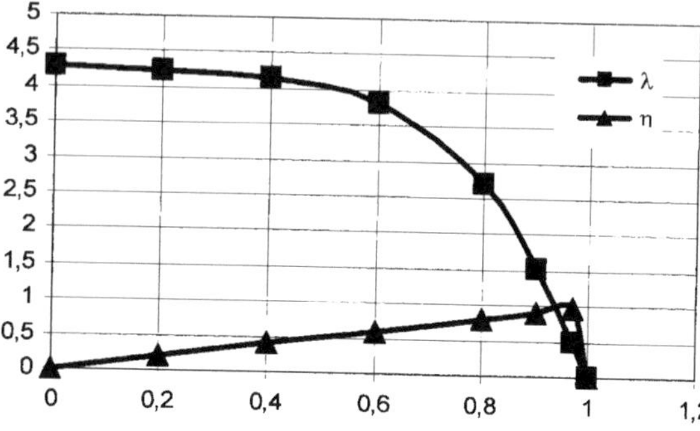

Fig. 11.17 Characteristic curves of a hydraulic clutch representing λ and η as function of ν. λ is multiplied by 10,000

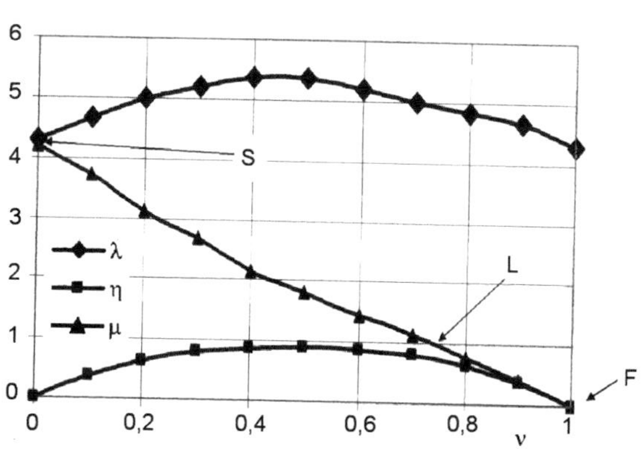

Fig. 11.18 Characteristic curves of a torque converter representing λ, η and ν as a function of ν. λ is multiplied by 10,000

Since there is a reaction element, torque transmission ratio isn't constant; the torque absorbed by the pump doesn't change very much as the speed transmission ratio increases, because of the reactor wheel.

Torque transmission ratio starts from a value of about 3–4 when the speed transmission ratio is zero, when the output shaft is stalling and goes to zero when the input and output shaft are synchronous; as a consequence, the efficiency that is the product of the two transmission ratio will be zero as $\nu = 0$ and $\nu = 1$.

On the same figure some typical operation conditions are shown:

- S, stands for *stall*, is the condition where the turbine is still and the pump is rotating.
- L, stands for *lock-up*, is the condition where the torque on the pump and the torque on the turbine are equal, so torque transmission ratio is one.
- F, stands for *free flow* point, is the condition where torque on the turbine is zero.

All curves can be justified by the construction of the speed triangles and by the principles of similarity.

The Trilok torque converter, named after the American research consortium that was in charge of its development, provides for a reactor wheel mounted on a free wheel that can react to torques only, who have the same direction as the input engine torque. With this provision, at the lock-up condition the reactor wheel is free; in fact, from this point on, the turbine should receive a torque lesser as the pump and, as a

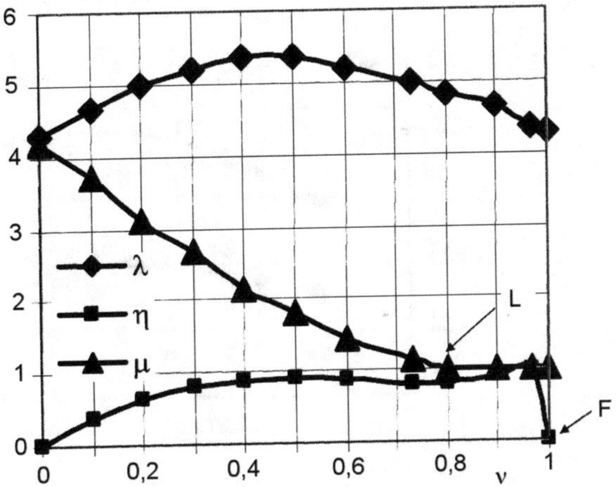

Fig. 11.19 Characteristic curves of a Trilok torque converter representing ■ λ, μ and η as a function of ν. λ is multiplied by 10,000

consequence, the reactor wheel should withstand with a torque opposite to the engine torque.

If the reactor is free wheeling, the torque converter will behave as a hydrodynamic clutch; therefore the efficiency, as it is shown on Fig. 11.19, will become linear at the point L, with a steep decrease to zero at the point F. From the point L on, also torque transmission ratio will remain constant, instead of going to zero.

As a first approximation, Trilok torque converters show, at any operation condition, the best efficiency value between the hydrodynamic clutch and the torque converter.

11.2.3 *Torque Converter Performance on a Vehicle*

Matching a torque converter to an engine and to a vehicle implies the performance calculation of a system including engine, torque converter and vehicle; it can be made taking into consideration a diagram where engine torque curve and pump torque curve are represented. Because pump and engine are rotating together, the engine speed is also representing the pump speed.

Once the torque converter diameter is chosen, λ has been defined; the torque absorbed by the pump is a parabola.

We can start drawing diagrams represented on Fig. 11.20. They represent the torque absorbed by the pump at different values of the speed transmission ratio, the engine output torque at wide open throttle M_{max} and a torque curve M_{reg} at partially open throttle.

As a value for the speed transmission ratio is assumed, for the equilibrium of the pump and engine system, engine speed will be forced at the value at the intersection point of the engine torque and of the pump torque diagrams.

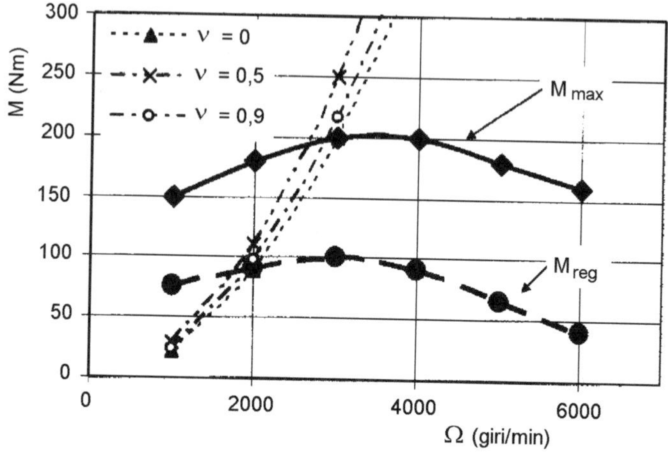

Fig. 11.20 Matching curves to the engine of a torque converter at different values of the speed transmission ratio ν. It must be noticed that the operation speed is determined by the engine torque and by the speed transmission ratio only

In order to change the matching point, engine torque must be changed.

Let us remember that the fact that engine can deliver the maximum torque at stall is bound to the choice of the pump diameter.

From torque converter characteristic curves, gearbox input torque (torque converter output torque) M_2 can be calculated as the engine torque M_1 is known.

The gearbox input speed Ω_2 will be given by the product of the engine speed Ω_1 by the speed transmission ratio ν; the gearbox input torque M_2 will be instead given by the product of the engine torque M_1 by the torque transmission ratio μ.

The calculation can be done starting from the diagrams of Figs. 11.19 and 11.20, for the wide open throttle engine torque; the procedure can be repeated for any values of regulated engine torque.

The characteristic of the engine transmission system is shown on Fig. 11.21; it can be observed how the very small part of the engine torque curve is expanded to the gearbox input torque curve.

Two other cases are shown on the same diagram:

- the case of an ideal continuously variable transmission with mechanical efficiency equal to one and
- the case of a slipping clutch with a safety margin of 15%.

The converter performance lays between the two reference curves, closer to the ideal continuously variable transmission.

The traction curve of the vehicle T can now be drawn, as a function of the vehicle speed V, multiplying M_2 and Ω_2 by the different values of the gearbox transmission ratios, utilizing the well known formulae:

$$V = \Omega_2 R_r / \tau_i \text{ , with } i = 1, n,$$

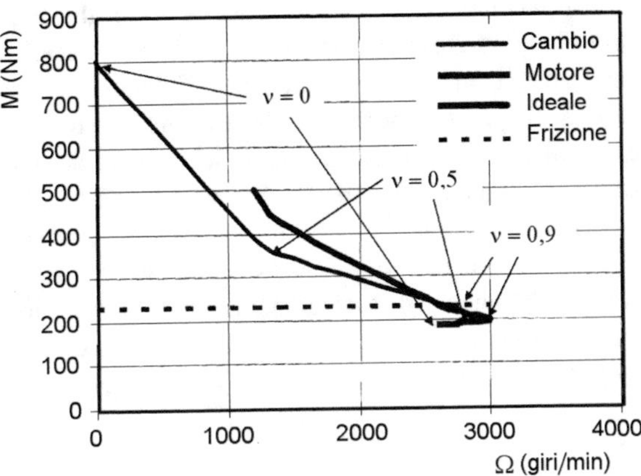

Fig. 11.21 Calculation of the gearbox input torque M_2 starting from the engine torque M_1. The values of the speed transmission ratio of the different curves are shown. The values of M_2 are also reported for an ideal continuously variable transmission and for a slipping friction clutch

$$T = M_2 \tau_i \eta_t / R_r \ , \text{ with } i = 1, n \ ,$$

where τ_i is the total transmission ratio of each speed of a gearbox with n speeds, R_r is the wheel rolling radius, η_t is the transmission mechanical efficiency, not including torque converter.

The time integration of the available power, according to the already known methods, can be made in order to calculate the performance of the vehicle with a torque converter.

By the repetition of this procedure for a number of values of partially open throttle torque, also the fuel consumption map of the engine and torque converter system can be calculated; for each value of the engine torque, the associated value of the fuel consumption can be divided by the corresponding value of the torque converter efficiency.

The use of this new fuel consumption map can lead us to calculate the consumption of the vehicle when the torque converter is working.

Sizing the torque converter in such a way as the engine converter system can stall at engine maximum torque doesn't always show as the best design praxis. It is convenient to look also for different sizing criteria, with torque converters with higher and lower diameters.

The first will force the engine to work at lower speeds, the second at higher speeds. The consequent change of the stall torque will also cause the traction curve to change at wide and partially open throttle. The start-up transients will show different duration and also different consumption. A trade-off between performance and consumption can be studied.

We can say that the contribution of a torque converter can be compared with that of an added range changer gearbox with variable efficiency and transmission ratio.

The poor efficiency of the torque converter has suggested using a *lock-up clutch*. This last is done with a multi disc oil clutch whose closure is controlled by the gearbox control system; when this clutch is closed the torque converter behaves as a rigid joint.

The closure of the clutch is made when the pump and the turbine speeds are very close; in these conditions the torque converter doesn't contribute much to the performance and is harmful for the efficiency.

It must not be forgot that by closing the lock-up clutch, also the mechanical damping function of the engine is lost; therefore a spring damper is necessary and sometime can be convenient to allow a controlled slip on the clutch, in order to avoid torsional vibrations.

The closure of the lock-up clutch must be therefore always gradual. In some automatic transmission of old generation this function is inhibited on lower gears; in modern transmission the clutch is also modulated in order to avoid vibration without jeopardizing the advantage on fuel consumption.

Chapter 12
Synchromesh Unit

12.1 Description

Synchronizers function is to enable changing of meshing gears couple, on a moving vehicle without negative consequences on mechanical integrity and on interior noise.

During synchronization the clutch must be disengaged.

12.1.1 Sincronizzatore Semplice

The most widely diffused synchronizer is the Borg Warner one, named after the manufacturer that developed it. Figure 12.1 shows a cross section of this synchronizer, in its single cone version; there are, as usual, two synchronizers, for two neighboring gears on the same shaft.

Gears are idle and are mounted on their shaft on needle bearings; they could be mounted either on the input or on the output shaft and are constantly meshed with those fixed on the other shaft.

The two idle gears show on their neighboring side the synchronizer hub 3, fixed to the relative gear. The joint between the *hub* and the wheel can be made in different ways: by laser welding, as in the figure, by spline shaft or teeth can be directly cut or stamped on the side of the gear.

In this last case, there must be a sufficient difference in radial dimensions between the gearing teeth of the wheel and the selector teeth of the synchronizer hub.

The different fabrication technologies of the joint between the hub and the wheel have a direct consequence on the length of the synchronizer assembly and of the gearbox itself.

The hub 3 shows a crown of teeth on his outside diameter; the *synchronizer ring* is free to move axially and matches with the hub through a tapered friction surface. It can also match with the selector teeth of the sleeve 4, in such a way as to block the rotation but leave the axial motion free.

G. Genta and L. Morello, *The Automotive Chassis*, Mechanical Engineering Series,
https://doi.org/10.1007/978-3-030-35635-4_12

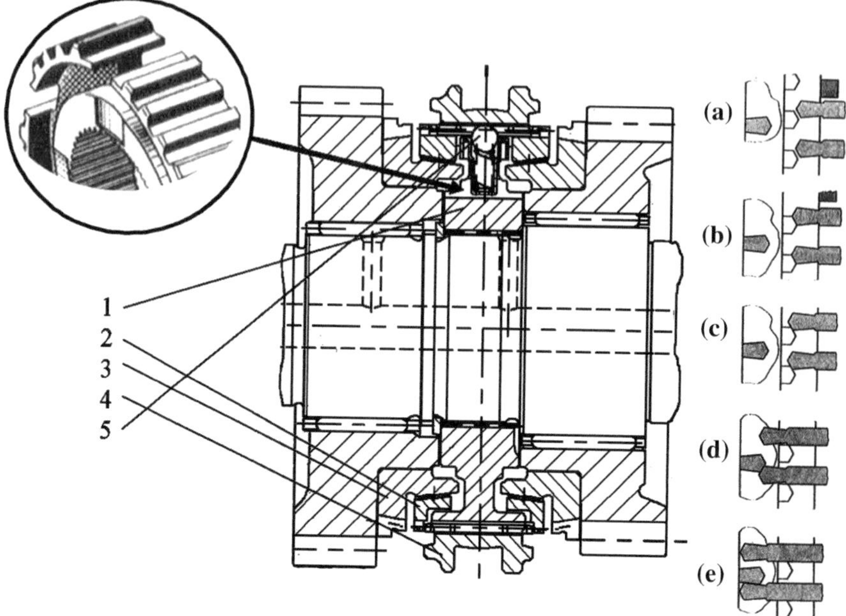

Fig. 12.1 Cross section of a single cone Borg Warner synchronizer (FIAT). In the circle on the upper part of the picture there is a prospective view of the synchronizer body 1. On the left the positions of teeth of the hub 3, of the ring 2 and the sleeve 4 are sketched, in the different phases of the synchronization process

The synchronizer ring 2, between the hub and the sleeve, acts as a friction clutch that can synchronize the elements before the engagement of the teeth; how this can be correctly made will be discussed on a following paragraph.

12.1.2 Multiple Synchronizers

Single cone synchronizers covered most applications until few years ago; in recent times the target of shortening shift time without increasing the manual force applied by the driver has been set. Multiple cone synchronizers have therefore been applied, particularly on low gears where the synchronization work is higher.

Figure 12.2 shows a double cone synchronizer; this arrangement can be appropriate to the mid gears of a large gearbox or for the lowest gears of a small one.

Let us look the different arrangement of the tapered rings of the synchronizer hub. The hub is reduced to the teeth crown only but there is a tapered friction ring connected to the hub in such a way ad to rotate with it with some freedom to move in other directions.

Friction surfaces are machined on the inside and outside faces of this ring.

The matching tapered surfaces are made with two different elements fixed to the selector sleeve in a particular way. The outside surface has the same shape as a single

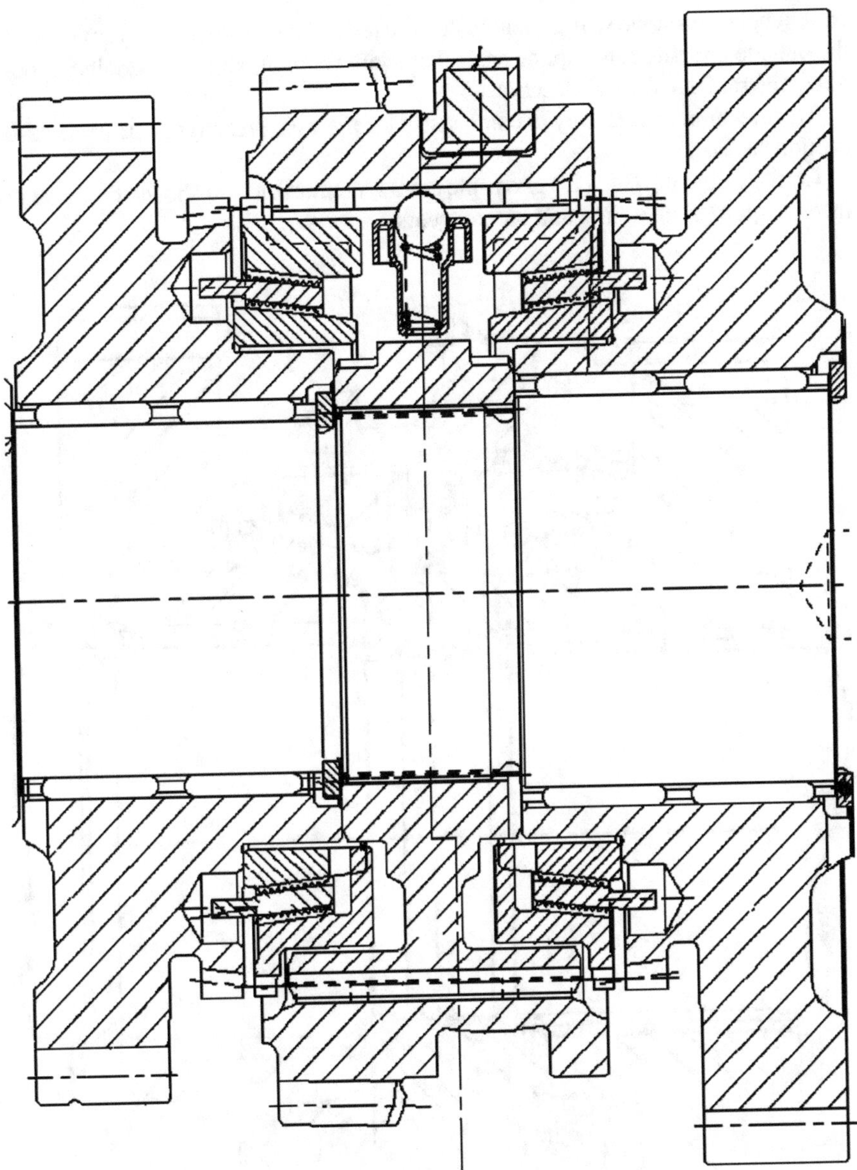

Fig. 12.2 Cross section of a double cone synchronizer (FCA)

cone synchronizer, we will explain in detail later on; the inside surface is matched to the outside one through a spline shaft that avoid rotation but leave certain freedom along the radial and axial directions.

This freedom is necessary to compensate for the wear and alignment errors of the friction surfaces.

On the following Fig. 12.3 two triple cone synchronizers are shown; this arrangement is appropriate to first and second gears.

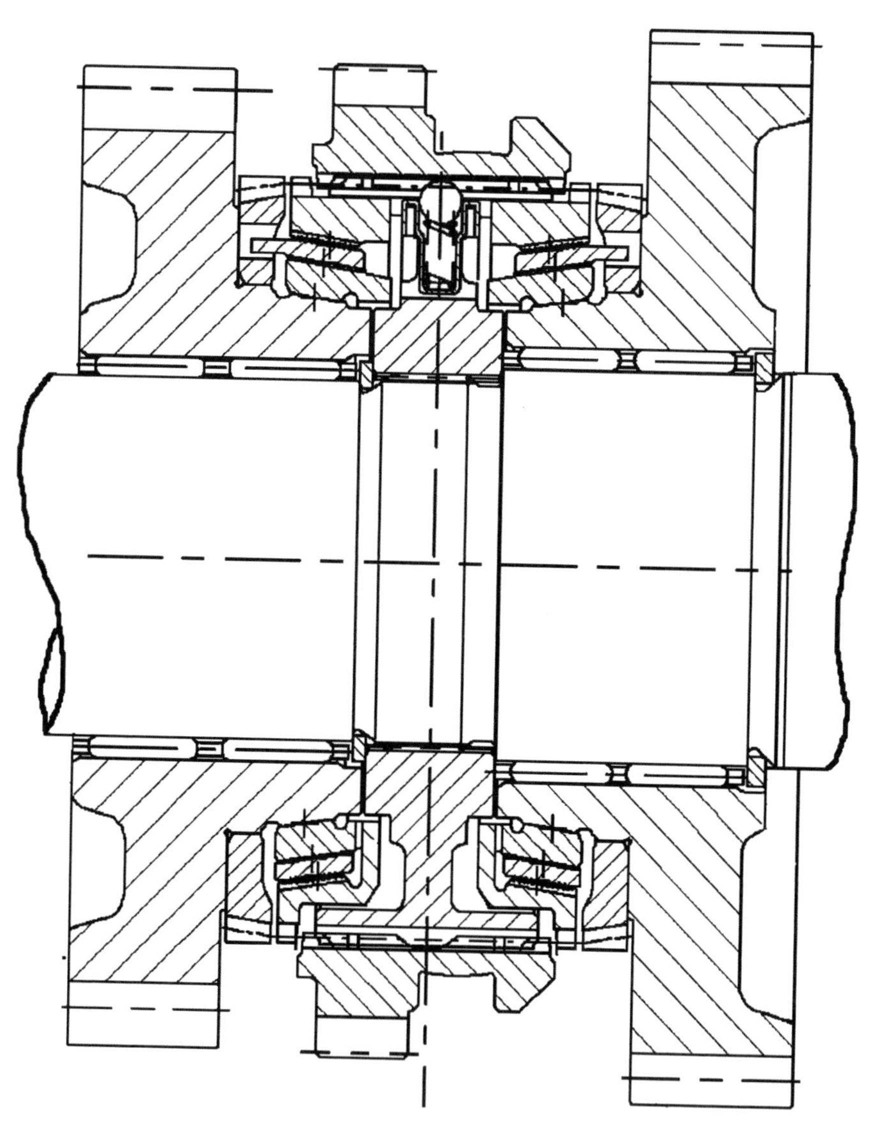

Fig. 12.3 Cross section of a triple cone synchronizer (FIAT)

In this case active friction surfaces of the hub are three; the innermost one is made as that of a single cone synchronizer, while the remaining two are similar to those of a double cone synchronizer.

Active sleeve surfaces are respectively arranged as a single cone synchronizer (the outside one) and mounted on a spline with some axial and radial clearance (the inside one).

The advantage shown by these arrangements is the multiplication of the synchronization torque made by the number of friction active surfaces present on the synchronizer; this allows the containment of the diameter of the ring, as compared with a single cone synchronizer of same performance.

It is possible to synchronize gearshifts without using synchronizers. Rotating parts should be accelerated during downshifts through the engine, after having engaged the clutch. It should also be possible to slow down rotating parts by means of a brake installed in the gearbox. This solution could prove of interest for the automation of manual transmissions by electro hydraulic electronic control systems.

12.1.3 The Gearshift Process

Let us examine the implications of synchronizing a gearbox during a speed shift; we will refer to a simple two speed single stage gearbox, shown on Fig. 12.4; power flows through the input shaft e to the output shaft u, trough two alternative gears couples.

To shift speed, the sleeve must alternatively engage with one of the wheels, in positions 1 and 2.

When couple 1 is engaged, if we call z_{1e} and z_{1u} teeth numbers on the two meshing wheels and Ω_e and Ω_u shaft rotation speeds, we will have:

$$\Omega_u = \frac{\Omega_e z_{1e}}{z_{1u}} , \tag{12.1}$$

while the speed of the idle gear of couple 2 will be:

$$\Omega_{e,2} = \Omega_e \left(\frac{z_{1e}}{z_{1u}}\right)\left(\frac{z_{2u}}{z_{2e}}\right) . \tag{12.2}$$

If we shift the sleeve to position two immediately, also after clutch disengagement, a number of shocks will occur between dog teeth of the two wheels; noise and structural damages could derive.

Synchronizing parts means bringing speed difference of moving parts to zero, before positive engagement. Because it is obviously impossible to change the speed of the u shaft, or the vehicle, without the help of the engine (the clutch is disengaged), synchronizing means also changing rotation speed of parts bound to shaft e in such a way as to delete the speed difference of the synchronizer body and the tooth wheel, before dog clutch engagement.

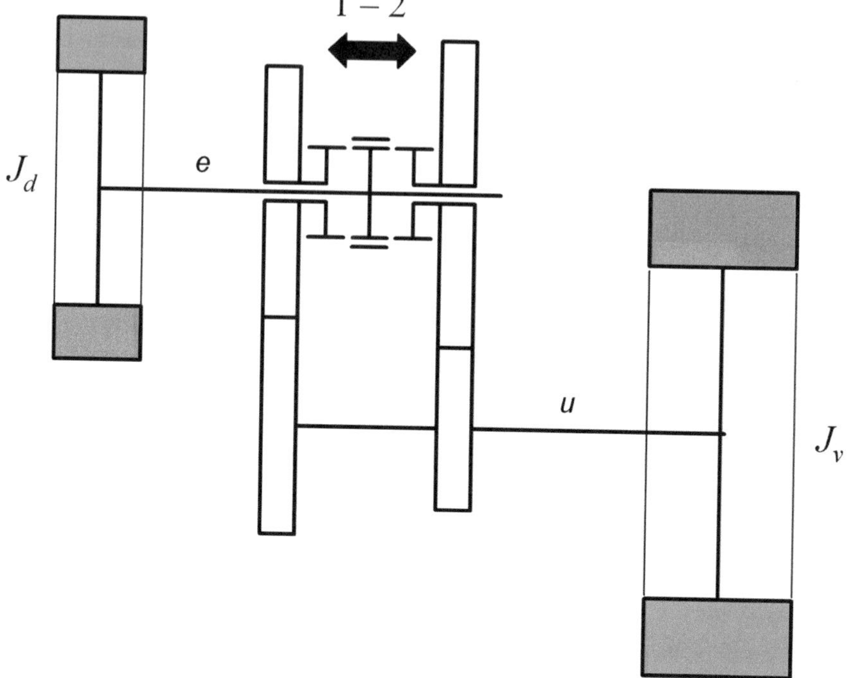

Fig. 12.4 Scheme of the masses to be synchronized during the speed shift of a two speed single stage gearbox

In our example, shifting from first to second speed implies reducing the speed of e, while during down shifts the same speed must be increased.

This operation needs a certain time delay, during which some energy must be added or subtracted to the system. In the mean time, taking into account the total driving resistance, the vehicle speed and the rotation speed of the shaft u will decrease. In a downhill road the opposite may occur.

Rotating masses, whose speed must be changed, are represented by a flywheel of inertia J_d; to its value contribute: driven clutch plate, gearbox input shaft with its fixed wheels (the constant gear wheels and the complete countershaft with its fixed wheels, in case of double stage gearboxes).

During a shift from second to first, some energy must be introduced in the system, while in the case of the shift from first to second, some energy must be subtracted; synchronizers perform this function through friction rings between the to masses to be brought to the same speed.

The necessary energy to this manoeuvre in subtracted or added to the apparent rotating mass of the vehicle, represented by a flywheel of inertia J_v rotating at the speed Ω_u. A vehicle deceleration or acceleration will therefore supply the synchronization work. Since the inertia J_v is much greater then J_d the vehicle speed variation will be negligible.

We can therefore summarize functions performed by the synchronizer as follows:

- Adapting speed of parts to be synchronized by an energy transfer from one to the other. The measure of this function is the synchronization time; it is a significant part of shifting time during which the vehicle is without traction.
- Positively joining of synchronized parts as to transmit the necessary torque; joint must be stable on time with no danger of gear self disengagement.
- In addition to the above, the following additional functions must be assured:
- Measuring rotating parts speed difference, to identify the suitable moment to engage the speed.
- Enabling positive engagement only when speeds are equal.

Let's make again reference to Fig. 12.1 and examine additional details.

A magnification of the synchronizer body is shown on the upper left of the figure. The body 5 shows a spline surface with 6 interruptions at 60°; these interruptions are replicated on the synchronizer ring 2.

Ring interruptions match with a defined angular clearance with 3 big teeth on the inside diameter of the sleeve.

The remaining three interruptions at 120° match with 3 push elements 5 that are free to slide inside the three said elements, that are the sleeve, the body and the ring; push elements show a radial hole in which a plunger with spherical head is fitted and is pushed by a coil spring against a seat on the inside of the sleeve.

Let's assume that the gearbox is in neutral, with both engine and vehicle in motion and the left wheel must be engaged.

We can identify five different phases of the shifting manoeuvre of the sleeve, actuated by the shift stick; for each of these phases a schematic cross section is drawn, showing a teeth view of sleeve, ring and hub developed on a plane.

1. The sleeve is pushed to the left against the reaction force of coil springs that try to keep push elements 5 in neutral position. Elements 5 and big teeth push the ring 2 and bring to contact the two matching tapered friction surfaces of ring and hub. Since the two parts have different speed, a friction torque will arise against ring and hub relative speed. The circumferential clearance between big teeth and ring will allow a small rotation of this last; the clearance is designed in such a way as the ring teeth stop the axial motion of the sleeve. See the detail a on the upper right. This phase is called *presynchronization* because it doesn't imply a real energy exchange between moving parts.
2. Friction torque between ring and hub transfers energy from the heaviest part of the system (the vehicle) to the lightest, in order to slow or to accelerate rotating parts fixed to the wheel to be shifted. Friction torque *measures* the relative angular speed because it will go to zero with relative speed. The same torque provides stopping sleeve motion until synchronization is completed. This phase is called *synchronization*.
3. Before describing next phase, let us consider the shape of dog teeth on hub and sleeve. They show a tapered end and a counter tapered surface. When ring and hub are synchronous the friction torque is zero and dog teeth tapered ends rotate

the hub of the angular thickness of half tooth, under the pressure exerted by the shift stick.

4. The sleeve is now free to move across the ring and to match the hub; again teeth tapered end will rotate the hub if necessary. In this phase the push elements plungers are completely retracted and do not withstand anymore sleeve motion.
5. Now a *positive engagement* is made; at the end of this phase the driver will stop pushing shift stick; the counter tapered shape of dog teeth will retain the sleeve in the engaged position under the action of the engine torque. The same shape will exert a certain reaction to the shift stick, when the driver will start to disengage the gear for the next speed shift.

12.2 Design Criteria

Synchronization torque determines shifting time duration which contributes to vehicle acceleration time.

This torque is proportional to the force applied the shift stick; this last must respect certain design conditions.

The force on shift stick knob must be kept below 80–120 N on a car, while 180–250 N can be allowed on an industrial vehicle; synchronization time must be lower than 0.15–0.25 s on a car, while on an industrial vehicle must be lower then 0.25–0.4 s.

Transmission ratios between shift stick knob and sleeve must be kept below 7–12:1, to avoid too long shifting strokes.

When calculating the force available at the synchronizer, starting from the lever force, mechanical efficiency of shifting mechanisms must be taken into account; refer to what was said about the clutch mechanism.

Figure 12.5 reports synchronization times versus shifting force applied to stick knob, for some manual single stage gearboxes, for shifts from fourth to fifth and vice versa; the values for up and down shifts are not the same, because the synchronization works are different. The more effective the design, the shorter the synchronization time at the same shifting force.

The effort on the knob isn't the only judgment parameter. The shape of shift force diagram versus lever stroke is very important too; Fig. 12.6 shows two diagrams of two different gearboxes.

A good behavior is shown on the upper part of the figure by a force diagram generally increasing with the stroke, featuring a first peak m_1 associated to the synchronization phase and a second peak m_2 at the end of the engaging phase.

With this lever mechanical response, drivers can perceive through a force increase the beginning of the synchronization phase; the end of this phase is announced by a sudden decrease of the reaction force. Drivers understand when shift stick must be kept in place and when shifting manoeuvre can be completed quickly.

A second positive feature of the first diagram is the good statistic consistency between the results of different manoeuvres; this consistency allows the driver to get quickly trained to the gearbox use.

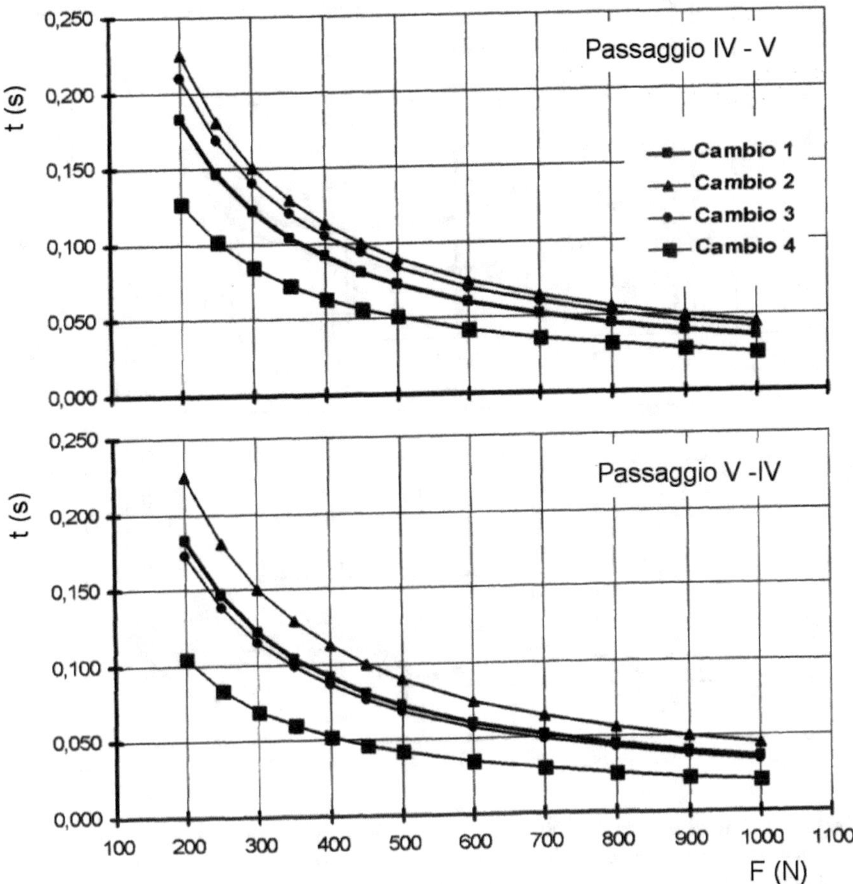

Fig. 12.5 Diagrams representing synchronization time (on the vertical axis) versus knob shifting force (on the horizontal axis), for different single stage gearboxes and for shifts from fourth to fifth and vice versa

A bad behavior is shown on the lower part of the figure, where the phenomenon of the *double shock* is also shown.

The double shock is measured by a third peak m_3; this occurs when the ring must turn at the synchronization phase end and must align his teeth with those of the sleeve.

This phenomenon may be due to a too high friction coefficient between the tapered ends of teeth of ring and sleeve or of sleeve and hub. An unsuitable opening angle of the tapered end or a different opening angle on different teeth can be the root cause of this inconvenience. The opening angle difference between teeth can also cause different behavior in different manoeuvres.

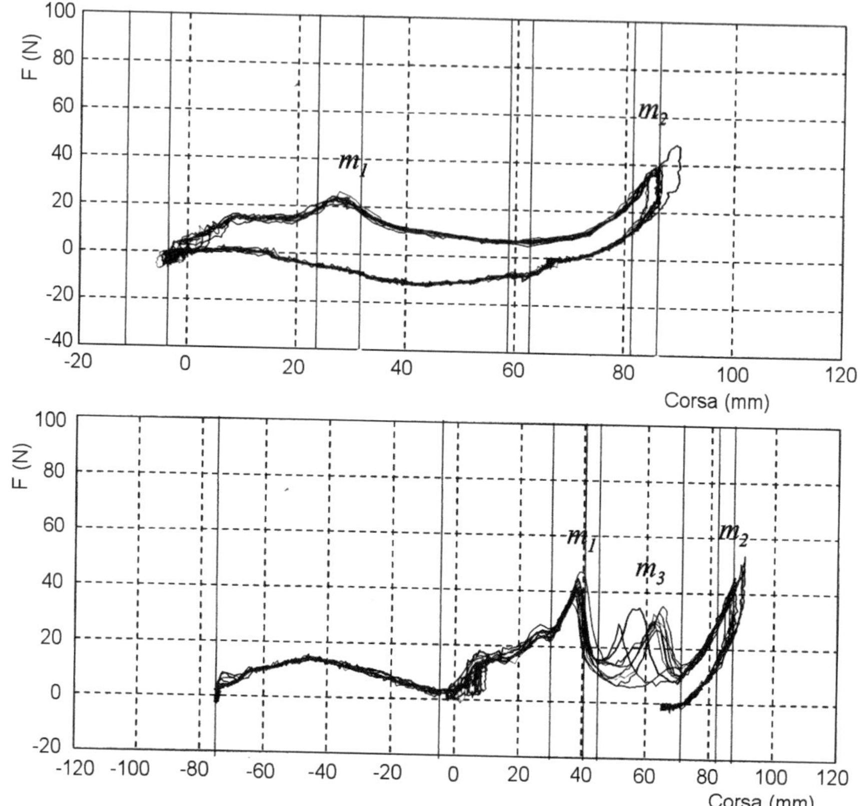

Fig. 12.6 Diagrams representing shift stick knob force (vertical axis) versus lever stroke (horizontal axis) on a single stage gearbox with good and poor behavior (good on the upper side, poor on the lower side): on the second diagram the double shock phenomenon is announced by the third peak m_3

The double shock can lead drivers to wrong manoeuvres characterized by an incomplete shift. The poor statistical consistency of the shifting force is particularly disagreeable because makes the gearbox behavior unpredictable.

Reconsider what was said about gearbox shifting quality at the end of chapter about shifting mechanisms.

12.2.1 Geometric Criteria

Ring and hub are represented with a simplified drawing on Fig. 12.7. Mean diameter d and opening angle α of the tapered surface are shown.

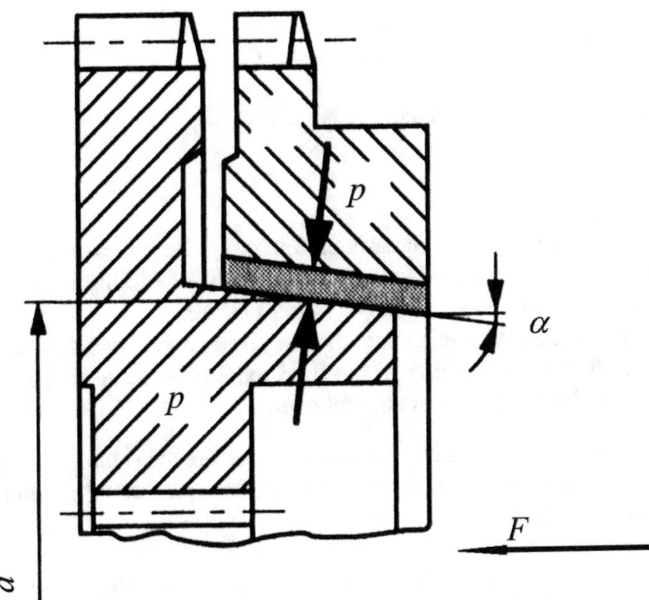

Fig. 12.7 Scheme of the force F acting on the synchronizer ring and of the pressure p between conical surfaces

Ring and hub must only contact on their conical surfaces; it is necessary to provide an axial clearance between parts to compensate for the allowed surface wear during gearbox design life time.

Any force F applied to the sleeve originates a normal pressure p on contacting surfaces; according to wear theory, this pressure is constant and if we call b the conical surface width, it will be:

$$pb\pi d \sin \alpha = F\ . \tag{12.3}$$

If we call with f the friction coefficient between surfaces and with M_s synchronization torque, we can write:

$$M_s = 1/2\ pb\pi d^2 f = \frac{Fdf}{2 \sin \alpha}\ . \tag{12.4}$$

Therefore synchronization torque will increase, at the same shift stick knob effort, if friction coefficient will increase and opening angle will decrease.

But opening angle can't be reduced indefinitely, to avoid irreversible sticking of cones.

This happens when the self unlocking force is lower than the friction force along the unlocking direction; the self unlocking force is done by the normal pressure resultant along the axial direction. Boundary sticking occurs when:

$$p \tan \alpha = pf \,, \tag{12.5}$$

therefore rings can unlock spontaneously if:

$$\tan \alpha > f \,.$$

The only effective mean of increasing available synchronization torque is increasing ring diameter or active surfaces number; this last possibility is exploited on the multiple cones synchronizers, as shown in this chapter.

Figure 12.8 represents a schematic drawing of a single cone synchronizer, who shows the different contribution to the total axial dimension of the assembly a; contributions to be considered are the following:

- Accepted wear on working friction surfaces; a reference value can range between 0.1 and 0.15 mm; for a conicity of 1:10 the consequent axial clearance must be greater than 1–1.5 mm (Δs).
- Gear torque that determines hub dog teeth width.
- Synchronization torque that determines ring dog teeth width.
- The geometrical condition of sleeve engagement and disengagement; sleeve width must be at least double of dimension l coming out from ring and hub teeth width and from dimension Δs.
- Synchronizer body width must be such as sleeve doesn't engage any synchronizer ring in neutral position.

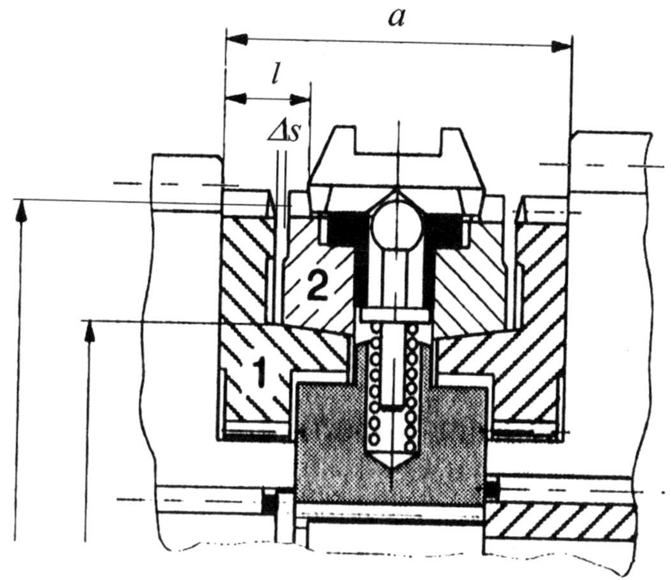

Fig. 12.8 The axial dimension a of a double single cone synchronizer is determined by the design wear projected on the axial direction and by dog teeth width of hub, ring and body

12.2.2 Functional Criteria

If we assume that vehicle apparent rotating mass is much greater than that of parts to be synchronized, the synchronization transient will be completely similar to that of the clutch during the vehicle start-up and will be described with a single equation, because with this assumption vehicle speed will remain constant.

If we call with M_s the synchronization torque, M_a the friction torque of the mechanical losses of parts to be synchronized, Ω the rotation speed of elements upstream the synchronizer, $J_{r,eq}$ the rotating mass of parts to be synchronized, we can write:

$$M_s + M_a = -J_{r_{eq}} \frac{d\Omega}{dt} . \tag{12.6}$$

Because M_s and M_a are almost constant with speed, angular acceleration will be constant on time while rotation speed will decrease constantly on time; during up shifts the acceleration is negative, while during down shifts it will be positive.

Synchronization torque will always show an opposite sign as that of angular acceleration, while friction torque will show the same sign as that of synchronization torque during up shifts, when rotating masses must be slowed down.

In the above hypotheses thermal power $P_{\max}$ and thermal energy E wasted by the synchronizer are:

$$P_{\max} = M_s \left(\Omega_2 - \Omega_1\right) , \tag{12.7}$$

$$E = \frac{1}{2} J_{r_{eq}} (\Omega_2^2 - \Omega_1^2) + \frac{1}{2} M_a \left(\Omega_2 - \Omega_1\right) \Delta t , \tag{12.8}$$

where subscripts 1 and 2 apply to initial and final conditions.

Let's now assume that we have to design the first gear synchronizer of a two stage gearbox, as far as the second to first shift is concerned; the scheme of the gearbox is shown on Fig. 12.9.

We call with z_i gear teeth numbers (i refers to the gear number on the scheme) and with J_i the related rotating mass. The equivalent rotating mass $J_{r,eq}$ to put in the above equation is that of all rotating masses up steam the synchronizer, reduced to the gear 3:

$$\begin{aligned} J_{r_{eq}} &= J_3 + (J_d + J_1) \left(\frac{z_3}{z_8}\right)^2 \left(\frac{z_7}{z_1}\right)^2 + (J_{ca} + J_7 + J_8 + J_9) \left(\frac{z_3}{z_8}\right)^2 + \\ &+ J_2 \left(\frac{z_3}{z_8}\right)^2 \left(\frac{z_8}{z_2}\right)^2 + J_4 \left(\frac{z_3}{z_8}\right)^2 \left(\frac{z_9}{z_4}\right)^2 . \end{aligned} \tag{12.9}$$

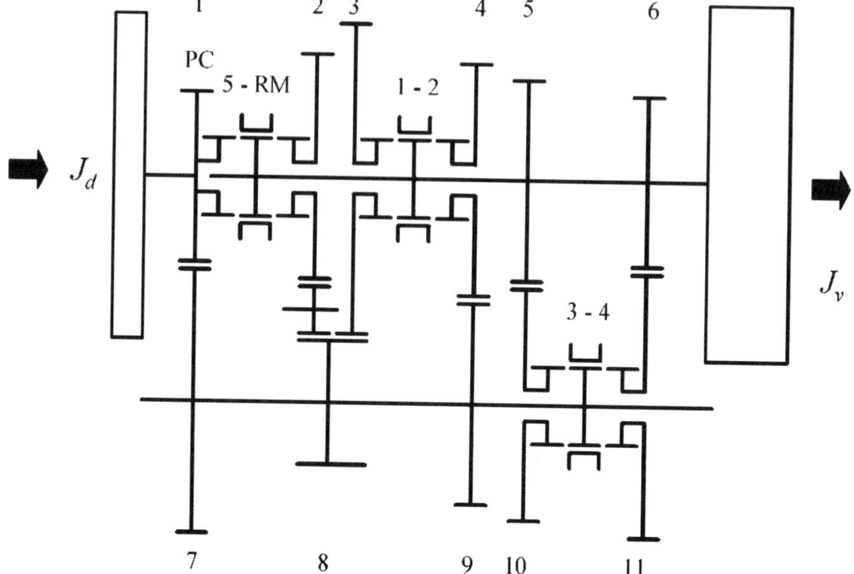

Fig. 12.9 Scheme for the calculation of equivalent rotating mass to be synchronized, during a gear shift manoeuvre

Masses of synchronizers and of reverse idlers are neglected; J_{ca} is the countershaft's rotating mass. Idle wheels on the countershaft mustn't be taken into account, because they are already synchronous with the output speed.

On the same gearbox, if the synchronizer is put on the countershaft, the rotating mass will be:

$$J_{r_{eq}} = J_8 + J_{ca} + J_9 + (J_d + J_1)\left(\frac{z_7}{z_1}\right)^2 + J_2\left(\frac{z_8}{z_2}\right)^2 , \tag{12.10}$$

because the remaining wheels will be rotating with the output shaft.

The advantage with this second configuration can be noticed; it can be also noticed that the rotating mass is decreasing notably while the speed order is increasing. On a single stage gearbox the calculation will be simpler and confirm the above conclusions.

After rotating masses evaluation it is possible calculating synchronization time as a function of the shifting force; in the same way wasted power can be calculated and compared, as a design parameter, with other gearboxes of success.

The synchronizer reference speed is defined by:

$$V = \frac{d}{2}(\Omega_2 - \Omega_1) , \tag{12.11}$$

Table 12.1 Reference design values for synchronizers made by steel/brass and steel/steel with molybdenum coating

Material	f	V (m/s)	E_s (J/mm^2)	P_s (W/mm^2)	p (N/mm^2)
St./Br.	∽ 0.1	5	0.09	0.45	3
St./St,m.	∽ 0.1	7	0.53	0.84	6

showing the relative speed at the synchronizer reference diameter at the beginning of the shift manoeuvre.

The empirical reference values, reported on Table 12.1, can be considered for the above magnitudes.

Specific power and energy are derived of wasted power and energy by dividing by the synchronizer active surface.

Because a friction coefficient as high as possible is wanted, a lubricated hydrodynamic film should be avoided.

Nevertheless synchronizers are splashed with lubricating oil; with reference to what we said above synchronizers active surface is grooved or threaded; threads should unwind at the relative speed.

If surfaces are grooved or threaded the reference surface area must be consequently reduced.

Something more must be said about dog teeth tapered ends; tangential force consequent to the opening mustn't rotate the ring before synchronization has been accomplished.

If we represent some of these teeth on Fig. 12.10, we can see that, under the action of the shifting force F, the opening angle of the end sets up a torque of opposite direction as the synchronization one; if the first were greater than the second, the shifting force would be as big as to engage the dog clutch before synchronization is accomplished.

If β is the opening angle of the tooth and d_s is the teeth mean diameter, the torque acting on the ring will be:

$$\frac{1}{2} d_s F \cot\left(\frac{\beta}{2}\right),$$

if we neglect friction on the end of the tooth; this torque mustn't be greater then the synchronization torque:

$$\frac{d_f F}{2 \sin(\alpha)}.$$

Materials matching for synchronization ring and hub must be determined taking into account the following needs:

- Wear must be small also without the help of hydrodynamic lubrication.
- Materials must be easily machined.
- Friction coefficient must be as constant as possible in production.

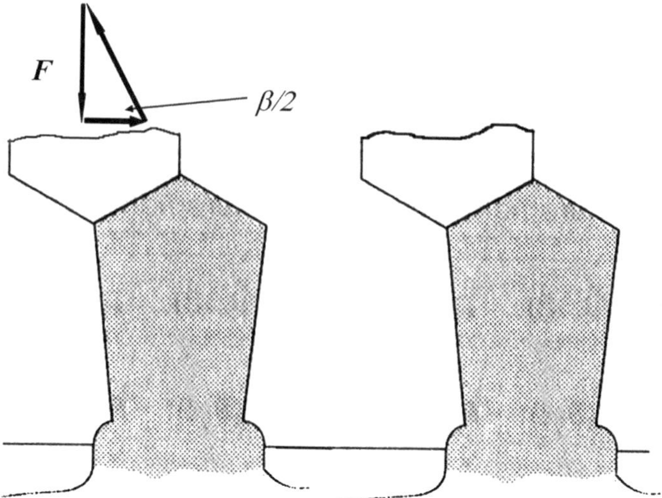

Fig. 12.10 Scheme for the calculation of the condition of possible engagement before the accomplishment of the synchronization

- Friction coefficient must be insensitive to wear and temperature.
- Materials must withstand possible overloads.

As a matter of fact the available options are very few; hubs are usually made of Cr Mn steel or Cr Mo steel. Rings can be made of special shot peened brasses or steel coated with a molybdenum layer. Also a paper layer could be considered in future, with advantage on the friction coefficient.

Chapter 13
Differentials and Final Drives

Sometime, the names of differential, transfer box and final drive are used imprecisely; in addition to that these mechanisms are sometime integrated in the same subsystem in almost all possible combinations. We try to introduce some definition to better clarify this issue.

The *differential* is a mechanism that allows dividing the torque coming from an input shaft in two predetermined parts flowing through two output shafts; torques ratios are independent of speed ratios of the same shafts.

This mechanism can be either used to divide the torque coming out of the final drive in to equal parts acting on the traction wheels of the same axle, or the torque coming out of the gearbox in two predetermined parts acting on different axles of the same vehicle; this second application is named sometimes *transfer box differential* or *central differential*.

The *final drive* is a gear train that further reduces the speed of the gearbox output shaft to adapt it to the traction wheels; this gear train is usually integrated with a differential mechanism. Under this name are also sometime included those speed reducers that are put on the transmission line after the differential final drive and are integrated in the wheel hub.

The *transfer box* is a mechanism that provides to move two or more drivelines with the only output shaft of the gearbox; this is used on vehicles with multiple traction axles. When multiple axle traction is permanent also a differential is necessary to allow different mean rotation speeds on the axles; in such case the differential train is usually integrated in the transfer box.

G. Genta and L. Morello, *The Automotive Chassis*, Mechanical Engineering Series,
https://doi.org/10.1007/978-3-030-35635-4_13

13.1 Differentials and Final Drives

There are different schemes according to vehicle drive kind; in all of them the differential mechanism shows usually the same configuration.

Look at the schemes shown from Figs. II.1 to II.3 to identify all possible configurations.

13.1.1 Rear Wheel Driven Cars

On rear wheel driven cars final drives carry out the task of rotating the driveline axis of 90°, from the longitudinal gearbox centreline, to the transversal axle centreline. On Fig. 13.1 the cross section of this subsystem is shown. Centreline rotation and speed reduction are achieved through a couple of bevel gears with spiral teeth; the wheel is bolted to a hub in order to allow the easy adaptation of the same production line to different final transmission ratios.

Shafts are supported by bevel roller bearings, because the axial thrust is relevant. The differential train is made with straight teeth bevel gears, because their rotation isn't continuous and their speed is small. *Planetary wheels* are fixed to half axles through splines, while *satellites* are idle on a short shaft fixed to a carrier through a pin; the bevel gear wheel of the final drive is bolted on this carrier.

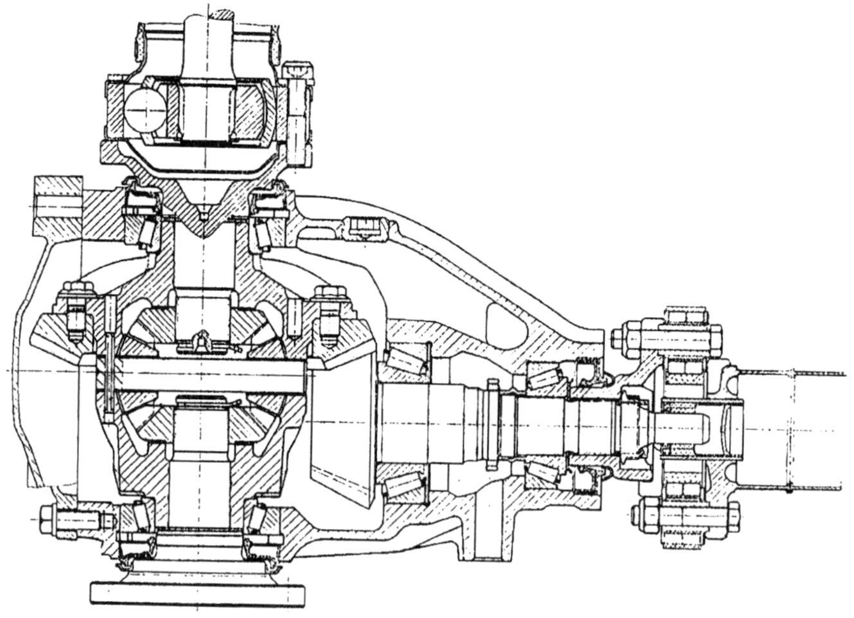

Fig. 13.1 Cross section of a differential assembly for a rear wheel driven car with rear independent suspensions (Mercedes)

The differential assembly is a suspended one, used in connection with independent suspensions: therefore half axles will be moved through constant speed joints, as in a front wheel driven car. It should be remembered that a suspended differential casing must react to the vectorial resultant of the wheel traction torque, acting on xz plane and to the gearbox output torque, acting on yz plane; for this reason the casing features a robust fixation interface, to suspend it to the car body, usually through an auxiliary frame.

It can be noticed, on the same figure, the small elastic tube on the bevel pinion shaft; it allows controlling the axial preload on bevel roller bearings, compensating for axial dimension tolerance.

In this application final drive bevel gears feature *hypoid* teeth, because their centre lines lie on different planes; this configuration allows having transmission line on a lower position with advantage for the interior roominess.

Planetary wheels and satellites thrust bearings are simple wear resistant washers, because of the small relative speed value.

13.1.2 Front Wheel Driven Cars

On front wheel driven cars differential and final drive are integrated in the gearbox; the reaction torque is therefore acting on the powertrain suspension. We can identify two cases where engine is transversal or longitudinal.

In the first case, shown on Fig. 13.2, gearbox output shaft and wheels centre lines are parallel; a couple of spur gear with helical teeth is sufficient. The pinion is cut on the output shaft and meshes with a wheel, which also in this case is flanged to the differential carrier.

The differential mechanism is very similar to the previous one. In this case planet wheels thrust bearing are made with needle cages to improve mechanical efficiency. A following paragraph will explain the influence of this mechanical efficiency on the vehicle dynamic behavior.

For the longitudinal engine, shown on Fig. 9.9, there are no relevant differences in comparison with the rear wheel driven case; usually hypoid teeth are unnecessary. On Fig. 13.3 we see the differential assembly corresponding to the traction scheme on Fig. II.1; the bevel pinion is fixed on a very short transmission shaft that allows shifting the front axle in front of the gearbox output shaft end.

Also the driving wheel of this short shaft is a bevel gear and a simple damper for the flexural vibration has been provided. This kind of architecture is necessary when front wheel overhang must be limited or when automatic epicycloidal gearboxes are used, where input and output shafts are aligned.

13.1.3 Industrial Vehicles

On industrial vehicles final drive and differential train (Fig. 13.4) are integrated in the same rigid structure supporting wheels and half axles, called rigid axle.

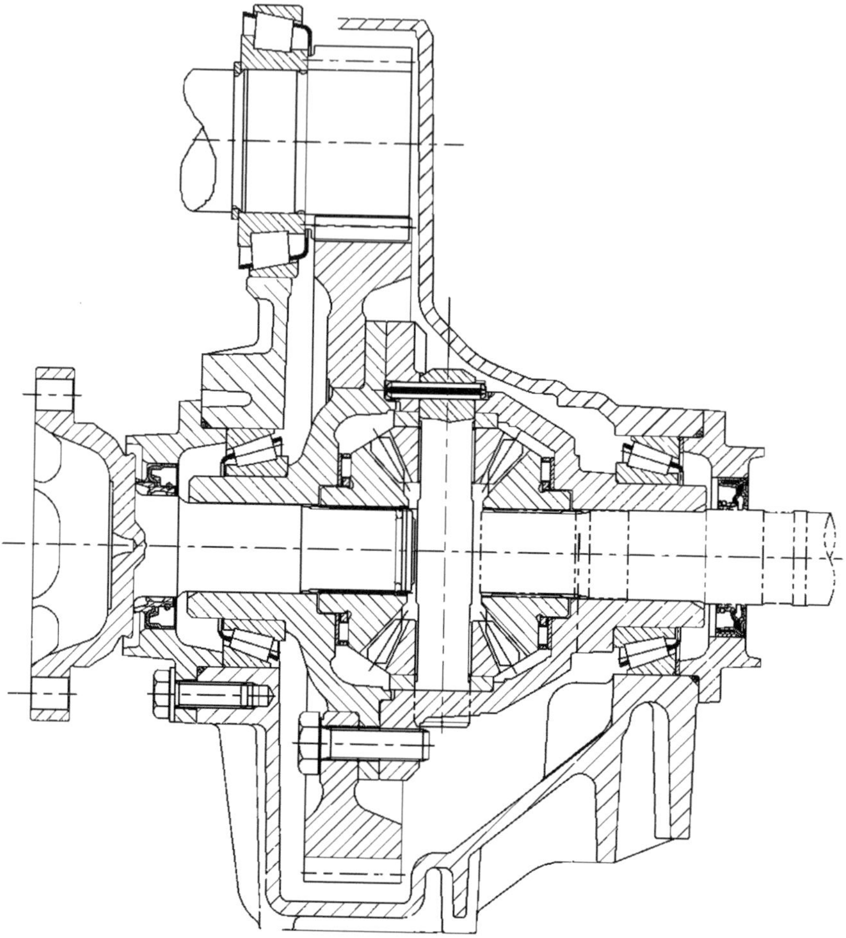

Fig. 13.2 Differential and final drive for a front wheel driven car with transversal engine (FCA)

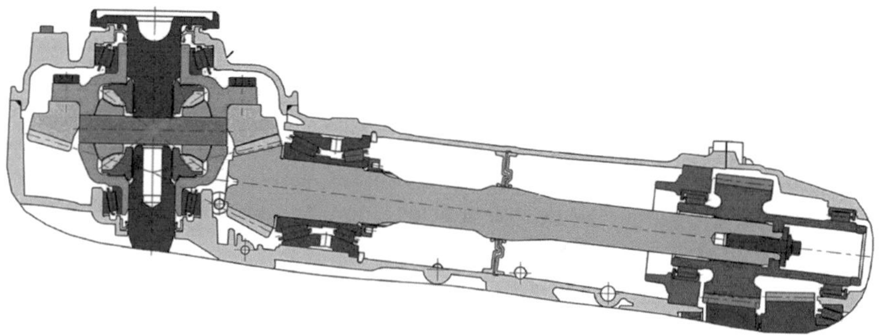

Fig. 13.3 Differential and final drive for a front wheel driven car with longitudinal engine; in this case wheel axis is shifted forward in order to reduce front overhang (Audi)

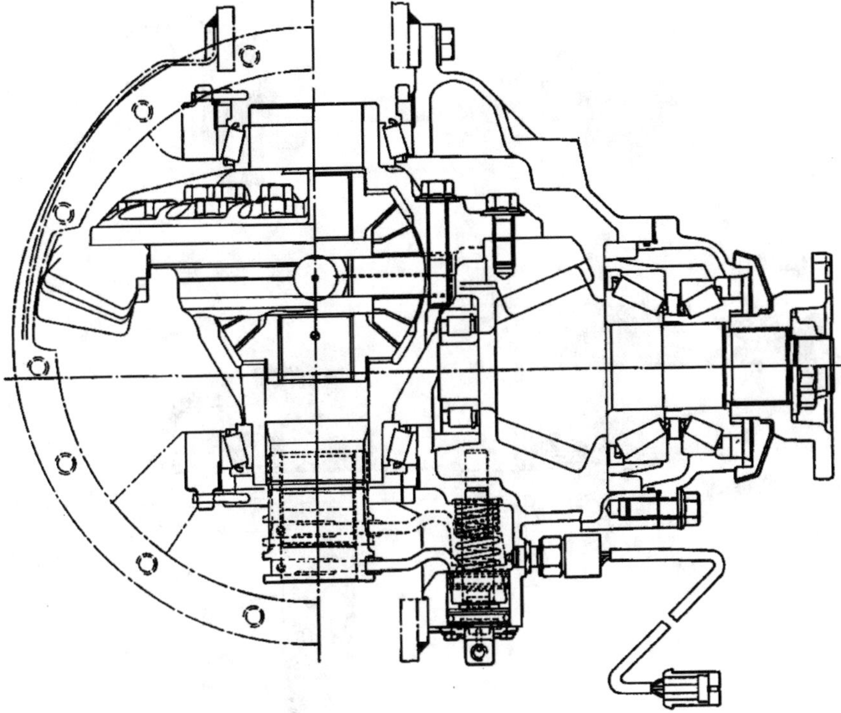

Fig. 13.4 Rigid axle differential assembly for a heavy duty truck (Iveco). It can be seen on the right half axle the differential lock dog clutch

Not considering the increased robustness of the group, there are no noticeable differences in comparison with rear wheel driven cars: final drive is made with bevel gears with hypoid teeth. Wheels offset is used in this case in the opposite way as before, to increase transmission shaft ground clearance.

On the right half axle a locking sleeve can be noticed; the lock is particularly useful to improve start-up on a slippery road.

For construction or off-road vehicles transmission ratios obtained with a single stage final drive may not be sufficient; transmission ratio is limited by the allowed bevel gear dimension and by the minimum teeth number which can be cut on the pinion once the pitch has been defined.

This problem can be solved by installing an additional epicycloidal final drive. In this case the differential carrier can be integrated with the epicycloidal drive carrier, where annulus gear is fixed on the bevel wheel shaft; this way, the additional drive doesn't change rotation direction.

On urban busses is quite important to limit the ground height of the aisle floor. Floor height and tire rolling radius define the maximum diameter of bevel gear wheel; therefore, also in this case represented on Fig. 13.5, a spur gear final drive is necessary on the output shaft of the differential. This choice is motivated by the need of having a bridge shaped rigid axle with the central part compatible with a low height aisle floor.

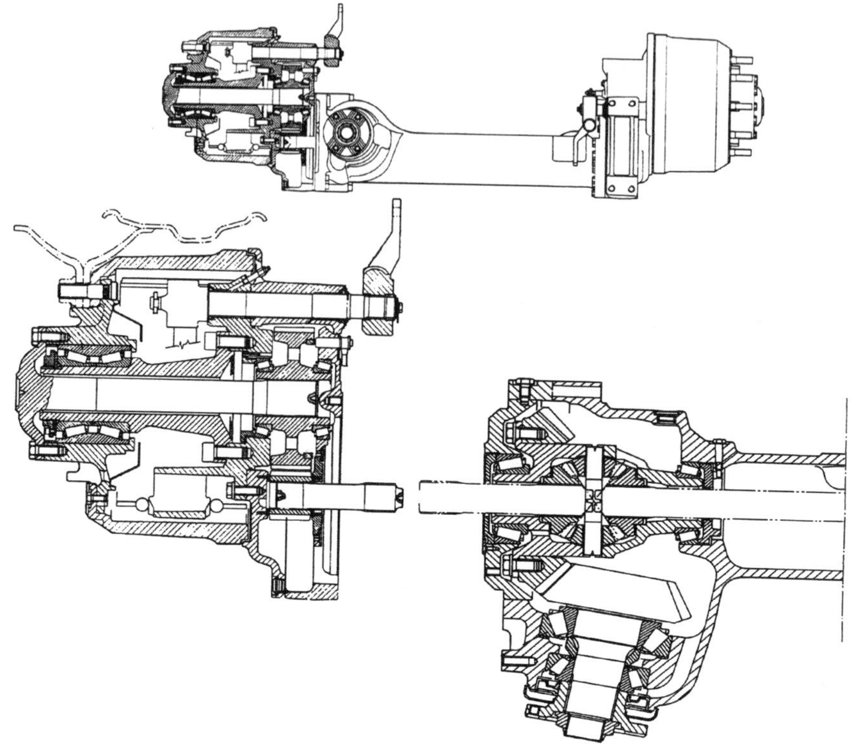

Fig. 13.5 Rigid rear axle for urban busses (Iveco). The lower detail shows the additional final drives on the wheel hubs that shapes the axle like a bridge. Note the inclination of the input shaft, caused by the transversal powertrain lay-out

The inverted bridge configuration can be also used to increase ground clearance on off-road vehicles; in this case will hubs will be lower as the half axles.

13.2 All Wheel Drive Transfer Boxes

On all wheel driven vehicles transfer box architecture is conditioned by two factors:

- The vehicle single axle traction configuration the all wheel drive is derived of, with reference to the architectures shown on Fig. II.3.
- The fact that the drive is *permanent* (can be used at any vehicle speed) or *non permanent* (can be used occasionally at low speed on bad roads). In the first case a central transfer differential is also necessary, to avoid unnecessary tire wear and additional rolling resistance when wheel speeds are different; in the second case the differential can be avoided.

We consider drivelines derived of conventional rear wheel driven vehicles and front wheel driven vehicle, either with transversal or longitudinal engine.

13.2.1 Modified Rear Wheel Drives

The architectures in use are many; a possible classification can derive of the vehicle mission. We can identify off-road vehicles, where the main objective is to obtain a good mobility on dirty and slippery roads and road vehicles, where the objective is to reach superior stability and handling on paved roads at high speed also.

On off-road vehicles (scheme c, Fig. II.3), the transmission line is offset in comparison to the engine crankshaft to increase ground clearance. The transfer box is also working for rear axle drive only; the transfer box lay-out is visible on Fig. 13.6.

Other functions are present; as a matter of fact the gearbox output shaft moves a synchronized two speeds range reducer. The reducer output shaft moves a third shaft featuring a bevel gear differential that provides splitting the torque for the two axles. On the same figure a dog clutch is also shown; it can fix the differential carrier with the front planet gear, locking in this way the differential for slippery road at low speed.

Fig. 13.6 Transfer box for a professional off-road vehicle (Mercedes). The transfer box integrates the two axles drive through a bevel gear lockable differential and a range speed reducer with two synchronized speeds

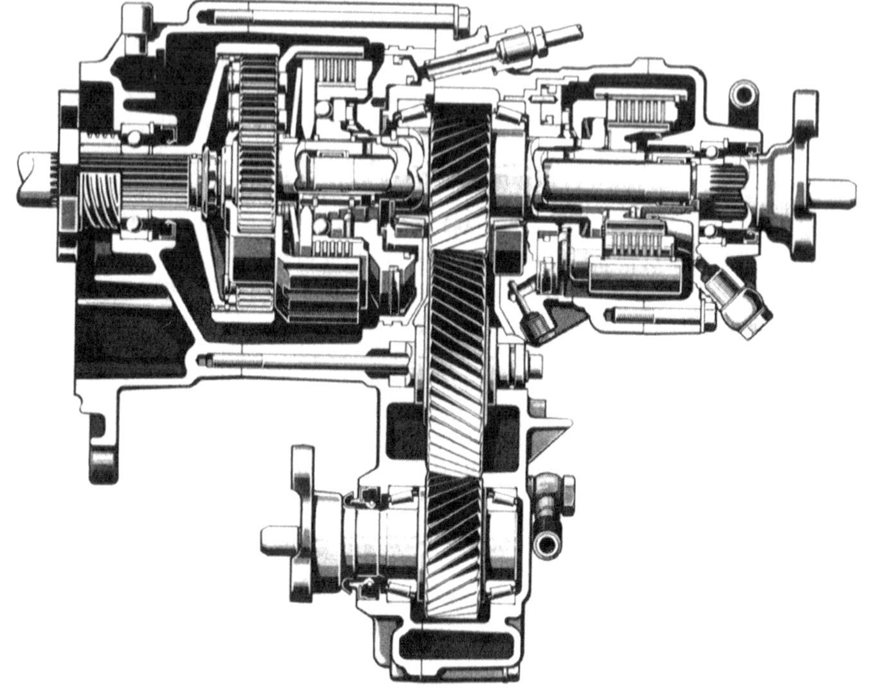

Fig. 13.7 Transfer box for an all wheel drive on-road vehicle with longitudinal front engine (Mercedes). The box integrates the axle drive with an epicyclical differential with slip control

The offset between crankshaft and transmission line is useful because it allows using rigid axle suspensions without interference with the oil sump.

This type of transfer box can have a simple dog clutch to insert the front wheel drive on low speed slippery road, instead of the differential.

On on-road vehicles, the original rear wheel drive line position is conserved; a transfer box is added to move the front axle too, using always, in this case, a central differential.

An example is shown on Fig. 13.7, where the motion to the front axle is given through an idler. The differential is of the epicycloidal spur gears type, where the carrier is fixed to the rear axle transmission shaft, the sun gear moves the front drive transmission shaft, the annulus gear is driven by the gearbox output shaft.

There is a multi disc wet clutch that controls the two axles speed difference; on the function of this clutch we will comment later on.

A second clutch, on the right of the figure, can put the front axle to idle. Three operating modes are available: rear wheel drive, constant rate all wheel drive, locked differential all wheel drive.

It should be noticed that the carrier shows twin satellites, to allow the correct rotation direction for the front axle.

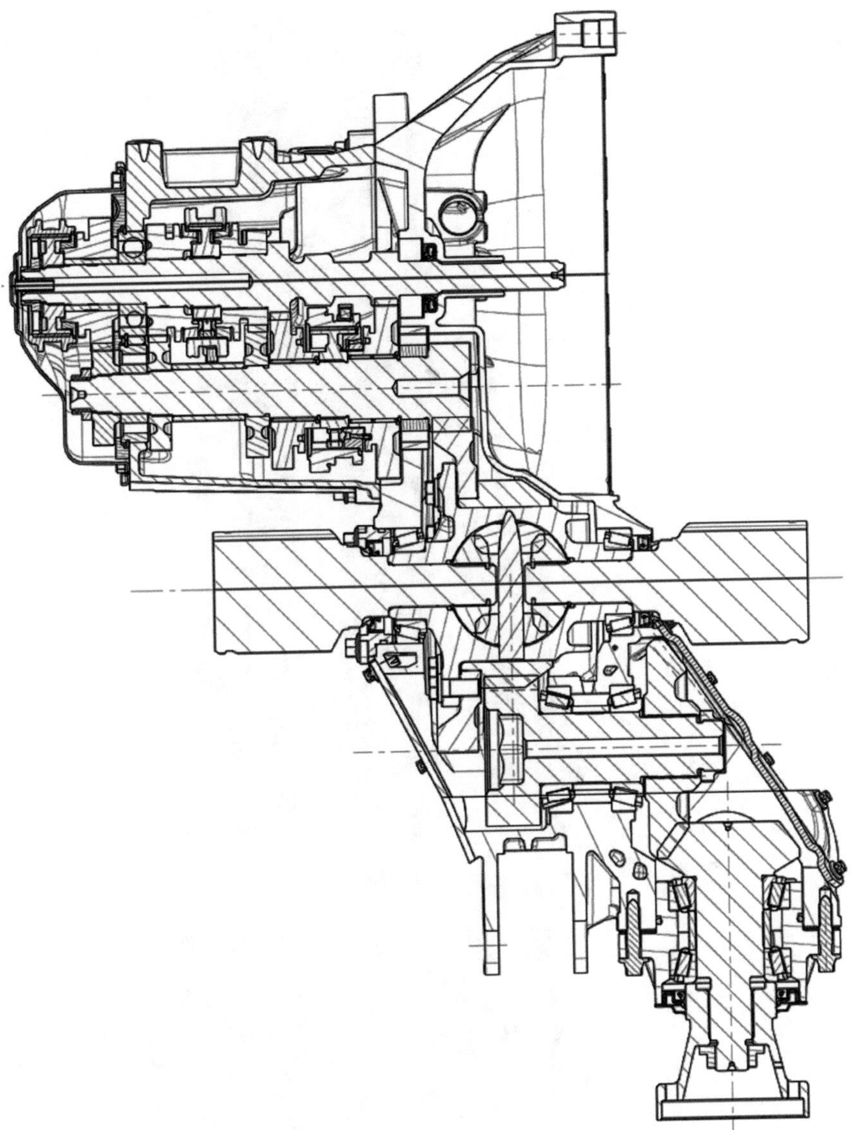

Fig. 13.8 Gearbox, differential and transfer box of an all wheel drive car with transversal front engine (FCA)

The use of a spur gear epicycloidal differential allows any torque breakdown rate different than 50/50; this is useful when axle loads are different and, in any case, to correctly control the vehicle static margin, peculiar on vehicle stability.

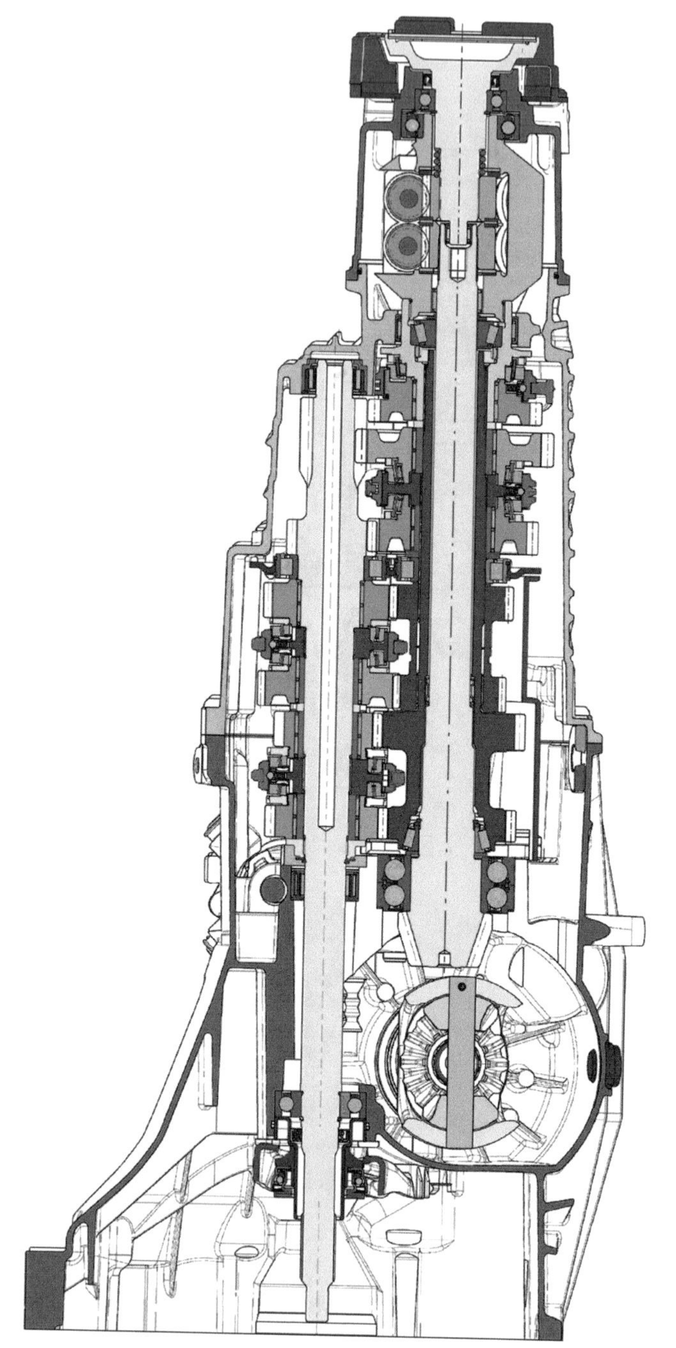

Fig. 13.9 Gearbox, differential and transfer box of a permanent all wheel drive with longitudinal front engine (Audi). The gearbox output shaft is hollow, in order to install the front wheel axle drive shaft

13.2.2 *Modified Front Wheel Drive Vehicles*

It may be useful, in this case, to distinguish between transversal and longitudinal engines.

In the most widely diffused case of transversal front engine, shown on Fig. 13.8, the front axle differential is coupled to a simple bevel gear drive featuring a transmission ratio close to one. The driving bevel gear is fixed to the front axle differential carrier; the rear axle transmission shaft is moved by the driven bevel gear. There can be a simple dog clutch, not shown on this figure, to move the rear axle, in case of non permanent all wheel drive.

The rear differential group is similar to that of a rear traction drive; in case of permanent traction, a viscous coupling is usually fit on the rear axle drive line; this joint provides subtracting part of the torque available to the front axle, only when this axle is showing an average speed greater than that of the rear axle. In a next paragraph we will comment about this joint and his way of working.

A self locking transfer differential can be installed in the transfer case; in this case the viscous coupling is avoided.

The case of the longitudinal engine is shown on Fig. 13.9. All wheel drive looks simpler; the gearbox output shaft is hollow; the shaft moving the front differential bevel pinion is left free to turn in the cavity. The gearbox output shaft moves a Torsen type differential, we will explain later on. Planet wheels of this differential are respectively moving the front and the rear axle; this differential transfers and controls the torque to the two axles.

13.3 Differential Theory Outline

In these paragraphs the influence of differential mechanical efficiency on transferred torque value is examined and compared with the case of an ideal friction free differential.

The friction torque is sometime deliberately increased with different devices to control the value of the output torque. These devices will be examined and explained and their effect on the vehicle dynamics will be investigated.

13.3.1 *Friction Free Differential*

The differential is a two degrees of freedom mechanism that can be idealized as a black box where an input shaft enters at a speed Ω and with a torque M; from this black box, two output shafts exit at Ω_1 and Ω_2 speeds, with M_1 and M_2 torques. The two degrees of freedom could be the input shaft rotation angle and the difference in angular displacement of the two output shafts.

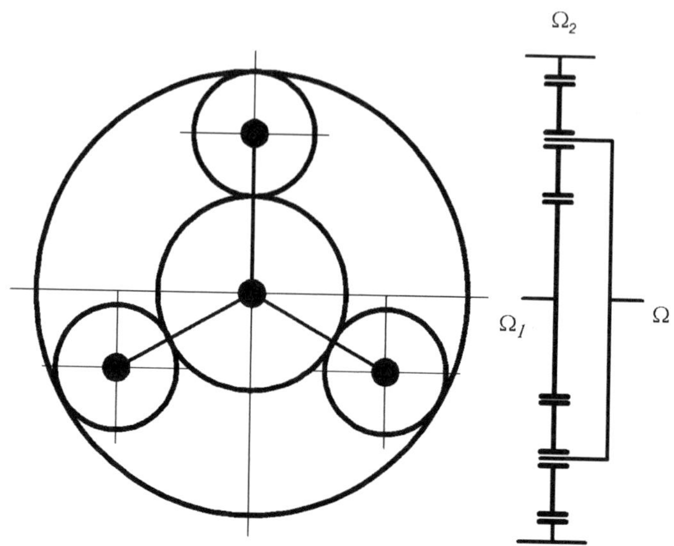

Fig. 13.10 Kinematic scheme of an epicyclical train, used as a differential

The properties of a differential must be the three following:

1. There must be only one relationship between the three speeds; in this way the speed difference between output shafts is undetermined (the mechanism is named after this property).
2. The input torque is split in two output torques, in a constant ratio independent on speed.
3. The two output torques must have the same direction.

These properties are fulfilled by an epicycloidal gear as that sketched, with the usual rules, on Fig. 13.10. If r is the ordinary transmission ratio, that is the transmission ratio of the one degree of freedom mechanism, obtained by locking the carrier, we can write:

$$r = \frac{\Omega_2 - \Omega}{\Omega_1 - \Omega}, \tag{13.1}$$

and:

$$\Omega_2 = \Omega + r(\Omega_1 - \Omega), \tag{13.2}$$

or:

$$\Omega_2 - \Omega_1 = (1 - r)(\Omega - \Omega_1). \tag{13.3}$$

From the last formula we see that the speed difference of the two output shaft can assume any value. The sign of r is negative, because when the carrier is locked output speeds are opposite.

For the system equilibrium it must be that:

$$M_1 + M_2 = M\,; \tag{13.4}$$

if there is no friction it must also be that:

$$M_1\Omega_1 + M_2\Omega_2 = M\Omega\,. \tag{13.5}$$

If we substitute one of the speed equations in the last, we obtain:

$$M_1 = -\,M\,\frac{r}{1-r}\,, \tag{13.6}$$

$$M_2 = M\,\frac{1}{1-r}\,, \tag{13.7}$$

$$\frac{M_1}{M_2} = -r\,. \tag{13.8}$$

where we can see that the torque is splitted in constant parts, independent of speed.

The very well known case of conventional bevel gear differentials can be described by the above equations where the ordinary transmission ratio is set to -1; in this case, output torques are always equal to half input torque.

In a transfer box differential, if a split ratio different than 50/50 is desiderated, it is necessary to use an ordinary transmission ratio different than -1.

13.3.2 *Differential with Internal Friction*

Let us assume now that mechanical losses do exist and that they can be expressed through the concept of mechanical efficiency, in other words, the power loss is proportional to the useful power.

We define η as the mechanical efficiency of the ordinary gear train associated to the differential; if the shaft 2 rotates in the same direction as the torque M_2, this torque will be called input torque and M_1 will be therefore output torque; in this assumption:

$$M_1 = -M_2 r\eta\,; \tag{13.9}$$

if, instead M_1 is input and M_2 is output, it will be:

$$M_2 = -\frac{M_1 \eta}{r} . \tag{13.10}$$

The efficiency η doesn't change value, as a first approximation, on the two different direction of power flow; if Ω is different than zero, the first formula is valid when the shaft 2 is slower and the second formula when the shaft 2 is faster.

Therefore, if $\Omega_1 > \Omega > \Omega_2$ it will be:

$$M_1 = \frac{r\eta}{1 - r\eta}, \tag{13.11}$$

$$M_2 = M\frac{1}{1 - r\eta}, \tag{13.12}$$

$$\frac{M_2}{M_1} = -\frac{1}{r\eta}. \tag{13.13}$$

Instead, if $\Omega_1 < \Omega < \Omega_2$ it will be:

$$M_1 = -M\frac{r}{\eta - r}, \tag{13.14}$$

$$M_2 = M\frac{\eta}{\eta - r}, \tag{13.15}$$

$$\frac{M_1}{M_2} = -\frac{r}{\eta}. \tag{13.16}$$

If the three speeds are equal, the output torque will be undetermined within a range included between the values of the above formulae, as an analogy with the friction phenomena, where there is no relative motion, if friction torque (or force) is still below the static friction value. The case of the axle differential can be again obtained by setting $r = -1$.

On the contrary of the ideal no friction case, if one wheel can bear a lower traction force because of a lower traction coefficient, there will be no relative motion between wheels if the difference in torque is included in the above limits.

In the case of a bevel gear differential, we can refer to the scheme on Fig. 13.11. Forces exchanged by planet wheels and satellites teeth are reported. The scheme represents the three reference planes of the bevel gears as if they were on the same plane; planet gears drawing planes are turned over those of satellites.

On the satellite tooth flank will act tangent forces T_2 and T_1, inclined according to the pressure angle ϑ.

But, because of the friction, this forces will cause also normal components N_1 and N_2, who must be oriented in such a way, as to oppose the relative motion; on this scheme we assume that:

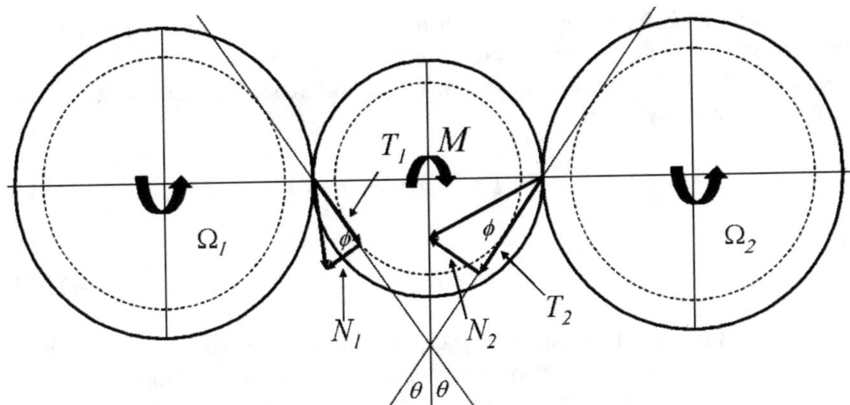

Fig. 13.11 Scheme of forces acting on a bevel gear differential; reference planes of planet wheels are turned over the reference planes of satellites

$$\Omega_1 > \Omega > \Omega_2 \, .$$

If R is the reference radius of the primitive satellites cone, we can write the rotation equilibrium equation of satellites, bearing in mind that the friction angle ϕ is bound to the friction coefficient f; it must be that:

$$\begin{aligned} & f = \tan(\phi) \ , \\ & T_1 \tan(\phi) = N_1 \, , \\ & T_2 \tan(\phi) = N_2 \, , \\ & (T_2 - T_1) R \cos(\vartheta) = T_1 + T_2 \, . \end{aligned} \tag{13.17}$$

If we call with R' the reference diameter of the planet wheels primitive cone, it must be that:

$$(T_1 + T_2) R' = M \, . \tag{13.18}$$

Therefore, we obtain the final equation:

$$\eta = \frac{T_1 R'}{T_2 R'} = \frac{1 - f \tan(\vartheta)}{1 + f \tan(\vartheta)} \, . \tag{13.19}$$

13.3.3 Self Locking Differential

A *self locking* differential is a particular mechanism where torque difference between the two output shafts can be limited to preset values; the limit case is when the differential is completely locked where the speed torque is undetermined (planet wheels are locked together).

We define as *locking coefficient b* of a differential the maximum difference of torque (referred to the total axle torque) between the two output shaft that can be sustained with no difference in speed; the torque difference is called locking torque.

We have by definition that:

$$b = (M_2 - M_1)/(M_2 + M_1)\,, \tag{13.20}$$

or:

$$\eta = (1 - b)/(1 + b)\,. \tag{13.21}$$

The b coefficient can be constant or can depend upon speed difference or input torque values; we will see in the next paragraph some practical examples.

We remember finally that the name self locking differential is given to devices where the locking torque is determined by the mechanical propertied of the system; the name *controlled differential* is given instead to devices where the locking torque is determined by a controlled clutch, for instance by electronic means.

13.4 Types of Self Locking Differentials

13.4.1 ZF System

According to the ZF system, shown on Fig. 13.12, the bearings where the thrust of the planet wheels 6 is loaded, are not made by anti friction material, but by a multi disc wet clutch.

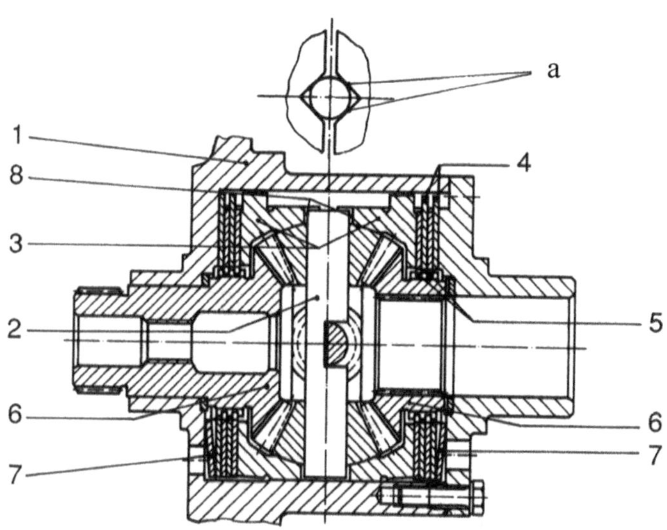

Fig. 13.12 Self locking ZF differential. The multi disc wet clutches apply a constant (upper version) or a variable (lower version) locking coefficient

On the above figure, two discs 5 are rotating with planet gears, other two discs 4 and the pressure plate 3 are rotating with carrier. Pressure plates 3 show a V shaped groove (detail a) that match with satellite shafts 3 where total input torque is applied.

By means of this device, the discs pressure force is about proportional to the total torque. The locking torque is therefore proportional to the total torque and will always oppose the relative rotation of the half axles. The locking coefficient b is constant.

On the lower part of the same figure, a different version of this differential is presented; the pressure plates are pushed by a diaphragm spring 7. In this case the locking torque features a minimum locking value also at zero total torque; the blocking coefficient won't be constant anymore.

13.4.2 Torsen System

A cross section of a Torsen differential (Torsen stands for torque sensitive) is represented on Fig. 13.13; planet wheels 5 and satellites 3 are made with worm gears. Satellites centre line lies on a plane that is perpendicular to the centre line plane of planet gears.

There are three satellites for each planet gear. The satellite teeth don't mesh with those of the neighboring satellite; small spur gear wheels 4 can counter rotate each couple of neighboring satellites.

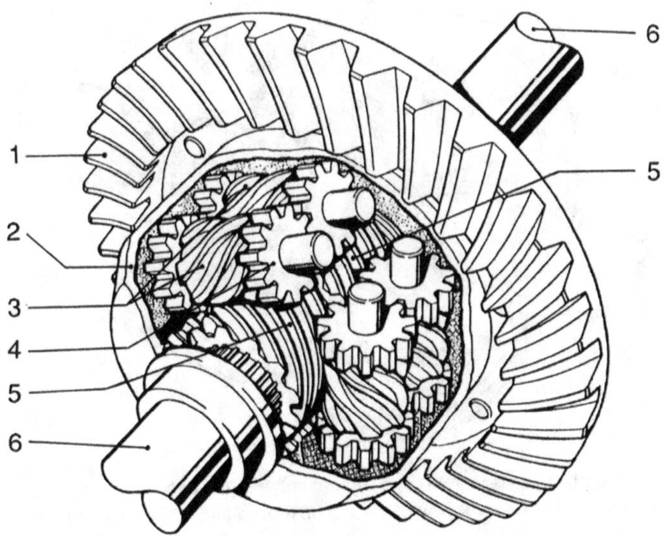

Fig. 13.13 Scheme of a Torsen differential. The locking torque is due to the high friction existing between the worm gear planet wheels and satellites

Satellites shafts are blocked in their radial and axial direction; their rotation only is free. It should be noted that, in order to have the mechanism working correctly, wheel threads must have same direction.

With these conditions, the ordinary transmission ratio is again – 1; as a matter of fact, if we lock the carrier 2, a planetary rotation will cause the corresponding rotation of the satellite that will be reversed in the neighboring satellite, which, finally, will cause the other planet wheel to rotate in the opposite direction, at the same speed.

The relevant friction existing on helical surfaces causes the mechanical efficiency to be very poor; a simplified mathematical model can lead to the same conclusions as those of bevel gear differential. In this case also, the blocking coefficient will be constant.

13.4.3 Ferguson System

The Ferguson system is made by applying a viscous joint to two different elements of a differential; the joint exploits the oil (silicon oil) viscosity to brake two elements of the differential with a torque depending on the elements relative speed.

Figure 13.14 shows the scheme of a joint suitable for the transmission shaft of an all wheel drive gearbox similar to that on Fig. 13.8; this joint can also be integrated in a differential according to different schemes.

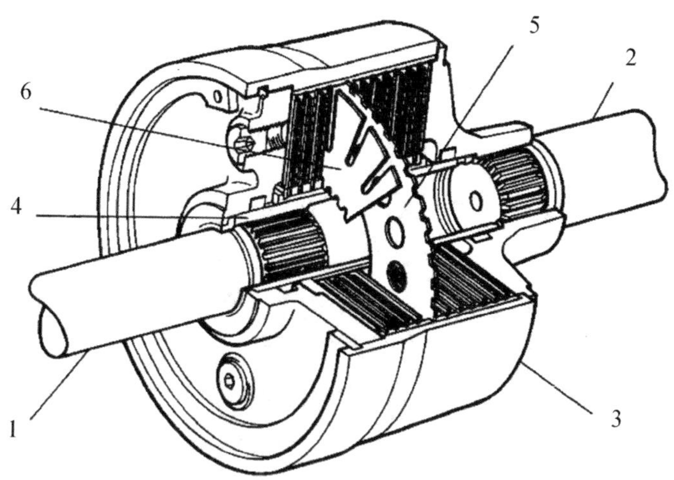

Fig. 13.14 Scheme of a Ferguson joint. The friction between the two shafts 2 and 3 is made by the viscosity of the silicon oil inside the joint, working on discs 5 and 6; the friction torque depends on relative speed between 2 and 3

The joint is made by a cylindrical reservoir 3 fixed to shaft 2, on the right of the figure. A second shaft 1, on the left, is free to rotate in the reservoir; a rotating seal avoids oil leakage outside the reservoir.

The left shaft has a spline where a series of metallic discs 6 is invested; another series of discs 5 is connected to the inside diameter of the reservoir, again through a spline. The two discs families are stacked with each of the discs 6 inserted between two discs 5.

The discs are drilled or cut to activate viscous forces between the disc families, when there is a relative rotation speed; axial forces are unnecessary, because the braking force is controlled by the oil speed gradient in gaps between facing discs. The torque value is determined by disc number (increasing with the number) and by the apparent oil density; this last can be easily modified by changing the quantity of air emulsified in the oil (decreasing with the quantity of air); this two magnitudes, together with discs diameter, are the design parameter of the joint.

The Newton law, applied to this case, says that:

$$\tau = \frac{\mu \Delta V}{d} , \tag{13.22}$$

where τ is the shear on the surface element dS of two facing discs, d is the gap dimension between discs, ΔV is the local difference in speed and μ is the coefficient of dynamic viscosity; the force acting on the surface element is:

$$\tau r^2 d\alpha \, dr , \tag{13.23}$$

where r is the local radius and $rd\alpha$ and dr are the circumferential and radial dimension of the surface element.

If we call r_i and r_e the inside and outside radii of the facing surfaces of two neighboring discs, the total braking torque of the joint bM, at a speed difference $\Delta\Omega$ will be:

$$bM = n \frac{\pi \Delta\Omega \nu \rho}{2d} r_e^4 (1 - \frac{r_i}{r_e})^4 , \tag{13.24}$$

where n is the number of facing surfaces of the discs.

Usually the cinematic viscosity coefficient ν is used in place of the dynamic one. The relationship between the two coefficients is:

$$\nu = \frac{\mu}{\rho} .$$

The value of obtainable viscosity, depending on the quantity of emulsified air in the oil, can range between 30,000 and 100,000 cSt (mm^2/s).

In the Ferguson differential the blocking coefficient is therefore depending upon the relative differential speed.

As we have said, the joint can be integrated either in the central or in the axle differential.

In the first case the differential could be a spur gear epicycloidal differential with simple or double satellites. If the input shaft is connected to the annulus gear, one half axle will be connected to the sun gear and the other to the carrier; the Ferguson joint must be sensitive to the difference in speed between sun gear and carrier.

13.4.4 Active System

An active differential is able to convey the transmission torque on each wheel without wasting a significant part of the input power as friction and independently from the difference of speed of the driving wheels; for this feature a system like this is also called torque-vectoring differential.

As Fig. 13.15 shows, each side of a differential of this kind is equipped with an epicycloidal transmission and a clutch able to transfer part of the torque from the final bevel gear to the half shafts directly.

In our example, each epicycloidal gear is made with two equal sun gears, two equal satellites, cut on the same shaft rotating in the casing in a fixed position, contained in a single annulus gear; therefore the transmission ratio between the two sun gears is 1.

The inside sun gears are rotating with the satellites case of the differential, while the outside ones are idle on the half shafts and can be engaged with them with a clutch. A stack of wet discs can be closed by the oil pressure, modulated by an electronic control and determines the torque transferred from the final gear to each half shaft.

Fig. 13.15 Torque vectoring differential (FCA)

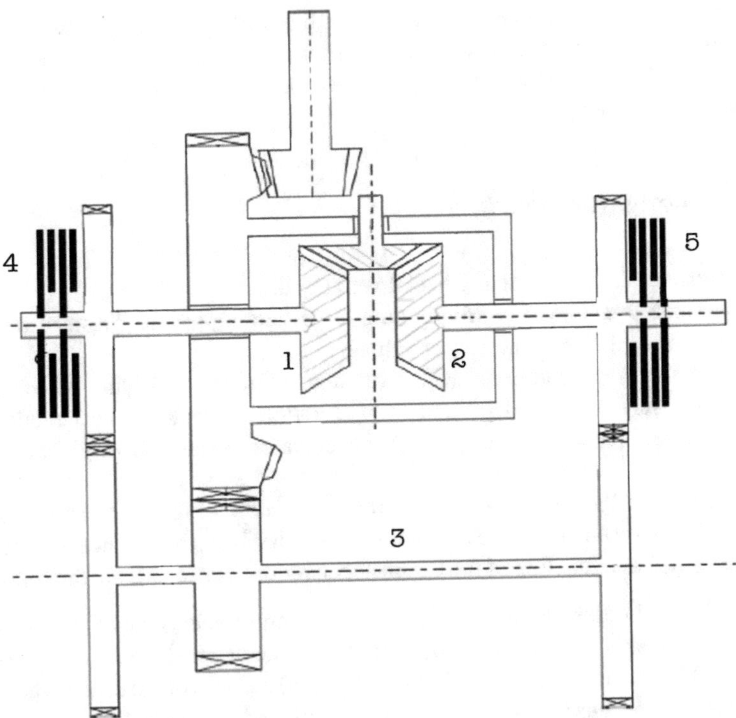

Fig. 13.16 In a torque vectoring differential, part of the torque is applied to the half shafts from the planetary wheels 1 and 2 and part is applied to auxiliary shaft 3, according to the total pressure acting on the two clutches 4 and 5

When clutches are completely closed, the differential is locked-up, when they are completely open the differential works normally.

The functional scheme in Fig. 13.16 explains this mechanism that, for sake of simplicity, is represented with conventional gears with the same function; part of the torque is applied to the half shafts from the differential planetary wheels 1 and 2 and part is applied to auxiliary shaft 3, according to the total pressure acting on the two clutches 4 and 5.

The difference in pressure determines the difference in torque applied to the two shafts; it is theoretically possible to convey the entire torque to a single half shaft, neglecting the effect of the friction between the rotating parts of the differential.

13.5 Differential Effect on Vehicle Dynamics

A real differential or a self locking differential have a beneficial effect on vehicle mobility, because they allow a wheel or an axle on a high traction coefficient ground to receive a traction force greater than that acting on the coupled wheel or axle on low traction coefficient ground.

This advantage has an energetic cost due to the work of friction forces on rotating elements and implies variations on traction forces as compared with the ideal reference values. We will see on the next paragraphs the effect of these variations on the vehicle dynamic behavior.

13.5.1 Driving Axle Differential

Let us consider the case or rear wheel drive at first.

We assume that the car is driving on a large steering-pad of 200 m of radius at steady speeds of different increasing values.

The car working conditions, with particular reference to driving wheels speed and torque, were calculated by means of a multibody mathematical model. The calculation parameter is the centrifugal acceleration, measured in g. The overall vehicle mass is about 1,300 kg.

We start from the case of an ideal differential and on Figs. 13.17 and 13.18 the diagrams of the speed difference between driving wheels and of driving wheels torque are represented as a function of the lateral acceleration. The results are the following:

- If mechanical efficiency is one, torque on driving wheels is equal, by definition. Torque increases quickly with speed (0.7 g of lateral acceleration correspond to about 130 km/h of speed on a 200 m steering-pad) because of the driving resistance increase and because of side slip forces longitudinal components.

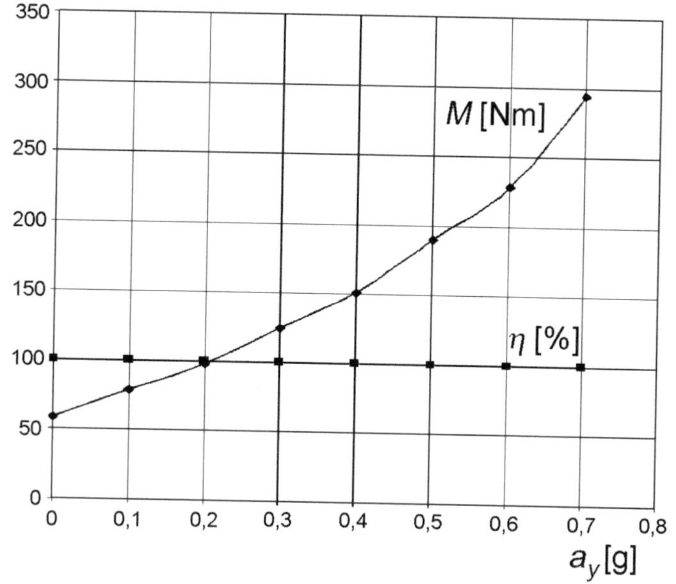

Fig. 13.17 Diagram of driving wheel torque M, for a rear wheel driven car, on a 200 m radius steering-pad, as a function of centrifugal acceleration a_y, in case of an ideal differential mechanical efficiency η of 100%

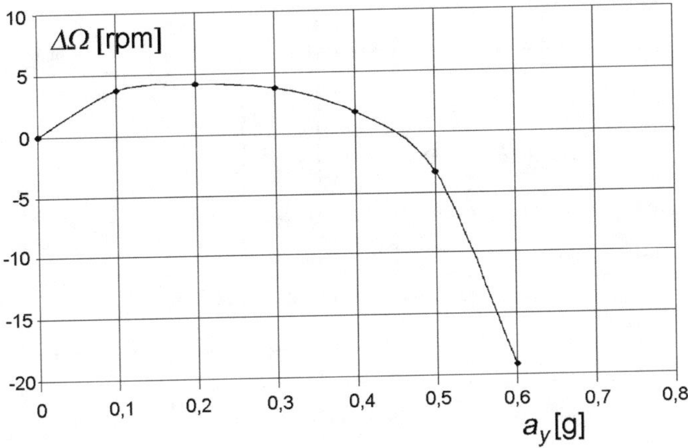

Fig. 13.18 Diagram of the speed difference $\Delta\Omega$ between the inside driving wheel and the outside driving wheel in the case of the previous figure

- Let us remember that vertical load transfer on driving wheels increases about with the square of the speed. The speed difference $\Delta\Omega$ increases because of the different wheel path, at first, and decreases after, because of the fact that the inside wheel, initially slower, will see its longitudinal slip increased, because of the vertical load loss at the same traction force; speed difference is negative and infinite when the wheel looses its grip.

In the assumption that the differential has a constant efficiency smaller than one, angular speed difference will allow a torque difference too.

The inside wheel, initially slower, will receive a larger torque until speed difference becomes zero; torque difference is proportional to total driving torque. Around the point where speed difference is zero, the driving wheels torque diagrams will cross, towards a situation where inside wheel receives a smaller torque, because its speed is faster.

The efficiency diagram will show a cusp with value one, where the speed difference goes to zero. Speed difference is zero as long as the differential is locked, because torque difference is too small.

Let's assume that on the two planet wheels of the differential there is a constant friction force (preload). Torque diagrams will cross where longitudinal slip assumes such a value as the speed difference becomes zero (Fig. 13.19).

The efficiency diagram shows the already commented cusp and is shaped like a hyperbole, that takes into account the constant value of the friction losses compared to an increasing value of the total torque.

The Torsen and the ZF differential may be simulated, with a good approximation, as constant efficiency differentials. The shape of the curve is similar to the previous one and is not reported.

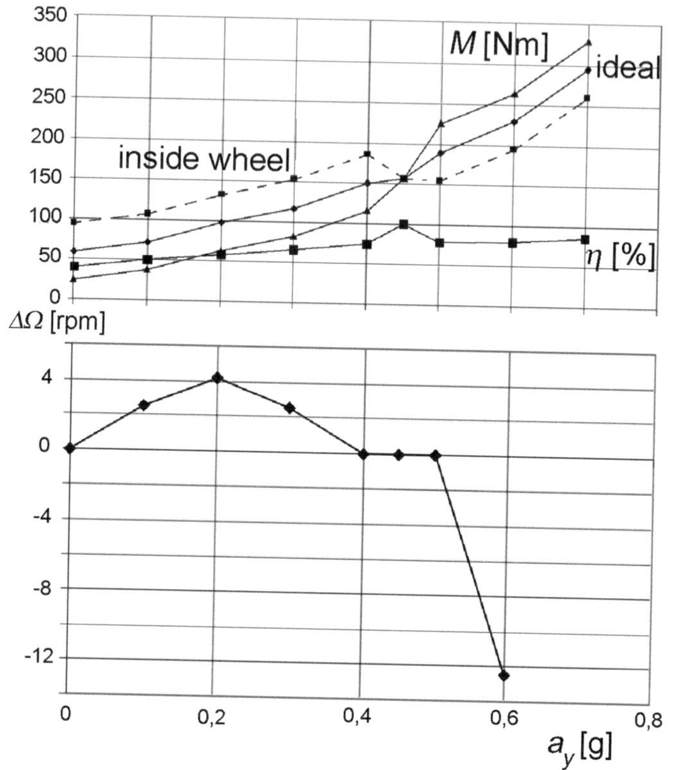

Fig. 13.19 Upper figure: driving wheels torque M diagram and efficiency η diagram in the case of a preloaded differential; torque on inside (dotted line) and outside wheel can be compared with the ideal differential value without mechanical losses. Lower figure: speed difference $\Delta\Omega$ between wheels

In the case of a viscous differential, shown on Fig. 13.20, the difference between driving wheels torques is not relevant between the origin of the diagram and the point where the speed difference becomes zero, because the locking torque is very much dependent on the value of the speed difference.

After the zero point speed difference increases and therefore the difference in torque increases till the inside wheel spin point.

The efficiency diagram shows two cusps with a value of one, if we neglect the mechanical losses, at zero acceleration and at the point where the speed difference is zero.

Let's study now the effect of these phenomena on the understeer curve of this car; we will assume as understeer parameter the difference between the actual steering wheel angle and the steering wheel angle at zero speed (about correspondent to the Ackermann steering wheel angle). We limit the study to the case of the ideal differential and of the Torsen differential. The understeer diagram is on Fig. 13.21.

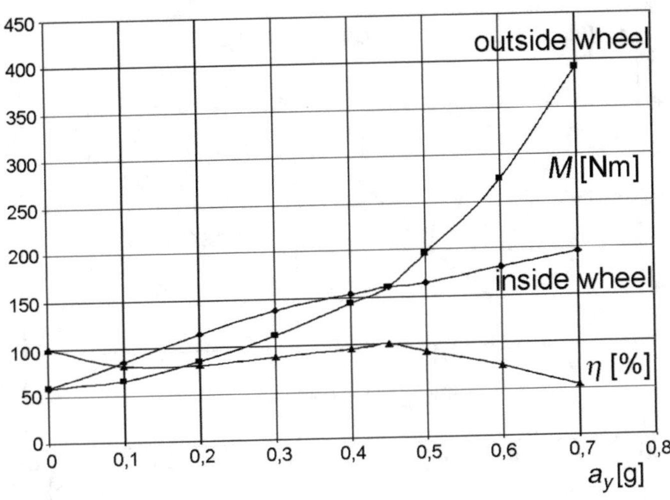

Fig. 13.20 Diagram of the wheel torque M and of the efficiency η of a differential of a rear driven car with Ferguson joint

We can outline the following facts:

- The zero speed steering angle (the value is shown on the caption of the figure) increases because of the differential friction torque. As a matter of fact, the non symmetric traction torque induces a yaw moment (understeering direction) at positive speed difference; the driver must increase steering wheels side slip angle, in order to equilibrate this torque.
- This additional yaw torque goes to zero where the speed difference becomes zero and becomes, after this point, an oversteering yaw torque.
- The acceleration limit where the rear wheels side slip angle becomes too big and the steering angle goes to zero (power oversteer) shifts toward higher values because of the lower traction force on the inside wheel.

For small curvature radii steering-pads the diagrams are not very different. The zero speed difference point moves to higher accelerations because the speed difference contribution of the curvature of the path is predominant on that of the longitudinal slip. The torque difference induced by the differential locking torque is more relevant because of the higher longitudinal slips. Steering wheel will become much heavier at low speed; from this stand point the constant locking coefficient differential are more penalized.

In the case of front wheel driven cars, these calculations can be repeated.

The additional yaw torque (driving torque unbalance) is now applied to the front wheels, with a more relevant effect on the steering torque; in addition to that the torque unbalance induces additional steering angle because of the suspension and steering mechanism compliance.

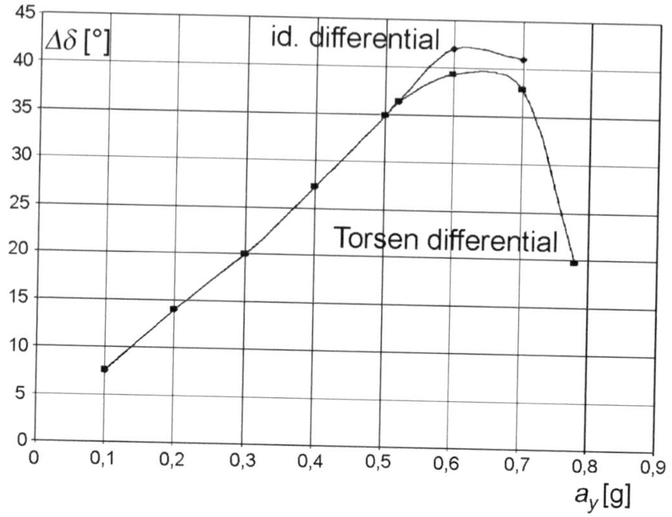

Fig. 13.21 Understeer index diagram (the understeer index is assumed to be the difference $\Delta\delta$ between the actual steering wheel angle and the zero speed steering angle) of a rear wheel driven car on a 200 m radius steering-pad in the ideal differential case and in the Torsen differential case. The zero speed steering angle is 13.6° for the ideal differential and 14.9° for the Torsen differential

Therefore the application of self locking differential to front wheel driven cars is very seldom, with some exception of Ferguson joints with modest self locking torque.

We can conclude that on a rear wheel driven vehicle (the case of the rear axle of an all wheel driven vehicle may be included) the application of a self locking differential will increase understeer while entering in the curve, but with an ampler exploitation of the road grip.

The worse handling is more evident on small curves or when accelerating the vehicle.

An ideal differential controlled by means of a controlled clutch should satisfy the following conditions:

- Simulate the behavior of a no friction differential for high curvature curves at low speed, or when the rolling speeds difference is only due to pressure or when a high grip isn't required.
- Simulate the behavior of a locked differential on high accelerations or low grip roads.

We shouldn't forget that the interconnection torque affects braking with an ABS system. As a matter of fact the prediction of the traction coefficient through the interpretation of the speed difference is affected by the effects on the locking torque; a third requirement of the above differential should, therefore, to act as a free differential during braking.

13.5.2 Transfer Box Differential

An all wheel drive transmission system allows higher longitudinal acceleration at a given traction coefficient value.

Let's recall the equilibrium equations of a longitudinally accelerating vehicle; said:

h	the height from the round of the centre of gravity,
l	the wheel base,
F_{z1}, F_{z2}	the vertical loads on front and rear axles,
μ_x	the peak value of the traction coefficient,
P	the vehicle weight,
a_x	the longitudinal acceleration,
F_{x1}, F_{x2}	the traction forces at the adhesion limit on the front and rear axles,

on a front wheel driven vehicle there will be a vertical load transfer due to the longitudinal acceleration:

$$\Delta F_z = a_x \left(\frac{Ph}{lg} \right) , \tag{13.25}$$

therefore:

$$F_{x1} = \mu_x (F_{z1} - \Delta F_z) , \tag{13.26}$$

$$a_x = g \frac{F_{x1}}{P} = \mu_x g \frac{F_{z1}}{P\,(1 + h\mu_x / l)} ; \tag{13.27}$$

in the same way on a rear wheel driver vehicle, it will be:

$$a_x = g \left(1 - \frac{F_{z1}}{P} \right) \frac{\mu_x}{1 - \mu_x \frac{h}{l}} . \tag{13.28}$$

In the case of an all wheel driven vehicle, we introduce the parameter i, identifying the characteristics of the transfer differential:

$$i = \frac{F_{x1}}{F_{x1} + F_{x2}} . \tag{13.29}$$

The former formulae become:

$$a_x = g \frac{F_{x1}}{P} \frac{\mu_x}{i + \mu_x \frac{h}{l}} ; \tag{13.30}$$

$$a_x = g \left(1 - \frac{F_{x1}}{P} \right) \frac{\mu_x}{1 - i - \mu_x \frac{h}{l}} , \tag{13.31}$$

according to the fact that the front or the rear axle is spinning at first.

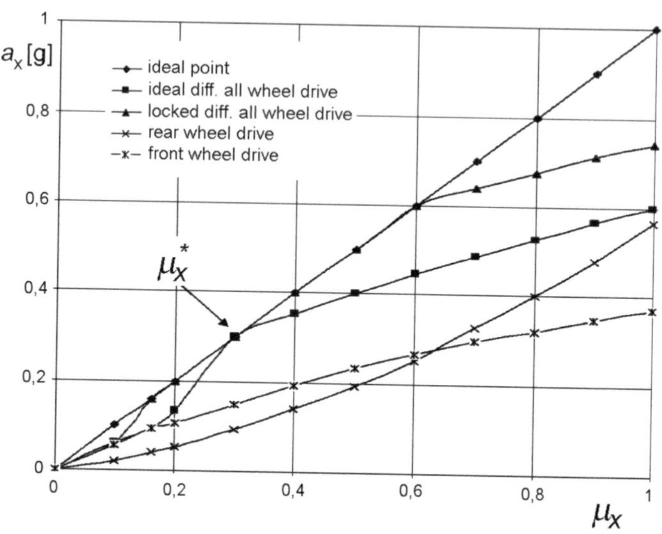

Fig. 13.22 Diagram of the maximum longitudinal acceleration a_x (on the vertical axis, measured in g), as a function of the longitudinal traction coefficient μ_x (on the horizontal axis), in the case of an ideal material point, for a front wheel drive, for a rear wheel drive and for an all wheel drive with no friction differential an for a locked differential with constant locking coefficient. The higher the locking coefficient, the wider the extension where ideal material point response is copied

Figure 13.22 compares the diagram of a_x as a function of μ_x in the three considered cases; the advantage of the all wheel drive is evident, where, in any case, the ideal condition of:

$$a_x = g\mu_x,$$

is reached in one point only.

The value of the traction coefficient μ_x^* where the traction force is the same as in a material point with no load transfer can be derived when the two previous equations are equal:

$$\mu_x^* = \frac{l}{h}\left(\frac{F_{x1}}{P} - i\right). \tag{13.32}$$

For acceleration lower than the $g\mu_x^*$ value, rear wheels will spin, because the load transfer is not sufficient to guarantee the necessary traction force; vice versa for higher accelerations.

If the transfer differential can be modelled as a device with locking coefficient b, we can substitute in the previous formulae a fictitious i' coefficient equal to:

$$i' = i - \frac{b}{2}$$

and i'' equal to:

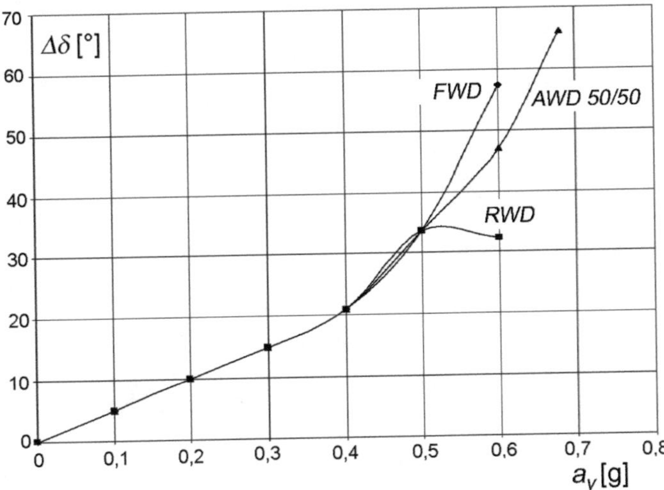

Fig. 13.23 Diagram of the understeer index $\Delta\delta$ as a function of the lateral acceleration a_y for the same car of the previous cases, with a front wheel FWD, rear wheel RWD and all wheel drive AWD (50% of torque on each axle) with no friction transfer differential

$$i'' = i + \frac{b}{2},$$

to take into account the driving torque increase or decrease on axles caused by the locking torque of the transfer differential; on the diagram representing the maximum acceleration as a function of the longitudinal traction coefficient, the curve will match the ideal straight line with a wider extension; the larger b, the wider the extension.

Let's examine finally the effect of all wheel drive on the understeer coefficient, again with the steering-pad simulation.

An ideal transfer differential splitting the traction torque into equal parts is considered; on Fig. 13.23 the understeer curves relative to this case are compared with those of the front wheel drive and of the rear wheel drive.

On rear wheel drives, at high acceleration when the traction force is relevant, the response tends to oversteer because of the high value of the rear side slip angles; on front wheel drives, the high value of the side slip angles raises the natural vehicle understeer to unacceptable values.

The acceleration value where the vehicle response is degenerating is almost the same for front and rear wheel driven cars. All wheel drive reduces side slip angle values and is characterized by a better response. In the reported case, an acceptable understeering behavior is present also at high acceleration value, with an extension of the manoeuvrability of the car.

Chapter 14
Shafts and Joints

Transmission shafts are used to apply torque to those driven components whose rotation axis cannot be perfectly aligned with their driving counterpart.

The following connections are in this situation:

- gearbox output shaft or reducer output shaft with final drive, for conventional drive vehicles either in the case of rigid axle or independent suspensions; this transmission shaft is also called *propeller shaft*;
- gearbox output shaft with reducer or transfer box input shaft, when they are not rigidly connected, as being separately mounted on the chassis structure;
- engine shaft and gearbox input shaft, when they are separated;
- differential output shaft with driving wheels in all independent suspensions; they are also called *half shafts*.

It is evident in this case that transmission shafts must be capable of working with sensible offsets, because of the suspension stroke and, for front wheels, because of steering angles.

Also transmission components directly connected to the chassis structure, as, for example, gearbox and differential box by conventional drive vehicles with independent suspensions, cannot be joined by rigid flanges, because of the chassis flexibility and of dedicated elastic suspensions, addressed to vibration dampening.

The types of shaft can be classified under two different categories: propeller shaft and half shafts.

14.1 Propeller Shafts

A typical propeller shaft can be exemplified by an industrial vehicle application, as shown in Fig. 14.1.

G. Genta and L. Morello, *The Automotive Chassis*, Mechanical Engineering Series,
https://doi.org/10.1007/978-3-030-35635-4_14

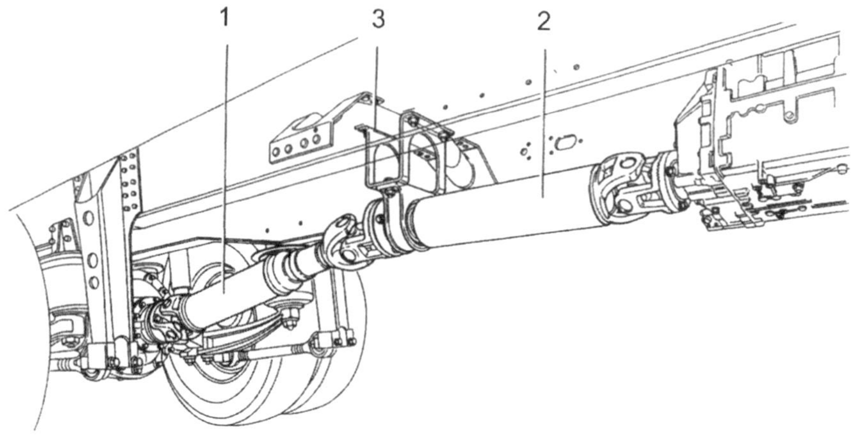

Fig. 14.1 Two pieces propeller shaft for the rigid rear axle of a heavy duty industrial vehicle (Iveco)

The final drive on the rigid axle is connected with the gearbox output shaft through a two pieces propeller shaft: the first swinging shaft 1 and the fixed shaft 2. An intermediate bearing 3 holds the fixed shaft aligned with the gearbox output shaft and the connection with the swinging shaft.

The swinging shaft is subject to change its inclination angle with the side beams because of the rear suspension bounce; to evaluate the offset to be covered, lateral displacement due to suspension and chassis flexibility should not be forgot.

Since the reaction to the driving or braking torque applied to the axle is taken by the suspension elastic elements too, we can expect also axle rotation around the y axis; see Part I above S deformation of leaf springs.

The two piece design is cause by the relevant distance between the flanges to be connected, considering the natural bending frequencies of the shaft; it is desirable that these frequencies are far away the chassis bending natural frequencies to avoid annoying booms.

Also for this purpose an elastic suspension is provided for the intermediate bearing.

For shorter vehicles, as semitrailer tractors or light duty industrial vehicle, a single piece propeller shaft can be used.

Also the natural torsion frequency of the system, including powertrain, transmission shaft and vehicle has a fundamental importance on vehicle comfort and drivability; this natural frequency is directly influenced by the propeller shaft.

A very simple model of the system could include two flywheels of inertia J_m and J_v representing respectively the powertrain inertia (engine and gearbox, in a given speed) and the inertia of the vehicle; these flywheels are connected through a transmission shaft, whose length is L.

Immaginiamo, per semplificare al massimo il problema, un sistema costituito da due volani J_m e J_v, rappresentanti rispettivamente il momento d’inerzia equivalente del motopropulsore (motore e cambio, in un certo rapporto) e la massa apparente rotante del veicolo; tali volani siano collegati mediante un solo albero di lunghezza L.

If G is the shear module of the shaft material, I_p the polar inertia of its cross section, we will have, at resonance condition:

$$\omega = \sqrt{\frac{GI_p}{J_m L} + \frac{GI_p}{J_v L}}, \tag{14.1}$$

where ω[1] is the torsional natural frequency of the transmission shaft.

As we know, internal combustion engines deliver a torque that is changing over time, because of the shape of the pressure cycle and because this pressure is transferred to the shaft through a crank mechanism; also the engine rotating and reciprocating masses introduce a further cause for driving torque variation over time.

If we decompose the resulting torque in harmonics, it will be possible to identify engine speeds introducing resonance between the driving torque on the gearbox output shaft and the natural frequencies of the transmission shaft; the desired result is that this resonance condition is beyond the vehicle maximum speed or at speeds so low that they don't affect practically the vehicle use conditions.

We should remember that other torsional excitations are coming from the joints connecting the different pieces of the propeller shaft.

If we now model the propeller shaft with a constant section beam with section S and inertia moment I_f, and we call with E the elasticity modulus and with ρ the density of the shaft material, we can also calculate with:

$$\omega_c = \frac{\pi^2}{L^2}\sqrt[2]{\frac{EI_f}{\rho S}}. \tag{14.2}$$

the bending natural frequency of the propeller shaft on its bearings.[2]

This approach is oversimplified, because it doesn't take into account the contribution of other elastic elements of the transmission system as clutch torsional dampers, tires and suspensions, for torsion or bearing and powertrain suspensions for bending; a more complete model will be introduced in the Part V of the second volume.

In any case, the issue of natural frequencies suggests to size transmission shafts with very rigid cross section and very light design; tubular section are, as consequence, the best choice.

High resistance steel or aluminum are used; sometime on sport cars, also filament wound composite epossidic plastic materials are applied, with kevlar fibers.

The propeller shaft of conventional drive cars with suspended differential is designed with the same criteria.

By these applications a universal Hooke joint is applied, as shown in Fig. 14.2.

---

[1] In this book vibration frequency is indicated by ω, while Ω is used for angular speeds. Frequencies will be always measured in coherent units (rad/s) and no difference will be made between frequency and pulsation.

[2] A more complete approach to the issue of natural frequencies and critical speeds, can be found, for example, on the book G. Genta, *Dynamics of Rotating Systems*, Springer, New York, 2005.

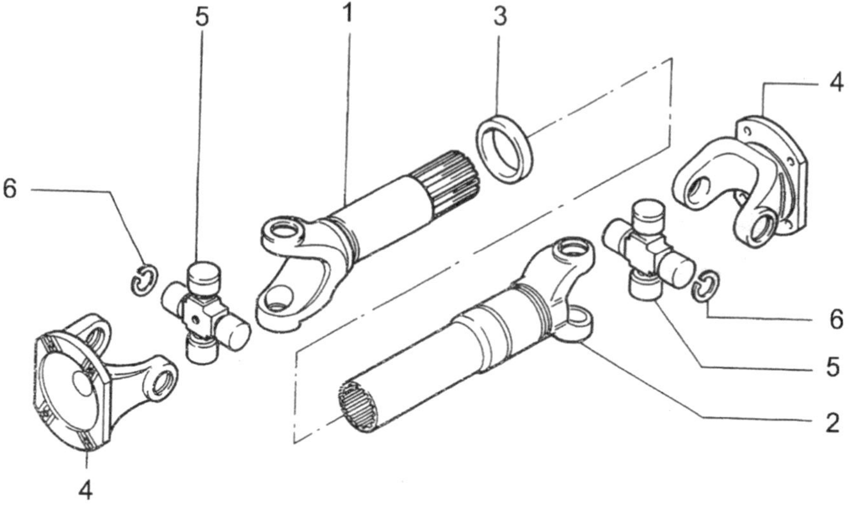

Fig. 14.2 Exploded view of the moving part of a propeller shaft with universal Hooke joints

The moving piece of the propeller shaft is made by two tubes 1 and 2, connected by a spline that allow length variations due to the axle suspension motion. The spline is lubricated by grease sealed by the rubber ring 3. The two ends of the moving piece show the *yokes* of the universal joints.

Other two yokes 4 are flanged to the gearbox output shaft and to the differential input shaft or fixed transmission piece, if any.

Two *trunnion crosses* 5 connect the yokes; the connection between crosses and yokes is made usually by sealed needle bearings, kept in place by Seeger rings 6: we comment in a next paragraph the universal joint operation.

14.2 Half Shafts

The half shafts layout of an independent wheels suspension for a front wheel driven car is shown in Fig. 14.3.

Being the driving axle also a steering one, joints must be designed for big working angles. The transversely mounted powertrain implies always half shafts of different length.

The different length can be a problem, if because of this fact half shafts feature a different torsional stiffness; this fact occurs by wide cars with powerful engines. In fact, under high values of driving torque, self steering moments could be applied to the steering mechanism, because the stiffer half shaft is transmitting higher torque. The problem can be solved by using bigger cross sections on the longer half shaft.

Fig. 14.3 Complete transmission for a mid size front wheel driven car with transversal engine; the transmission is seen from its back (FCA)

By this design choice, the different geometry of the two half shaft could cause the natural bending frequency of the longer half shaft to be too low; to avoid this inconvenience, as by propeller shafts, the longer half shaft is divided in two pieces, the first of them is fixed to the powertrain assembly.

What was written in the previous paragraph about torsional and bending natural frequencies could be repeated in this case, but the application of tubular sections in this case could be sometimes impossible because of the limited space.

It can therefore happen that a half shaft natural frequency is coupling with a tire vertical natural frequency, with resonance. This problem can be solved by applying to the half shaft additional masses; they are made by an iron ring press fit on a rubber bushing with suitable elasticity: this systems works as a damper tuned on its natural frequency.

The study of torsional and bending vibrations must take into account the entire powertrain suspension, now interested also by the reaction to the driving torque on the front axle; again we address to the Part V to approach this study.

Because of the limited space and wide angles, universal joints are not applied to half shafts, but constant velocity joints based upon the Rzeppa principle.

By independent rear suspensions technical solutions are similar to front driving suspensions.

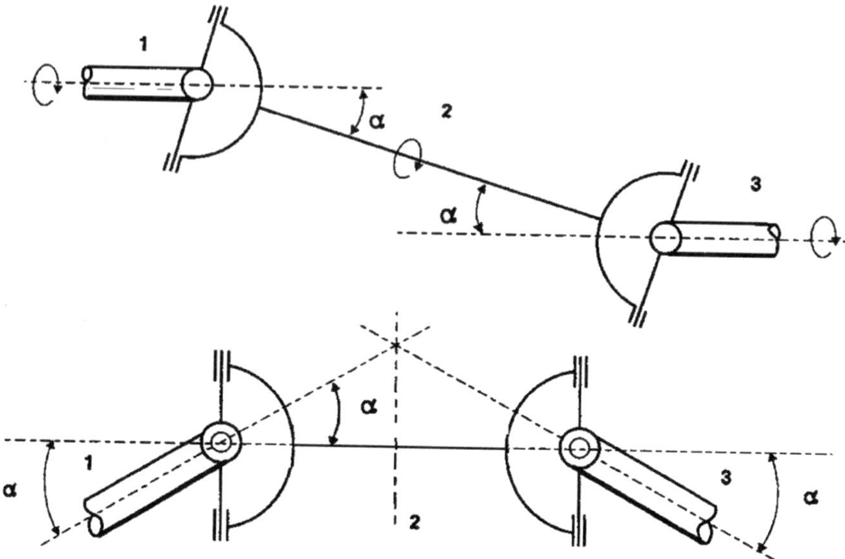

Fig. 14.4 Universal joints layout necessary to obtain a constant speed transmission; the inertia moments of the the shaft mass between the joints will generate, in any case, a periodic output torque

14.3 Universal Joints

The design of a universal joint has been already described in the previous Fig. 14.2; it is easy to understand why a universal joint alone is not a constant speed[3] joint.

Let us consider the scheme in the upper part of Fig. 14.4 and in particular the left joint.

If θ is the rotation angle of the first shaft piece 1, before the joint, at the time t, φ is the corresponding rotation angle of the second shaft piece 2, after the joint and α is the angle between the center line of the shafts 1 and 2, we can write that:

$$\tan(\varphi) = \frac{\tan(\theta)}{\cos(\alpha)} . \tag{14.3}$$

If we derive the two members of the above equation, we obtain:

$$\frac{d\varphi}{dt} = \frac{\cos(\alpha)}{1 - \sin^2(\alpha)\cos^2(\theta)} \frac{d\theta}{dt} . \tag{14.4}$$

Therefore if the input speed

[3]In a constant speed joint, the ratio between input and output speeds is always independent of angular position; in particular, if the input speed is constant, also the output speed will be constant and equal.

$$\frac{d\theta}{dt}$$

is constant, the output speed

$$\frac{d\varphi}{dt}$$

will not and there will be a periodic function of θ expressing the transmission ratio between input and output shafts; this function will depend on α, and constant speed conditions are obtained only for $\alpha = 0$.

But if we install on the second shaft piece a second universal joint, connecting the third shaft piece 3, we can, for the same equation, obtain a constant speed transmission if the second joint has the same working angle α and its input yoke is on the same plane as the output yoke of the first joint. This task can be accomplished in two ways, as in the upper and lower part of the figure.

This sentence is only true in the ideal case where the rotating mass of the second shaft piece 2 is negligible in comparison with the total rotary mass of the system; if not the non constant speed of the shaft 2 will introduce torque variations in the system that will not conserve its constant velocity.

The configuration applied to propeller shafts is that shown in the upper part of the figure and implies that the shafts 1 and 3 are always parallel; this condition must be obtained by a suitable elasto-kinematic behavior of the rear axle suspension. This condition is only approximately obtained by leaf springs suspension because of the S deformation of springs because of the driving torque.

The second layout in the lower part of the figure is used particularly by half shafts of front driving axles of some off-road vehicles, featuring rigid axle suspension, where the high value of the torque doesn't allow to use constant speed joints.

Figure 14.5 shows an application of this kind. The stub and the axle are articulated through the king-pin for steering; the king-pin axis must be coincident with the middle of the shaft 2; this is very short and is reduced to the minimum necessary to contain the needle bearing of the trunnion crosses.

In this arrangement, the steering motion induces two equal working angles in the two universal joints, as requested by the previous scheme. The axle and the stub have a particular shape building up a sealed rotula to protect the transmission.

Sizing universal joints is not difficult; usually they are selected in a catalog, taking into account transmission torque and working angles. It is important to take into account that torque is limited by working angles and there are also geometric limits, due to interference between yokes and cross that must be verified before the application.

It also important to avoid applying transmission shafts with too small working angles; in fact small working angles can wear locally needle bearings because of too small rotation.

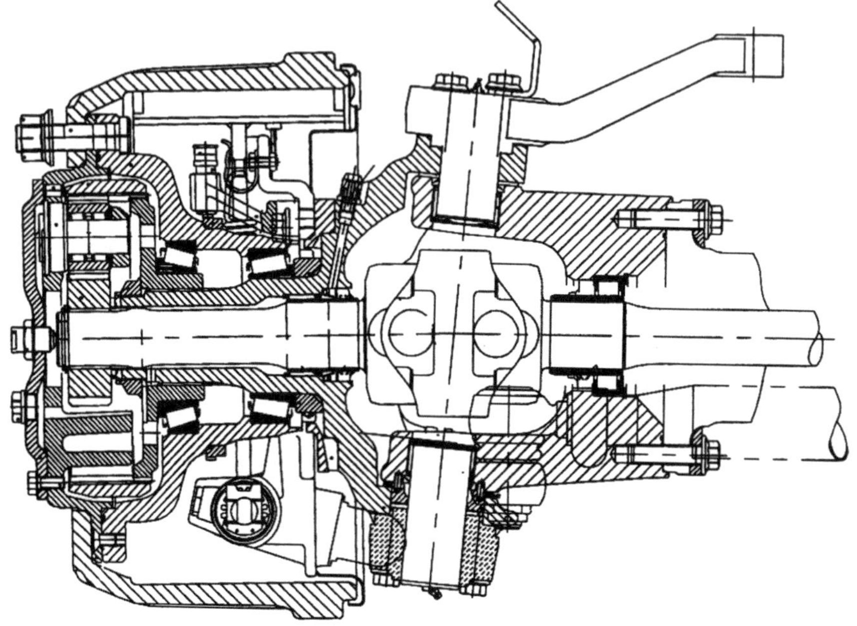

Fig. 14.5 Scheme for a universal joints transmission for a front rigid steering axle of an off-road industrial vehicle; the two yokes of the shaft 2 are almost coincident (Iveco)

14.4 Constant Speed Joints

The double universal joint cannot be used in the transmission to the steering wheels of an independent suspension; this solution has been applied in the past for non steering rear wheels: also this solution doesn't appear today acceptable because of the limitations on camber recovery, in order to keep parallel the first and third shafts of the transmission.

By independent wheel suspensions only constant velocity joints are applied of the Rzeppa type, as those shown in the upper part of Fig. 14.6; the scheme represents a section of a complete half shaft for a front wheel driven car; the wheel, not represented, is on the right of this scheme.

The joint on the wheel side must be able to rotate because of two components: the steering angle, in an almost horizontal plane and the angle imposed by the suspension motion (stroke and camber) in an almost vertical plane; the total working angle allowed by these joints is about 45°.

The other joint on the gearbox side is only subject to the angles caused by the suspension stroke in an almost vertical plane and by the steering angle in an almost horizontal plane; in fact, because the joint on the wheel side cannot be set on the king-pin axis, the wheel steering motion will move the joint center on an almost horizontal plane; the total working angle allowed for these joints is about 20°.

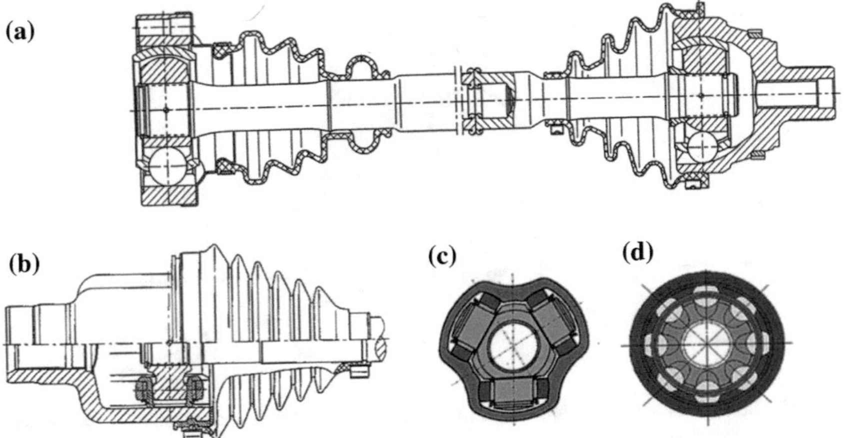

Fig. 14.6 Complete half shaft for a front wheel driven car (**a**) with a fixed joint on the right and a sliding joint on the left. Scheme **b** represents a tripod joint that can be used in place of a sliding Rzeppa joint. Details **c** and **d** show the cross sections of these joints

The wheel steering and the suspension stroke impose to the joint on the wheel side a trajectory not coincident with a circle with its center on the joint on the gearbox side; for this reason a sliding joint must be applied that allow till 50 mm of displacement.

Rzeppa sliding and fixed joints are made with a crown of balls engaged in corresponding grooves on the inside nut and on the outside cup where the wheel or the differential output shafts are connected. Balls are kept on the same plane by a cage, made with a ring with holes.

The scheme d on the figure shows a section of the fixed joint, not very different in this view from the sliding joint one.

The shape of the ball grooves on the nut and cup is such as to determine four different contact points where the surface in contact is almost perpendicular to the force to be exchanged.

The intersection of the plane containing the ball centers with the joint rotation axis determines the joint center or the half shaft articulation point; the cup of the joint on the wheel side has its grooves with a curvature that avoids ball displacements along the rotation axis, while the cup on the gearbox side has straight grooves to allow half shaft sliding freedom.

It is important to verify that no suspension position (usually the most dangerous one is at rebound with full steering angle) can bring any ball to exit its groove.

Cups are sealed with rubber bellows, rotating with the half shaft, that keep balls lubricated for life.

The constant speed behavior of these joints can be justified by means of Fig. 14.7. In this scheme only one ball 3 is represented, engaged with two grooves in the same time; these grooves are respectively fixed to the two parts of the joint 1 and 2; only

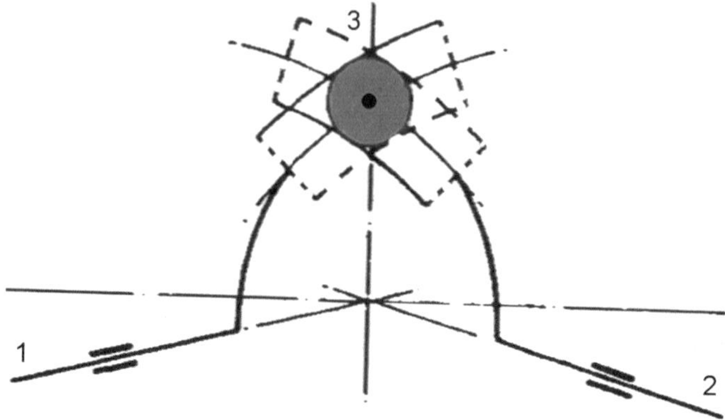

Fig. 14.7 Scheme of a constant speed Rzeppa joint; only one ball 3 is represented, engaged with the two grooves on shafts 1 and 2

the part of these grooves represented with full line is really existing on the nut and the cup.

The shape of the groove is such as to impose to any ball to have its center on the bisector plane of the angle made by the two rotation axis of the shafts of the joint; in this condition the distance between each ball and the two rotation axis will be constant, determining the same rotation speed of the two axis. During the joint rotation balls will be subject to displacements along their grooves proportional to the joint working angle.

From the description of their operation appears that friction must be carefully considered by ball joints; in fact in a complete revolution of the joint each ball is displaced along its groove in proportion to the working angle; this displacement cannot be covered by pure rolling of the balls in their grooves.

The friction is causing a small energy loss in the transmission and an exchange of forces in the transversal direction, because of suspension and powertrain motions; these forces can cause vibrations and harshness that is not efficiently filtered by the powertrain suspension.

For this reasons the limit of the force necessary to have the joint sliding under torque must be carefully specified and taken under control; tripod joint try to solve this problem with a different architecture.

A scheme of this joint is represented on the scheme b in Fig. 14.6. The balls are substituted by three rollers; they are mounted on pins with needle bearings and can roll on the straight grooves of the cup. This design is suitable to contain friction by joint sliding motion under torque.

Chapter 15
Automatic Gearboxes

15.1 General Issues

Opposite to manual gearboxes, automatic gearboxes show a wide diversity of technical solutions; this situation could demonstrate the fact that this gearbox doesn't have reached his technical maturity yet.

Automatic gearboxes have been born, in fact, for high displacement engine comfortable family cars. They are increasing now their minicars and sport cars share, in a market where consumption and emissions are particularly relevant; these issues have proliferated technical solutions, with no final winners, for the time being.

Technical solutions will be described in this chapter that have reached the highest diffusion share for cars market; some solution specific to industrial vehicles will be also outlined.

A very important topic for automatic gearbox study is their control system; modern cars automatic gearboxes have mainly electronic control systems. Their study is usually undertaken in a course on electric and electronic systems.

Nevertheless we will examine automatic gearbox control strategies, which can be defined as rules and methods that are adopted in control systems to decide actual speed and to control speed shift. Hybrid vehicle automatic transmissions are also not considered here; they feature a further function of deciding, which is the most convenient energy source in a given situation (prime engine, energy storage system, wheels).

15.1.1 *Automation Level*

In a historical perspective, the first goal to be reached by automatic gearboxes has been driver and passengers comfort improvement.

As far as the driver is concerned it is evident that he is relieved of the physical effort of operating clutch pedal during start-up and speed changes and of gears shifting;

G. Genta and L. Morello, *The Automotive Chassis*, Mechanical Engineering Series,
https://doi.org/10.1007/978-3-030-35635-4_15

also the need of identifying the most suitable shifting moment and of coordinating operations, especially in an uphill start, is eliminated.

As far as passengers are concerned the objective has been the limitation of the *jerk* (derivative of the vehicle acceleration) during gear shifts and starts; as a matter of fact the human body is quite sensitive to this parameter and it can be assumed as relevant to travel comfort.

If we consider automatic gearboxes from the level of automation standpoint, we can identify the following types:

- *Full automatic* gearboxes; shifting speed and starting-up functions are both carried out automatically, with mechanisms that are developed for this purpose. Between different available working modes, there is usually a semi automatic one where speed ratio choice is left to the driver, being shifting sequence made entirely automatically.
- *Semiautomatic* gearboxes; one of the above automatic functions is partly or totally missing. For example start-up device operation is automatic, but not gearshift; or this second function is automatic, but decisions about the most suitable shifting moment are left to the driver. In this second version accelerator pedal must be operated by the driver in coordination with clutch operation, to avoid engine over speed or stall or any kind of shock. This kind of gearbox has lost interest on cars, because it doesn't represent a favorable cost benefit trade-off. It is vice versa adopted on some industrial vehicle, because of the relevant effort of moving manual commands; sometime these gearboxes share with their manual counterparts most components.
- *Robotized* or *automated* gearboxes; mechanisms are deliberately derived of manual gearboxes, including friction clutch. They can't be defined as semiautomatic gearboxes, as previous ones, because their control system can handle all automatic functions. The decision of using a modified manual gearbox can be justified by other targets, such as reducing production and operation cost. The first is reduced because existing production facilities can be used. The effect on the second is due to reduction of mechanical losses in comparison with automatic gearboxes. We should not forget, finally, a positive impact on improving car sport performance, because shifting and start-up time can be reduced to figures not easily available also to professional drivers.

This last category of gearboxes is reaching a significant market share and could have a further development in next years.

15.1.2 Gearshift Mode

A very important aspect conditioning automatic gearbox configuration is gearshift mode; we usually distinguish between gear shifts with power interruption and gear shifts without power interruption or *powershifts*.

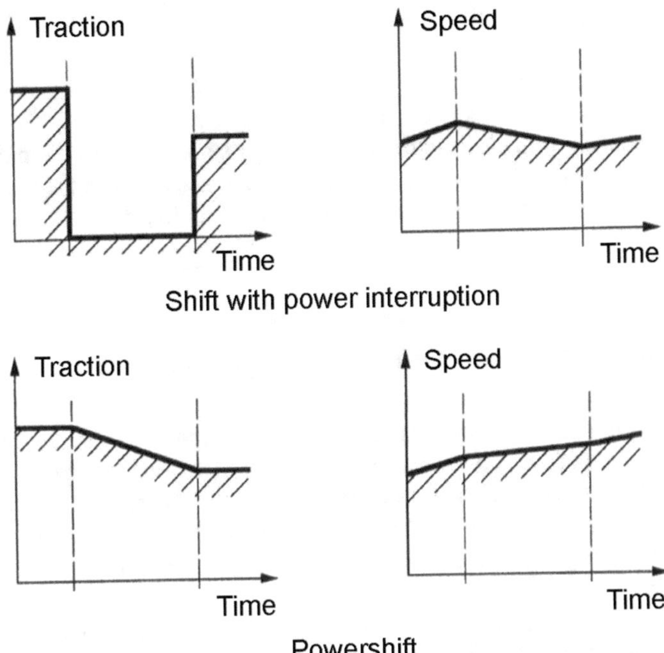

Fig. 15.1 Qualitative diagrams of traction force and speed as a function of time for gear shift with interruption of power and powershifts

In the first case gears are shifted as in a manual gearbox using synchronized or non synchronized dog clutches; during gear shift the power flow going through the gearbox must be interrupted to allow the disengagement of the existing speed and the engagement of the next one.

The power available to move the vehicle goes to zero during this manoeuvre and the vehicle slows down; the jerk is significant also if it human body is not sensitive under certain duration values.

Figure 15.1, in the upper part, shows a qualitative diagram of traction force and of vehicle speed during a gear shift with power interruption.

The same figure shows on the lower part the case of a powershift gearbox. It is possible to bring the jerk to a minimum and to continue to accelerate the vehicle if the manoeuvre occurs without interruption of power. This result can be achieved on stepped gearboxes if each synchronizer is substituted with a clutch and disengagements and the engagements are overlapping; it can be easily understood that gear shift sequence implies a more significant power dissipation than a conventional synchronizer; with this procedure shifting time is not so important and can be compromised in favour of a better driving comfort.

We should notice that on automatic gearboxes with power interruption synchronizers may be avoided, if an external power source is used: this can be the engine that can speed up itself and rotating parts to be synchronized with a double clutching manoeuvre, during down shifts, and a brake on the gearbox that slows down same parts during up shifts.

This option isn't only conditioned by costs but also by shifting time reduction; in this case a drive-by-wire accelerator system is mandatory.

On continuously variable gearboxes the shifting manoeuvre is, by definition, powershift type, in consideration of the very small difference between the previous and the next transmission ratios; also the wasted energy during gear shift can be very small.

15.1.3 Stepped and Continuously Variable Gearboxes

Automatic gearboxes are said stepped if they have available a limited number of transmission ratios (*steps*) as in manual gearboxes. The total number of speeds is seldom larger than six in consideration on the increased mechanical complexity.

When designing the number of speeds on an automatic gearbox with torque converter, it is useful to remember that this start-up device is a continuously variable transmission.

In this case it is usual to avoid the first speed of corresponding manual gearbox; its function is performed by the torque converter from stall to lock-up and therefore the number of available speeds is reduced by one unit without affecting vehicle performance. A four speed torque converter automatic gearbox can be compared with a five gear manual gearbox with clutch and so on.

Stepless or *continuously variable transmissions*, shortly CVT, feature an unlimited number of transmission ratios between a lower and an upper limit.

Mechanisms that perform this function are developed for the purpose and will be explained later on. An important design parameter for CVT is the so called *range*, which is the ratio between the highest and the lowest available transmission ratio. When designing CVT range, it must be kept in mind that it should be the same as for stepped gearboxes for the same vehicle, taking into account what was said about torque converter application. A CVT with very large range (with very short low ratio) could avoid start up device.

Gear trains used on stepped automatic gearboxes can be classified according to two categories:

- Gear wheels with fixed rotation axis, similar to those adopted on manual gearbox; the solution is simple but seldom adopted on powershift gearboxes because of the installation space to be left for clutches.
- Epicycloidal gear trains of different configuration that in association with band brakes or multi disc clutches can realize different ratios including reverse speed, featuring always coaxial input and output shafts.

15.2 Car Gearboxes with Fixed Rotation Axis

15.2.1 Synchronizer Gearboxes

Automatic gearboxes with synchronizers are often derived from manual transmissions.

In the past, manual gearboxes with torque converter received a limited diffusion. In these gearboxes the torque converter was coupled to a clutch. It worked as an automatic CVT from stall to lock-up. The clutch, with electromagnetic actuator, was open automatically at any speed change.

Selection and engagement manoeuvres were manual, acting on the shift stick; the electromagnet of the clutch was switched by a sensor, able to measure the effort applied to the shift stick knob.

This kind of semi automatic gearboxes eliminated the clutch pedal only, being shift stick still manual. They don't exist anymore on cars but could be reconsidered in the future.

Automatic gearboxes, whose mechanisms are shared with those of manual gearboxes with hydraulic clutch, are present now on the market with growing volumes; their shifting mechanisms are on the contrary very specific.

External shifting mechanisms are electrically or electrohydraulically servo actuated. Firsts look less appropriate to a quick response. As a matter of fact hydraulic systems can feature higher power peaks, by use of pressure accumulators; electric 12 V actuation is, on the contrary, limited by battery capacity and wiring harness dimension.

On hydraulic systems, the external shifting mechanism is made by two categories of components: a group providing for hydraulic energy generation, regulation and distribution and a set of actuators. Actuators are as many as the displacements to be made: one for the clutch and two, at least for the gearbox (one for selection and one for engagement).

Figure 15.2 presents a system of this kind; a hydraulic actuator 1 engages and disengages the clutch, while a second double actuator provides for shifting and turning the shaft 2, for engagement and selection motions; the hydraulic group 3 includes an oil pump, a pressure accumulator and actuation electric valves. The selection mechanism is done by a rack which can raise the shaft 2 matching the finger with the three selection gates (first and second speed, third and fourth speed, fifth and reverse speed).

The second actuator can rotate the same shaft to obtain engagement and disengagement; it must feature three different positions: the engagement of the two neighboring speeds and the intermediate idle position.

On the next Fig. 15.3 actuators are illustrated with more detail. The clutch actuator 1 is moved by pressure acting on his back chamber (coil spring side): oil pressure operates disengagement, while engagement is obtained by opening a controlled exhaust; the coil provides for actuator return.

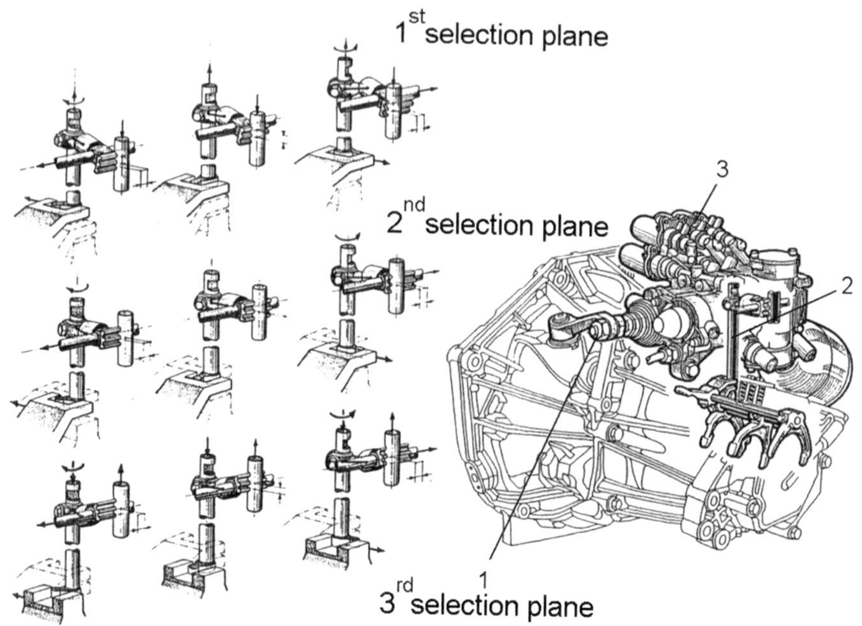

Fig. 15.2 Scheme of an electro-hydraulically actuated internal shifting mechanism; a hydraulic actuator 1 disengages and engages the clutch, while a second combined actuator provides for rotation and shift of the shaft 2; a hydraulic group 3 includes an oil pump, a pressure accumulator and actuation electric valves (Marelli)

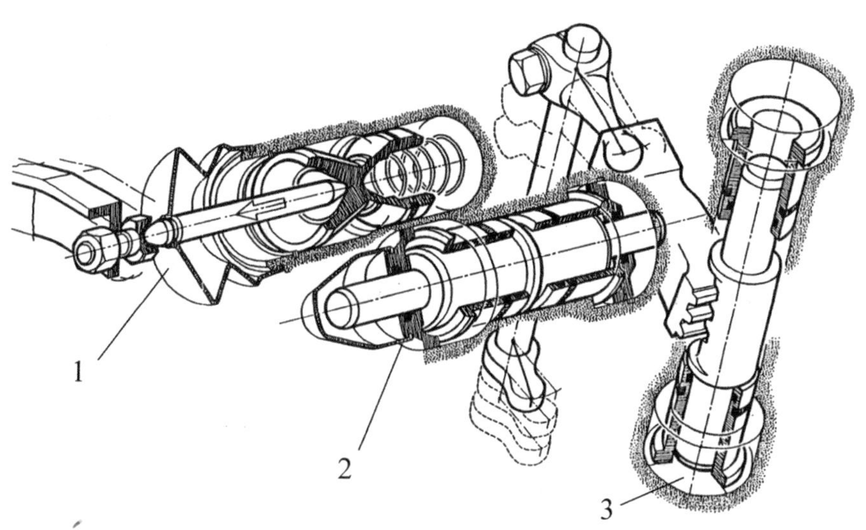

Fig. 15.3 Hydraulic actuation group for a gearbox; 1 is the clutch actuator; 2 and 3 the engagement and selection actuators (Marelli)

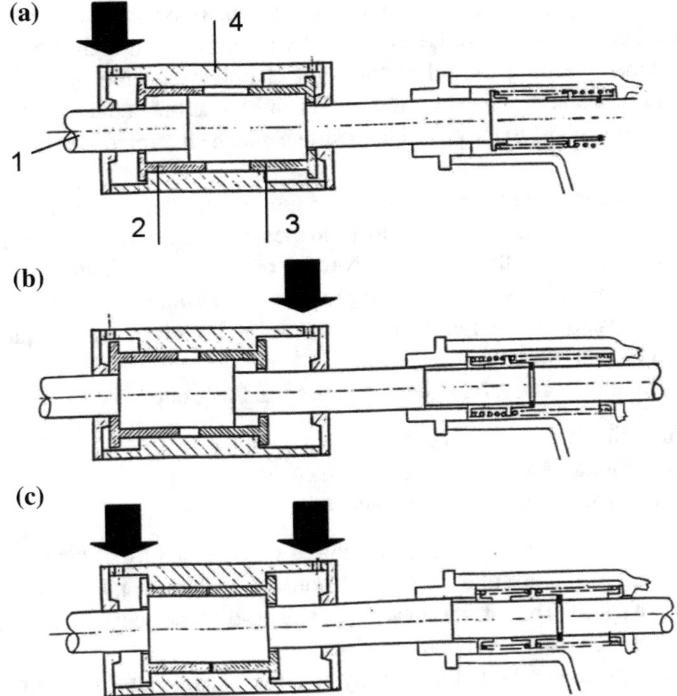

Fig. 15.4 Actuator piston with three stable positions. Two bushings 2 and 3 are set inside the cylinder; stroke limiters on bushings determine an intermediate stable position that can be obtained. While feeding the oil to the left chamber (scheme **a**), a stable position with right full stroke can be obtained; feeding the oil to the right chamber (scheme **b**), a the stable position with left full stroke can be obtained. Feeding both chambers (scheme **c**) brings the two bushings to the end of their stroke and an intermediate stable position is obtained (Marelli)

Pressure control must be made through a proportional valve, suitable to control actuation and exhaust pressure to obtain a suitable actuation speed (i.e.: slow start-up, quick start-up, speed change, etc.). Clutch actuators must feature two stable rest positions (disengaged and engaged clutch) made by stroke limiters; the control system must recognize the driven plate wear and adjust the incipient engagement point accordingly.

Selection actuator 3 and engagement actuator 2 must feature, vice versa, three different stable rest positions (for three selection planes and for idle and two engaged neighboring speeds). These positions remain almost unchanged during gearbox useful life; they must be adjusted at the assembly only. Contrary to clutch actuators, engagement and selection actuators don't have return coil springs and must therefore receive oil pressure on two piston sides.

Figure 15.4 shows the design feature which allows also the obtainment of an intermediate stable piston position. Two bushings 2 and 3 are invested on the double effect piston 1, inside the cylinder 4; they are free to move inside the cylinder. Their stroke is limited by a collar that determines the intermediate stable position too (scheme c). Feeding the oil to the left chamber (scheme a) the right stable position can be obtained; feeding the oil to the right chamber (scheme b) the left stable position can be obtained. Feeding the two chambers at the same time (scheme c) brings the two bushings to the end of their stroke and the intermediate stable position is obtained.

In the generator, regulator and distributor group there is a pressure accumulator that regulates pressure at a value almost independent on the flow rate required by actuators. To obtain this, the pump is driven on-off by an electric motor, controlled by a pressure sensor. In the accumulator pressure oil is available also when motor is stopped; it is possible to design the pump according to the averaged required flow, instead than on peak values.

Three different electrovalves are also present in the group:

- One, proportional type, for clutch actuation;
- Two, proportional type, for engagement mechanism;
- Two, on-off type, for selection mechanism.

Valve type choice for engagement mechanism is decided by of the need of controlling thrust in order not to damage synchronizers and in the mean time to obtain high actuation speed. The selection valves can be on-off type because isn't necessary to control actuation force with high precision.

Clutch actuator is similar to that already seen for manual clutch hydraulic mechanism; shift control system includes also an axial position sensor (engagement stroke) and an angular position sensor (selection stroke); through them selected gears can be identified.

When selection planes number is increasing, for example on a six speeds gearbox, actuation system could become quite complex for the added fourth selection plane; in this case a cam actuation system is used, named *S-cam*, after the shape of his cam.

With reference to the Fig. 15.5 shift actuator is made by a single bistable piston 1 that moves the finger of the internal shift mechanism. The piston can rotate in the cylinder, in order to reach different selection planes.

The two piston stroke ends correspond to the two engagement positions of each sleeve; an intermediate idle position couldn't be obtained with sufficient precision. Because of this limitation, the internal mechanism is completely reorganized, by bringing neutral and reverse on the first selection plane, first and second on the second plane and so on. See the scheme on the lower left of Fig. 15.6.

How can be seen on the same figure, the piston 1 engages through a pin 4 a cam cut on a sleeve 3; this cam is shaped like an S. The sleeve can rotate or can be fixed by a magnetic actuator 5. Since there is a spring plunger that can match one of the axial grooves on piston 1 (one groove for each selection plane), piston 1 engages alternatively two neighboring speeds, until actuator 5 is switched on; if this occurs, when the piston goes to the left, the finger tip lowers of one selection plane, or rises of one, when the piston moves to the right.

Working on the piston and on the actuator 5 all speeds can be selected and engaged in a sequential way only, according to the order R, N, 1, 2, 3, 4, 5, 6 and vice versa.

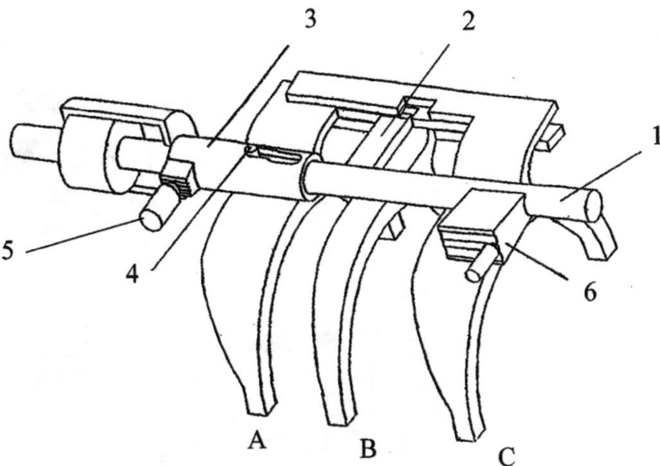

Fig. 15.5 S-cam actuator (Marelli). See the working details on the next figure

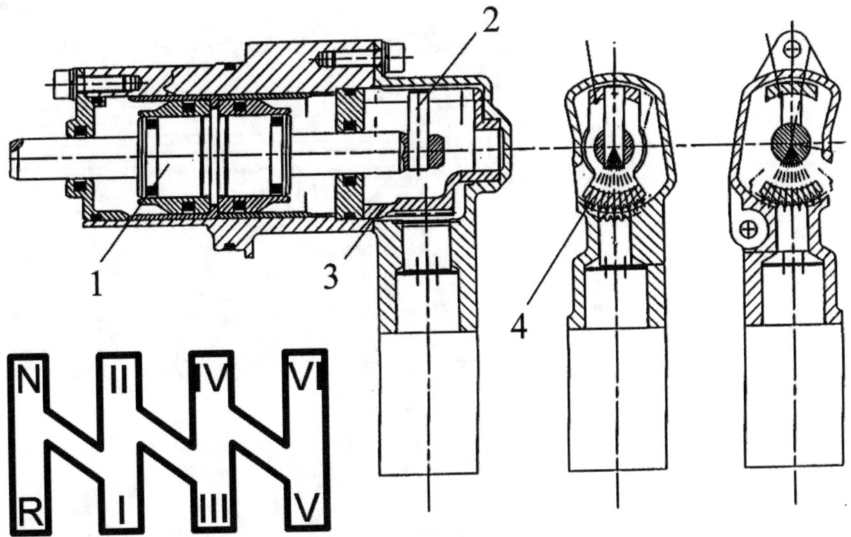

Fig. 15.6 Working details of an S-cam actuator; on the lower left there is the selection scheme of the relative gearbox (Marelli)

15.2.2 Multi Disc Clutch Gearboxes

The configuration of this automatic gearbox kind is still similar to one of a manual gearbox. Synchronizers only are substituted by multi disc wet clutches and speed shift can be power-shift type.

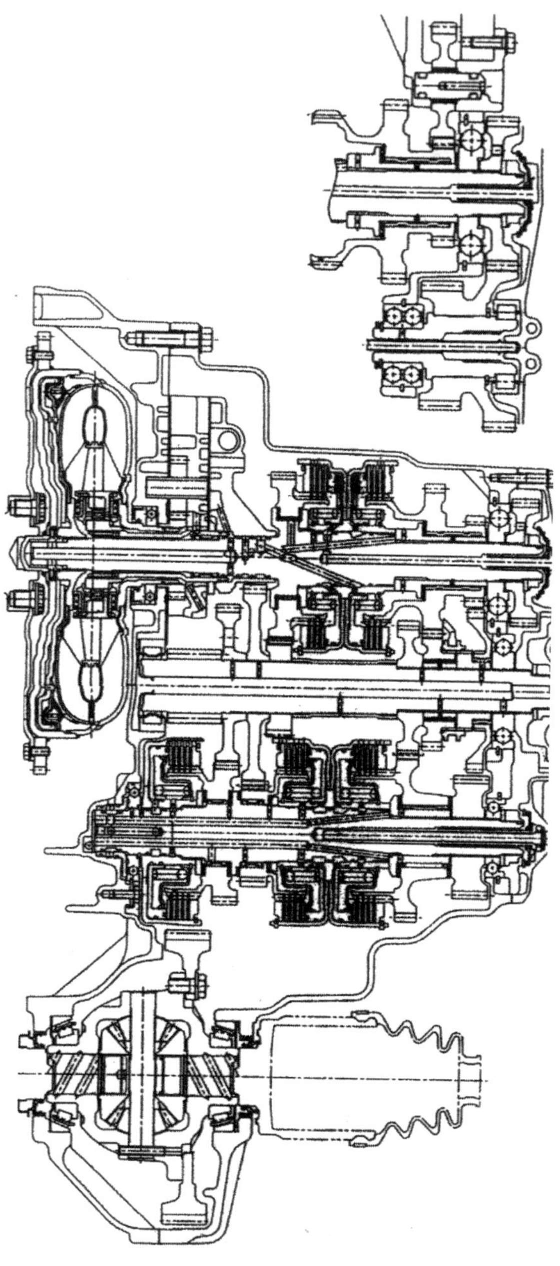

Fig. 15.7 Countershaft powershift gearbox with multi disc wet clutches (Honda)

Reverse speed too is made with an idler, in a similar way to a manual gearbox. With this kind of gearbox there are no limits, how in epicycloidal ones, to choose desired transmission ratios.

The simplicity of the basic concept should not suggest that this kind of gearbox can be really derived of a manual one. As a matter of fact, clutch dimension notably increases centre distance between gearbox shafts and shafts are complicated by drills for oil abduction to the different clutches; a gearbox with five speeds only needs to have three transmission stages.

A construction drawing of a five speed gearbox for a front wheel driven car with transverse engine is showed on Fig. 15.7. On this drawing shafts are represented on the same drawing plane; input shaft gears (the first from top) match with those of the output shaft (the second from top) and with those of a counter shaft (the third from top), through a gear set drafted on the left.

This double gear set is necessary because the rotation of the countershaft must not be inverted as compared with the input shaft; the design complexity is due to space limitation.

Let's look, on the upper right of the same figure, the lock-up torque converter whose hydraulic clutch is fed through a hole on the input shaft; the converter pump moves a gears oil pump for gearbox lubrication and pressure supply to different actuators.

Gears for third, first and second are on the countershaft, starting from the right. On the input shaft, always from the right, there are gears for fourth and fifth; gears radial dimensions are conditioned by clutch diameters. The reverse speed gears are on the right of the input and of the output shaft and are in constant mesh with an idler; a sleeve dog clutch is shared with the fifth speed.

15.2.3 Dual Clutch Gearbox

Since the first dual clutch gearboxes or transmissions, shortly called DCTs, have appeared at the beginning of the new century, they have constantly increased their market penetration thanks to their design simplicity and high mechanical efficiency, showing very few limitations in gearshift quality, as compared with conventional epicycloidal gearbox with torque converter.

They are extensively applied to small and medium size cars but there are no theoretical reasons against their adoption in large high-class cars also.

Their working principle is quite simple: a double clutch C_e and C_o with coaxial output shafts can operate two different sets of gears, engaging each with two different countershafts, as shown in Fig. 15.8.

Both countershafts end with a common output gear that is operating the differential.

The gears on one countershaft are dedicated to even speeds, while those on the other are dedicated to odd speeds; obviously, there is also a third countershaft, not shown in the scheme, for the reverse speed, engaging with the even speeds shaft.

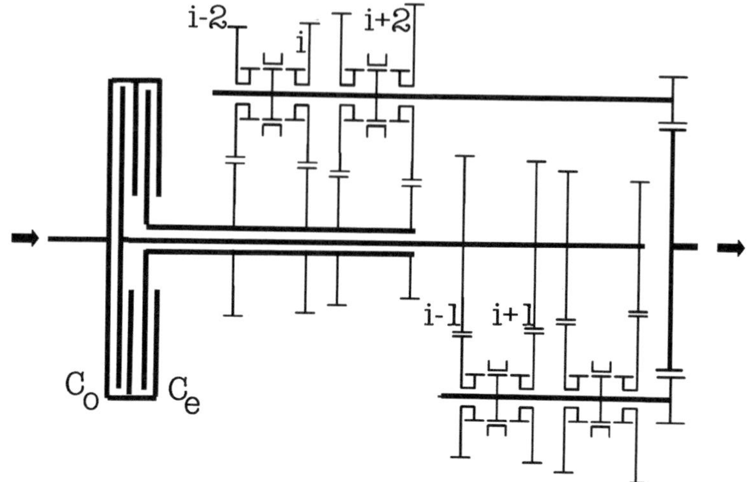

Fig. 15.8 An eight speeds dual clutch gearbox includes two clutches C_e and C_o with coaxial output, operating two single stage transmissions, where each output gear is moving the differential

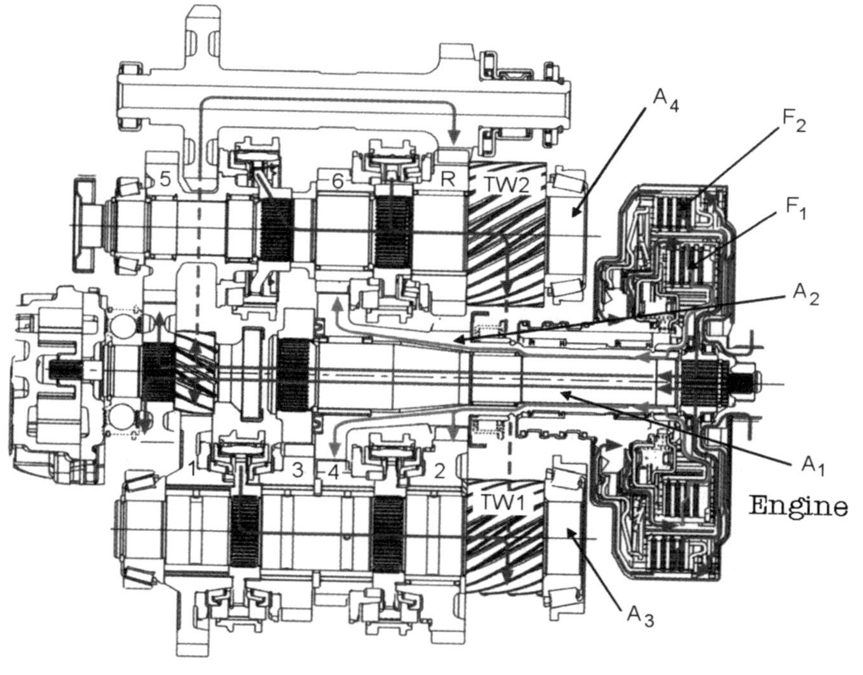

Fig. 15.9 Working scheme of a powershift dual-clutch six speeds gearbox (Audi)

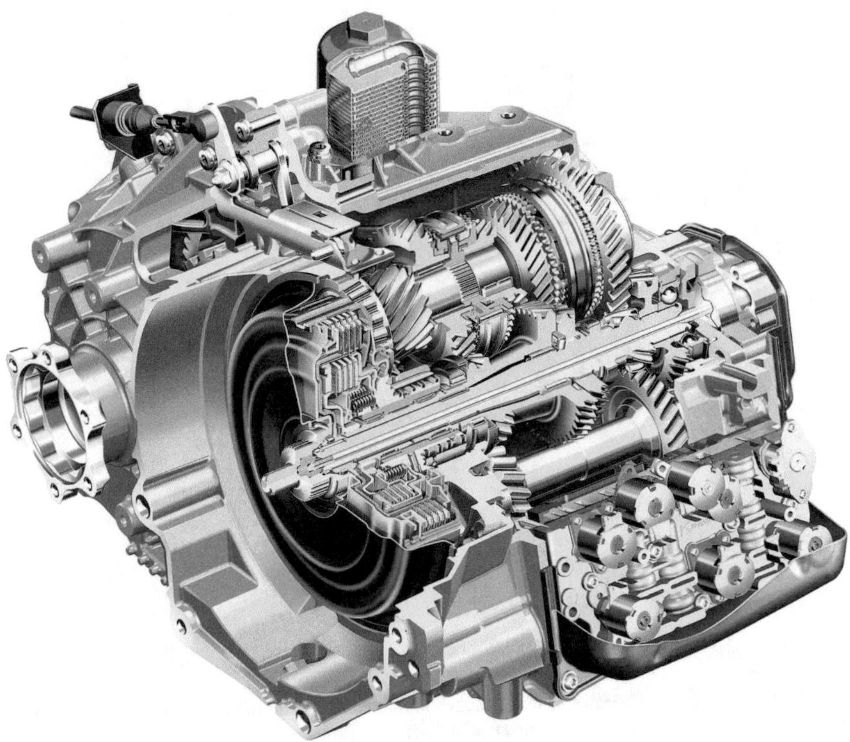

Fig. 15.10 Cut away of a dual speed powershift six speeds gearbox (Audi)

Each idle gear can be engaged with its countershaft through a conventional synchronizer, moved by an electro hydraulic actuator.

Let us suppose that the gear i is presently engaged and the clutch C_e of even speeds is closed; according to the accelerator pedal position and the values of vehicle speed and acceleration, the control system is able to predict if the next ratio to be engaged will be gear $i+1$ or $i-1$: as a consequence the selected synchronizer will be engaged, with no consequence on car motion because the odd speeds clutch C_o is still open.

Only at the shifting time, the even speeds clutch will be open and the odd speeds clutch will be contemporarily closed, with a pressure cycle suitable to the continuity of power transmission. Obviously, gears $i+2$ and $1-2$ cannot be selected without applying gears $i+1$ or $i-1$ before.

On Fig. 15.9 a DCT configuration configuration scheme is reported; the scheme is suitable for a front wheel drive with transverse engine, but could be easily adapted for different traction configurations.

Clutches F_1 and F_2 can move two coaxial shafts, where one is fixed to the odd speed input wheels and the other to the even speed input wheels. In first speed, for example, clutch F_2 is engaged, while clutch F_1 is disengaged; the second gear shaft

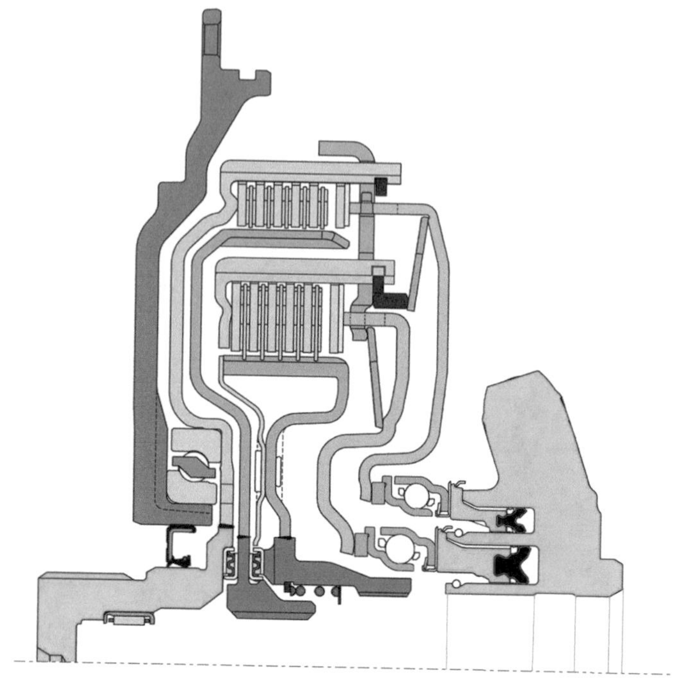

Fig. 15.11 Multi disc concentrical wet clutches suitable to a dual clutch gearbox; two oil pistons operate the clutches trough ball bearings (Schaeffler)

is idle and doesn't receive torque. If the sleeve of the second speed is engaged, the internal shaft starts moving but will be still idle; at any time clutch F_1 can be engaged and the clutch F_2 disengaged, according to a powershift change of speed procedure.

In the same way first gear can be disengaged, to engage the third, in preparation for the next shift. As we have already said, peeds must be used sequentially; double up or down shifts are impossible.

The clutch F_2 is used for the start-up of the vehicle too. The particular architecture of this gearbox features a reduced number of idle turning clutches and no torque converter. The effect on mechanical efficiency is surely positive; the advantage is paid by an increased design complexity.

The scheme is drafted on the same plane, as usual; on the engine shaft are installed the two clutches F_1 and F_2 (start-up device) that move shafts A_1 and A_2; shaft A_1 moves the input wheels of first, third and fifth speeds, while the shaft A_2 moves the input wheels of second, fourth and sixth (one single wheel).

Two countershafts A_3 and A_4 match the final drive wheel and are respectively fixed with the driven wheels of fifth and sixth speeds and with the driven wheels of first, second third and fourth speeds.

The reverse speed shows a third idler countershaft, not shown on the scheme; it can be seen on the front of the following Fig. 15.10. It meshes with the driving

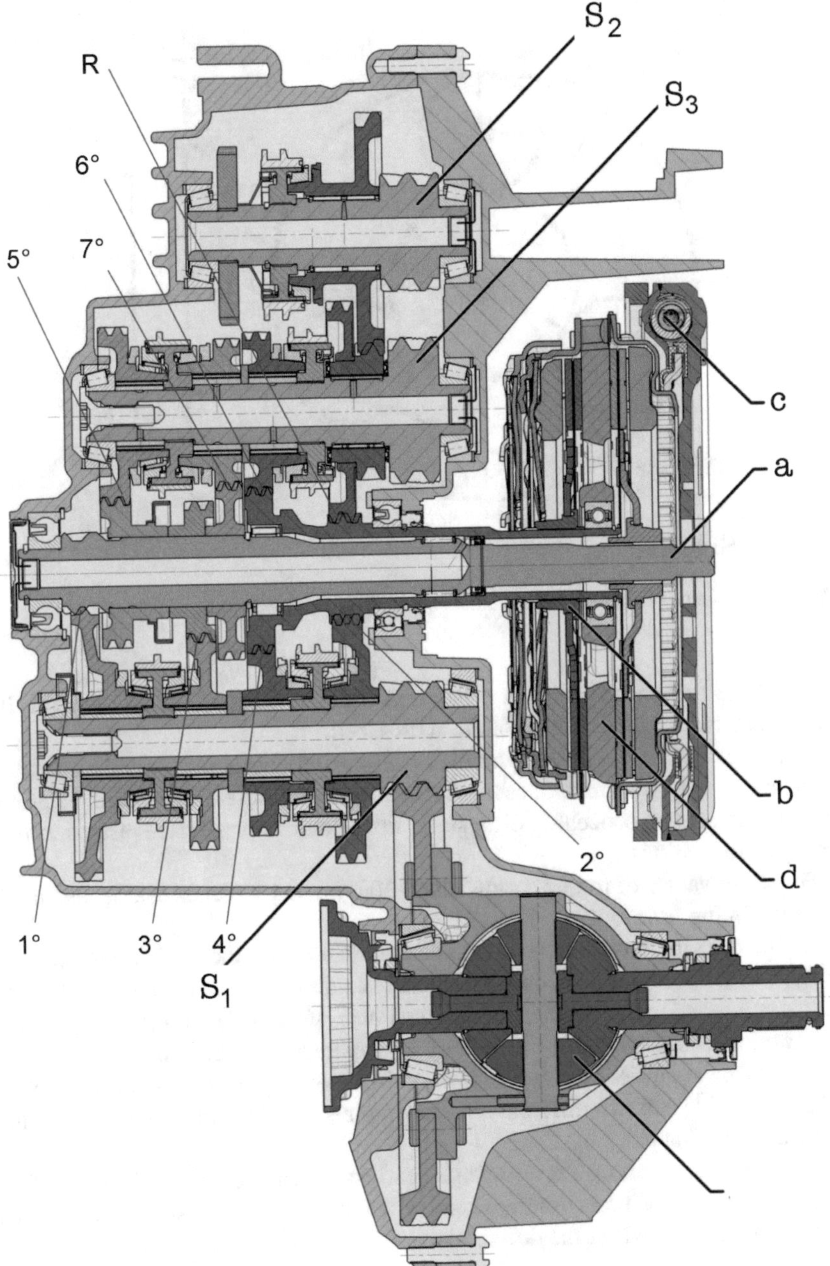

Fig. 15.12 Dry clutch seven speeds multidisc transmission; note that each countershaft is bearing odd and even gears (Volkswagen)

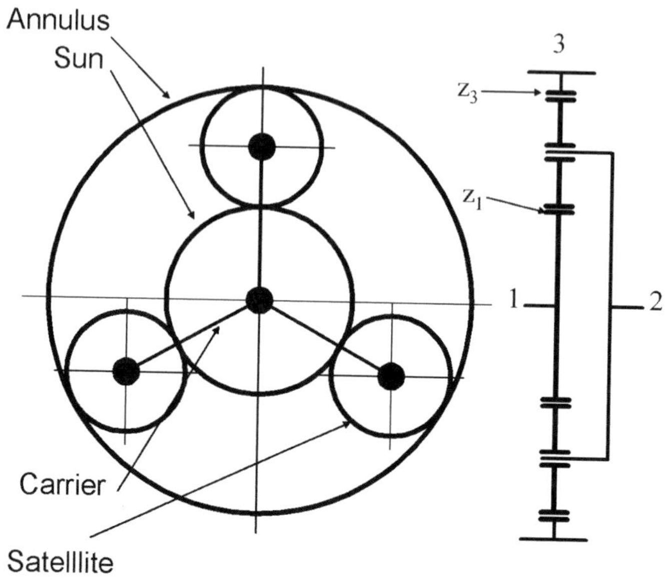

Fig. 15.13 Scheme of a simple epicycloidal gear train

wheel of the first speed on the A_2 shaft and with the R wheel on the A_3 shaft; its synchronizer is neighbor of that of the sixth speed.

In this gearbox the clutches are multi disc oil wet type, because of the relevant value of the torque to be transmitted; as Fig. 15.11 shows, two different pressure plates are pressed, against two concentric discs stack, by hydraulically actuated pistons, acting through ball bearings.

For lower values of torque, below about 300 Nm, dry clutches can be also used, as shown in Fig. 15.12.

In this case the clutch is more complicate than usual; there are three pressure plates, one at the engine side, one at the gearbox side and one in the middle: this one only is connected to the engine, through a number of straps, as shown in Fig. 11.3.

The odd speed pressure plate in pulled by a diaphragm spring, while the even speed pressure plate is pushed; the two necessary diaphragm springs are both installed at the gearbox side and two different disengaging mechanisms are provided, operated by hydraulic actuators.

The drawing, reporting a seven speeds gearbox, is again represented with the four shafts on the same plane as usual; remember that the three output shafts are engaged with the final ratio gear of the differential.

15.3 Epicycloidal Car Gearboxes

15.3.1 Epicycloidal Trains

We will examine the most important gear trains used on automatic gearboxes; we admit that the configurations in use are many more.

These trains are always made with spur gears; in the past some example with bevel gears was available.

The simplest configuration for an epicycloidal gear train is shown on Fig. 15.13; the same was presented for gear speed reducers, splitters and transfer differentials. The wheel train is made by an internal tooth gear (*annulus*) and by an external tooth gear (*sun*) with concentric centre lines; the two wheels mesh with other wheels (*satellites*) whose hubs are supported by shafts mounted on a rotating structure coaxial to the sun, called *carrier*.

Satellites are normally three and are put on the carrier at constant angles; radial components of gearing forces are, therefore, self equilibrated.

This systems shows three degrees of freedom; in order to transmit a torque from an input element to an output element, the third element must be fixed to the gearbox casing (stator); three transmission ratios that can be obtained with a train of this kind (as many as the elements that can be stator: sun, carrier, annulus), multiplied by two (the number of possibilities for a given element to be input or output shaft). To this six possible transmission ratio, one more should be added as a direct drive, by fixing the input and the output shafts together.

If we call with:

$$\tau_0 = -\frac{z_3}{z_1}$$

the ordinary transmission ratio that can be obtained by fixing the carrier, we can obtain the six different value of the transmission ratio τ reported on Table 15.1.

It should be noticed that τ_0 is negative, because, when the carrier is locked, the output shaft is counter rotating as compared with the input shaft; not all ratios can be utilized on the same gearbox, for the obvious difficulty of having the same element working as input and output in the same mechanism. With this mechanism two ratios (one reduced and one direct drive) can be obtained in practice.

To obtain a larger number of transmission ratio, more epicycloidal gear train must be combined together; an obvious solution is putting more simple gears in series, where each of them makes available one of the desiderate ratio alone or in combination with the remaining ones.

A second way is utilizing compound gear trains, where different elements are integrated together. This second way, preferred on not very modern configurations, allows design simplification but with limited flexibility in determining transmission ratios.

Table 15.1 Possible values of the transmission ratio of a simple epicycloidal gear, in function of the constraints and of the ordinary transmission ratio τ_0, with reference to Fig. 15.13

Input	Output	Stator	Transmission ratio τ
1	3	2	$\tau = \Omega_1/\Omega_3 = \tau_o$
3	1	2	$\tau = \Omega_3/\Omega_1 = 1/\tau_o$
1	2	3	$\tau = \Omega_1/\Omega_2 = 1 - \tau_o$
2	1	3	$\tau = \Omega_2/\Omega_1 = 1/(1 - \tau_o)$
2	3	1	$\tau = \Omega_2/\Omega_3 = 1/(1 - 1/\tau_o)$
3	2	1	$\tau = \Omega_3/\Omega_2 = 1 - 1/\tau_o$

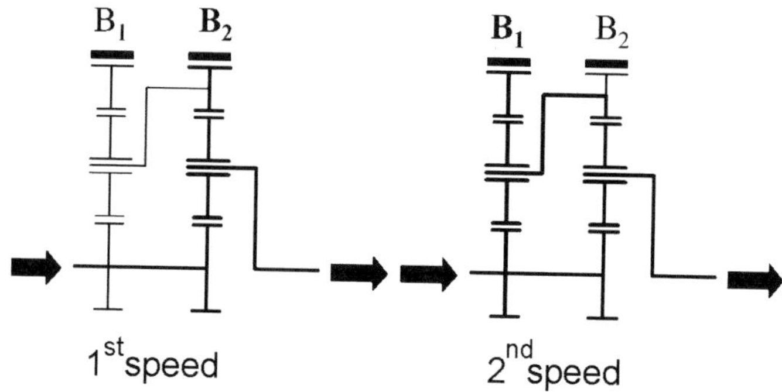

Fig. 15.14 Scheme of a Wilson compound epicycloidal gear train; two speeds can be obtained by stopping one of the two annulus wheels with B_1 and B_2 band brakes

The Wilson epicycloidal gear train, shown on Fig. 15.14, is a first example of compound train. In this mechanism two simple trains are matched is such a way as the carrier of the first train is fixed to the annulus of the second one; by stopping one of the annulus wheels at a time, two different not inverted transmission ratios can be obtained.

By fixing the two carriers together, a direct drive can be obtained.

An alternative scheme of the same mechanism is shown on Fig. 15.15; on this figure B_1 and B_2 are band brakes, while C_1 and C_2 are two multi disc clutches. Three forward and one reverse speed can be obtained by locking the different elements according to the table shown on the same figure (X designates locked elements).

On first speed, when the B_2 brake is locked, we can write for the left mechanism with blocked carrier:

$$\frac{\Omega_{As}}{\Omega_S} = -\frac{z_S}{z_A} . \quad (15.1)$$

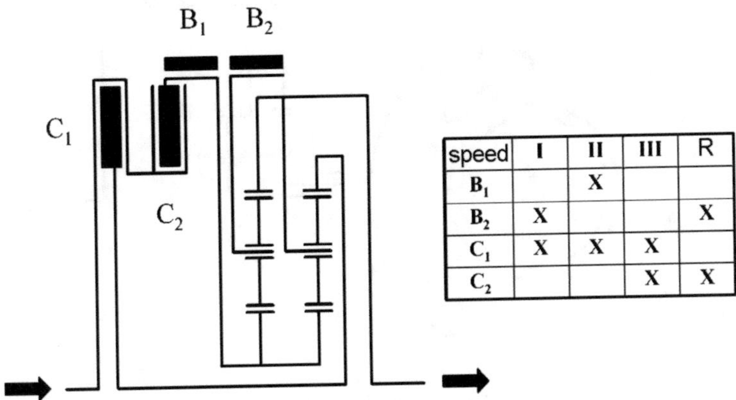

speed	I	II	III	R
B1		X		
B2	X			X
C1	X	X	X	
C2			X	X

Fig. 15.15 Scheme of a second Wilson compound epicycloidal gear train with brakes and clutches; the table shows statuses corresponding to different speeds

In the above formula, Ω are the angular speeds, z the gear teeth number, subscripts d and s show the right and the left gear train, where A and S stay for annulus and sun gear. It should be noticed that teeth numbers don't show any d or s subscript, because the corresponding wheel of the two simple trains are equal.

For the right train, distinguishing again with the 1 and 2 subscripts the input and the output shafts:

$$\frac{\Omega_1 - \Omega_2}{\Omega_S - \Omega_2} = -\frac{z_S}{z_A}\,. \tag{15.2}$$

let's remember that Ω_{AS} and Ω_2 are equal, because their corresponding elements are fixed together.

From the two equations derives the first gear transmission ratio:

$$\tau_1 = 2 + \frac{z_S}{z_A}\,. \tag{15.3}$$

For the second speed the right sun gear is locked and the left simple gear train is idle; the right gear train has the annulus fixed to the input shaft and the carrier fixed to the output shaft; therefore:

$$\tau_2 = 1 + \frac{z_S}{z_A}\,. \tag{15.4}$$

For the third gear the contemporary lock of clutches C_1 and C_2 puts the mechanism in direct drive, with:

$$\tau_3 = 1. \tag{15.5}$$

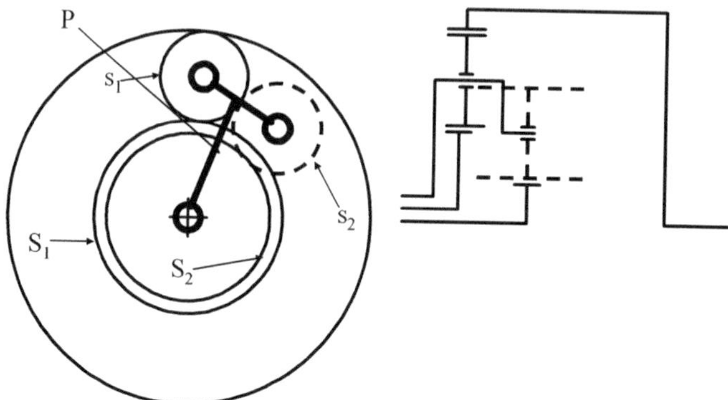

Fig. 15.16 Scheme of a compound Ravigneaux epicyclical gear train; the satellite s_1 is represented conventionally with dotted lines, because its drawing plane is not on the sketch plane

For the reverse gear the carrier of the left simple train is locked, while the right train is idle. The sun gear of the left train is fixed to the input shaft and the annulus gear is fixed to the output shaft; therefore:

$$\tau_{RM} = -\frac{z_A}{z_S} . \tag{15.6}$$

If, for example, $z_S/z_A = 1/3$, the first speed ratio will be 2.333, the second speed 1.333, the third 1 and the reverse – 3; if, instead, $z_S/z_A = 2/5$, the first speed ratio will be 2.4, the second 1.4, again 1 for the third and – 2.5 for the reverse speed.

If we remember what we said about torque converters, obtainable ratios are similar to those of a reference manual four speed gearbox (the stall torque converter torque ratio can be about 2.5); the possibility of obtaining a desired speed step or a suitable reverse speed ratio is unfortunately compromised.

A different compound gear train configuration is that of the so called Ravigneaux train, shown on Fig. 15.16; it is very much diffused on automatic gearboxes. In this train a single annulus wheel is provided with two satellites s_1 and s_2 on the same carrier P; satellites mesh with two different sun wheels S_1 and S_2. Satellites s_2 mesh with the smaller sun S_2; they don't mesh with the annulus but with the satellites s_1 of the larger sun S_1.

Since satellites s_2 have their centre line on a different drawing plane, they are conventionally sketched with dotted lines.

On the following Fig. 15.17, different gearbox statuses are shown with the already known procedure. With this scheme, with three multi disc clutches and two band brakes, four different forward speeds and one reverse speed can be obtained.

We adopt the same symbols of the previous example; we have, in first speed, the right sun gear fixed to the input shaft and the annulus gear fixed to the output shaft, while the carrier is the stator element; therefore:

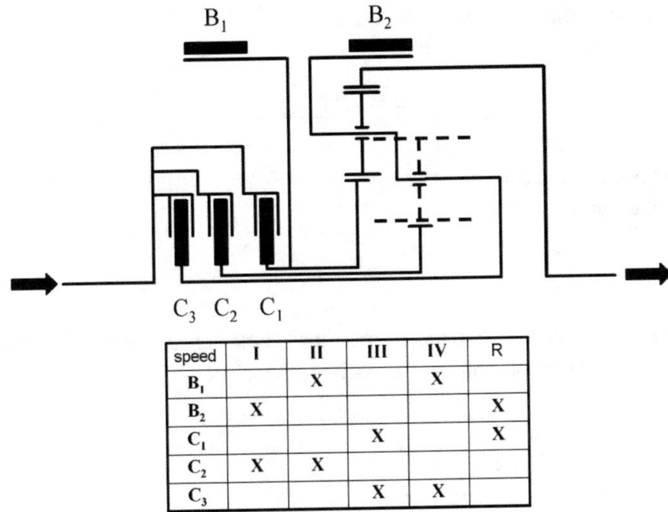

speed	I	II	III	IV	R
B_1		X		X	
B_2	X				X
C_1			X		X
C_2	X	X			
C_3			X	X	

Fig. 15.17 Scheme of an automatic gearbox with a Ravigneaux epicycloidal gear train; the table shows statuses of clutches and brakes in different speeds

$$\tau_1 = \frac{z_A}{z_{Sd}}; \tag{15.7}$$

it represents the ordinary transmission ratio of the right gear train; the sign is positive because the double meshing satellites don't change the rotation direction.

In second speeds the annulus wheel is fixed to the output shaft, while the input shaft is fixed to the right sun gear, being the left one locked. For the left gear train, we can write:

$$\frac{\Omega_{Ss} - \Omega_P}{\Omega_A - \Omega_P} = -\frac{z_A}{z_{Ss}}; \tag{15.8}$$

because the left sun gear is locked and the annulus is fixed to the output shaft:

$$-\frac{\Omega_P}{\Omega_2 - \Omega_P} = -\frac{z_A}{z_{Ss}}. \tag{15.9}$$

For the right epicycloidal gear train, we can write:

$$\frac{\Omega_{Sd} - \Omega_P}{\Omega_2 - \Omega_P} = -\frac{z_A}{z_{Sd}}. \tag{15.10}$$

If we remember that the right sun gear is fixed to the input shaft, we can also write:

$$\frac{\Omega_1 - \Omega_P}{\Omega_2 - \Omega_P} = \frac{z_A}{z_{Sd}}. \tag{15.11}$$

If we compare the equations, we obtain:

$$\tau_2 = \frac{1 + \frac{z_{Ss}}{z_{Sd}}}{1 + \frac{z_{Ss}}{z_A}} . \tag{15.12}$$

In third speed the carrier and the right sun gear are fixed, therefore the gearbox is in direct drive, and:

$$\tau_3 = 1 ; \tag{15.13}$$

in fourth speed, the carrier is fixed to the input shaft and the annulus is fixed to the output shaft, while the left sun gear is locked; therefore:

$$\tau_4 = \frac{1}{1 + \frac{z_{Ss}}{z_A}} . \tag{15.14}$$

In reverse speed the carrier is locked, the sun is fixed with input shaft and the annulus with the output shaft; the transmission ratio is that of the ordinary left gear train and therefore:

$$\tau_{RM} = -\frac{z_A}{z_{Ss}} . \tag{15.15}$$

Also in this case remarks made for previous case about the possibility to obtain a given range and ratio step do apply.

If the number of available speed increases, applied gear train gets more complex; for example a scheme suitable for a five speed gearbox may be conceived by adding a simple epicycloidal gear train to the Ravigneaux gear train previously described.

The scheme could be compared to the application of a range speed reducers that we described for the industrial vehicles.

The total speed number should be eight, by multiplying the four speeds of the compound gear train by the two of the added speed reducer (reduced and direct drive); in a real example many obtainable speed ratios should be too close and therefore useless.

The simplest scheme, considered till yesterday, are abandoned in favour of compound gear trains that look unnecessarily complicated, but obtain transmission ratio values closer to theoretical specification.

Figure 15.18 shows the example of a Ravigneaux compound gear train joined to two simple epicycloidal gear trains; the number of theoretically available speeds is well over the five practically implemented. Since many possible combinations were discarded, the implemented transmission ratios can assume almost the desired values and therefore allow a better optimization of performance, fuel consumption and emission.

On the same figure is represented with the same procedure brakes and clutches status for different speeds; this scheme has been designed and applied for longitudinal engine conventional driven cars.

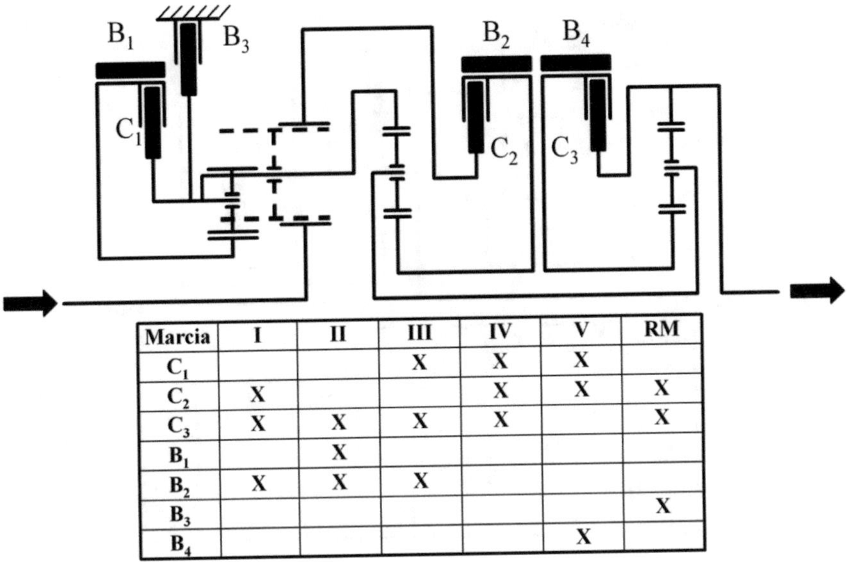

Marcia	I	II	III	IV	V	RM
C_1			X	X	X	
C_2	X			X	X	X
C_3	X	X	X	X		X
B_1		X				
B_2	X	X	X			
B_3						X
B_4					X	

Fig. 15.18 Scheme of an automatic powershift gearbox featuring three epicycloidal gear trains, conceived by Mercedes; five different speeds are obtained

Quite recently, this scheme has been modified in a very simple manner, to obtain what represented on Fig. 15.19; substantially an additional annulus wheel has been added to the Ravigneaux gear train and the lay-out of brakes and clutches has been modified, without increasing their number. The result is seven forwards speeds and two reverse speeds, haw the status table shows on the picture.

The transmission ratio calculation will be omitted for the two previous schemes for sake of simplicity.

This overview on gearbox schemes doesn't pretend to be a complete report on what is or can be produced. Possible combinations of simple and compound epicycloidal gear trains are many more.

The epicycloidal gear solution is, for the time being, the most diffused one for powershift automatic gearboxes.

15.3.2 Production Examples

Conventionally driven cars benefit of the alignment of input and output shafts.

The example on Fig. 15.20 corresponds to the scheme on Fig. 15.18 and adopts, as we have seen, a Ravigneaux gear train 3 and two epicycloidal speed reducers 2 and 4, downstream of the first; it features five forward speeds and a reverse speed.

We draw the attention on the fact that on torque converter automatic gearboxes is impossible to keep the vehicle braked at parking, by engaging only the lowest speed, how is possible on manual gearboxes.

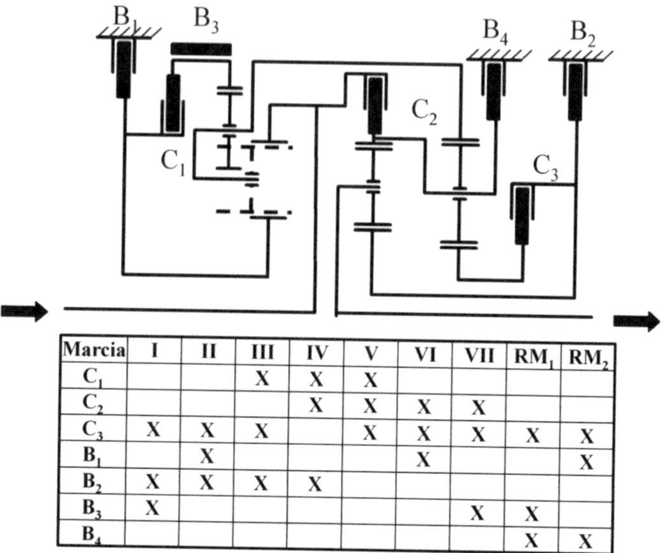

Marcia	I	II	III	IV	V	VI	VII	RM_1	RM_2
C_1			X	X	X				
C_2				X	X	X	X		
C_3	X	X	X		X	X	X	X	X
B_1		X				X			X
B_2	X	X	X	X					
B_3	X						X	X	
B_4								X	X

Fig. 15.19 Scheme still conceived by Mercedes derived of the previous one; it allows seven forward speeds and two reverse speeds

On these gearboxes engine compression and friction torque is used to withstand vehicle tendency to move. The torque converter, inserted on the transmission line, acts as free wheel joint at low speed; also the lock-up clutch, if any, couldn't keep the car braked, because actuation pressure is not available when the engine is stopped.

To keep the vehicle stopped, a manually actuated park pawl is added, matching a sprocket wheel on the output shaft; in the figure the pawl is the element 15 and the corresponding sprocket wheel is the element 14.

Also the torque converter can be seen (without lock-up clutch), the oil pump 6 and the different multi disc wet clutches used for gear engagement.

A recent evolution of this scheme is represented on Fig. 15.21; the operation principles are explained on the previous Fig. 15.19. Epicycloidal gear trains are almost the same as in previous case. A lock-up clutch can be noticed on the torque converter.

Automatic gearboxes for front wheel driven cars, with longitudinal or transversal engine, are complicated by the necessity of having an output shaft at a convenient distance of the input shaft, not coaxial as on epicycloidal gearboxes; in addition, particularly critical on transversal engines are space limitation caused by the front steered wheel.

Nevertheless, automatic gearboxes have gear trains that are substantially similar to those of conventional drives. Additional single stage trains are added to locate the output shaft on the desiderated final drive pinion position.

One example of this architecture is reported on Fig. 15.22, for a front wheel driven car with longitudinal engine.

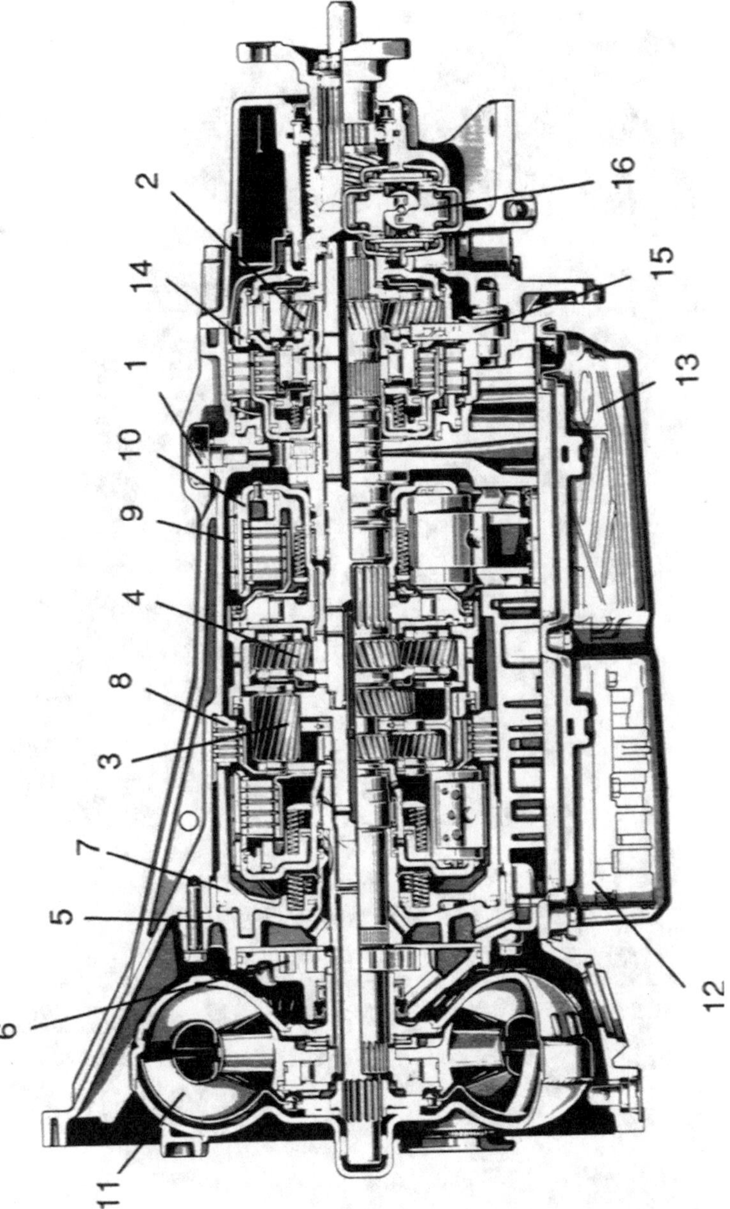

Fig. 15.20 Scheme for a five speed automatic powershift gearbox for conventional drive (Mercedes)

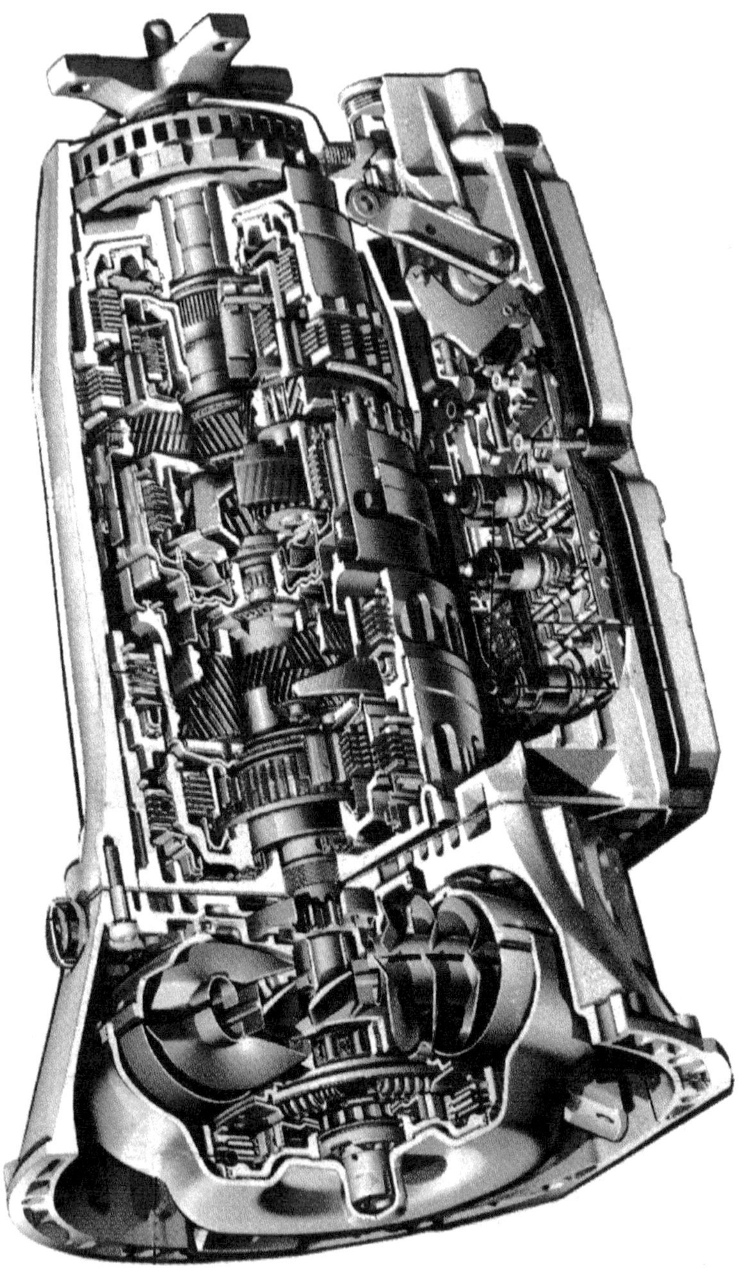

Fig. 15.21 Scheme for a seven speed automatic powershift gearbox for a conventional drive (Mercedes)

Fig. 15.22 Six speeds automatic gearbox with epicycloidal gear trains; the lateral shaft connects the gearbox output shaft with the differential pinion, behind the engine (Audi)

The scheme includes a simple and Ravigneaux gear trains that in conjunction with three clutches and two brakes make six speeds available. The Ravigneaux gear train output shaft moves the differential final drive through an auxiliary shaft. The presence of the all wheel drive doesn't influence the lay-out.

The possibilities available to front wheel driven cars with transversal engine are two.

A first solution provides for torque converter and oil pump in line with the engine; the torque converter output shaft is fixed to a sprocket wheel that, through a silent chain, moves the input shaft of the real gearbox on a parallel centre line. Every geometrical problem on the transversal direction of the car is eliminated; on the contrary there are consequences on the front overhang of the car that becomes relevant.

A second solution provides dividing the epicycloidal gearbox in two sections.

The first section could include a Ravigneaux gear train coaxial to the engine; its output shaft moves through a single stage train a simple epicycloidal gear train coaxial to the differential pinion. It is therefore possible to have five speeds within the space available on a front wheel drive with transversal engine.

15.4 Car CVTs

15.4.1 Motivations

Continuously variable transmissions (CVTs) are mostly diffused on front wheel drives. They are particularly appreciated for having available an unlimited number of transmission ratios, with benefits for comfort, performance and fuel economy.

A simple explanation of this sentence can be supplied by Fig. 15.23. On this figure the specific fuel consumption map is reported using engine torque and speed

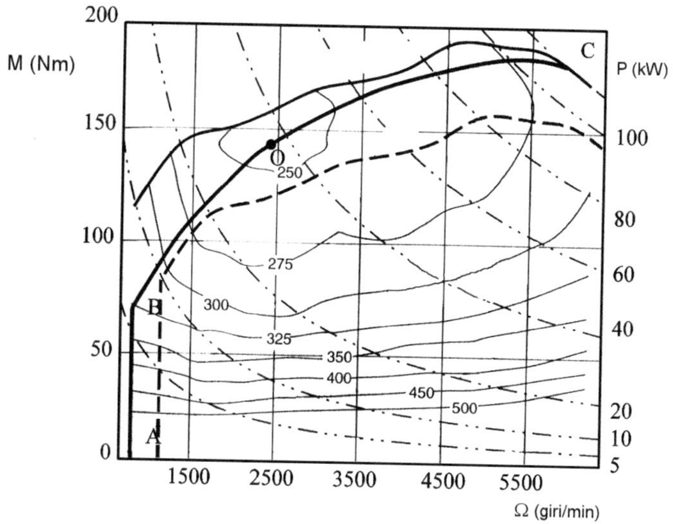

Fig. 15.23 Engine specific fuel consumption map; the minimum fuel consumption line ABC and the area of practical use for a stepped gearbox are shown. This area is included between the wide open throttle curve and the dotted curve showing the torque obtained on the higher speed, when existing

as parameters; for each output power value, one and only one minimum fuel consumption working point does exist.

These points are represented by the thick line below the wide open throttle curve; it can be drawn by connecting all tangency points of equal fuel consumption curves (quoted on the figure) with the hyperbola (drafted with a line and dots) representing the constant power curves.

This curve is made by a vertical straight line AB on the left of the map that represents the minimal stable working speed of the engine; this speed depends on engine characteristics, on elastic and damping properties of the transmission line, made of clutch, gearbox, on car suspensions and on tires.

Other notable points of this curve are point C, coincident, by definition, with full power point and point O that represents the minimum absolute fuel consumption value.

Let's assume changing plane coordinates in such a way as they represent the wheel torque and vehicle speed at the actual speed $i - 1$ of a stepped gearbox; the dotted curve, similar to the maximum torque curve, represent the wheel torque at the next i speed.

A very efficient usage of a stepped gearbox allows specific fuel consumption to fluctuate between the maximum torque curve at the $i - 1$ speed and the maximum torque curve at the i speed, at a given power level; actual fuel consumption can be higher than minimum.

At the last speed, fuel consumption can fluctuate between the maximum torque and the road load torque at the last speed; in this case fuel consumption increase could be substantial.

It is possible to have the engine working at the specific minimum fuel consumption only if an unlimited number of speeds are available.

This fact explains why in these last years so many efforts have been made to increase the speeds number either for manual or for automatic gearboxes.

It should be said that only on these lasts, when a sufficiently sophisticated control system is available, the instantaneously optimum speed is guarantied to be engaged.

Similar statements can be made about acceleration and pollutant emissions.

Possible systems suitable for a CVT include in principle:

- Electric transmissions;
- Hydrodynamic transmissions (among them the torque converter);
- Hydrostatic transmissions;
- Variable geometry mechanical transmissions.

The first three categories have been considered many times in the past, but have been abandoned on conventional vehicles for the poor transmission efficiency; this penalty jeopardizes the advantage theoretically available by a CVT.

These kinds of transmissions can be reconsidered for hybrid vehicles, where the intermediate form of energy generated by the transmission (electric or hydraulic pressure energy) can be used to recover vehicle kinetic energy or as a temporary source alternative to fossil fuel.

Variable geometry mechanical transmissions can show similar efficiency as that of gear transmissions. At this time belt or chain drives are available and roller traction drives are at the developmental stage.

On special purpose vehicles, such as tractors, earth moving machines, airport boarding busses, etc. some hydrostatic CVT are used.

15.4.2 Production Examples

Steel belt transmissions represent the most diffused technical solution for CVT. Steel belts that are known today are the VDT belt and the LuK chain, which is used as a belt anyway. Their working principles present many common points but some differences are presents mostly as far as component design is concerned; this issue will be explained later on.

The cross section of a CVT with a Van Doorn Trasmissie steel belt (VDT) is shown on Fig. 15.24 with the usual drawing system. The gearbox on the figure doesn't include a torque converter, because start-up functions are performed by an electronically controlled multi disc clutch on the output shaft.

The reason for this choice is twice:

- While towing the car, any damage on the belt is avoided; the belt in fact is loose without oil pressure on pulleys;
- During emergency braking manoeuvres, transmission ratio can be changed independently of vehicle speed, making the next acceleration quicker.

The engine is fixed to the carrier 8 of an epicycloidal gear train with twin satellites that is used for forward and reverse speed; the gearbox input shaft 1 is instead fixed to the sun gear 9 of the same train. By using the clutch 4 is possible to set the train in direct drive; by engaging instead the brake 5, the annulus is locked and sun gear rotation is opposite to that of the carrier, obtaining reverse speed.

With this design it would be possible, if necessary, to change transmission ratio in both forward and reverse drives.

The transmission ratio variation is obtained through two variable gouge V pulleys which match the belt (see detail on Fig. 15.25).

Each pulley is made by two steel cones that can slide axially but are fixed together when turning. Looking at the drawing, the left input semi pulley and the right output one are also fixed in the axial direction; the other two semi pulleys are instead free to slide.

Because the developed belt length cannot change, by narrowing the two driving semi pulleys the driving pulley primitive diameter will be increased and the two driven semi pulleys must be accordingly widen, to reduce their primitive diameter; the belt will be shifted to the right and the transmission ratio will be decreased. With the opposite motion the transmission ratio will be increased.

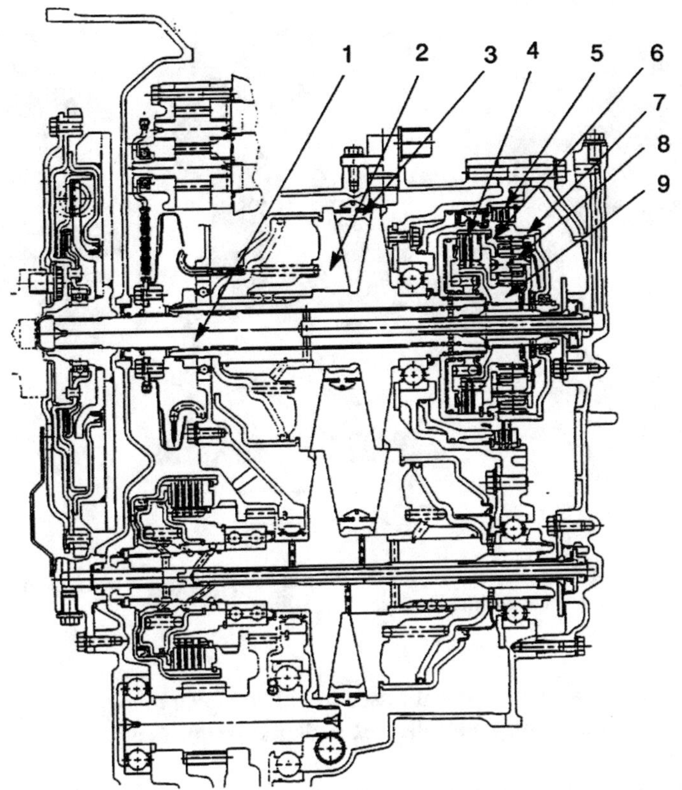

Fig. 15.24 VDT steel belt CVT; the gearbox is characterized by the start-up multi disc clutch on the output shaft (Honda)

A given transmission ratio range will be obtained with a minimum overdrive ratio and a maximum reduced ratio.

The two limit ratios are reciprocal numbers; this situation where the 1:1 value is in the middle of the ratio range must be adapted to the vehicle use by an additional final drive bringing higher ratio to be close to one.

For this reason there is an idler gears final drive between output pulley and the differential pinion; it should be necessary, in any case, to position different centre lines appropriately to the engine compartment lay-out. This final drive is also used to adapt same gearbox to different cars.

Semi pulleys active surfaces are moved by hydraulic pistons; this motion can change their primitive diameter. Oil pressure determines semi pulleys contact force with the belt.

Controlling this parameter is vitally important, because contact force determines torque transmission through friction between the pulley and the belt; if this force is insufficient parts would slip, with lack of efficiency and surface damage; if this

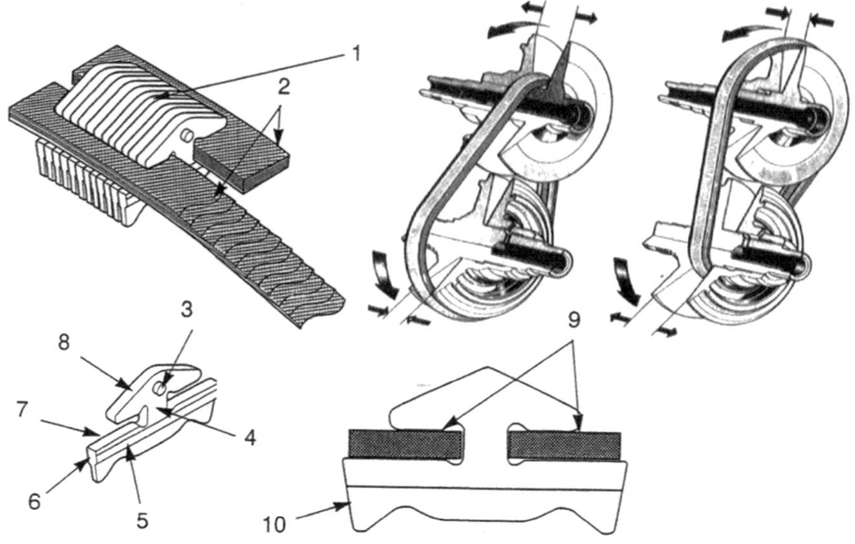

Fig. 15.25 Detail of the pulley motion in a VDT steel belt CVT (upper part); detail of the belt (lower side). The torque is transmitted by belt compression

force is excessive, also mechanical losses would be higher than necessary, because contact area between belt and pulley exceeds the cinematically correct primitive circumference.

VDT belts include the use of many thrust elements as that on the lower side of Fig. 15.25.

Each of them contacts the next and is set along the belt centre line; they are kept aligned with a series of flexible steel ring bands, one inserted inside the other with no clearance. Thrust elements have a trapezoidal shape with a median appendix; the two inclined sides match pulleys sides and exchange the friction force necessary to torque transmission.

The median appendix of each thrust element matches the two sets of concentric rings; ring contact surfaces are inclined in such a way as the rings don't come in contact with pulley sides.

Rings must be very as the curvature stress will be contained within reasonable values; the necessary tension is supported by an adequate number of rings.

Ring developed length must allow an assembly without clearance do uniformly distribute tension between rings.

Thrust elements can only exchange compression forces with neighboring ones; therefore VDT belt will never work with positive tension and will transmit the torque through the compression of the driven part of the belt circuit.

Rings only purpose is keeping elements at the correct radial position in the pulleys and aligned when outside of pulleys.

Friction forces between elements and pulleys wouldn't waste energy if there weren't relative motions. They are very limited because there isn't macroscopic slip between parts; nevertheless there is local microscopic relative motion because of two facts:

- Thrust element contact area must have a certain radial extension to limit contact pressure; on each contact area there will be a small slip on points outside primitive radius.
- Thrust elements entering and leaving pulleys must slide because they change their path from straight to circular.

The power wasted by these small relative motions will be in any case determined by pressure acting between belt sides and pulleys; this pressure must be limited to the minimum necessary to avoid a macroscopic slip. Contact pressure must be therefore carefully adjusted as a function of transmitted torque.

There are also belt CVT with torque converter between engine and speed variation unit, as that shown in Fig. 15.26.

Fig. 15.26 VDT steel belt CVT with start-up torque converter (Mercedes)

Other CVT apply different steel belts that could be better named as chains.

Figure 15.27 shows a chain CVT lay-out, not very different as compared with the first as far as pulleys are concerned; start-up clutch is instead before speed variation unit. The gearbox shows the usual final drive before differential drive, but differential shows a bevel gear drive, to adapt the gearbox to a longitudinal engine.

The LuK chain is represented on the upper part of the same figure.

The links of this chain are made by a stack of plates mounted on pins, rolling on their contacts without slip; pin ends are inclined to match pulley sides. The friction between pin tips and pulley transmits efforts from belt to pulleys and vice versa; in this case the belt is in tension on the driving side of its circuit.

Plates length can show a limited difference in different links, to avoid whistles and vibrations.

A good point in favour of the chain is the possibility of adopting a smaller minimum radius on the pulley, reducing dimensions at the same performance; mechanical efficiency should also be slightly better.

Chain or belt design has presented many problems; steel has been chosen as most suitable material because of stress necessary level and for oil compatibility.

Many solutions of reinforced elastomeric materials have been also proposed for belts; they must be used outside of gearbox lubricated areas and have found their application for limited torque, particularly on scooters.

A different working principle usable for CVT, not present on mass production yet, is that of rolling bodies able to exchange significant tangent forces with rather limited contact areas. This is possible thanks to special synthetic lubrication oils having the property of viscosity increasing with contact pressure (*traction fluids*).

The advantage of this principle consist in a high mechanical efficiency, that can be reached because of very limited slips of contacting parts, working very closely to pure rolling conditions; on the contrary, the disadvantage is due to the very high mechanism weight, necessary to obtain a sufficient contact pressure, for adequate traction at lubricated conditions.

A typical configuration according to this principle is that of the so called toroidal gearbox; the essential parts of this gearbox are shown on Fig. 15.28. The gearbox shows two identical traction bodies each of them made by two symmetric toroidal surfaces.

These toroidal surfaces are made by two revolution surfaces obtained by turning a circumference arc around the rotation axis; on rotating facing surfaces, a number of rollers are set whose main curvature is slightly higher than that of the toroidal surface.

By changing the rotation axis angular position of these rollers, it is possible changing transmission ratio between toroidal bodies; when rollers rotation axis is perpendicular to the toroidal surface rotation axis, transmission ratio will be 1:1; by turning the rotation axis in the two directions on the same plane it will be possible to decrease or to increase transmission ratio.

In order to limit reaction loads on casing, two symmetric mechanisms are employed whose thrusts can be self equilibrated in a simple way though a central

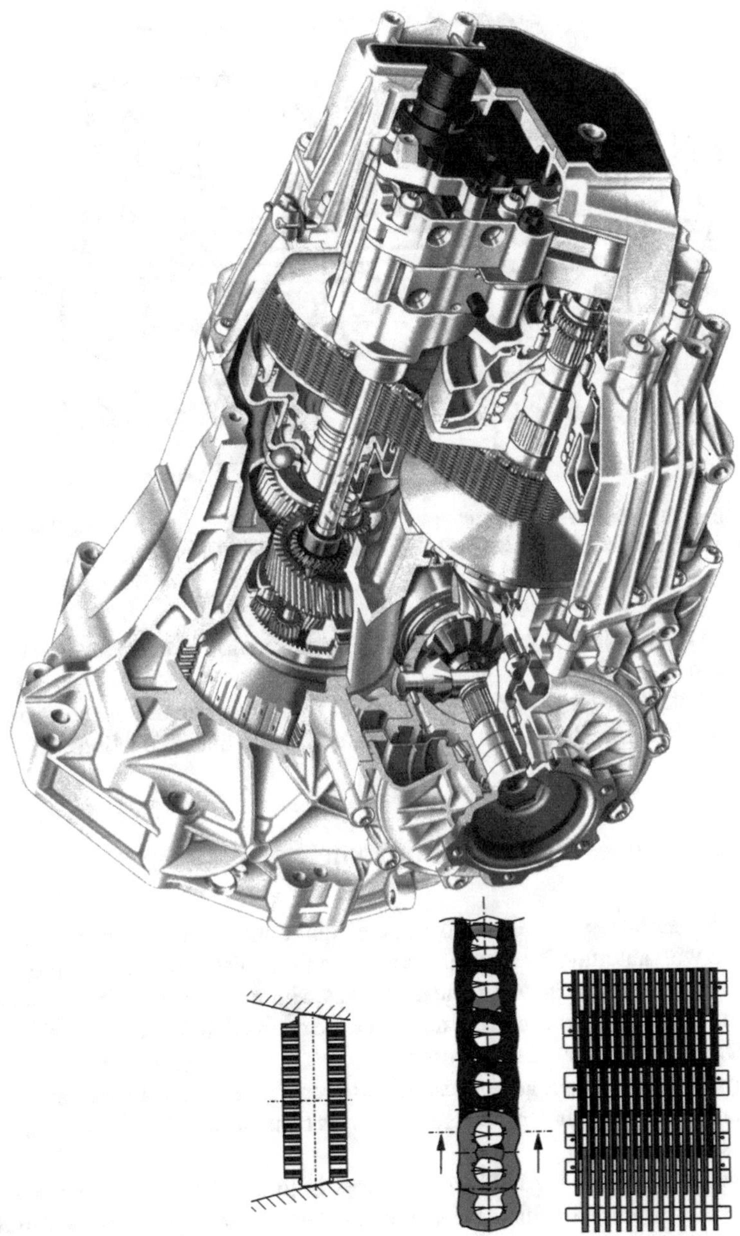

Fig. 15.27 On the lower side: chain CVT suitable to front wheel drive with longitudinal engine (Audi). On the upper side: view and cross section of a LuK chain

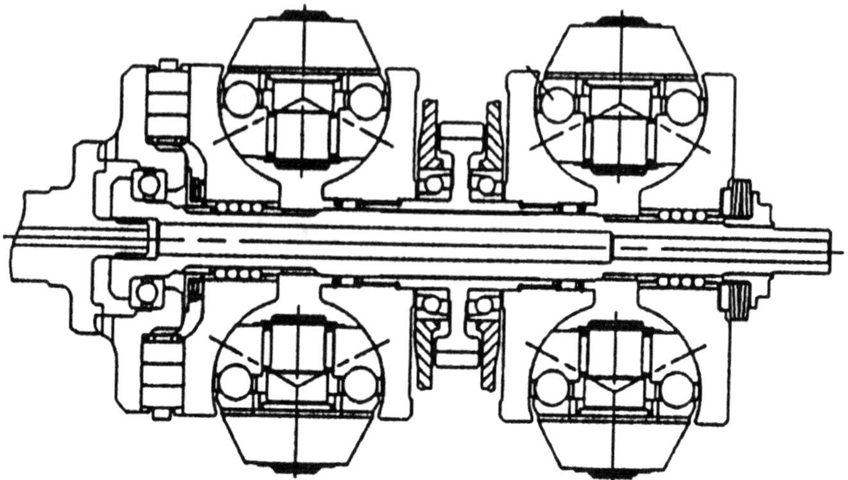

Fig. 15.28 Rolling bodies toroidal CVT; by steering appropriately satellites it is possible changing transmission ratio between input (on the left) and output pinion (in the middle) (Torotrak)

shaft. The output shaft is meshing with the central pinion fixed to the two toroidal driven surfaces.

Three rollers are employed to limit contact forces; a particular mechanism must be used to obtain on all of these rollers the same centre line inclination.

15.5 Gearboxes for Industrial Vehicles

Automatic gearboxes for industrial vehicles don't show substantial differences if compared with automobiles ones; the countershaft architecture is diffused on semi-automatic gearboxes or in automatic gearboxes with many speeds, where an epicycloidal train configuration could be difficult to be adopted. In these applications both clutches and torque converters are used.

Epicycloidal gear train gearboxes, not different as those considered on cars, are particularly applied on busses, where comfort is a major priority and the number of speeds is limited.

A relevant problem for industrial vehicles is torque converter behaviour. The mechanical efficiency of this device is particularly poor when pump and turbine exchange their function; this happens when reaction torque inverts because the engine is expected to brake the vehicle with its mechanical friction or with dedicated devices (*engine brake*).

This fact is justified by blades angles that can't be optimized for two opposite flow directions.

Engine brake effect at these conditions is particularly decreased, as compared with a manual gearbox. This fact can be accepted and is sometimes beneficial on cars fuel economy; is unacceptable vice versa on a heavy vehicle.

Some specific devices (*retarder*) are used to solve this problem; they will be explained later on.

15.5.1 Semiautomatic Gearboxes

Together with the already commented pre selection semiautomatic gearboxes, there automatic and automatized gearboxes are on the market with clutch and torque converter in series.

Torque converter is set at first with the purpose to increase start-up torque and to smooth transmission output torque; the torque converter features a lock-up clutch to increase the transmission efficiency in cruise drive.

Torque converters include an additional free wheel between pump and turbine; this free wheel is designed to be able to transmit only negative torque, i.e. it is like an open joint when the engine is driving, it is like a locked joint when the engine is braking; in this way the desiderate engine braking effect is granted.

After torque converter, an additional component is fit, not necessarily only because of gearbox automation; this is the retarder that helps the vehicle brakes on long downhill drives.

Retarders can be assimilated to high diameter hydrodynamic clutch, with a limited radial blade dimension; this shape is dictated by the needs of limiting longitudinal gearbox length and obtaining a high stall torque.

The pump is fixed to the torque converter output shaft; the turbine blades are directly cut on gearbox casing and are still.

During vehicle slow-down, the hydrodynamic clutch absorbs a braking torque equal to stall torque and wastes the corresponding power to heat; to adjust or delete braking torque, clutch oil quantity is adjusted or eliminated with dedicated valves.

Remaining gearbox elements are those already seen.

15.5.2 Automatic Gearboxes

Totally automatic gearboxes are also available to industrial vehicles and they are almost similar to those of conventional drive cars with torque converter and epicycloidal gear trains.

Speed number is limited to 5 or 6; for this fact this kind of gearbox is suitable to busses or limited speed trucks, subject to frequent stops.

Figure 15.29 shows an example of this gearbox, with six forward and one reverse speeds; there are three simple epicycloidal gear trains that can be used separately or in combination.

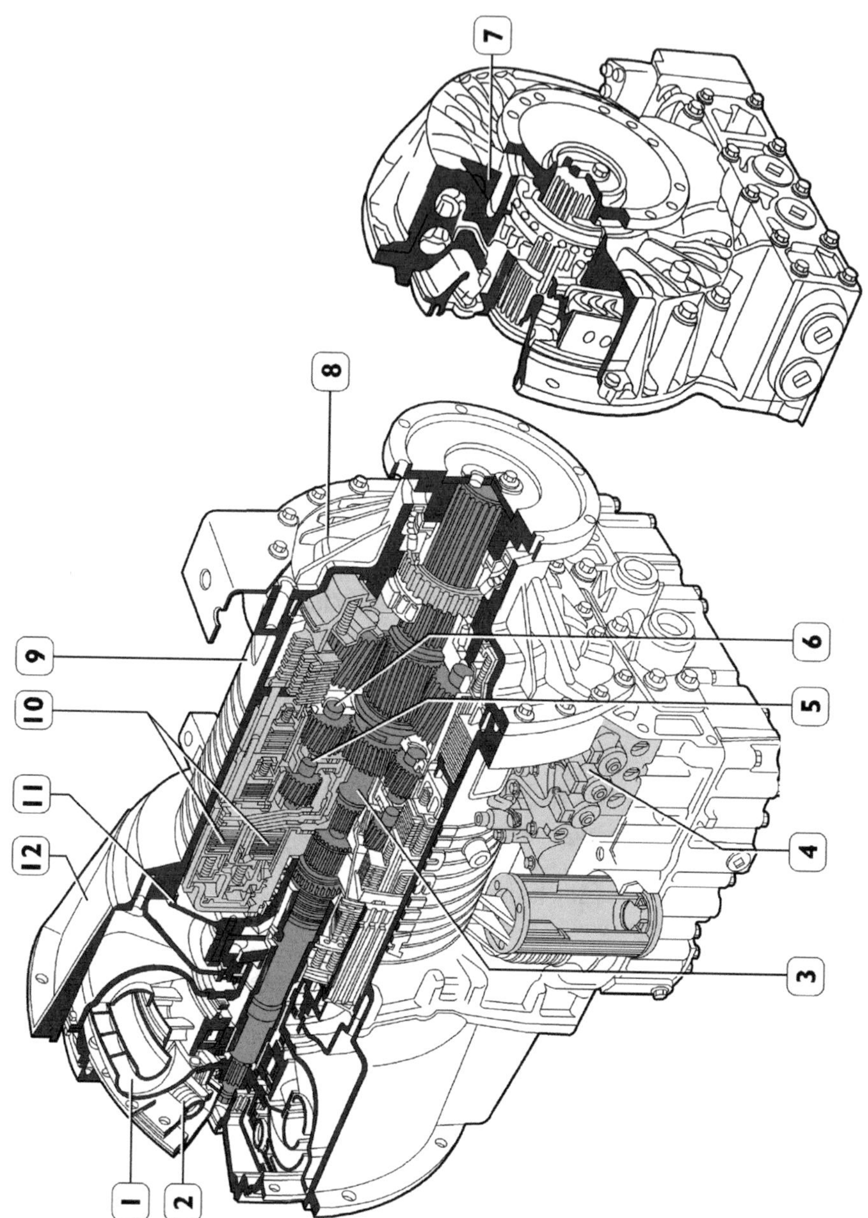

Fig. 15.29 Epicycloidal gear trains automatic gearbox for industrial vehicle; on the right the retarder unit 7 is shown that can be optionally mounted in place of the back cover 8 (Iveco)

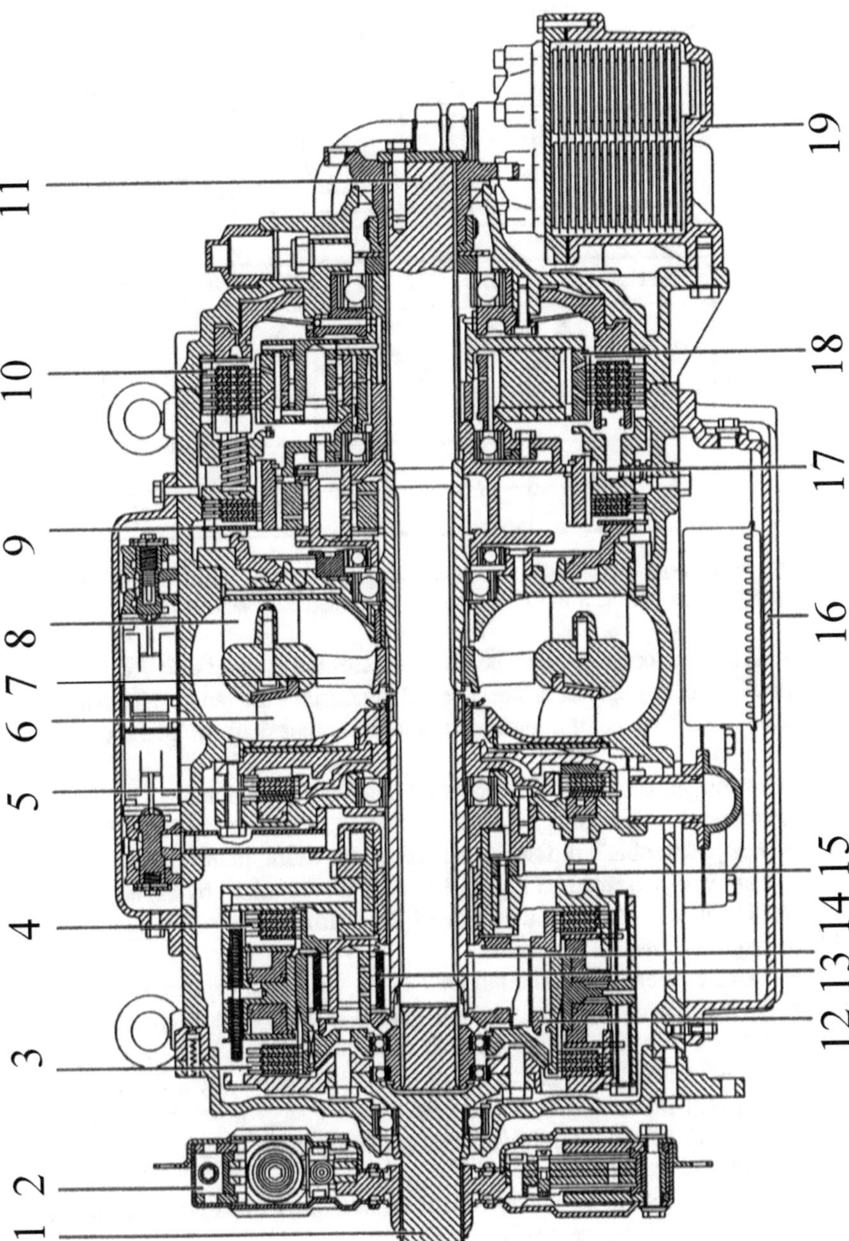

Fig. 15.30 Automatic gearbox for urban bus; the architecture is characterized by a torque converter in the central section of the gearbox; it is used for vehicle start-up as well as a retarder (Iveco)

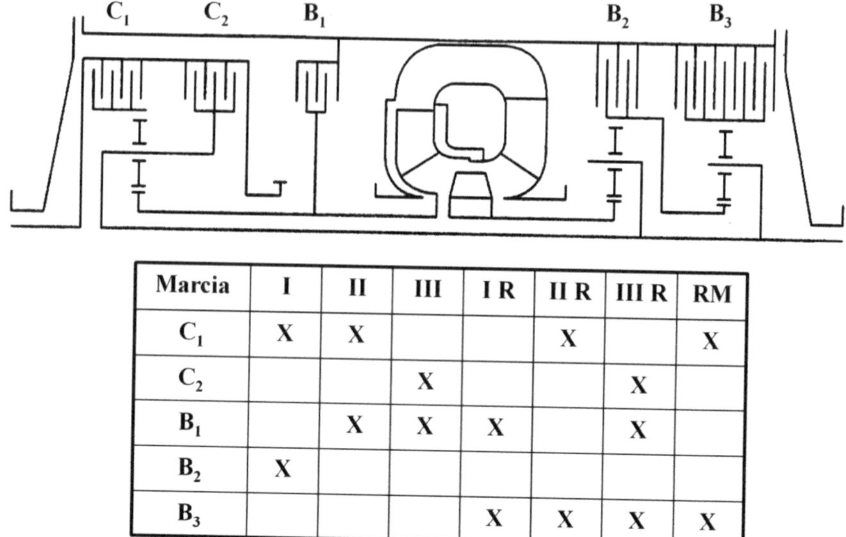

Marcia	I	II	III	I R	II R	III R	RM
C_1	X	X			X		X
C_2			X			X	
B_1		X	X	X		X	
B_2	X						
B_3				X	X	X	X

Fig. 15.31 Working scheme of the previous automatic gearbox; three normal forward speeds, three retarded forward speeds (I R, II R, III R) and one reverse speed are available

The back gearbox cover 8 can be changed with the retarder unit 7 we described in the previous paragraph. In this case radial dimensions are reduced by using a two side bladed wheel; the central rotor is fixed to the output shaft, while the stator is cut in the casing.

Braking torque regulation is made by oil quantity in the clutch.

We introduce finally a very particular automatic gearbox, suitable for urban busses; it was developed to obtain different retarded speeds useful in hilly towns.

The gearbox is presented on Fig. 15.30 and is characterized by having the torque converter installed in the central section of the gearbox, instead of the joint flange with the engine, as usual.

The torque converter (6 is the pump, 7 is the turbine, 8 the stator) is used as a start-up device as well as a retarder; input shaft 1 is fixed to the engine through the joint 2 that acts as a torque damper. Two different clutches 3 and 4 are provided for a first stage epicycloidal gear train 13 and for the direct drive. The two output epicycloidal gear trains 17 and 18 are used to mix energy flows through the input shaft and the turbine and for the reverse speed. The heat exchanger 19 can waste the heat generated by the retarder.

The scheme on Fig. 15.31 and the related status table of brakes and clutches allow us to understand how the mechanism works.

In first speed, power flows through two parallel ways. The first way is made by a rigid connection with the epicycloidal gear train carrier 13; the second way flows through the torque converter, whose pump is fixed to the sun gear of the same train and turbine is fixed to the sun gear on train 17. The annulus gear is fixed to input shaft.

In second speed, the revolution speed is reduced by the epicycloidal gear train 13 whose sun gear is locked; the torque converter isn't loaded.

In third speed the gearbox is a direct drive.

In retarded speeds the third brake involves the epicycloidal gear train 18 the counter rotates the turbine which becomes pump; the torque converter pump is locked by the B_1 brake and acts like a stator.

The torque converter is now an effective retarder.

15.6 Control Strategies

Next paragraphs are going to explain the main rules for speed choice and speed shift actuation that automatic gearbox control systems must observe to interpret driver's intentions in the best way, taking into account the power requested by the road profile and speed.

We assume that the gearbox adopts an electrohydraulic control system where engagements and disengagements are made by hydraulic actuators fed through electronically controlled electrovalves. This technology is common to new automatic gearboxes, even if some applications exist on the market where also control functions are performed by hydraulic logic circuits.

A typical electronic control system for an automatic powershift gearbox with epicycloidal gear trains receives the following magnitudes as inputs:

- speed selector position; typically positions P, R, N, D exist which indicate situations for parking, reverse speed, neutral position and automatic forward drive; selectors can also show:
  - fixed speed positions (i.e. 1, 2, 3, 4) to indicate driver's will to lock the gearbox in one of the speeds or, alternatively, to limit automatic shifts up to one of these speeds;
  - selectors to choose different automation programs such as W for winter (when ice and snow are on the road the first speed is inhibited and shifts are slower), E for economy to show the driver's will to drive at low fuel consumption, S for sport driving, etc.
- +/– positions to use the gearbox sequentially, increasing or decreasing speeds of one unit at a time; in this case additional commands may exist on the steering wheel;
- engine speed;
- throttle position in petrol engines or pumped quantity on diesel engines; more recent engine that have a drive-by-wire regulation system and this signal can by accelerator pedal position or some calculated magnitude indicating the required power;
- cruise control position;
- torque converter turbine speed;

- gearbox output shaft speed;
- accelerator kick-down position;
- oil brake pressure or stop light switch, indicating a braking situation;
- gearbox oil temperature; this sensor is particularly important because oil actuators response is influenced by viscosity and therefore temperature;
- engine coolant temperature, showing the engine readiness to deliver maximum performance;
- driver's door opening to inhibit the gear selection from the outside or other safety signals.

The control system manages normally the following output actuators:

- on/off electrovalves, as many as hydraulic lines for brakes and clutches actuation; valve number may be reduced combining more functions on the same valve;
- proportional valves, as many as the functions to be contemporarily actuated with regulated pressure; at least one for the next speed engagement and one for the torque converter lock-up clutch, on powershift gearboxes;
- engine control system communication line, to regulate engine torque during speed shifts when critical to clutches life;
- instrument panel communication line to show selected gear and possible malfunctions;
- selector interlock to inhibit N/D or N/R or P/R shifts if the brake pedal isn't depressed; the vehicle could in fact move if isn't braked when torque converter is applied.

Input and output list will be surely implemented as control system functions will increase for number and complexity. Many inputs used by gearbox control system could be already available for other vehicle controls; in this case signals could be gathered by a serial communication bus.

15.6.1 Speed Selection for Minimum Consumption

The first function, to be performed by control systems, is speed selection; speed selection doesn't have a unique definition because available speeds in certain engine operation regions could be many.

Let's for instance consider the diagram on Fig. 15.32, where available power curves at different gearbox speeds and power necessary to motion at different slopes are shown as a function of car speed; at zero slope, between 15 and 20 km/h the engine can work in first and second speed; between 40 and 60 km/h all speeds are usable.

Speed selection must take into account the requested acceleration too; to deliver for instance 50 kW at 50 km/h the first speed is necessary, but to deliver 40 kW only first and second can be utilized. To define gearbox speed in an unambiguous way another condition must be set, that we assume to be minimum fuel consumption.

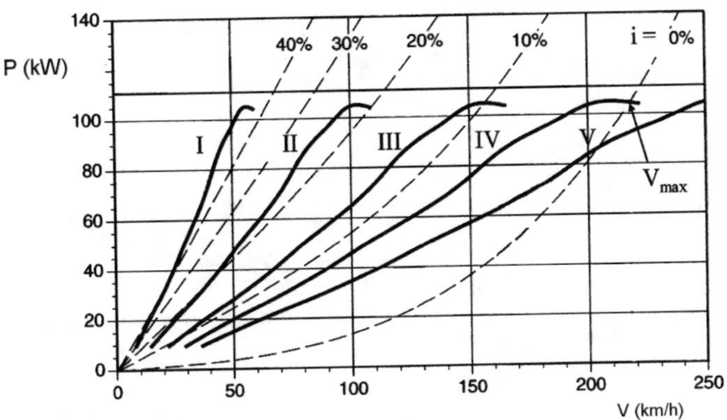

Fig. 15.32 Diagram representing available power at different gearbox speeds to accelerate the vehicle on an assigned slope at a given vehicle speed; available power is the distance between one of the thick lines and the selected slope curve

If we recall the specific fuel consumption map on Fig. 15.23, we see that the minimum consumption curve is a bit lower than the maximum power one. At minimum engine speed, if constant versus load, the minimum consumption curve is a vertical segment and the engine must be regulated on load only.

If we imagine drawing the map of Fig. 15.32 on the diagram of Fig. 15.23 and if we consider the already defined transmission ratios, we can say that first speed is suitable for minimum consumption within the working area limited:

- on the left by maximum power available in first and by minimum feasible engine speed in first and
- on the right, by minimum feasible speed in second, by maximum available power in second and by maximum engine speed in first;

same conclusions can be drawn for all speeds till the fifth, where one of the boundaries is given by the power absorbed by driving resistance.

Desired power isn't in reality an available input to the control system; in addition to that drivers aren't used to think in terms of power, but act on accelerator pedal by increasing or decreasing its stroke after comparison of actual performance with desiderate one.

In fact, when drivers think about accelerating vehicle, they depress the pedal until the desiderated feeling is obtained; if maximum acceleration is desiderated they push the pedal all the way down.

Let's consider, for example, a petrol engine with throttle valve and mechanically linked accelerator pedal; the torque delivered by the engine isn't very linear on the engine speed as we can observe on Fig. 15.33 on different diagrams at different throttle angles α. The same torque can be offered at different speeds and different throttle angles.

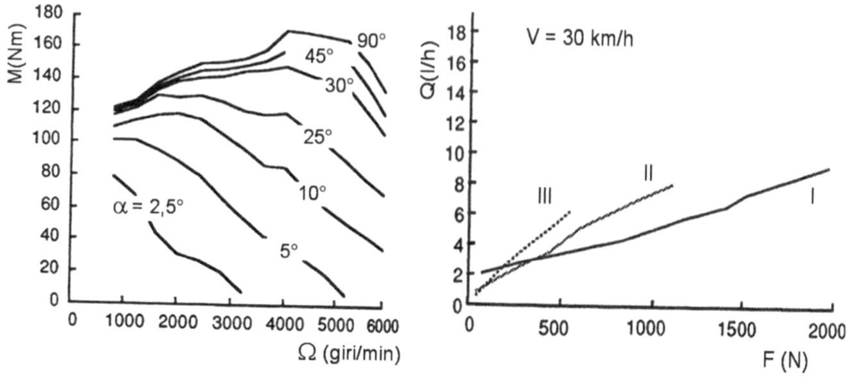

Fig. 15.33 Map of the torque delivered at different throttle angles (on the left); with simple elaborations it is possible to calculate the fuel rate curves necessary to obtain a given traction. Starting from the intersection of these curves it is possible to obtain throttle angles at a given speed where shifting speed is convenient

If we imagine to elaborate the fuel consumption map as to build up, for each vehicle speed (on Fig. 15.33, on the right, the example for 30 km/h is shown), the traction diagram as function of fuel rate, we can draw a set of curves for each of the available transmission ratio at that vehicle speed.

Intersections between traction curves show shift points to keep the consumption at a minimum.

Shift point throttle angle can be easily calculated.

From fuel consumption map and throttle valve map, by repeating this process for a sufficient number of vehicle speeds, it is possible to draw the curves of Fig. 15.34 which limit the optimum utilization regions of each gearbox speed on the throttle valve versus vehicle speed plane; both magnitudes are known to the control system.

To avoid, finally, any oscillation between two neighboring gearbox speeds, the obtained curves are now interpreted as upshift curves; other curves are moved to lower speeds for downshifts (in grey). Within the area between curves actual speed is maintained.

15.6.2 Speed Selection for Comfort

If we consider, instead, the problem of defining the most suitable shift speed for comfort, we must request that traction at a given gearbox speed is the same as for the next speed, at the same accelerator pedal position; as a matter of fact, with an automatic gearbox, the driver doesn't foresee shifting time and doesn't move, usually, accelerator pedal during speed shifts.

Let's consider again a petrol engine; the case of a diesel engine isn't substantially different. We transform now throttle angle map into a family of traction curves at the

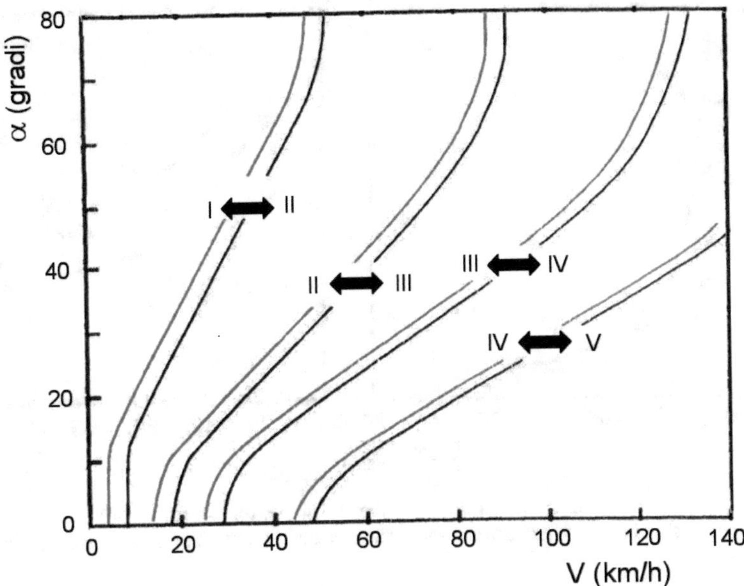

Fig. 15.34 Qualitative diagram of optimum shift curves for minimum consumption; the black curves are for upshifts, the grey ones for downshifts

same throttle angle (we produce, in other words, the traction map as a function of speed, having throttle angle as parameter); we can have two different situations that are explained in the following examples.

On the left side of Fig. 15.35 we see the curves family for a small throttle angle, of 5° (usually the wide open position is at 90°); curves show an intersection that represents the most suitable point for shifting.

In the case of an opening of 45° (right side), the curves don't have any intersection; in this situation the optimum shift point should be set where curves have minimum distance; at wide open throttle shift point is positioned at the maximum engine speed.

If we group the optimum shift points on a throttle angle diagram as function of vehicle speed, we obtain a curves family similar to that of previous paragraph.

The minimum consumption curves family is usually shifted to lower speeds as compared with that of maximum comfort, except maximum power point (wide open throttle) where the two families are coincident.

15.6.3 Definition of a Compromise Choice

Engaged speed choice on a gearbox can be made according to different approaches.

A first simple approach is compromising between the two criteria, defining a curve family in between. But, at very low vehicle speed the jerk produced by lowering the

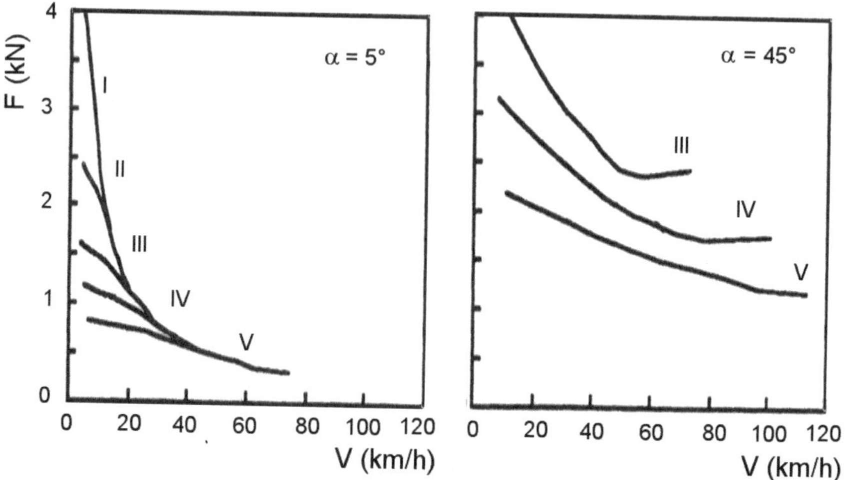

Fig. 15.35 Traction curves family at the same throttle angle, for two different values of this last parameter. At small angle values curves show an intersection, at higher angle values intersection may not exist

shift speed may be disappointing, because acceleration is modest; in addition to this, damage on fuel consumption may be of some percent if leaving optimum condition.

The most diffused approach was, in the past, assigning the choice responsibility to the driver, by introducing a program selector which determines the two alternatives. The economy option can be selected, for instance, in city traffic or slow suburban traffic, while comfort option is for different situations.

This approach is questionable because introduces an additional command without immediate feed-back, which can be easily forgot in the wrong position.

On more modern control systems, algorithms are implemented that recognize driver's intention and driving environment by quick statistic elaborations on operation parameters already known to the control system.

15.6.4 Speed Choice in Real Driving Conditions

A fixed shift speed curve at a given vehicle speed and throttle angle doesn't represent an acceptable choice in conditions different of driving on a flat road with no traffic.

Let's assume that the vehicle is moving on a flat road on a high gear ratio. The operating conditions are represented on Fig. 15.36 by the point 1 of a diagram; the dotted curves represent the traction force F_x at the wheel, as a function of the throttle angle α.

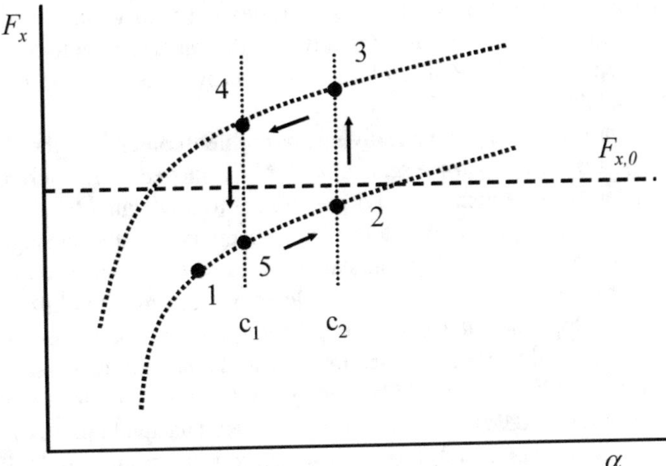

Fig. 15.36 Typical gearbox speed oscillation cycle occurring when desiderate traction value is between feasible values of two neighboring speeds at the same throttle angle

Let's assume also that the car is facing a sudden slope, rising traction to $F_{x,0}$, as shown on the diagram; the driver will feel the car slowing down and react increasing throttle angle up to the corresponding value of point 2, where a downshift is started.

When the shift has been completed the driver must probably release the accelerator pedal, in the attempt to reach the desiderated value, starting a new upshift that reduces again traction force.

A series of not requested speed shifts will occur (cycle 2, 3, 4, 5), very well known to people having driven a car on hilly roads at imposed low speed.

The solution of the problem should be to temporarily move the shifting line to a lower throttle angle as to reach the desiderated traction value on the lower speed, as it is accomplished on a manual gearbox.

A similar situation occurs when on an uphill road a sudden short descent is met or when a curve is met on a flat road. In the first case the accelerator pedal is released and the next speed may be shifted with a following upshift when the uphill road restarts. With a manual gearbox the speed wouldn't be changed. On the sudden curve, the same situation could occur.

The above inconvenience can be avoided when setting the selector to a manual condition as to maintain the low gear.

The solution isn't satisfactory because implies additional operations and because drivers may forget the selector in the wrong position, particularly if a high gear is chosen.

As an alternative, drivers expert of automatic gearbox increase distance with preceding cars as to avoid any action on the accelerator pedal; power reserve for acceleration will decrease, making overtaking more difficult and sometimes reducing average speed.

Manual gearbox drivers can see the oncoming descent or curve and decide whether to maintain or to change speed; this possibility isn't available to automatic gearbox control systems; the diffusion of satellite navigation systems could, in future, leave this option available.

A simpler and effective system of solving this problem comes from the observation that on a hilly or bending road the accelerator pedal is moved much more frequently and with ampler displacements, if the driver desires to drive quickly.

The presence and the value of a slope may be estimated of throttle angle average value and variance. By comparing calculated values with reference one that can be achieved empirically, is possible deciding to lower the downshift points.

Another road slope indicator may be engine torque; it can be guessed by control systems by a statistical analysis of injection time and other engine parameters.

A sport driving behavior can be detected by the throttle angle first derivative.

If after accelerator depression, vehicle response isn't adequate (the existing gearbox speed has been maintained) a second actuation step on the pedal will follow; this event can be an indicator of the driver's lack of satisfaction.

This situation may suggest adopting a higher upshift curve; in the same way the persistence of low throttle angle speeds may suggest to return to lower speed shifts more suitable to fuel consumption.

Another particular situation may occur when the accelerator pedal is released to obtain weak slow down, strong slow down or slow down with brakes usage. In the first case it is better to leave the actual gearbox speed or to select a higher one to lower fuel consumption; in the second and third case a downshift may be necessary to enhance the engine braking force.

The first behavior is already implemented in the shifting lows of the previous paragraphs.

The second behavior can be implemented by considering brake pressure and throttle angle first derivative; a desire for stronger slow down is characterized by higher accelerator release speeds.

15.6.5 Brakes and Clutches Actuation

A last problem to be tackled with the control system is clutches or brakes actuation, involved during speed shifts and torque converter lock-up clutch actuation.

In previous paragraphs we have defined shift conditions for maximizing comfort or minimizing consumption; above conditions still don't consider shifting transient generated by clutches.

It is necessary that clutches disengagement and engagement, involved in speed shifts, follow certain criteria to avoid disturbing jerk during the manoeuvre.

Let's consider the example of Fig. 15.37. An epicycloidal gear train is represented with four different speeds; we consider only the first to second speed shift in order to understand what is going on during this period of time.

In first gear the band brake B_I is closed.

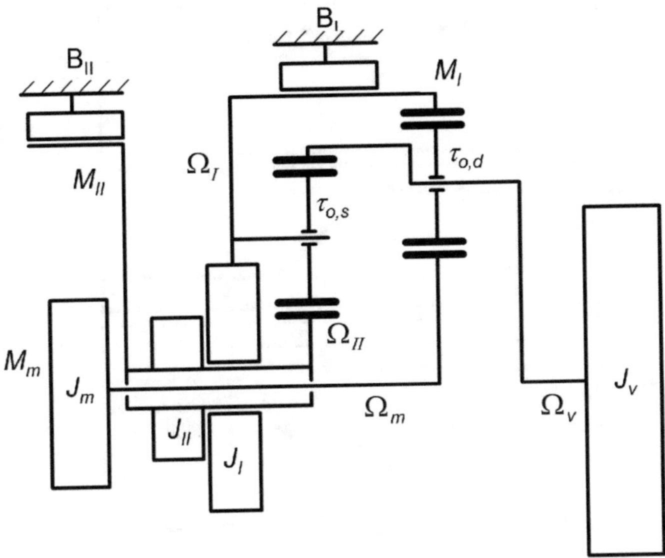

Fig. 15.37 Epicycloidal gear train scheme suitable for four speed; the rotating masses involved in the shift from first to second speed are shown

Let us consider the ordinary transmission ratios of the left and right trains, called $\tau_{o,d}$ and $\tau_{o,s}$ respectively.

Subscripts m and v show elements turning at engine speed and at vehicle transmission shaft speed; for sake of simplicity torque converter is ignored.

In first gear the band brake B_{II} bell that will be actuated to obtain the second speed is turning at a speed given by:

$$\Omega_{II} = -\Omega_v \tau_{o,s} \ . \tag{15.16}$$

In second gear the brake B_{II} is closed why the bell of the band brake B_I will be left free to turn.

The shift procedure from first to second will start working on brake B_{II}; the engine, gearbox and vehicle system will be slowed down by braking moment application but will receive a positive contribution of the partial transformation of brake B_{II} rotating parts kinetic energy. They are modelled with flywheel J_{II} while the engine is modelled with flywheel J_m; in the same way a positive contribution will come by the torque reduction of brake B_I, and a negative contribution will come of the absorbed work to accelerate flywheel J_I to the speed:

$$\Omega_I = \Omega_v + \frac{1}{\tau_{o,s}(\Omega_m - \Omega_v)} \ . \tag{15.17}$$

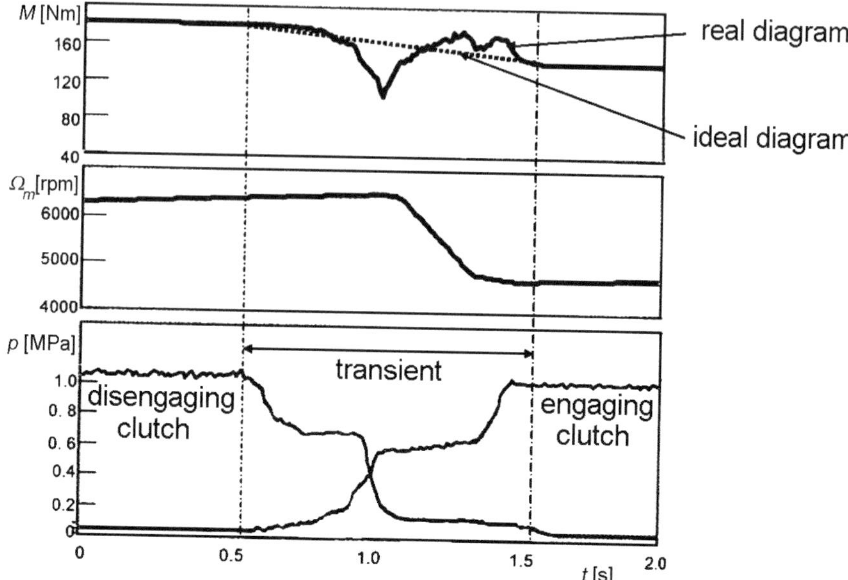

Fig. 15.38 Diagram versus time t of the most important magnitudes involved in an upshift; from the top gearbox output torque M (real and ideal values), engine speed Ω_m and oil pressure p on brakes are represented

Input magnitudes for a good quality upshift are therefore many; on Fig. 15.38 is reported as an example a typical diagram of most relevant magnitudes as a function of time, for the first to second upshift at wide open throttle. Engine speed is increasing before and after shift, if supplied power is greater than resistant one; engine speed will decrease as a consequence of transmission ratio decrease.

Gearbox output torque will be reduced as a consequence of the upshift; an ideal solution, with the aim of reducing jerk, will be to have a torque diagram made with a straight line joining the torque values before and after shift point; the real value is represented by the oscillating line.

In the two hydraulic circuits for clutch or brake actuation there will be:

- a decreasing pressure versus time on elements locked on first speed;
- an increasing pressure versus time on the element to be locked for the second speed; the two diagrams should overlap with a precise criterion to minimize traction variation.

The curve shape of pressure rise and fall is assigned to proportional control valves to obtain a traction curve continuous shape.

Optimum actuation lows for brakes and clutches must be determined for each speed shift as a function of torque and speed. Mapping actuation functions can be a too heavy work in terms of experimental activity and of control system memory occupation; mathematical models could be more convenient.

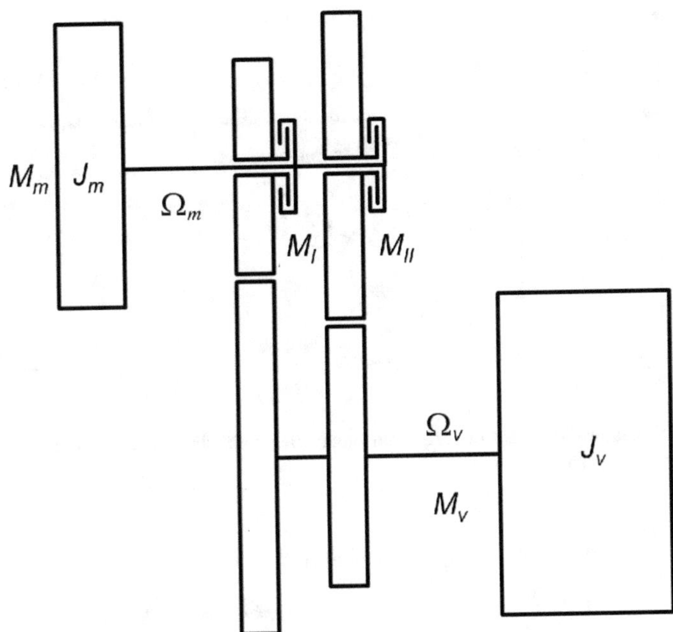

Fig. 15.39 Scheme for the study of speed shift transient in a powershift gearbox with fixed rotation axis

While designing these models it must be taken into account that gearbox output torque is not only determined by input torque, but also by gearbox rotating elements inertia, which can't be neglected, particularly for epicycloidal gear trains.

The best result, in terms of jerk reduction, can be reached when clutches (or brakes) pressure can be regulated proportionally and independently; acceptable results, such that shown on the figure, can be obtained with a continuous regulation of the closing element and with a pulsed discharge of the opening one.

Also in case of non epicycloidal gear train gearboxes (fixed rotation axis and dual clutch) the inertia contribution isn't negligible.

Let's consider a two speed gearbox, as represented by the scheme in Fig. 15.39; M_I and M_{II} are the instantaneous slip torques (function of time) of clutches B_I and B_{II} involved in the transmission ratio variation from τ_I to τ_{II}. The apparent vehicle rotating mass is represented by J_v while J_m is the equivalent inertia of the engine, M_m is the engine torque and M_m' the driving torque at the gearbox output shaft.

We assume that during shifts there is no important change in resistant torque. Engine torque can be sometimes considered as being constant too; in most general case, M_m is a known function of Ω_m, if the throttle valve angle remains unchanged during the manoeuvre.

We assume finally that the clutches actuation forces are linear on time.

On Fig. 15.40 we can draw a schematic diagram of the phenomenon; we can write:

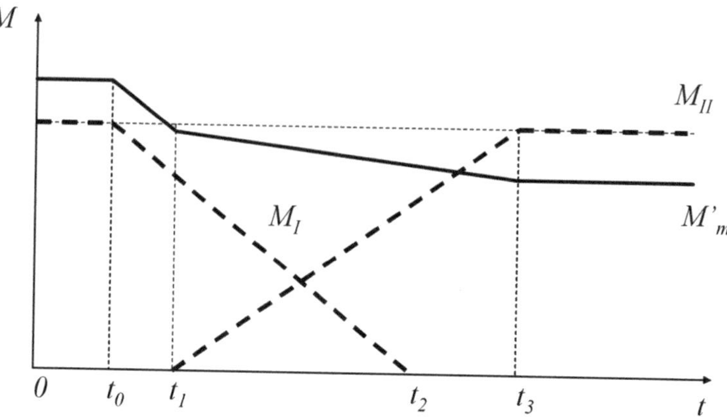

Fig. 15.40 Schematic diagram of torque on engagement clutches of a countershaft powershift gearbox during an upshift

$$\begin{cases} J_m \frac{d\Omega_m}{dt} = M_m(\Omega_m) - M_I(t) - M_{II}(t) \\ J_v \frac{d\Omega_v}{dt} + M_v = \tau_I M_I(t) + \tau_{II} M_{II}(t) = M'_m \end{cases} . \tag{15.18}$$

The derivative of the second equation is the jerk; it could never be zero because the M'_m torque will change at least because of the change of the transmission ratio. We can understand that the minimum jerk is achieved if M_I and M_{II} are linear on time and when t_1 and t_0 are equal.

Piston filling and emptying times are not only determined by their volumes and by orifices dimensions, but also by:

- oil circuit pressure, which is determined by pump speed (rotating with the torque converter pump) and by oil viscosity, function of temperature; pressure is in fact determined by the oil internal spillage in the pump;
- clutches rotational speed, because, very often, the centrifugal pressure field is not negligible;
- friction coefficient on clutch working faces, dependent on wear and on temperature;
- oil viscosity and temperature that condition filling and emptying time at a given pressure;
- dimension tolerances of the gearbox.

For this reasons the development of a reliable mathematical model can be particularly difficult. It is a good practice to use a first approximation mathematical model, integrated by feed-back parameters; they can be considered from time to time to compensate for the tolerance or, more frequently, for temperature effects.

The ideal feed-back parameter should be output torque, but at this time viable torque sensors don't exist; but torque can be estimated by accelerations, once inertia masses are known.

A further parameter to be used to minimize acceleration variations is controlling engine torque too, through spark advance or, if possible, directly on the throttle.

It is also important to eliminate some of the variables of the system, by regulating oil pressure on a constant value, eliminating part of the effects of temperature and viscosity.

Some few words must be written on lock-up clutches; to their actuator apply all considerations made for clutches and brakes.

The lock-up clutch must be engaged when converter slip ratio goes to small values, indicating that torque multiplication is no more necessary.

The clutch engagement eliminates damping capacity too, which is quite useful particularly for low speeds where the vehicle equivalent rotating mass is high.

On old gearboxes lock-up clutch was only actuated on top gear. A higher attention given on more recent models to fuel consumption, forced to use lock-up clutch in all gearbox speeds; the benefit on consumption is of some percent.

In this case the actuation pressure must be carefully controlled as a function of torque also during normal operation and not only at the engagement, to allow a small slip when torsional vibrations may occur.

A damper similar to that of a conventional start-up clutch is in this case recommended.

Chapter 16
Design and Testing

In this chapter we will discuss about design rules and procedures for those transmission components, we have not considered in previous chapters yet, and about test methods applied for validation and qualification.

We shouldn't forget, looking at the essential nature of reported information, that the objective of this manual is only to offer some complement to the knowledge gathered in engineering courses of machine design and calculation; we refer back to them for the fundamental know-how.

16.1 Transmission Mission

We will describe in this paragraph the operating condition of the transmission during the vehicle's life, to understand qualitatively which loads are applied and how long they must be withstood without damage.

The transmission plays a top importance role on the vehicle system operation and, therefore, must have an average useful life at least as long as that of the vehicle itself; a suggested maintenance program must be obviously attended.

Transmissions must not only preserve their structural integrity but also their functional characteristics, where ease of use, we have described writing on shifting mechanisms and synchronizers, mechanical efficiency and generated noise are the most relevant.

Transmission life, as other vehicle systems one, cannot be described by deterministic terms, but only by statistic terms, because loads coming from road and driving stile have statistic nature.

Therefore endurance specification also, normally adopted by manufacturers, is assigned statistically through the magnitude B_{10} that corresponds to the endurance which isn't reached by the 10% of produced transmissions.

G. Genta and L. Morello, *The Automotive Chassis*, Mechanical Engineering Series,
https://doi.org/10.1007/978-3-030-35635-4_16

The *endurance* is the maximum distance traveled according to the foreseen mission, without any major damage.

If we take into account the recent product evolutions, we shouldn't only include in the major damage category the failures that interrupt vehicle motion, but also all those that are usually claimed by customers at the service shop, as:

- noise increase,
- shifting loads increase,
- lubricant spills, etc.

As reference values for B_{10}, we can assume:

- automobile transmissions $>$ 150,000 km,
- construction trucks $>$ 300,000 km,
- urban busses $>$ 400,000 km,
- long haul trucks $>$ 800,000 km.

The final result it should be obtained is clear enough; it must be said that the quantitative specification of transmission life is made difficult by working conditions which are always changing and sometimes unforeseeable.

If we only cover the automobile case, we can say that in Europe the life is shared in these four typical environments:

- 40–70% on motorways,
- 15–30% on suburban roads,
- 15–20% on urban roads,
- 3–10% on mountain roads.

Each manufacturer promoted independently test campaigns addressed to obtain reliable date on usage of cars; the results of these tests are an important part of the company know-how and are published very seldom.

These data acquisitions must be made for different groups of homogeneous customers, for different car types and must be repeated at different times, to take into account evolutions in living habits, in road network, in traffic and in car preferred types.

It is usual to employ data gathered of special costumers' fleets, such as taxis, company cars, etc., which allow data measurement on their cars (through inboard digital recorders), in change of favorable selling prices.

A typical data acquisition includes time histories of car speed, gear shift position, clutch position, accelerator position and road slope; they are statistically synthesized to obtain information easier to be used. From these many other data are derived, such as those related to transmissions; transmission operation is identified by input torque, engaged speed and clutch pedal position.

The time history of these parameters can be used as reference mission for design calculations and experimental endurance validations.

As an example, in Fig. 16.1 data elaborations results are reported, for a car and for a medium duty truck, showing the percentage of use of available gearbox speeds.

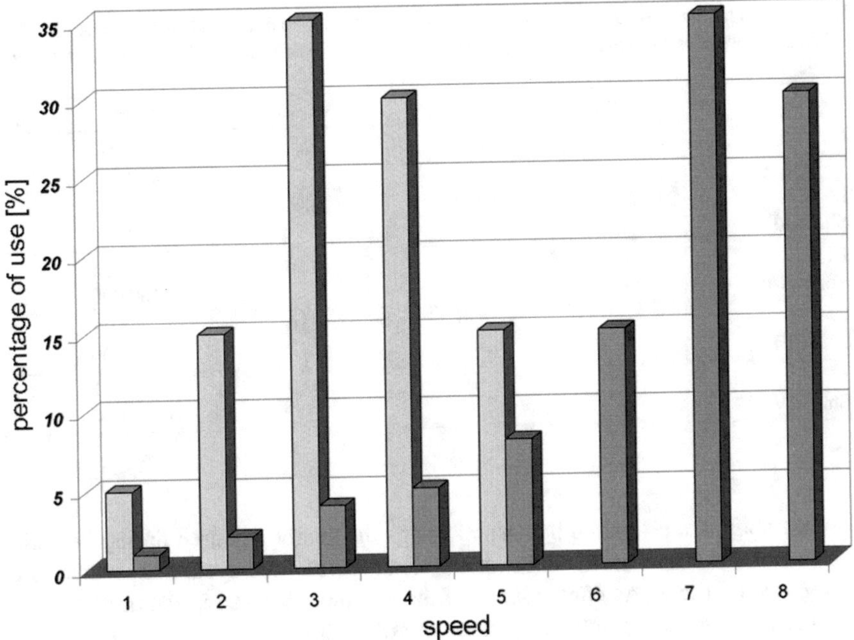

Fig. 16.1 Qualitative diagram of use percentage of the different speeds, in the average life of a car (dotted bars) and of a medium duty industrial vehicle (dashed bars)

The input torque is strongly changing versus time and can also be negative; all structural transmission components service life is therefore limited by fatigue.

Gears and other rotary elements introduce additional load variations versus time.

Table 16.1 shows a qualitative description of the transmission life, after having accepted some simplifications.

As a first assumption, to stay on the safe side, only mechanisms stressed by I and II speed are designed for a limited life; mechanisms stressed by III, IV and V speed are designed for unlimited life; since they are basically made by steels a life of 100 million cycles is assumed.

As far as the reverse speed is concerned a life as long as one quarter of the I speed life is usually assumed.

The endurance requested for the I and II speeds are simply derived by multiplying the target life of 200,000 km by the percentages shown in Fig. 16.1.

Load conditions are simplified and made much more severe by assuming that the car is working:

- 70% of the time, at maximum torque and at the corresponding engine speed;
- 20% of the time at maximum power and at the corresponding engine speed;
- for the remaining 10% of the time with the engine braking the car with a torque as big as half of the maximum torque and at maximum engine speed.

Table 16.1 Qualitative description of the average use condition of a car transmission. The life spent in I and II speeds is measured in (km), that spent in III, IV and V speeds is measured in million cycles

Speed	Max torque	Max power	Braking	Total
I Shifts N/I	5,600	1,600	800	8,000 260,000
II Shifts I/II	21,000	6,000	3,000	30,000 180,000
III Shifts II/III	70	20	10	100 100,000
IV Shifts III/IV	70	20	10	100 20,000
V Shifts IV/V	70	20	10	100 20,000

Gear shifts are estimated by assuming that the gearbox is used during its entire life of 200,000 km as in the fuel consumption and emission cycle that is, therefore, driven 20,000 times. As consequence of the assumption made in the third points a number of downshifts is added equal to 10% of upshifts.

These load conditions can be used for designing gears, shafts, bearings, housings and seals, while the number of up and downshifts can be used to design synchronizers, shifting mechanisms and clutch.

This table should be improved with an accurate statistic study, for the application under exam, but can be considered as a starting point for a preliminary design of a car transmission.

Input parameters acquired during real driving conditions are useful to estimate component average life; as a good praxis, we recommend to consider also some exceptional condition that may occur when transmissions are used in a non correct but possible way; in this conditions also, catastrophic failures are not accepted.

Also these test procedures are part of manufacturers' know-how; they are performed driving the car according to very heavy driving schedules that condense the car useful life in short time; a typical test of this kind consists, for example, in releasing the clutch pedal suddenly as it may happen while wearing muddy and slippery shoes.

The transmission components are classified for design process, according to three different categories:

- A category, including components that are critical to correct transmission operation and can be calculated following reliable and validated procedures (shafts, gears, bearings);
- B category, including components that are critical to correct transmission operation, but can't be calculated following reliable procedures (synchronizers, lubricants, seals, etc.);

- C category, including non critical components, because previous experience doesn't record failures (circlips, venting valves, housings, etc.).

B category components are designed in similitude to previous successful examples; the prototypes of these components must be extensively tested on bench and vehicle, to assure their reliability before building the first transmission prototype.

It is therefore very important, for an effective management of development plans, that all components accomplish the experimental validation process with the same level of confidence on success.

All components, including those which benefit of a reliable design process, will be tested on a real transmission on a bench and finally on a real vehicle, as soon as they have reached their final design, to demonstrate the achieved reliability.

Reliability is the probability of surviving a certain mission without failures.

Reliability demonstration must be made on a sufficiently wide prototype sample; these must be built with their critical characteristics on the border of their tolerance field in a real production process.

The design of these tests, oriented to guarantee a sufficient confidence level can be studied on more specific handbooks.

16.2 Gears

16.2.1 Endurance

Gear life is limited by four different tooth failure kinds:

- bending;
- pitting;
- scuffing;
- wear.

Figure 16.2 shows the qualitative extension of survival fields to these phenomena, on a diagram of transmitted torque versus circumferential speed.

Following issues are very relevant to determine the gear life:

- working conditions (transmitted torque, circumferential speed, oil temperature),
- tooth material,
- tooth geometry,
- surface treatment,
- lubricant formulation.

For spur gear wheels calculation, the standard procedure ISO 6336 is used by almost all car manufacturers.

Tooth bending failure occurs when a substantial part of gear wheel is removed; it is useful to separate failures due to overload from those due to fatigue.

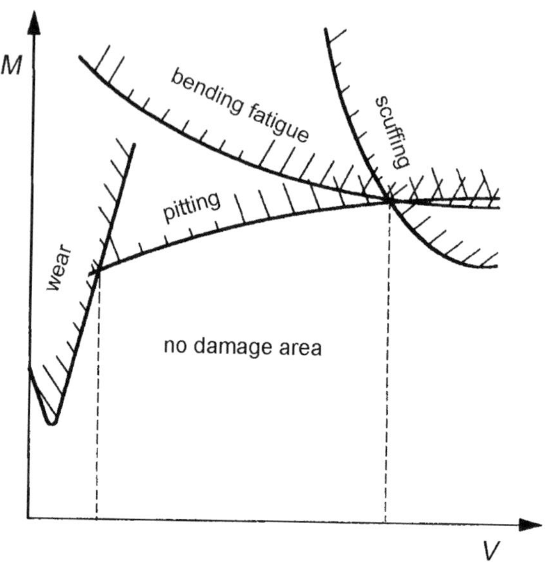

Fig. 16.2 Qualitative diagram of the transmitted torque M as function of circumferential speed V showing the typical shape of operation boundary curves for bending fatigue, pitting, scuffing and wear; the area below curves envelope corresponds to operation without damage danger; in consideration of the most diffused speed values, pitting is the most important limiting condition

Failures of second kind have the aspect shown on Fig. 16.3 and show very clearly a point where the failure started (a surface defect, a notch or a non catastrophic overload failure).

A set of lines which are almost concentric to the starting point, show the propagation direction of the failure. The failure surfaces are polished by the continuous shock of the two parts, due to load pulsation.

A failure terminal part can be seen, which is characterized by a very irregular surface with crystal aspect, due to a sudden rupture when the resistant section is too much decreased.

Overload failures have the aspect of the terminal part of fatigue failures.

The stress variation at the origin of the fatigue phenomenon is due to the following causes:

- low frequency torque variation; torque can also change direction;
- cyclic teeth mesh and number variation of teeth in contact, at medium frequency;
- engine torque harmonic content, at high frequency.

By their nature, idler are working under alternate loads.

Pitting damage is shown by the appearance of growing craters which can join each other. They are due by failures under the tooth surface caused by too high hertzian contact pressure.

The aspect of this failure can be characterized by very small craters, which give the surface a velvety aspect, as in Fig. 16.4 or by big and irregular craters as those on Fig. 16.5; the aspect difference is caused by the severity of the phenomenon in question.

Fig. 16.3 Aspect of a tooth flank damaged by bending fatigue

Fig. 16.4 Aspect of a tooth flank damaged by pitting

It must be remembered that pitting phenomenon occurs on lubricated surfaces only; therefore lubricant characteristics play a major role.

The low speed wear is characterized by abrasions on tooth flanks, according to relative speed direction; it is very seldom on automotive transmissions, except with evident geometric errors or by lack of lubricant.

Scuffed surfaces aspect is quite similar two the previous one; the damage can be caused by a rupture of lubricant film, due to high temperature or high pressure. It

Fig. 16.5 Aspect of a tooth flank damaged by heavy pitting

can cause metal on metal contact; the aspect is caused by micro welds and rips or by chemical action.

The quoted ISO norm includes calculation procedures addressed to verify that none of the above phenomena occur, except scuffing.

On the diagram of Fig. 16.2 relative to automotive transmissions, the safe tooth operation area is essentially limited by pitting.

Because of applied load time variation, we must consider the phenomenon of fatigue at the design stage.

Therefore the so called material Wöhler curves must be considered; one of these curves is reported on Fig. 16.6.

The two lines (inclined and horizontal) on the right of the figure represent, on a bilogarithmic scale, stress amplitude (on the vertical axis) in function of cycle number leading to failure; there is, as known, a horizontal limit, the fatigue limit, below of which, according to a simplified theory, failure will never occur.

It is a good praxis, to consider not only material probe average failure curves, but also B_{10}, B_{50}, B_{90} curves, if not only average life, but also reliability has to be predicted.

In order to compare working stress with Wöhler limit, it is useful to transform, as a first step, torque time history into stress time history.

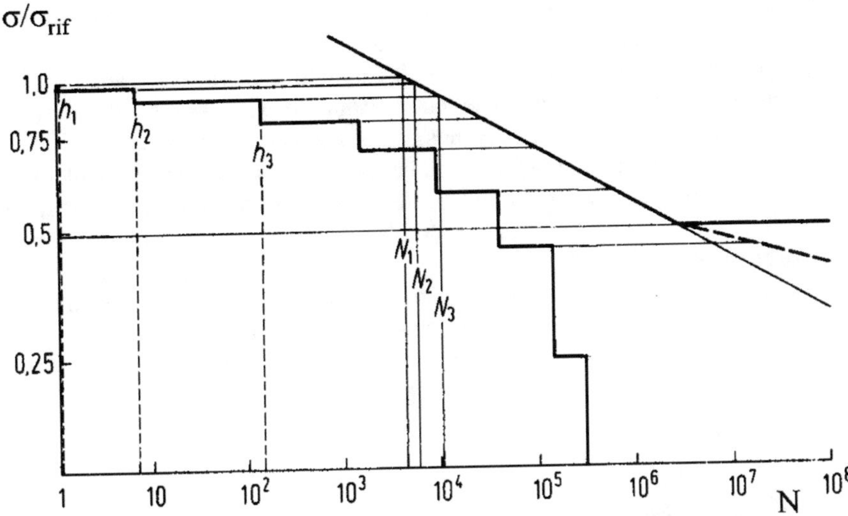

Fig. 16.6 Comparison of a Wöhler curve with an assigned load profile (staircase line). Stress is divided by a reference value; N_1, N_2, N_3 represent cycles number during which relative loads h_1, h_2, h_3 are applied

It is possible, at this time, to count cycles number at the same stress amplitude (h in the diagram) in such a way as to define a histogram in which each stress amplitude class is represented with the corresponding cycle number (N in the diagram); this histogram is the so called *load profile*.

Road profile, reordered in decreasing load amplitude classes, is represented by the staircase line below the Wöhler curve.

At this time the comparison of actual stress with allowed one is possible.

The rule that is normally applied in this comparison was developed by Miner; according to this rule, each group of same amplitude loads produces on the material a partial damage, equal to the ratio between actual load cycles and no damage cycles, on the Wöhler curve.

The same rule says that failure occurs when the sum of partial damages (less than one, by definition) is equal to one.

Miner's rule implies no damage for load amplitudes below fatigue limit; this hypothesis should be put under discussion, because if a partial damage already occurred, cracks have been already started; it is possible that also stress levels below fatigue limit are causing the damage to proceed.

To take into account this fact, some empirical corrections are used; according to those corrections the horizontal fatigue limit line is changed to an inclined one, whose slope is defined by the bisector line (on the bilogarithmic representation) between inclined and horizontal lines. It is reported on the diagram with dots.

While choosing in literature the appropriated Wöhler curve for gear material, it is important to take into account for the actual heat treatment (case hardening or nitriding, this last applied usually to epicycloidal gears) and the finishing of surface in contact that can be ground, shaved or honed.

A common characteristic of countershaft automotive gearboxes is adopting the same centre distance for all gears. The centre distance of a gearbox is decided by the most stressed wheels couple, usually those of first and reverse speeds.

The smaller stress for other speeds will be taken into account with a smaller face width; the ratio between face width and diameter is conditioned by the available space to install the transmission: for this reason this ratio isn't very different in different gearboxes (between 0.3 and 0.6).

The centre distance is therefore an engineering parameter indicating design value; it may be used to compare same torque capacity gearboxes.

16.2.2 Noise

Many different noise qualities produced by the gearbox are judged unpleasant and can draw driver's attention; they can be classified according to different categories and precisely:

- gearing *whistle*;
- gearing *rattle*;
- speed shift *grating*;
- bearing *whine*.

Whistle can be caused by three different reasons:

- pitch errors in contacting teeth and pitch variations due to contacting teeth deformation;
- tooth bending stiffness variation along the contact line; under this aspect helical spur gears show a better behavior as straight ones;
- undulations of contacting teeth surfaces.

Rattle noise is particularly disagreeable and is produced by idle gear wheels in constant mesh and by sleeves and synchronizers hubs; in Fig. 16.7 is reported, on the left, the case of a single stage gearbox and on the right the case of a double stage countershaft gearbox.

The engaged gear wheels are shown by their sleeves position; circles mark meshing points of rotating idle gears. On the figure top what happens on a conventional drive gearbox in idle speed is shown.

Constant gear puts countershaft in rotation. Because engine speed isn't constant, the countershaft will have torsional oscillations that will cause teeth flanks to hit in the two angular directions, left free by clearance.

A similar phenomenon may occur between sleeve spline and synchronizer hub and may be perceivable on idle couples, in direct drive or other speeds.

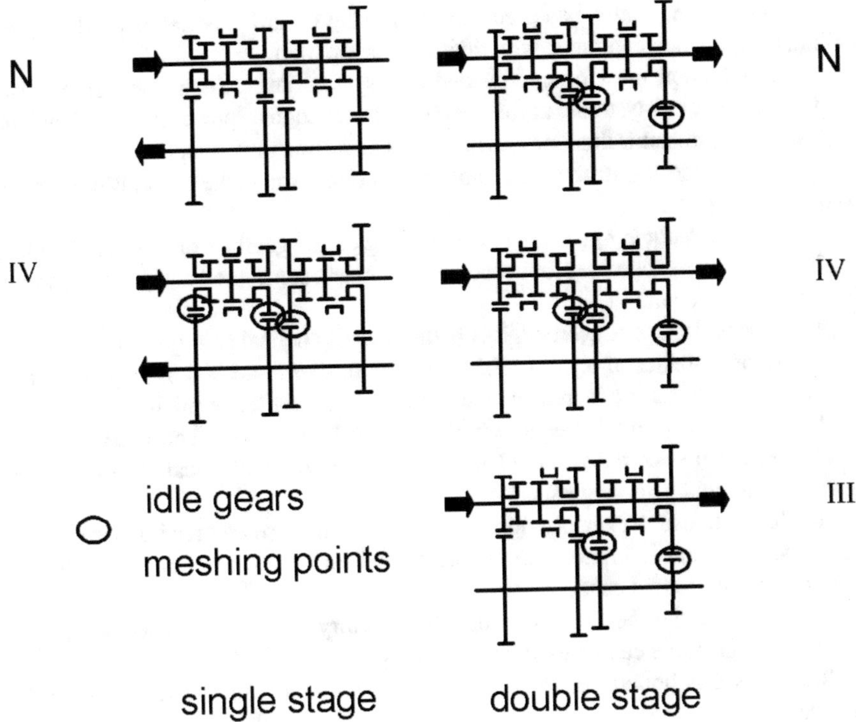

Fig. 16.7 Schemes of a single stage and a double stage countershaft gearboxes; circles mark meshing points of rotating idle gears, responsible for rattle noise generation

Single stage gearboxes, on the left figure, present the same kind of phenomena, except at idle speed and for wheel couples that have sleeves on the driving shaft; these will be in fact still.

Rattle noise at idle is similar to rattle noise in acceleration, but, due to the limited speed may produce a metallic like sound; during acceleration the sound has many more harmonic frequencies and is similar to the flow noise of fluids in pipes.

The probability to perceive this noise is reducing consistently at medium engine speed, because many harmonic frequencies exit the audible field and the noise is covered by other sources.

The influence of oil viscosity is notable, because when high, for example at low temperature, the vibration of the aforesaid parts is damped.

Grating noise is produce during speed shifts only, if the thrust force has overcome the synchronizer reaction to a premature completion of the shift manoeuvre.

This noise is caused by an inadequate synchronizer dimension or to a violent manoeuvre; it can be detected at slow speed and low ambient temperature, when the synchronizer cone doesn't succeed in removing oil film.

Bearing whine, finally, is a precursor of bearing failure only and is produced by rolling bodies when clearance with their runs is too high.

Noise reduction must be approached at a system level. To transmission noise contribute other subsystems, at the source, such as engine, powertrain suspension, elastic parts of transmission line and half axles and, sometime, suspensions and tires.

Air borne noise and structurally transmitted noise through the car body contribute to noise perception.

Thinking of reducing gearbox noise working on the gearbox only, could be misleading and cause excessive product cost; nevertheless some good design rules should be taken into account.

As far as whistle is concerned, it is better to apply helical spur gears only, with a coverage ratio greater or equal to 2.5; it can be further increased by applying high contact ratio profiles, by increasing the ratio of teeth height and pitch.

A profile correction oriented to achieving a slim tooth tip could be also beneficial.

Entire transmission ratios should be avoided; the preferential wear of ever coupled gears can modulate gearing sound.

It is better to use reasonably tight tolerances, such those included under IT5 and IT7 classes; tighter tolerance may be prescribed for gear wheels of most frequent use, to contain product cost.

Teeth surface must be smooth enough. Satisfactory results are reached by shaving and honing; grinding can be used exceptionally.

Rattle noise is bound, as we have seen, to circumferential clearance; it can be reduced by:

- reducing clearance;
- reducing inertia rotary mass of passive parts;
- improving lubrication; the better washed are hitting parts, the higher the damping action;
- by a more suitable tuning of the clutch damper, natural frequencies can be moved outside of a critical region;
- by adoption of a double mass damping flywheel.

Because noise is irradiated partly through the air and partly through the structure, it is a good practice to stiffen housing panels as much as possible, by suitable ribs and to increase local stiffness of interface points with the powertrain suspension.

16.3 Shafts

Gearbox shafts are made with many diameter changes, as they must allow fitting many different parts; it may also happen that some tooth profile is directly cut on them, because of small diameter.

Figure 16.8 shows the details of a shaft for a front wheel drive gearbox with transverse engine. The many diameter variations must be smoothened with chamfers

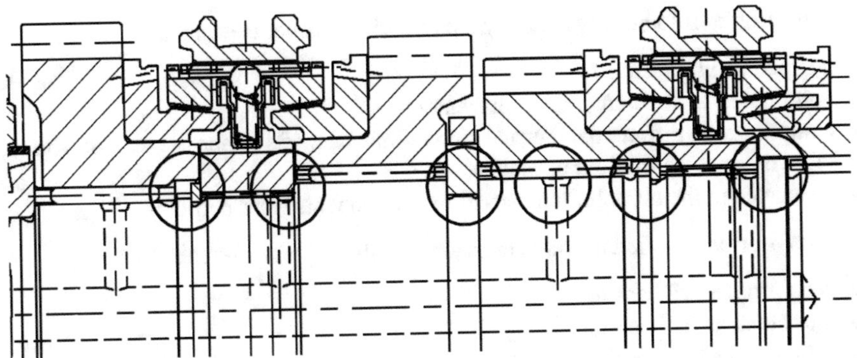

Fig. 16.8 Some example of notching effect reduction on diameter transitions and transverse drills of a shaft of a single stage gearbox for front wheel drive (FIAT)

or with edges rounding, to reduce torsion and bending stress concentration, even reducing resistant section.

Machine design manuals report static form factor for the most widely diffused section transitions; these factors are coefficients by which the stress obtained by the De Saint Venant theory application must be multiplied, to obtain real stress.

Shaft calculation can be performed following the same procedure we have suggested for gear wheels, to take into account fatigue phenomenon and changes in load amplitude.

While designing shafts, it is vitally important to introduce the displacement calculation due to load application; as a matter of fact, displacements (both linear and angular) may change teeth working conditions, with negative impact on noise and useful life.

From displacement study, some good design rules arise than can be summarized as follows:

- to reduce the shaft span between bearing, by limiting as much as possible gear wheels and synchronizers width;
- to install the wheels subject to highest loads as close to the bearings as possible;
- to avoid too steep diameter transition, organizing components by increasing or decreasing diameters;
- to avoid feather keys, preferring spline connections;
- to smoothen diameter transitions and drilling, as shown on Fig. 16.8;
- to use circlips at the ends of the shaft only.

16.4 Bearings

Gearbox bearings are normally roller bearings; bush bearings are limited to lowly stressed idle gear wheels and to sliding bearings for internal shift mechanisms.

If possible, bearings are limited to two per shaft, avoiding hyperstatic mountings, too sensitive to machining tolerances.

The problems to be considered when choosing a roller bearing are:

- adequate component life;
- compliance to shaft angular displacements;
- compliance to differential thermal elongations of shafts (made of steel) and housings (made of aluminium or magnesium);
- resistance to the oil pollution caused by components wear particles.

The bearings applied to shafts ends are mainly of four different types:

- deep groove ball bearing;
- four contacts ball bearings;
- cylindrical roller bearings;
- tapered roller bearings.

Deep groove ball bearings are widely applied because they can withstand both radial and axial loads. They are easily assembled on the shaft, don't request position adjustment and are reasonably cheap; on the contrary they have the disadvantage of large dimension and oil pollution sensitivity.

In consideration of this point, sometime self lubricated sealed ball bearings are preferred, also if gearbox splashed oil is abundantly available.

Four contact ball bearings are almost equivalent to previous ones, with the advantage of smaller dimensions at a slightly increased cost.

Cylindrical roller bearings have a high radial load capability but can't withstand axial loads; they are usually coupled to ball bearings at the other end of the shaft. They have the disadvantage of a higher cost and not working correctly, when substantial angular displacements are foreseen.

Tapered roller bearings are more and more extensively applied because of their optimum radial and axial load capacity (see for reference Fig. 9.11). They have limited dimensions; their assembly on the gearbox is difficult, because inner race with roller cage and outer race are separable components.

They require, for a correct operation, an appropriate axial preload, which must be maintained at every work condition. The axial position of at least one the two bearing must be adjusted, to compensate for the length tolerance of shaft and housing; in addition to that, one of the two shoulders must be conveniently made, in such a manner as to be insensitive to thermal displacements. This result can be achieved by a spacer made with high thermal expansion material.

Bearings on idle gear wheels are mostly needle bearings; sometimes the races are the outer surface of the shaft and the inner surface of the gear wheel, to reduce radial dimensions.

16.5 Lubricants

The correct lubricant specification is vitally important for gearbox endurance, considering the manifold functions it is expected to perform; the lubricant functions are the following:

- reducing friction and wear of metallic and non metallic (rotary and sliding seals) parts;
- distributing to colder areas the heat generated in hot ones, contributing to generated heat dissipation;
- building up lubricated hydrodynamic films;
- protecting components against corrosion;
- deterging residual particles produced by wear;
- performing all of the above functions for long time and at any possible working temperature.

Gearbox components lubrication is usually made by splashing and spraying oil by rotating gears. It is therefore necessary to correctly exploit moving parts in order to grant presence and renewal of oil to all couples which need lubrication.

It is therefore necessary to design oil channels into the housing in such a way as they pick up the oil projected by wheels and distribute it to less exposed parts; transversal drilling on shafts may contribute to lubricate bearings of idle gear wheels.

On heavy duty manual gearboxes and on automatic gearboxes pressure lubrication is used; a gear wheels pump is moved by the input shaft and dedicated pipes and channels distribute the oil to the utilization points.

The lubricant distribution study would be rather difficult if tackled by mathematical models; in a simpler way, experimental analyses are set up, using modified gearboxes with transparent windows.

When teeth flanks mesh together different phenomena can be identified:

- *boundary lubrication*, when surfaces are in contact without interposition of lubricant oil; their protection is solely granted by their nature; in this situation only lubricant additives modify surfaces chemical nature and prevent micro welding (friction modifiers);
- *mixed lubrication*, when a partial separation of surfaces takes place, by effect of hydrodynamic forces generated by lubricant;
- *hydrodynamic lubrication*, when a completely separated lubricant film is set up.

These three different situations take place according to the position of the contact point on the tooth flank and to the peripheral speed; on the tooth part which is closer to the primitive circumference mixed lubrication will take place at low speed, while hydrodynamic lubrication will take place at high speed; on parts near to the first and last point of contact there will be boundary and mixed lubrication.

These facts should be taken into account while choosing lubricant, as well as maximum temperature; it can reach 90–100 °C, in the bulk of the lubricant, and 150–160 °C locally.

On present cars the expected lubricant life is as long as gearbox life.

Lubricants which satisfy above conditions are mineral oils mixed to synthetic ones; a suitable additives package must be provided to:

- prevent corrosion and oxidation products build up;
- deterge and disperse pollutant particles;

- modify chemical nature of surfaces in contact, to prevent micro welds in boundary lubrication conditions.

Viscosity grade depends on operation temperature; multigrade oils are widely applied in Europe to standardize product throughout the market and to avoid unaccepted seasonal oil changes.

16.6 Housings and Seals

Functions of housings are the following:

- reacting to forces and torques applied by the contained parts and distributing resultant forces to interfaces with engine and powertrain suspension;
- maintaining the exact position of part contained inside;
- wasting generated heat;
- insulating generated noise;
- allowing simple gearbox assembly and disassembly.

Housing lay-out can be classified according to three alternative architectures:

- through housing, when bearings seats are cut on the same housing element which results particularly stiff and simple to be machined; there are openings, closed by removable covers which allow assembling and disassembling interior parts;
- end loaded housings; the housing is cut transversely to shafts in two halves, therefore the bearing seats of the same shaft rest on different housing parts;
- top loaded housings; they are cut along shafts in two halves, therefore each bearing rests on two different half seats; also in these last cases, additional covers must be provided, to make assembly and disassembly possible.

If the housing is divided in two halves each part is machined separately during most of the cycle. Final boring of bearing seats will be made on assembled parts, to grant necessary tolerance; there are therefore no clearance location pins that allow unambiguous half housings assembly.

The most diffused architecture is the second one; it shows the advantage of an easier assembly and, on industrial vehicles gearboxes, allow organization by modules (clutch, splitter, gearbox, reducer, accessories, etc.) assembled in different versions.

Housings are usually made in aluminium and, sometime, in magnesium for weight reduction; they show a large number of local reinforcements, such ribs and webs to reach maximum stiffness with contained weight.

The example in Fig. 16.9 shows inclined ribs to increase torsional stiffness to shafts reaction forces.

Housings must have breathers. As a matter of fact, lubricant doesn't occupy all available interior volume, to contain weight and the friction losses; without breather, the air in the free space would change its pressure, because of temperature variation, with problems on seals.

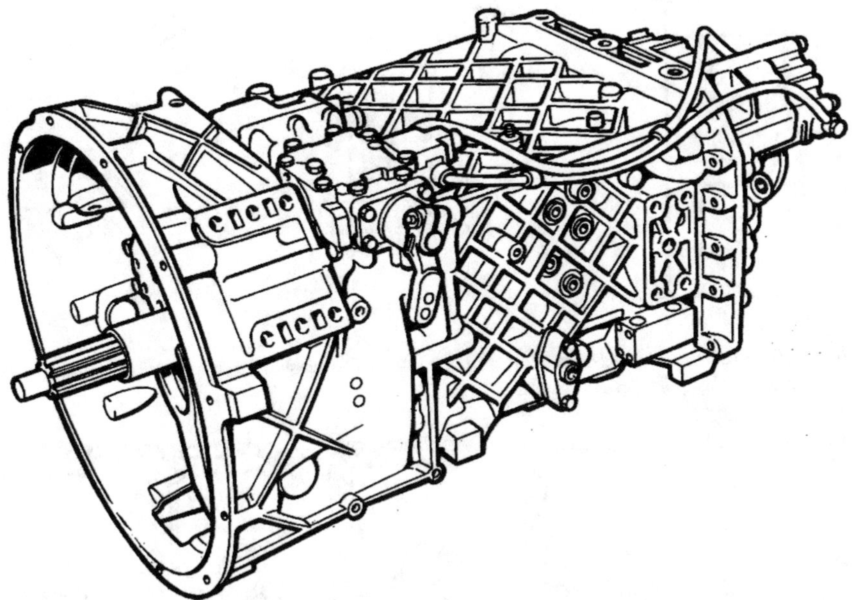

Fig. 16.9 Industrial vehicle gearbox housing, characterized by considerable inclined ribs; they increase global torsional stiffness and panel bending stiffness (Iveco)

Air must exit during vehicle operation and re-enter at stops; dust and other pollutants must be kept away. The breather is like a cap; suitable openings with separation labyrinths and filters, these lasts made with low density sintered metal are provided on the cap.

Rotary and sliding seals must be carefully designed. Seals must be completely tight; even small leakages are now unaccepted for environment pollution and consequent refills.

To seal fixed parts are used preformed gaskets (Fig. 16.10, middle), or in situ polymerized gaskets (Fig. 16.10, right). In this case, gaskets are made with synthetic materials that are distributed on parts as paste and polymerize, becoming solid, after assembly; for this reason, covers must have suitable teeth (pointed by an arrow in figure) to avoid paste intrusion, after bolt tightening. Covers must also have suitable projections to make the gasket rupture easier at disassembling.

For limited dimension and round covers, O-rings are also used (Fig. 16.10, left). In case of preformed gaskets the number of bolts and the cover plate stiffness must grant an almost constant pressure contact.

To verify this fact a photographic pressure sensitive film can be useful.

Rotary seals are lip type with coil spring that clamps rotary shaft (Fig. 16.11, top). Seals must be assembled with their springs inside the housings, to improve tightening to pressure increase due to temperature; a second lip can be added on the seal, to protect the sealing circle of dust contamination.

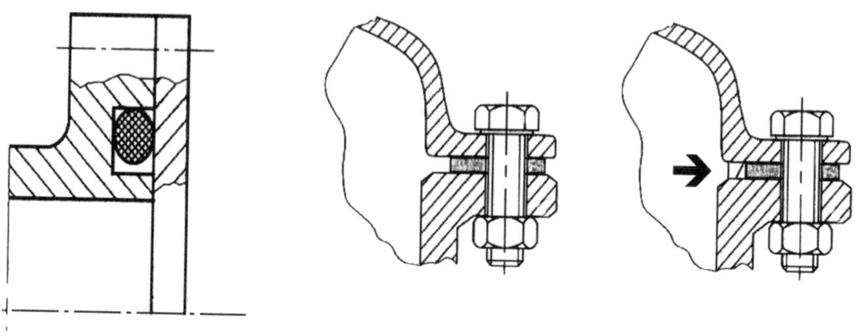

Fig. 16.10 Examples of seals on covers; from the left, an O-ring seal for round covers, a preformed gasket and a gasket polymerized in situ

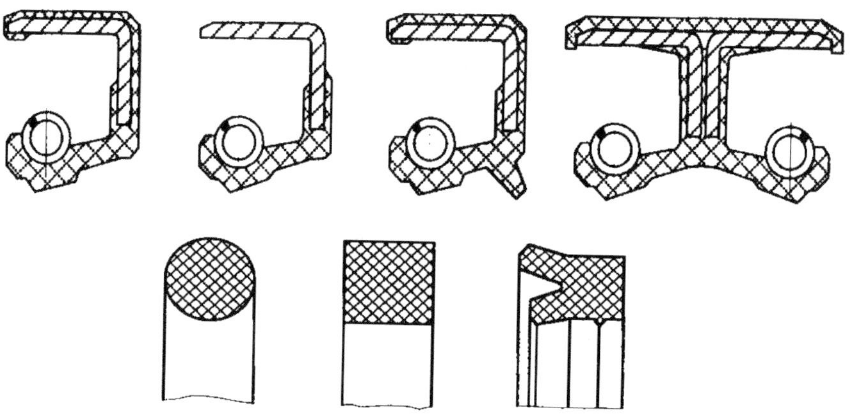

Fig. 16.11 Examples of rotary seals (top) and of sliding seals (bottom)

Sliding seals or small angle rotary seals (selector rods, Fig. 16.11, bottom) are made with O-ring or square rings.

Seals for multi disc clutch actuation pistons are subject to high pressure and are made with rectangular rings with pressure sensitive lip (Fig. 16.11, bottom right); these seals must be correctly oriented at assembled.

16.7 Test Technologies Outline

To verify functions and reliability of a gearbox, suitable test activities are conveniently performed after calculation ones; these tests must be made on different prototypes, to grant the suitable confidence level.

Test activities can be classified according to their time position, with reference to the development process of a new gearbox.

There will be test demonstrating design adequacy, performed on a limited number of prototypes, manufactured with experimental tools.

After these tests there will be a second series of them, suitable to demonstrate manufacturing process adequacy, performed on a significant number of prototypes manufactured with mass production tools.

These tests are performed on benches and prototype vehicles; they are followed by a vehicle reliability demonstration program that should confirm transmission reliability and identify residual problems.

Same test cycle must be separately performed on all supplied parts.

Test activities can also be classified according to expected results; from this point of view we identify *functional* and *reliability tests*.

The fundamental characteristic of functional tests is that they are performed in short time, because the life doesn't imply sudden result changes.

Some functional test can be repeated on the same prototype at different times of its life. For example, mechanical efficiency measurements should be repeated after run-in in time, to verify the consequent improvement. In the same way leakage tests must be performed on new gearboxes to verify design and production process adequacy and at the end of useful life to find out unacceptable variations due to wear.

Typical functional test results are characteristic measurements that must be compared with project objectives.

Functional tests include the following:

- Lubrication tests, where it is verified that oil reaches all points to be lubricated, also when the gearbox is inclined in three directions, according to the vehicle mission.
- Leakage test of lubricant oil.
- Power absorption tests, to be performed at any possible input torque, engine speed and gearbox speed.
- Selection and engagement forces, to be performed at different gearbox speed, vehicle speed and different meaningful oil temperatures.
- Noise emission tests at idle and at different working conditions at different speeds.
- Operation temperatures measurements.
- Misuse and abuse tests.

All these tests can be performed on few prototypes that must be machined and assembled with those dimensions that are relevant to phenomena in cause, as close to tolerance limits as possible; for instance rattle noise tests should be performed with the widest angular clearance, allowed by drawing specifications.

Endurance tests consist in having the component working for the expected life according to different possible mission profiles. Expected results consist of failures that must occur after useful life; if they occur prematurely, they must be analyzed to design corrective countermeasures. Repetition of all scheduled tests is in any case requested, until success is reached.

Reliability can be demonstrated by repeating endurance tests on a statistically significant prototype number.

Almost all functional and endurance tests can be performed on a bench or on a vehicle; it is useful to test a vehicle only for result confirmation, when success has

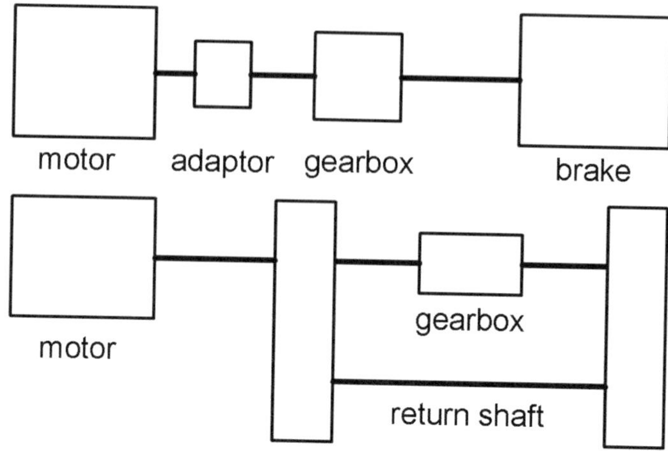

Fig. 16.12 Schemes for transmission test benches; on the lower part of the figure there is the scheme of a recirculation power bench

been reached in an adequate number of bench tests. These are, in fact, easier to be supervised and failures are easier to be analyzed.

The stand-by time of a vehicle on test and the kind of damage consequent to a gearbox failure are unacceptable, in consideration of the high prototype fabrication cost.

Transmission test benches are particularly simple and include a foundation block on which a complete transmission can be installed; it can be put in rotation by an actual engine or by an electric motor; this last is more appropriate for long endurance tests.

When an electric motor is used, for some kind of test, a control circuit is necessary to produce an input torque with the periodic irregularity of the internal combustion engine. The same result can be obtained by connecting motor and transmission through a torque pulsator.

According to the kind of test, bench schemes drawn on Fig. 16.12 can have a brake to simulate the vehicle resistance.

Brakes can be coupled to variable inertia flywheels when vehicle inertia must be reproduced; brake and flywheel assembly can be substituted by a suitably controlled motor/generator, able to emulate vehicle resistance and to recover part of the wasted energy.

Sometimes (Fig. 16.12, at bottom) and when tests imply constant input torque, shafts are connected through a transmission line. The transmission line is preloaded through a constant torque, which stresses the gearbox at the desiderate level. In this case motor power must pay for friction resistance only.

The picture in Fig. 16.13 shows the interior of a typical modern test cell for complete transmissions.

Fig. 16.13 Interior of a typical modern test cell for complete transmissions. This kind of test cell can be adapted to all kind of tests, including acoustic emission measurement: it is acoustically reverberant, but it can be changed to anechoic, by encapsulating the only transmission in a suitable cabin

This kind of test cell can be adapted to all kind of tests, including acoustic emission measurement: it is acoustically reverberant, but it can be changed to anechoic, by encapsulating the only transmission in a suitable cabin.

We can see on the left the electric motor and its control system, the transmission test bed and the torsiometer shaft to measure the output torque. In the small picture below, taken from the control console, we can see the electric brake.

The electric brake has a maximum power of 220 kW and a maximum torque of 600 Nm and can operate up to 7,000 rpm, simulating the torque fluctuation of an internal combustion engine in the field of frequencies between 0 and 500 Hz.

The brake can absorb up to 200 kW, with a maximum torque of 3,000 Nm at 650 rpm; a maximum transmission ratio of 5 at maximum torque can be simulated, sufficient to test gearbox and final drive separately, for the maximum torque or together for reduced torque.

References of Volume I

Part I

1. L. Baudry de Saunier, *L'automobile théorique et pratique* (Omnia, Paris, 1900)
2. O.C. Schmidt, *Practical Treatise on Automobiles* (The American Text-book, Philadelphia, 1909)
3. W. Neubecker, *Antique Automobile Body Construction and Restoration* (Post Publication, Arcadia, 1912)
4. M. Peter, *Der Kraftwagen* (R. C. Schmidt, Berlin, 1937)
5. M. Serruys, *La suspension et la direction des véhicules routiers* (Dunod, Paris, 1947)
6. M. Boisseaux, *L'automobile, méthodes de calcul* (Dunod, Paris, 1948)
7. J.C. Maroselli, *L'automobile et ses grands problèmes* (Larousse, Paris, 1958)
8. M.G. Bekker, *Off-the-Road Locomotion* (University of Michigan Press, Ann Arbor, 1960)
9. J.P. Norbye, *Sports Car Suspension* (Sport Car Press, NY, 1965)
10. J. Pawlowski, *Vehicle Body Engineering* (Business Books, London, 1969)
11. G. Oliver, *Cars and Coachbuilding* (Sotheby Parke Bernet, London, 1981)
12. I.S. Ageikin, *Off-the-Road Mobility of Automobiles* (Balkema, Rodderdam, 1987)
13. D. Giacosa, *Progetti alla FIAT* (Automobilia, Torino, 1988)
14. T.D. Gillespie, *Fundamentals of Vehicle Dynamics* (SAE, Warrendale, 1992)
15. M. Mitschke, *Dynamik der Kraftfahrzeuge* (Springer, Berlin, 1995)
16. W.F. Milliken, D.L. Milliken, *Race Car Vehicle Dynamics* (SAE, Warrendale, 1995)
17. K. Newton et al., *The Motor Vehicle* (SAE, Warrendale, 1996)
18. J. Reinpell, H. Stoll, *The Automotive Chassis* (Arnold, London, 1996)
19. P.L. Bassignana et al., *Storia fotografica dell'industria automobilistica italiana* (Boringhieri, Torino, 1998)
20. J. Fenton, *Handbook of Automotive Body and Systems Design* (Professional Engineering Publishing, London, 1998)

G. Genta and L. Morello, *The Automotive Chassis*, Mechanical Engineering Series,
https://doi.org/10.1007/978-3-030-35635-4

21. J. Fenton, *Handbook of Automotive Powertrain and Chassis Design* (Professional Engineering Publishing, London, 1998)
22. H. Heisler, *Vehicle and Engine Technology* (Arnold, London, 1999)
23. J. Fenton, *Handbook of Vehicle Design Analysis* (SAE, Warrendale, 1999)
24. J. Appian-Smith, *An Introduction to Modern Vehicle Design* (SAE, Warrendale, 2002)
25. J. Brown et al., *Motor Vehicle Structures: Concepts and Fundamentals* (SAE, Warrendale, 2002)

Part II

1. L. Baudry de Saunier, *L'automobile théorique et pratique* (Omnia, Paris, 1900)
2. O.C. Schmidt, *Practical Treatise on Automobiles* (The American Text-book, Philadelphia, 1909)
3. A. Seniga, *Il meccanismo di trasmissione negli automobili* (Biblioteca d'automobilismo e d'aviazione, Milano, 1912)
4. E.B. Butler, *Transmission Gears* (Griffin, London, 1917)
5. M. Peter, *Der Kraftwagen* (R. C. Schmidt, Berlin, 1937)
6. M. Boisseaux, *L'automobile, méthodes de calcul* (Dunod, Paris, 1948)
7. W.H. Crouse, *Automotive Transmissions and Power Trains* (Mc Graw-Hill, New York, 1955)
8. P. Patin, *Les transmissions de puissance* (Eyrolles, Paris, 1956)
9. D. Thirlby, *The Chain Driven Frazer Nash* (Mc Donald, London, 1965)
10. G. Rogliatti, C. Valier, L. Giovanetti, *La frizione nel tempo* (Valeo, Torino, 1980)
11. D. Giacosa, *Progetti alla FIAT* (Automobilia, Torino, 1988)
12. H.J. Schöpf, G. Jürgens, J. Pickard, Das neue Fünfgang-automatikgetriebe von Mercedes-Benz. ATZ **91** (1989)
13. A. Vari, *Design Practices: Passenger Car Automatic Transmissions* (SAE, Warrendale (PA), 1994)
14. J. Fenton, *Handbook of Automotive Powertrain and Chassis Design* (Professional Engineering Publishing, London, 1998)
15. H. Heisler, *Vehicle and Engine Technology* (Arnold, London, 1999)
16. G. Lechner, H. Naunheimer, *Automotive Transmissions, Fundamentals, Selection, Design and Application* (Springer, Berlin, 1999)
17. R.K. Jurgen, *Electronic Transmission Control* (SAE, Warrendale (PA), 2000)
18. J. Happian-Smith, *An Introduction to Modern Vehicle Technology* (SAE, Warrendale (PA), 2002)
19. J. Greiner et al., 7-Speed Automatic Transmission from Mercedes-Benz. ATZ **105** (2003)

Index

G. Genta and L. Morello, *The Automotive Chassis*, Mechanical Engineering Series,
https://doi.org/10.1007/978-3-030-35635-4

Printed in the United States
By Bookmasters